The Physiology
and
Biochemistry *of*
Prokaryotes

The PHYSIOLOGY *and* BIOCHEMISTRY *of* PROKARYOTES

David White
Indiana University

New York   Oxford
OXFORD UNIVERSITY PRESS
1995

Oxford University Press

Oxford   New York   Toronto
Delhi   Bombay   Calcutta   Madras   Karachi
Kuala Lumpur   Singapore   Hong Kong   Tokyo
Nairobi   Dar es Salaam   Cape Town
Melbourne   Auckland   Madrid

and associated companies in
Berlin   Ibadan

Copyright © 1995 by Oxford University Press, Inc.

Published by Oxford University Press, Inc.,
198 Madison Avenue, New York, New York 10016

Oxford is a registered trademark of Oxford University Press

Library of Congress Cataloging-in-Publication Data
White, David, 1936–
The physiology and biochemistry of prokaryotes / David White.
p. cm. Includes bibliographical references and index.
ISBN 0-19-508439-X
1. Microbial metabolism. 2. Prokaryotes—Physiology. I. Title.
OR88.W48   1995
589.9—dc20   94-6328

5 7 9 8 6 4
Printed in the United States of America
on acid-free paper

PREFACE

The prokaryotes are a diverse assemblage of organisms that consists of the Bacteria (eubacteria) and the Archaea (archaebacteria). This text covers major aspects of the physiology and biochemistry of these organisms at the advanced undergraduate and beginning graduate levels. Topics include the structure and function of the prokaryotic cell (Chapter 1), growth physiology (Chapter 2), membrane and cytosol bioenergetics (Chapters 3 and 7), electron transport and photosynthetic electron transport (Chapters 4 and 5), metabolic regulation (Chapter 6), intermediary metabolism (Chapter 8 through 13), homeostasis (Chapter 14), solute transport (Chapter 15), protein export (Chapter 16), and responses to environmental signals (Chapter 17).

The diversity of the prokaryotes is due, in part, to adaptations to the different habitats in which they grow. The habitats range in pH from approximately 1 to 12, temperature from below 2°C to over 100°C, pressures that can be several thousands of atmospheres (at the bottom of the oceans), habitats without oxygen, and habitats with salt concentrations that can reach saturating levels. The physiological types among the prokaryotes mirror the diversity in the habitats and the sources of nutrient. Thus, there are aerobes, anaerobes, facultative anaerobes, heterotrophs, autotrophs, phototrophs, chemolithotrophs, alkaliphiles (alkalophiles), acidophiles, and halophiles. Some of these physiological types are found only among the prokaryotes, with rare exceptions. For example, among microorganisms that have been cultured in the laboratory, the ability to live indefinitely in the absence of air (anaerobic growth) is confined almost entirely to prokaryotes. Also, growth using the energy extracted from inorganic compounds (lithotrophy) seems not to have evolved in the eukaryotes and is restricted to certain groups of prokaryotes. The same can be said about the reduction of nitrogen gas to ammonia (nitrogen fixation), although it is more widespread among the prokaryotes than is lithotrophy. Another typically prokaryotic capability is the use of inorganic compounds such as nitrate and sulfate as electron acceptors during respiration (anaerobic respiration).

On the other hand, it is also true that most of the metabolic pathways that exist in the prokaryotes are the same as those in all living organisms, reflecting a unity in biochemistry dictated by principles of chemistry and (with respect to bioenergetics) physics. This text emphasizes the underlying principles of chemistry and physics in explaining the various physiological and metabolic features of prokaryotic cells, and thus provides the background for further advanced studies in this area. A summary at the end of each chapter emphasizes the major points in the chapter, and is followed by a set of study questions.

Chapter 1 is an overview of the parts of prokaryotic cells and the functions of these parts. The chapter also introduces the student to the two evolutionary lines of prokaryotes, Bacteria (eubacteria) and Archaea (archaebacteria).

Chapter 2 is a general treatment of bacterial growth. It includes an explanation of the growth equations and their use and applications, as well as selected topics, such as responses to different growth conditions, the stringent response, and adaptive responses by stationary phase cells.

Chapter 3 is devoted to membrane bioenergetics. It explains the principles of the chemiosmotic theory, including the basic tenets of the theory, and the different ways that prokaryotes generate the proton and sodium ion potentials that drive membrane activities such as solute transport, motility, and ATP synthesis. This chapter, as well as Chapter 7, which explains bioenergetics in the cytosol, brings together principles of physics, thermodynamics, and organic chemistry to explain how prokaryotes use and interconvert light, chemical, and electrochemical energies. The first part of the chapter explains the chemiosmotic theory and provides an overview of how the proton potential is created by respiration, photosynthesis, and ATP hydrolysis. The details of respiration and photosynthesis are covered in the two subsequent chapters. The second part of the chapter includes detailed discussions of specialized processes that generate a proton or sodium potential in certain groups of prokaryotes. This includes light-driven proton pumping by bacteriorhodopsin in the halophiles, as well as sodium-dependent decarboxylases, and proton and sodium-coupled end-product efflux in fermenting bacteria.

Chapter 4 describes electron transport pathways. The first part is a general description of electron transport and coupling sites in both mitochondria and bacteria, and includes a discussion of the bc_1 complex and the Q cycle, as well as linear Q loops and proton pumps. The second part of the chapter describes electron transport pathways in certain well-studied bacteria. Electron transport pathways in particular bacteria are

also discussed in Chapter 11 (the lithotrophs and sulfate reducers).

Chapter 5 covers photosynthesis. It begins with an introduction to the various photosynthetic prokaryotes and gives an overview of electron flow in the different photosynthetic systems. Later sections of the chapter are devoted to a more detailed examination of photosynthesis, including a closer look at the reaction centers, the light-harvesting pigments, and energy transfer between pigment complexes.

Chapter 6 is an introduction to the principles of metabolic regulation and provides the necessary background for learning the metabolic regulation of the pathways discussed in Chapters 8 and 9. It includes a discussion of the patterns of metabolic regulation, the Michaelis–Menten equation, allostery, and covalent modification of enzymes.

Chapter 7 explains the principles of bioenergetics in the soluble part of the cell, the cytosol, and is the counterpart to Chapter 3 which is concerned with membrane bioenergetics. It includes a discussion of what is meant by the term "high-energy molecule" and how these molecules are synthesized, group transfer potentials, high-energy intermediates, substrate level phosphorylation, and coupled reactions. This chapter provides the background for understanding the energetics of the pathways discussed in Chapters 8 and 9.

Chapters 8 and 9 describe the central pathways of intermediary metabolism, including the metabolism of carbohydrates, organic acids, lipids, purines, pyrimidines, amino acids, and aliphatic hydrocarbons. The chapters on intermediary metabolism are by no means inclusive with respect to the pathways that are covered. They were written to provide an overview of metabolism in both aerobically and anaerobically growing prokaryotes, as well as to prepare the student for the chapter on sugar fermentations (Chapter 13). Major pathways found in most bacteria (as well as eukaryotes) are discussed, as well as some that are seen only in certain groups of prokaryotes (e.g., the modified Entner–Doudoroff pathway in some archaea and bacteria, and the nonphosphorylated Entner–Doudoroff pathway in some archaea). Special emphasis is given to the relationships between the pathways (i.e., how they interconnect) and their physiological roles.

Chapter 10 discusses the biosynthesis of the bacterial cell wall and covers peptidoglycan and lipopolysaccharide synthesis.

Chapter 11 describes the metabolism of inorganic molecules. This includes ammonia and sulfate assimilation, nitrogen fixation, nitrate and sulfate respiration, and lithotrophy.

Chapter 12 explains the pathways for the assimilation of single-carbon compounds as well as methanogenesis.

Chapter 13 describes the main classes of carbohydrate fermentations.

Chapters 14–16 cover solute transport, protein export, and pH and osmotic homeostasis, respectively.

Chapter 17 is concerned with responses to environmental signals such as chemotaxis effectors, and changes in nutrient and oxygen levels. Most of Chapter 17 is devoted to two-component regulatory systems.

As can be seen from this introduction, the organization of the text is according to topics rather than organisms, although the physiology of specific groups of prokaryotes is emphasized. This pattern of organization lends itself to the elucidation of general principles of physiology and metabolism.

A word about the chemical structures and the naming of compounds. Most of the carboxyl groups are drawn as nonionized and the primary amino groups as nonprotonated. However, at physiological pH, these groups are ionized and protonated, respectively. The names of the organic acids indicate that they are ionized (e.g., acetate rather than acetic acid).

There are two distinct evolutionary lines of prokaryotes commonly referred to as eubacteria and archaebacteria. However, this terminology implies a specific relationship between eubacteria and archaebacteria, whereas at the molecular level, the archaebacteria are not more closely related to the eubacteria than they are to the eukaryotes. In recognition of this fact, Woese *et al.* suggested that the eubacteria be referred to simply as bacteria and that the archaebacteria be called archaea.[1] That terminology is followed in this text.

REFERENCES AND NOTES

1. Woese, C. R., Kandler, O., and M. L. Wheelis. 1990. Towards a natural system of organisms. Proposal for the domains Archaea, Bacteria and Eucarya. *Proc. Natl Acad. Sci. USA* 87:4576–4579.

ACKNOWLEDGMENTS

I would like to express gratitude to the following individuals who thoughtfully reviewed all or portions of the text: Tom Beatty, Martin Dworkin, Howard Gest, George Hegeman, Alan Hooper, Arthur Koch, Jörg Overmann, Norman Pace, Carlos Ramirez-Icaza, and Palmer Rogers, as well as the anonymous reviewers who carefully read the manuscript and made valuable suggestions. Particular thanks to Martin Dworkin for his long-time support and encouragement. I am deeply indebted to Kirk Jensen of Oxford University Press for his editorial advice and confidence in the project. I would also like to thank Alan Chesterton and the production staff at Keyword Publishing Services Ltd. The text was illustrated by Eric J. White. His careful work and enthusiasm for the project are greatly appreciated, and it was a personally rewarding experience working with him. This book is dedicated to my family.

CONTENTS

SYMBOLS

c	Speed of light (3.0×10^8 m/s)
C	Coulomb
cal	Calorie (4.184 J)
E_0	Standard redox potential (reduction potential) at pH 0
E_0'	Standard redox potential at pH 7
E_m	Mid-point potential at a specified pH (e.g., $E_{m,7}$). For pH 7, also written as E_m', which is numerically equal to E_0'.
E_h	Actual redox potential at a specified pH. For pH 7, E_h' or $E_{h,7}$.
eV	Electron volt. The work required to raise one electron through a potential difference of one volt. It is also the work required to raise a monovalent ion (e.g., a proton) through an electrochemical potential difference of one volt. One eV = 1.6×10^{-19} J.
F	Faraday constant (approximately 96,500 C). The charge carried by one mole of electrons or monovalent ion. It is the product of the charge carried by a single electron and Avogadro's constant.
g	Generation time (time/generation)
ΔG_0	Standard Gibbs' free energy change at pH 0 (J/mol or cal/mol)
$\Delta G_0'$	Gibbs' free energy at pH 7
ΔG_p	Phosphorylation "potential" (energy required to synthesize ATP using physiological concentrations of ADP, inorganic phosphate, and ATP)
h	Planck's constant (6.626×10^{-34} J·s)
J	Joule. One coulomb volt (C × V). The work required to raise one coulomb through a potential difference of one volt.
k	Instantaneous growth rate constant (time^{-1})
kJ	Kilojoule.

K	Absolute equilibrium constant.
K'	Apparent equilibrium constant at pH 7.
K	Kelvin (273.16 + °C)
$\Delta\tilde{\mu}_{ion}$	Electrochemical potential difference between two solutions of an ion separated by a membrane. Units are joules.
$\Delta\tilde{\mu}_{ion}/F$	Electrochemical potential difference expressed as volts or millivolts.
$\Delta\tilde{\mu}_{H^+}/F$	The protonmotive force. Electrochemical potential difference in volts or millivolts of protons between two solutions separated by a membrane. Also written as Δp.
N	Avogadro's number (6.023×10^{23} particles/mol)
Δp	See $\Delta\tilde{\mu}_{H^+}/F$.
ΔpH	Difference in pH between the inside and outside of the cell. Usually, $pH_{in} - pH_{out}$.
R	The ideal gas constant ($8.3144 \text{ J K}^{-1} \text{ mol}^{-1}$ or $1.9872 \text{ cal K}^{-1} \text{ mol}^{-1}$)
V	Volt. The potential difference across an electric field.
$\Delta\Psi$	Membrane potential, usually $\Psi_{in} - \Psi_{out}$.

CONVERSION FACTORS, EQUATIONS, AND UNITS OF ENERGY

Electrode potential at pH 7

$E'_h = E'_0 + [RT/nF] \ln [(\text{ox})/(\text{red})]$ volts. The symbol n refers to the number of electrons, and (ox) and (red) refer to the concentrations of oxidized and reduced forms, respectively. When (ox) = (red), then $E'_h = E'_0 = E'_m$.

$E'_h = E'_0 + (60/n) \log [(\text{ox})/(\text{red})]$ mV at 30°C.

Electron, charge

1.6023×10^{-19} C

Gibb's energy

For the reaction $aA + bB \leftrightarrow cC + dD$,

$$\Delta G = \Delta G_0 + RT \ln [C]^c[D]^d/[A]^a[B]^b$$

Gibb's energy and equilibrium constant

$$\Delta G'_0 = -RT \ln K'_{eq} \text{ or } -2.303RT \log K'_{eq}$$

Gibb's energy for solute uptake and concentration gradient

$$\Delta G = RT \ln [S]_{in}/[S]_{out}$$

at 30°C, $\Delta G = 5.8 \log_{10}[S]_{in}/[S]_{out}$ kJ/mol

or

$$\Delta G/F = 60 \log_{10}[S]_{in}/[S]_{out} \text{ mV}$$

Growth

$g(k) = 0.693$, where g is the doubling time for the population and k is the instantaneous growth rate constant.

$x = x_0 2^Y$, the equation for exponential growth. Y is the number of generations. x is mass or any parameter that changes linearly with mass.

$Y = t/g$

Light, energy in a quantum

$E = h\nu = hc\lambda$, where ν = frequency (c/λ), h = Planck's constant, λ = wavelength

E (kJ) $= 1.986 \times 10^{-19}/\lambda$, where λ is in nm

E (eV) $= 1.24 \times 10^3/\lambda$, where λ is in nm

Light, energy in an einstein

$E = Nhv = Nhc/\lambda = 1.197 \times 10^5$ kJ/λ, where λ is the wavelength in nanometers, N is Avogadro's number, and c is the speed of light.

E (kJ) $= 1.196 \times 10^5/\lambda$, where λ is in nm

Nernst equation

$n\Delta\Psi = (RT/F) \ln [S]_{in}/[S]_{out}$ V, where S is the concentration of a diffusible ion of valency n or, $n\Delta\Psi = 60 \log [S]_{in}/[S]_{out}$ mV at 30°C

According to the equation, each 10-fold concentration difference of a permeant monovalent ion corresponds to a potential difference of 60 mV.

Phosphorylation potential

$\Delta G_p = \Delta G_0' + RT \ln [ATP]/[ADP][P_i]$

Proton potential

$\Delta p = \Delta\mu_{H^+}/F = \Delta\Psi - 60\Delta pH$ mV at 30°C

Proton potential and ΔE_h at equilibrium

$-n\Delta E_h = y\Delta p$, where n is the number of electrons transferred over a redox potential difference of ΔE_h V, and y is the number of protons translocated over a proton potential difference of Δp V.

$2.303RT$

5.8 kJ/mol or 1.39 kcal/mol at 30°C

$2.303RT/F$

0.06 V or 60 mV at 30°C

DEFINITIONS

Acetogenic bacteria	Anaerobic bacteria that synthesize acetic acid from carbon dioxide and secrete the acetic acid into the medium.
Acidophile	Grows between pH 1 and 4 and not at neutral pH.
Aerobe	Uses oxygen as an electron acceptor during respiration.
Aerotolerant anaerobe	Cannot use oxygen as an electron acceptor during respiration but can grow in its presence.
Anaerobe	Does not use oxygen as an electron acceptor during respiration.
Alkaliphile (alkalophile)	Grows at pH above 9, often with an optimum between 10 and 12.
Autotroph	Uses carbon dioxide as sole or major source of carbon.
Chemolithotroph	See lithotroph.
Cytoplasm	The fluid material enclosed by the cell membrane.
Cytosol	The liquid portion of the cytoplasm.
Dalton	All atomic and molecular weights refer to the carbon isotope (^{12}C) which is 12 D or 1.661×10^{-24} g. Daltons are numerically equal to molecular weights and can be used as units when molecular weight units of grams per mole are not appropriate (e.g., when referring to ribosomes).
Einstein	One "mole" of light (6.023×10^{23} quanta).
Facultative anaerobes	Can grow anaerobically in the absence of oxygen or will grow by respiration if oxygen is available.
Facultative autotroph	Can grow on carbon dioxide as sole or major source of carbon or on organic carbon.

Growth yield constant, *Y*   The amount of dry weight of cell produced per weight of nutrient used.

Halophile   Requires high salt concentrations for growth.

Heterotroph   Uses organic carbon as major source of carbon.

Lithotroph   Oxidizes inorganic compounds as a source of energy for growth.

Molar growth yield constant (Y_m)   Grams of dry weight of cells produced per mole of nutrient used.

Neutrophile   Grows with a pH optimum near neutrality.

Obligate anaerobes   Will grow only in the absence of oxygen but not necessarily killed by oxygen.

Phosphorylation potential   Energy required to phosphorylate one mole of ADP using physiological concentrations of ADP, P_i, and ATP.

Photon   Quantum. A particle of light.

Photosynthesis   The use of light as a source of energy for growth.

Photoautotroph   An organism that uses light as a source of energy for growth and carbon dioxide as the source of carbon.

Photoheterotroph   An organism that uses light as a source of energy for growth and organic carbon as the source of cell carbon.

Phototroph   An organism that uses light as the source of energy for growth.

Quantum   A particle of light (photon).

Standard conditions   All reactants and products are in their "standard states." This means that solutes are at a concentration of one molar (1 mol of solute per liter) and gases are at one atmosphere. Biochemists usually take the standard state of H^+ as 10^{-7} M (pH 7). By convention, if water is a reactant or product, its concentration is set at 1.0 M, even though it is 55.5 M in dilute solutions.

Strict anaerobes   Will grow only in the absence of oxygen and also killed by traces of oxygen.

Thermophile   An organism that can grow at temperatures greater than 55°C.

The Physiology
and
Biochemistry *of*
Prokaryotes

1

Structure and Function

Although prokaryotes[1] are devoid of organelles such as nuclei, mitochondria, chloroplasts, Golgi vesicles, and so on, their cell structure is far from simple. Despite the absence of organelles comparable to those found in eukaryotic cells, metabolic activities in prokaryotic cells are nevertheless compartmentalized, which is necessary for efficient metabolism and growth. For example, compartmentalization occurs: within multienzyme granules that house enzymes for specific metabolic pathways; in intracellular membranes and the cell membrane; within the periplasm in gram-negative bacteria; and within the cell wall itself. There are also various inclusion bodies that house specific enzymes, storage products, or photosynthetic pigments. In addition, prokaryotic cells display appendages such as fimbriae and pili that are used for adhesion to other cells, and flagella that are used for swimming. This chapter describes these structural cell components and their functions. There are actually two distinct types of prokaryotes—the Bacteria and the Archaea—and these are described later.

1.1 Phylogeny

Figure 1.1 shows a current phylogeny[2,3] of life forms based upon comparing ribosomal RNA nucleotide sequences.[4] Notice that there are three lines (domains) of evolutionary descent— Bacteria (eubacteria), Eucarya (eukaryotes), and Archaea (archaebacteria)— that diverged in the distant past from a common ancestor.[5] (The term *archaeon* may be used to describe particular archaea.) Archaea differ from bacteria in ribosomal RNA nucleotide sequences, in cell chemistry, and in certain physiological aspects, described in Section 1.2. Table 1.1 lists examples of prokaryotes in the different subdivisions within the domains Bacteria and Archaea. Notice that the gram-positive bacteria are a tight grouping. Although there is no single grouping of gram-negative bacteria, most of the well-known gram-negative bacteria are in the purple bacteria group.

1.1.1 Archaea

Phenotypes
The archaea commonly manifest one of three phenotypes (i.e., *methanogenic*, *extremely halophilic*, and *extremely thermophilic*) which also correspond to phylogenetic groups.
1. The *methanogenic* archaea (Euryarchaeota) are obligate anaerobes that grow in environments such as anaerobic ground waters, swamps, and sewage. They derive energy by reducing carbon dioxide to methane, or by converting acetate to carbon dioxide and methane. The metabolism of the methanogens is discussed in Chapter 12.

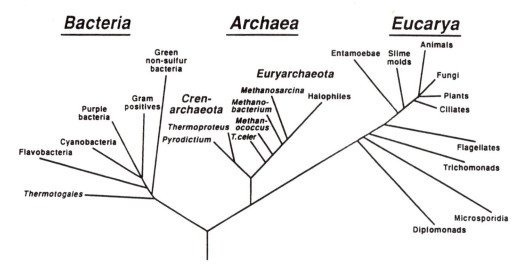

Fig. 1.1 Phylogenetic relationships among life forms based upon rRNA sequences. The lengths of the lines are proportional to the evolutionary differences. The position of the root in the tree is approximate. (From Woese, C. R. and N. R. Pace. 1993. Probing RNA structure, function, and history by comparative analysis, pp. 91–117. In *The RNA World*. R. F. Gesteland and J. F. Atkins (eds.). Cold Spring Harbor Press, Cold Spring Harbor, NY.)

2. The *extremely halophilic* archaea (also Euryarchaeota) require very high sodium chloride concentrations (at least 3–5 M) for growth. They grow in salt lakes and solar evaporation ponds. The halophilic archaea have unique light-driven proton and chloride pumps called *bacteriorhodopsin* and *halorhodopsin*, respectively (Sections 3.8.4 and 3.9). Most extreme halophiles are archaea. Some exceptions are the bacterium, *Ectothiorhodospira*, and the alga, *Dunaliella*.

3. The *extremely thermophilic* archaea (Crenarchaeota) grow in thermophilic environments (generally 55–100°C).[6] Some of these have an optimal growing temperature near the boiling point of water. They use inorganic sulfur either as an electron donor or as an electron acceptor in energy yielding redox reactions. (The pathways for sulfate reduction and sulfur oxidation are described in Sections 11.2.2 and 11.4.1, respectively.) For this reason they are also called *sulfur-dependent*. For example, some, including *Sulfolobus* and *Acidianus*, oxidize inorganic sulfur compounds such as elemental sulfur and sulfide using oxygen as the electron acceptor, and derive ATP from the process,

while others are anaerobes that oxidize hydrogen gas using elemental sulfur or thiosulfate as the electron acceptor. The latter include *Thermoproteus*, *Pyrobaculum*, *Pyrodictium*, and *Archaeoglobus*. [*Pyrobaculum* and *Pyrodictium* use $S°$ (elemental sulfur) as an electron acceptor during autotrophic growth (i.e., growth on CO_2 as the carbon source) whereas *Archaeglobus* uses $S_2O_3^{2-}$.] Archaea belonging to the genus *Pyrodictium* have the highest growth temperature known, being able to grow at 110°C. A few of the sulfur-oxidizing archaea are *acidophiles*, growing in hot sulfuric acid at pH values as low as 1.0. They are called *thermoacidophiles* in recognition of the fact that they grow optimally in hot acid. For example, *Sulfolobus* grows at pH values of 1–5 and at temperatures up to 90°C in hot sulfur springs, where it oxidizes H_2S (hydrogen sulfide) or $S°$ to H_2SO_4 (sulfuric acid). Although most of the extreme thermophiles are obligately sulfur-dependent, some are facultative. For example, *Sulfolobus* can be grown heterotrophically on organic carbon and O_2 as well as autotrophically on H_2S or $S°$, O_2, and CO_2. Interestingly, some of the sulfur-dependent

Table 1.1 Major Subdivisions of Bacteria

Bacteria and their subdivisions

Purple bacteria

  α subdivision
    Purple nonsulfur bacteria (*Rhodobacter,
    Rhodopseudomonas*) rhizobacteria,
    agrobacteria, rickettsiae, *Nitrobacter,
    Thiobacillus* (some), *Azospirillum, Caulobacter*

  β subdivision
    Rhodocyclus (some), *Thiobacillus* (some),
    *Alcaligenes, Bordetella, Spirillum,
    Nitrosovibrio, Neisseria*

  γ subdivision
    Enterics (*Acinetobacter, Erwinia, Escherichia,
    Klebsiella, Salmonella, Serratia, Shigella,
    Yersinia*), vibrios, fluorescent psedomonads,
    purple sulfur bacteria, *Legionella* (some),
    Azotobacter, Beggiatoa, Thiobacillus (some),
    Photobacterium, Xanthomonas

  δ subdivision
    Sulfur and sulfate reducers (*Desulfovibrio,*
    myxobacteria, bdellovibrios

Gram-positive eubacteria

  A. High (G + C) species
    *Actinomyces, Streptomyces, Actinoplanes,
    Arthrobacter, Micrococcus, Bifidobacterium,
    Frankia, Mycobacterium, Corynebacterium*

  B. Low (G + C) species
    *Clostridium, Bacillus, Staphylococcus,
    Streptococcus*, mycoplasmas, lactic acid bacteria

  C. Photosynthetic species
    Heliobacterium

  D. Species with gram-negative walls
    Megasphaera, Sporomusa

Cyanobacteria and chloroplasts
  *Oscillatoria, Nostoc, Synechococcus, Prochloron,
Anabaena, Anacystis, Calothrix*

Spirochaetes and relatives
  A. Spirochaetes
    Spirochaeta, Treponema, Borrelia
  B. Leptospiras
    Leptospira, Leptonema

Green sulfur bacteria
  Chlorobium, Chloroherpeton

Bacteroides, flavobacteria and relatives
  A. Bacteroides group
    Bacteroides, Fusobacterium

  B. Flavobacterium group
    *Flavobacterium, Cytophaga, Saprospira,
    Flexibacter*

Planctomyces and relatives
  A. Planctomyces group
    Planctomyces, Pasteuria
  B. Thermophiles
    Isocystis pallida

Chlamydiae
  Chlamydia psittaci, C. trachomatis

Radio-resistant micrococci and relatives
  A. Deinococcus group
    Deinococcus radiodurans
  B. Thermophiles
    Thermus aquaticus

Green nonsulfur bacteria and relatives
  A. Chloroflexus group
    Chloroflexus, Herpetosiphon
  B. Thermomicrobium group
    Thermomicrobium roseum

Archaea subdivisions

Extreme halophiles
  Halobacterium, Halococcus morrhuae

Methanobacter group
  *Methanobacterium, Methanobrevibacter,
  Methanosphaera stadtmaniae, Methanothermus
  fervidus*

Methanococcus group
  Methanococcus

"Methanosarcina" group
  *Methanosarcina barkeri, Methanococcoides
  methylutens, Methanotrhix soehngenii*

Methospirillum group
  *Methanospirillum hungatei, Methanomicrobium,
  Methanogenium, Methanoplanus limicola*

Thermoplasma group
  Thermolasma acidophilum

Thermococcus group
  Thermococcus celer

Extreme thermophiles
  *Sulfolobus, Thermoproteus tenax,
  Desulfurococcus mobilis, Pryodictium occultum*

Hodgson, D. A. 1989. Bacterial diversity: The range of interesting things that bacteria do, pp. 4–22. In: *Genetics of Bacterial Diversity.* D. A. Hopwood and K. F. Chater (eds.). Academic Press, London.

Table 1.2 Comparison between Bacteria, Archaea, and Eucarya

Characteristic	Bacteria	Archaea	Eucarya
Peptidoglycan	Yes	No	No
Lipids	Ester linked	Ether linked	Ester linked
Ribosomes	70S	70S	80S
Initiator tRNA	Formylmethionine	Methionine	Methionine
Introns in tRNA	No	Yes	Yes
Ribosome sensitive to diphtheria toxin	No	Yes	Yes
RNA polymerase	One (4 subunits)	Several (8–12 subunits each)	Three (12–14 subunits each)
Ribosome sensitive to chloramphenicol, streptomycin, kanamycin	Yes	No	No

archaea have metabolic pathways for sugar degradation not found among the bacteria. These are described in Section 8.5.4.

Comparison of domains archaea, bacteria, and eucarya

Listed below are several structural and physiological differences that distinguish bacteria from archaea. In addition, some similarities between archaeal and eukaryotic RNA polymerases, and archaeal and eukaryotic ribosomes are noted (Table 1.2).

1. Archaeal membrane lipids are long-chain hydrocarbons linked by *ether-linkage* to glycerol.[7] The lipids in the membranes of bacteria and eukaryotes are *fatty acids ester-linked to glycerol* (Figs. 1.16 and 9.4), whereas the archaeal lipids are methyl-branched, isopranoid *alcohols ether-linked to glycerol* (Fig. 1.17). Archaeal membranes are discussed in Section 1.2.5, and the biosynthesis of archaeal lipids is discussed in Section 9.1.3.

2. Archaea lack peptidoglycan, a universal component of bacterial cell walls. The cell walls of some archaea contain pseudomurein, a component absent in bacterial cell walls (Section 1.2.3).

3. The archaeal RNA polymerase differs from bacterial RNA polymerase by having 8–10 subunits, rather than four subunits, and by not being sensitive to the antibiotic rifampicin. The difference in sensitivity to rifampicin reflects differences in the proteins

of the RNA polymerase. It is interesting that the archaeal RNA polymerase resembles the eukaryotic RNA polymerase which also has many subunits (10–12) and is not sensitive to rifampicin.

4. Some protein components of the archaeal protein synthesis machinery differ from those found in bacteria. Archaeal ribosomes are not sensitive to certain inhibitors of bacterial ribosomes (i.e., erythromycin, streptomycin, chloramphenicol, and tetracycline). These differences in sensitivity to antibiotics reflect differences in the ribosomal proteins. In this respect, archaeal ribosomes resemble cytosolic ribosomes from eukaryotic cells. Other resemblances to eukaryotic ribosomes are the use of methionine rather than formylmethionine to initiate protein synthesis, and the requirement for an elongation factor, EF-2, that can be ADP-ribosylated by diphtheria toxin. In contrast, bacteria use formylmethionine to initiate protein synthesis, and their elongation factor EF-2 is not sensitive to diphtheria toxin. However, the archaeal ribosomes are similar to bacterial ribosomes in being 70S.

5. The halophilic archaea have light-driven ion pumps not found among the bacteria. The pumps are bacteriorhodopsin and halorhodopsin which pump protons and chloride ions, respectively, across the membrane. The proton pump serves to create a proton potential that can be used to drive ATP synthesis. This is discussed in Section 3.6.2.

6. The methanogenic archaea have several coenzymes that are unique to archaea. The coenzymes are used in the pathway for the reduction of carbon dioxide to methane, and in the synthesis of acetyl-CoA from H_2 and CO_2. These coenzymes and their biochemical roles are described in Section 12.1.5.

1.2 Cell structure

Much of the discussion of cell structure will refer to the well-studied bacteria because structural studies of archaea are not as common; however, there are well-known differences between archaea and bacteria with respect to cell walls and cell membranes, and these will be discussed. The structure and function of the major cell components will be described, beginning with cellular appendages (various filaments and flagella) and working our way into the interior of the cell.

1.2.1 Appendages

Numerous appendages, each designed for a specific task, extend from the surface of bacteria. There are three classes of appendages:

1. Flagella, used for motility

2. Fimbriae (sometimes called pili), used for adhesion.

3. Sex pili, used for mating by some bacteria.

Flagella

Swimming bacteria have one or several flagella[8-10] that are organelles of locomotion that protrude from the cell surface. The *flagellum* is a stiff (semirigid) helical filament (either left-handed or right-handed, depending upon the species) that rotates like a propeller. It is unrelated in composition, structure, and mechanism of action to the eukaryotic flagellum. The arrangement and number of flagella vary with the species. They can be located at one or both cell poles, or inserted laterally around the entire cell.⑪ The flagella that are studied in most detail are those of *Escherichia coli* and *Salmonella typhimurium*, and most of the following dis-

amphitrichous

cussion concerns these flagella. The flagella of other bacteria, except for the spirochaetes, are similar in general structure to those of *E. coli* and *S. typhimurium*. Much less is known about the flagella of archaea, but they appear to to be similar in morphology to bacterial flagella. Each bacterial flagellum has a small rotating motor at its base, built of protein, embedded in the cell membrane. The motor is attached to a helical protein filament of about 2 nm in diameter that extends approximately 5–10 μm from the cell surface (Fig. 1.2). The motor is actually an electrochemical machine. The energy to drive the motor comes from a current of protons that moves down a proton potential gradient through the flagellar motor from the outside of the membrane to the inside. (Some marine bacteria use a sodium ion current.) The passage of protons turns the motor in a way that is not understood. When the motor turns, the attached filament rotates and propels the bacterium through the medium, often quite rapidly (e.g. 20–80 μm/sec). Flagellar rotation and bacterial swimming are described in more detail in Chapter 17. The use of ion currents across membranes to do work is described more fully in Chapter 3.

1. General structure

The flagellum consists of three parts—a *basal body*, a *hook*, and a *filament*—plus additional proteins required for motor function. Intact flagella have been isolated and their components subjected to analysis.[12,13] The flagellum contains about 20 different polypeptides and requires approximately 40 different genes for its assembly and function. The motor can rotate either clockwise or counterclockwise (in most studied bacteria), which determines the directionality of swimming, and responds to chemotactic signals. Section 17.4 contains a detailed description of chemotaxis.

a. **The basal body** (Fig. 1.2). At the base of the flagellum there is a *basal body* embedded in the membrane. Although it frequently has been speculated that the basal body may be the motor, the isolated structure does not contain several proteins described later that are required for motor function. It is therefore probably not the entire motor. The basal

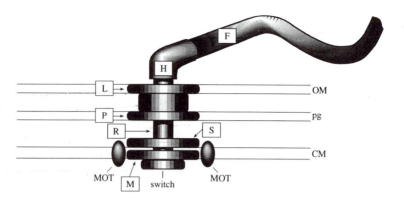

Fig. 1.2 Bacterial flagellum in a gram-negative envelope. The basal body is 22.5 × 24 nm and is composed of four rings, L, P, S, and M connected by a central rod. The M ring is embedded in the cell membrane and the S ring appears to lie on the surface of the membrane. Recently, it has been concluded that the S and M ring are actually one ring, the MS ring (see Wilson and Beveridge[9]). The P ring may be in the peptidoglycan layer and the L ring seems to be in the outer membrane. The P and L rings may act as bushings that allow the central rod to turn. Gram-positive bacteria have similar flagella but lack the L and P rings. The MotA and MotB proteins couple the proton potential to the rotation of the motor. The switch complex consists of three peripheral membrane proteins, FliG, FliM, and FliN, which are probably closely apposed to the cytoplasmic face of the M ring. Not shown are hook accessory or adaptor proteins called HAP1 and HAP3 between the hook and filament, and a protein cap called HAP2 on the end of the filament. The flagellum is assembled from the proximal to the distal end, with the filament being assembled last. It appears that the HAP1 and HAP3 proteins are required for the proper assembly of the filament on to the hook. Abbreviations: OM, outer membrane; pg, peptidoglycan; CM, cell membrane; R, central rod; MOT, MotA and MotB; H, hook; F, filament.

body consists of two stacked rings in gram-positive bacteria and four rings in gram-negative bacteria, through which a central rod passes, which is attached to the flagellum filament. The innermost ring (M) appears to be in the cell membrane. Next to the M ring is the S ring, which appears to be located on the external membrane surface. Electron microscopic evidence suggests that the M and S rings are actually one ring, called the MS ring. Gram-negative bacteria have an additional pair of rings (L and P rings) that may act as bushings, allowing the central rod to rotate in the peptidoglycan and outer membrane. The P ring is in the area of the peptidoglycan, and the L ring is in the outer envelope.

b. Additional proteins required for motor function. There are proteins that are not part of purified flagella and which are probably components of the motor because mutations in the genes that code for these protons affect motor function. For example, mutations in the *motA* and *motB* genes result in paralyzed flagella. The MotA and MotB proteins are in the membrane, and it is suggested that they surround the S and M rings in a ring of 8–12 proteins. Such rings of particles were seen in electron micrographs of freeze-fracture cytoplasmic membranes, but not when either MotA or MotB were missing.[14] The MotA and MotB proteins are believed to transduce the proton potential in an unknown fashion into mechanical rotation of the motor. Membrane vesicles prepared from strains synthesizing wild-type MotA were more permeable to protons than were vesicles prepared from strains synthesizing mutant MotA, indicating that MotA is likely to be a proton channel.[15] In bacteria such as *E. coli* and *S. typhimurium*, the motor changes its direction of rotation periodically. Mutants that fail to change flagellar rotation map in three genes, *fliG*, *fliM*, and *fliN*, which code for the switch proteins FliG, FliM, and FliN. The switch proteins seem to be peripheral membrane proteins closely associated with the basal body.

have them produce different single protein and not have HaPs so it's secreted, lots.

c. The hook. The central rod is attached to a curved hook which is made of multiple copies of a special protein called the hook protein (product of *flgE* gene). There are also two hook-associated proteins (HAPs), which are necessary to form the junction between the hook and the filament, and a third HaP (HaP2) which caps the filament. The HAPs are the products of the *flgK, flgL,* and *fliD* genes. Mutants that lack the HAPs secrete flagellin into the medium.

d. The filament. Attached to the hook is a semirigid helical filament that, along with the hook, protrudes from the cell. The protein in the filament is called *flagellin*, which is present in thousands of copies. Flagellin is not identical in different bacteria. For example, the protein can vary in size from 20 to 65 kD depending upon the species of bacterium. Furthermore, although there is homology between the C-terminal and N-terminal ends of most flagellins, the central part can vary considerably and is distinguished immunologically in different bacteria. In some cases, there is no homology at all. For example, nucleotide-derived amino-acid sequences for the flagellins from *Rhizobium meliloti* showed almost no relationship to flagellins from *E. coli, S. typhimurium,* or *Bacillus subtilis,* but were 60% similar to the N- and C-termini of flagellin from *Caulobacter crescentus.*[16] The flagellin subunits are arranged so that there is a central 60 Å hole, which may be important for transporting flagellin subunits from base to tip during growth.

2. Mechanics of motor function

The rotational force originates in the cell membrane, presumably in the M ring and its associated proteins, and is transmitted to the filament through the rod and hook. The P and L rings are assumed to be bushings through which the rod passes through the outer envelope. The flagellar motor must have both a rotor and a stator. Because of the large mass of the filament and the significant viscous drag that it encounters, the stator must be attached to a structural element sufficiently massive so as to preclude the stator from spinning in the membrane while the filament barely moves. It is usually assumed that whatever functions as the stator is firmly anchored to the peptidoglycan.

3. Growth of the flagellum

The filament grows at the tip as demonstrated by the use of the phenylanine analogue fluorophenylanine, or radioisotopes. Incorporation of the phenylalanine analogue fluorophenylalanine by *Salmonella* resulted in curly flagella that had only one half the normal wavelength. When the analogue was introduced to bacteria that had partially synthesized flagella, then the completed flagella were normal at the proximal ends and curly at the distal ends, implying that the fluorophenylalanine was incorporated at the tips during flagellar growth.[17] When *Bacillus* flagella were sheared off the cells and allowed to regenerate for 40 min before adding radioactive amino acids ([^{3}H]-leucine), radioautography showed that all of the radioactivity was at the distal region of the completed flagella.[18] The flagellin monomers are possibly transported through the central hole in the filament to the tip, where they are assembled.

growth goes towards tip

4. Differences in flagellar structure

Although the basic structure of flagella is similar in all bacteria thus far studied, there are important species-dependent differences. For example, some bacteria have *sheathed flagella,* whereas others do not. In some species (e.g., *V. cholerae*) the sheath contains lipopolysaccharide and appears to be an extension of the outer membrane.[19] There are also differences regarding the number of different flagellins in the filaments. Depending upon the species, there may be only one type of flagellin, or two or more different flagellins in the same filament. For example, *E. coli* has one flagellin and *R.meliloti* and *B. pumilis* each have two different flagellin proteins, but *C. crescentus* has three.[20–22] The data for *B. pumilis* and *C. crescentus* support the conclusion that the different flagellins reside in the same filament. It has been proposed that the filaments of *R. meliloti* are composed of heterodimers of the two different flagellins. Whereas most bacteria that have been studied have flagella that have a smooth surface as seen with the electron microscope, called *plain filaments* (e.g., *E. coli)*, certain bacteria (e.g., the soil bacteria, *Pseudomonas rhodos*, *R. lupini*, and *R. meliloti*) have "complex" filaments that have obvious helical patterns of ridges and grooves on the surface. Flagella with plain filaments rotate either clockwise or counterclockwise, whereas flagella with complex filaments rotate only clockwise with intermittent stops. It is thought that complex filaments are more rigid (because they are brittle) than are plain filaments, and are better suited for propelling the bacteria in viscous media. Although *E. coli* and *S. typhimurium* have four rings through which the central rod passes, other bacteria may have fewer or more. For example, gram-positive bacteria lack the outer two rings (the L and P rings). Additional structural elements of unkown function (e.g., additional rings or arrays of particles surrounding the basal body) have also been observed in certain bacteria.

The spirochaete flagella are called *axial filaments*. A major difference between the spirochaete flagella and those found in other bacteria is that the spirochaete flagella do not protrude from the cell; rather, they are wrapped around the length of the helical cell between the cell membrane and the outer membrane (i.e., in the periplasm). Spirachaetes have two or more flagella (some have 30 or 40 or more) inserted near each cell pole. The number inserted at each pole are the same. The flagella are usually more than half the length of the cell and overlap in the middle. Another difference is that the spirochaete flagellum is often surrounded by a *proteinaceous sheath*. The rotation of the periplasmic filaments are thought to propel the cell by propagating a helical wave down the length of the cell which causes the cells to "corkscrew" through viscous media. There are five antigenically related flagellins in the axial filaments but it is not known whether individual filaments contain more than one type of flagellin.[23] Despite these differences, the structure of the spirochaete flagella is the same as the flagella found in other bacteria in having a basal body composed of a series of rings surrounding a central rod, a hook, and a filament.

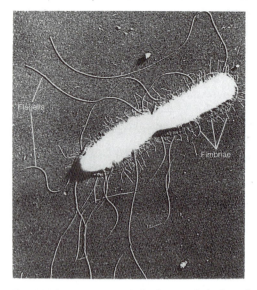

Fig. 1.3 Electron micrograph of a metal-shadowed preparation of *Salmonella typhi* showing flagella and fimbriae. The cell is about 0.9 μm in diameter. (Reprinted with permission of J. P. Duguid.)

Fimbriae, pili, filaments, and fibrils

Protein fibrils have been called various names, including fimbriae, pili, filaments, and simply fibrils. They extend from the surface of most gram-negative bacteria (Fig. 1.3).[24–27] Similar fibrils are rarely seen in gram-positive bacteria, but they are present in the gram-positive *Corynebacterium renale* and *Actinomyces viscosus*.[28–30] All such appendages are not alike nor are their functions known in all instances. The size varies considerably. They can be quite short (0.2 μm) or very long (20 μm), and differ in width from 3 to 14 nm or greater. Some originate from basal bodies in the cell membrane. However, most seem not to originate in the cell membrane at all and how they are attached to the cell is not known. Furthermore, they are not always present. Although freshly isolated strains frequently have fibrils, they are often lost during subculturing in the laboratory. They apparently are useful in the natural habitat but dispensable in laboratory cultures. What do they do?

1. Fimbriae

Many of the fibrils can be observed to mediate attachment of the bacteria to other cells, e.g., to other bacteria, animal, plant, or fungal cells. They are therefore important for colonization because they help the bacteria stick to surfaces. In nature, most bacteria grow while attached to surfaces where the concentration of nutrients is frequently highest. Attachment can be via fibrils that extend from the surface of the cells, and/or nonfibrillar material that may be part of the cell wall or glycocalyx (Section 1.2.2). It has been proposed that fibrils that mediate attachment be called *fimbriae*. That convention will be followed here, although many researchers instead refer to adhesive protein fibrils as adhesive pili. Fimbriae have *adhesins*, (i.e., molecules that cause bacteria to stick to surfaces). The adhesins are proteins in the fimbriae, often minor proteins at the tip, that recognize and bind to specific receptors on the surfaces of cells.

Fimbrial proteins (and other adhesins) are of important medical significance, as the following discussion will illustrate. Adhesion is studied in two types of experimental situations: attachment of bacteria to erythrocytes causing them to clump (*hemagglutination*), and attachment of bacteria *in vitro* to host cells to which they normally attach *in vivo*.[31] An example of the latter is the attachment of the causative agent of gonorrhea, *Neisseria gonorrhoeae*, to epithelial cells of the urogenital tract. When studying such attachments it has often been found that specific monosaccharides and oligosaccharides inhibit the attachment or hemagglutination when added to the suspension. The implication is that at least some, if not all, fimbriae bind to oligosaccharides in cell surface receptors, and that the added monosaccharides and oligosaccharides are inhibitory because they compete with the receptor for the adhesin. Receptors on animal cell surfaces include *glycolipids* and *glycoproteins*, which are embedded in animal cell membranes via the lipid and protein portions, and present their oligosaccharide moieties to the outer surface. It has been reported that many strains of *E. coli*, *Salmonella*, and *Shigellae* carry fimbriae whose hemagglutinin activity is prevented by D-mannose and methyl-α-mannoside. These fimbriae, therefore, are believed to attach to mannose glycoside residues on the cell surface receptors. They are called *common fimbriae*, or *mannose-sensitive fimbriae*, or (more frequently) *type 1 fimbriae*. Other fimbriae recognize different cell surface receptors and are sometimes called mannoseresistant fimbriae. For example, although most clinical isolates of *E. coli* from the urinary and gastrointestinal tracts possess type 1 fimbriae, many also (or instead) have galactose-sensitive fimbriae, called *P type fimbriae*.[32,33] P fimbriae bind the α-D-galactopyranosyl-(1-4)-β-D galactopyranoside [Galα-(1-4)Gal] in the glycolipids on cells lining the upper urinary tract.[34] Other known fimbriae include the *S*, *K88*, *F17*, and *K99* fimbriae of *E. coli*, the *type 4 fimbriae* produced by *N. gonorrhoeae*, *Moraxella bovis*, *Bacteroides nodosus*, *P. aeruginosa*, and the *Tcp fimbriae* produced by *Vibrio cholerae*.[35] The point being made here is that different types of fimbriae are specialized for attachment to specific receptors and can account for the specificity of bacterial attachment to hosts and tissues. They can be distinguished by inhibition of binding by mono- and oligosaccharides as well as by a variety of other methods, including morphology, antigenicity, molecular weight of the protein subunit, isoelectric point of the protein subunit, and amino acid composition and sequence. Since gram-positive bacteria generally do not have fimbrias, their adhesins are part of other cell surface components (e.g., the glycocalyx, described in Section 1.2.2).

2. Sex pili

Bacteria are capable of attaching to each other for the purpose of transmitting DNA from a donor cell to a recipient cell. This is called mating. Some bacteria (not all) use fimbriae for mating attachments. *The fimbriae that mediate attachment between mating cells are different from the other fimbriae and are called sex pili.* The requirement of sex pili for mating is found for enteric bacteria such as *E. coli*, and pseudomonads, but it is not universal among the bacteria. For example, gram-positive bacteria do not have sex pili. The sex pilus grows on "male" strains that donate DNA to recipient ("female") strains. In *E. coli*, it is coded for by a conjugative transmissible plasmid, the F plasmid, that resides in the donor strains.[36] An illustration of how the F pilus in *E. coli* works is shown

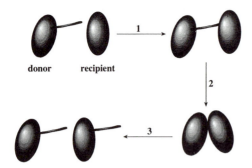

donor   recipient

Fig. 1.4 F-pilus mediated conjugation. Transfer of a sex plasmid. The donor cell has a plasmid and an F-pilus encoded by plasmid genes. 1. The F-pilus binds to the recipient cell. 2. The pilus retracts bringing the two cells together. This is due to a depolymerization of the pilus subunits. 3. The plasmid is transferred as it replicates so that when the cells separate each has a copy of the plasmid and is a potential donor.

in Fig. 1.4. The tip of the F pilus adheres to receptor molecules on the recipient cell surface. This is followed by a retraction of the pilus (depolymerization of the F-pilin subunits into the cell membrane) bringing the two cells closer together until their surfaces are in contact. DNA transfer takes place at the site of contact (not through the sex pilus). Sex pili seem to be designed for mating in cell suspension where the bacteria are not in intimate contact. The sex pilus presumably helps to stabilize the mating pairs of bacteria. They are not needed for many bacteria, perhaps because they mate in colonies or aggregates on solid surfaces where the cells are in close contact. However, it should be pointed out that the gram-positive bacterium *Streptococcus faecalis* (now called *Enterococcus faecalis*) forms efficient mating aggregates in liquid suspension and does not have sex pili. In other words, their adhesins are located on the cell surface rather than on pili. Mating in *E. faecalis* is induced by a sex pheromone secreted into the medium by recipient cells. The sex pheromones signal the donor cells to synthesize cell surface adhesins that promote cell aggregate formation (clumping) and subsequent DNA transfer.[37] Mating interactions mediated by surface adhesins is widespread in the microbial world. Other well-studied examples include *Chlamydomonas* mating and yeast mating.[38]

1.2.2 The glycocalyx

The term *glycocalyx*[39–41] is often used to describe all extracellular material *external to the cell wall*. All bacteria are probably surrounded by glycocalyces as they grow in their natural habitat, although they often lose these external layers when cultivated in the laboratory. The extracellular polymers may be in the form of S layers, capsules, or slime.

S layers
These are an array of protein or glycoprotein subunits on the cell wall. The S layers are found on the cell wall surfaces of a wide range of gram-positive and gram-negative eubacteria. They are also present in the archaea, where the S layer sometimes covers the cell membrane and serves as the cell wall itself. If the S layer is the wall itself, it is not called a glycocalyx.

Capsules
Capsules are composed of fibrous material at the cell surface (Figs. 1.5 and 1.6). They may be rigid, flexible, integral (i.e., very closely associated with the cell surface), or peripheral (i.e,. loose) material that is sometimes shed into the medium. Material that loosely adheres to the cell wall is sometimes called *slime or slime capsule.*

Chemical composition
The glycocalyces from several bacteria have been isolated and characterized. Although many are polysaccharide, some are protein. For example, some *Bacillus* species form a glycocalyx that is a polypeptide capsule. Also, pathogenic *Streptococcus* species have a fibrillar protein layer, the M protein, on the external face of the cell wall.

Role
An important role for the glycocalyx is adhesion to the surfaces of other cells or to inanimate objects. Such adhesion is necessary for colonization of solid surfaces or growth in biofilms (Figs. 1.5 and 1.6). An example is the complex ecosystem of several different bacteria maintained by intercellular adherence in human oral plaques.[42]

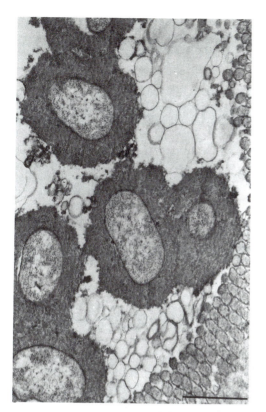

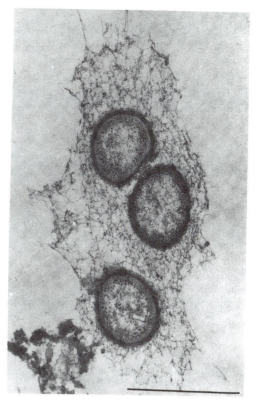

Fig. 1.5 Electron micrograph of a ruthenium red-stained thin-section of *E. coli* shown adhering to the neonatal calf ileum. Scale bar represents 1.0 µm. (From Costerton, J. W., T. J. Marrie, and K.-J. Cheng. 1985. Phenomena of bacterial adhesion, pp. 3–43. In: *Bacterial Adhesion* D. C. Savage and M. Fletcher (eds.). Plenum Press, New York, London.)

Fig. 1.6 Electron micrograph of a ruthenium red-stained thin section of bacteria adhering to a rock surface in a subalpine stream showing fimbriated glycocalyx. Scale bar represents 1.0 µm (From Costerton, J. W., T. J. Marrie, and K.-J. Cheng. 1985. Phenomena of bacterial adhesion, pp. 3–43. In: *Bacterial Adhesion*. D. C. Savage and M. Fletcher (eds.). Plenum Press, New York, London.)

Another role of the glycocalyx is protection from phagocytosis. For example, mutants of pathogenic strains of bacteria that no longer synthesize a capsule, such as unencapsulated strains of *S. pneumoniae*, are more easily phagocytized by white blood cells, making them less virulent.

1.2.3 Cell walls

Most bacteria are surrounded by a cell wall that lies over the external face of the cell membrane and protects the cell from bursting due to the turgor pressure that exists within the cell.[43,44] The turgor pressure is due to the fact that bacteria generally live in environments that are more dilute than the cytoplasm. As a consequence, there is a net influx of water resulting in a large hydrostatic pressure (turgor) of several atmospheres directed out against the cell membrane. Two different types of walls exist among the bacteria. Bacteria possessing one type of wall can be stained using the Gram stain procedure. They are called gram-positive. Bacteria possessing the second wall type do not stain and are called gram-negative. The Gram stain is described next, followed by a description of peptidoglycan, a cell wall polymer found in both gram-positive and gram-negative walls responsible for the strength of bacterial cell walls. This will be followed by descriptions of the gram-positive and gram-negative cell walls.

The Gram stain

The Gram stain was invented in the nineteenth century by Christian Gram to visualize bacteria in tissues using ordinary bright-field microscopy. When appropriately stained, bacteria with cell walls can usually be divided into two groups, depending upon whether they retain a crystal violet–iodine stain complex (gram-positive) or not (gram-negative) (Fig. 1.7). Gram-negative bacteria are visualized with a pink counter stain called safranin. During the staining procedure the complex of crystal violet and iodine can be removed with alcohol or acetone from gram-negative cells, but not from gram-positive cells. This is thought to be related to the thickness and composition of the gram-positive wall as compared to the much thinner gram-negative wall. The crystal violet–iodine complex is trapped within gram-positive cells by the thick cell wall. The difference in Gram staining is also related to the outer wall layer of gram-negative bacteria (also called the outer envelope or outer membrane), which is rich in phospholipids and thus made leaky by the lipid solvents, alcohol, and acetone. Archaea can stain either gram-positive or gram-negative, but their cell wall composition is different from that of bacteria, as will be described later. Therefore, the Gram stain cannot solely be relied upon to determine relationships among bacteria. Nor can one rely on the Gram stain as a probe for cell wall structure or composition.

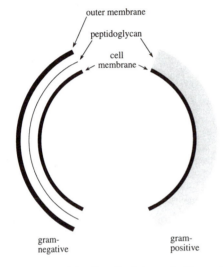

Fig. 1.7 Bacterial cell walls. Schematic illustration of a gram-negative wall and a gram-positive wall. Note the presence of an outer membrane (also called outer envelope) in the gram-negative wall and the much thicker peptidoglycan layer in the gram-positive wall.

Peptidoglycan

The strength and rigidity of bacterial cell walls is due to a glycopeptide called peptidoglycan or murein, which consists of glycan chains crosslinked by peptides (Fig. 1.8, Fig. 10.3). The structure and synthesis of peptidoglycan is discussed in more detail in Section 10.1. The glycan consists of alternating residues of N-acetylglucosamine (GlcNAc or G) and N-acetylmuramic acid (MurNAc or M) attached to each other via β-1,4 linkages. Attached to the MurNAc is a tetrapeptide that crosslinks the glycan chains via peptide bonds. The tetrapeptide usually consists of L-alanine, D-glutamate, a diamino acid, and D-alanine. The peptidoglycan forms a three-dimensional network surrounding the cell

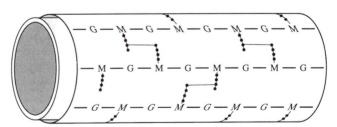

Fig. 1.8 Schematic drawing of peptidoglycan layer. The peptidoglycan surrounds the cell membrane and consists of glycan chains (–G–M–) crosslinked by tetrapeptides (closed circles). The thickness of the peptidoglycan varies from approximately a monomolecular layer in gram-negative bacteria to several layers in gram-positive bacteria. The direction in which the glycan chains are depicted to be running is not to be taken literally. M, N-acetylmuramic acid; G, N-acetylglucosamine.

membrane, covalently bonded throughout by the glycosidic and peptide linkages. Destruction of the peptidoglycan by the enzyme lysozyme (which hydrolyzes the glycosidic linkages) or interference in its synthesis by antibiotics such as penicillin, vancomycin, or bacitracin, results in the inability of the cell wall to restrain the turgor pressures. Under these circumstances, the influx of water bursts the cell in dilute media.

There are some important differences between the peptidoglycan in gram-positive and gram-negative bacteria. The peptidoglycan of gram-negative bacteria can be isolated as a sac of pure peptidoglycan which surrounds the cell membrane in the living cell. It is called the *murein sacculus*. The sacculus is elastic and believed to be under stress *in vivo* because of the expansion due to turgor pressure against the cell membrane.[45,46] In contrast, the peptidoglycan from gram-positive bacteria is covalently bonded to various polysaccharides and teichoic acids, and not isolatable as a pure murein sacculus. (See the discussion later of the gram-positive wall.) The peptidoglycan from gram-negative bacteria differs from the peptidoglycan from gram-positive bacteria in two other ways: (1) diaminopimelic acid is generally the diamino acid in gram-negative bacteria, whereas it is the diamino acid in only some of the gram-positive bacteria; and (2) the cross-linking is usually direct in gram-negative bacteria, whereas there is usually a peptide bridge in gram-positive bacteria (Fig. 10.3).

As described later, the peptidoglycan in gram-negative bacteria is attached non-covalently to the outer envelope via lipoprotein. Whether it is also attached to the cell membrane is a matter of controversy. The evidence for attachment to the cell membrane is that when *E. coli* is plasmolyzed (placed in hypertonic solutions so that water exits the cells), and prepared for electron microscopy by chemical fixation followed by dehydration with organic solvents, the cell membrane is seen to have adhered to the outer envelope in numerous places while generally shrinking away in other locations. It is reasonable to suggest that the zones of adhesion may be areas where the peptidoglycan bonds the cell membrane to the outer membrane, because

the peptidoglycan lies between the two membranes. However, this is presently a controversial subject. It has been argued that the zones of adhesion seen in electron micrographs of plasmolyzed, chemically fixed, and dehydrated cells are artifacts since they were not visible in cells prepared for electron microscopy by rapid cryofixation (also called freeze substitution) after plasmolysis.[47] In cryofixation, the water in and around the cells is rapidly frozen by liquid helium and the frozen water is substituted with osmium tetroxide in acetone.

Gram-positive walls

1. Chemical composition

The gram-positive cell wall in bacteria is a thick structure approximately 15–30 nm wide and consists of several polymers, the major one being peptidoglycan (Figs 1.8 and 1.9). The kinds and amounts of the other polymers in the wall vary according to the genus of bacterium. The nonpeptidoglycan polymers can comprise up to 50% of the dry weight of the wall. The ones most commonly found are usually covalently bound to the glycan chain of the peptidoglycan. These polymers include (Fig. 1.10):

Teichoic acids. Teichoic acids are polymers of either ribitol phosphate or glycerol phosphate.

Teichuronic acids. Teichuronic acids are acidic polysaccharides containing uronic acids, (e.g., some have N-acetylgalactosamine and D-glucuronic acid).

Neutral polysaccharides. Neutral polysaccharides are particularly important for the classification of streptococci and lactobacilli where they are used to divide the bacteria into serological groups (e.g., groups A, B, and C streptococci).

Lipoteichoic acids. A teichoic acid found in most gram-positive walls is lipoteichoic acid (LTA), which is a linear polymer of phosphodiester-linked glycerol phosphate covalently bound to lipid. The C2 position of the glycerol-phosphate is usually glycosylated and/or D-alanylated. Because of the negatively charged backbone of glycerol phosphate and the hydrophobic lipid, the molecule

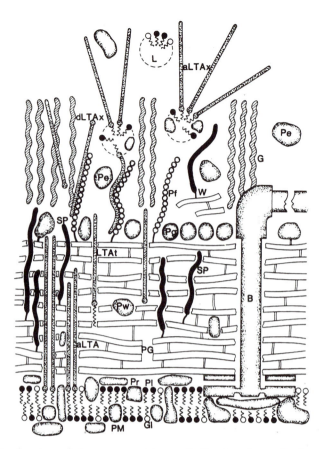

Fig. 1.9 Schematic drawing of the gram-positive wall. Starting from the bottom up, PM is the plasma cell membrane consisting of protein (Pr), phospholipid (Pl), and glycolipid (Gl). Overlaying the cell membrane is highly cross-linked peptidoglycan (PG) to which is covalently bound teichoic acids, teichuronic acids, and other polysaccharides (SP) and noncovalently attached protein (Pw). Acylated lipoteichoic acid (aLTA) is bound to the cell membrane and extends into the peptidoglycan. Some of the LTA is in the process of being secreted (LTAt). LTA that is already secreted into the glycocalyx is symbolized as aLTAx. Some of the LTA in the glycocalyx is deacylated and symbolized as dLTAx. (Deacylated means that the fatty acids have been removed from the lipid moiety.) Within the glycocalyx can also be found lipids (L), proteins (Pe), pieces of cell wall (W), globular protein (Pg) and polymers which are part of the glycocalyx proper (G). Also shown is the basal body of a flagellum (B). (From Wicken, A. 1985. Bacterial cell walls and surfaces, pp. 45–70. In: *Bacterial Adhesion*. D. C. Savage and M. Fletcher (eds.). Plenum Press, New York, London.)

is amphipathic (i.e., it has both a polar and a nonpolar end). The lipid portion is bound hydrophobically to the cell membrane, whereas the polyglycerol phosphate portion extends into the cell wall. Unlike the other polymers thus far discussed, LTA is not covalently bound to the peptidoglycan. Its biological role at this location is not understood, but in some bacteria it is secreted and can be found at the cell surface where it is thought to act as an adhesin. For example, LTA is secreted by *S. pyogenes* where it binds with the M protein and acts as a bridge to receptors on host tissues.[48] Under these circumstances, one should consider the secreted LTA as part of the glycocalyx along with the M protein.

Other glycolipids. There is a growing list of gram-positive bacteria that do not contain LTA but have instead other amphiphilic glycolipids that might substitute for some of the LTA functions, whatever they might be.[49] Bacteria having these cell-surface glycolipids (also called macroamphiphiles or lipoglycans) belong to various genera, including *Micrococcus, Streptococcus, Mycobacterium, Coryne*

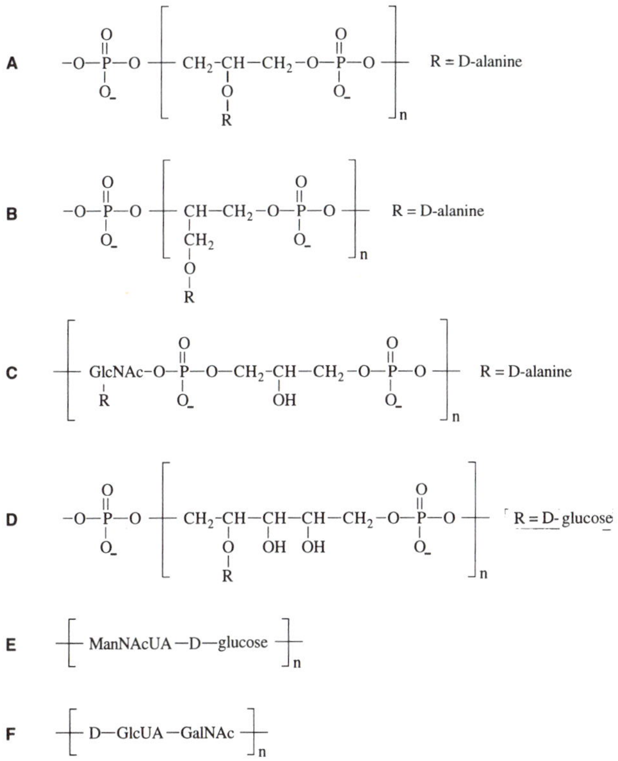

Fig. 1.10 Some teichoic and teichuronic acids found in different gram-positive bacteria. A, glycerol phosphate teichoic acid with D-alanine esterified to the C-2 of glycerol. B, glycerol phosphate teichoic acid with D-alanine esterified to the C-3 of glycerol. C, glycerol phosphate teichoic acid with glucose and N-acetylglucosamine in the backbone subunit. D-alanine is esterified to the C-6 of N-acetylglucosamine. D, ribitol phosphate teichoic acid, with D-glucose attached in a glycosidic linkage to the C-4 of ribitol. E, teichuronic acid with N-acetylmannuronic acid and D-glucose. F, teichuronic acid with glucuronic acid and N-acetylgalactosamine. It is believed that teichoic and teichuronic acids are covalently bound to the peptidoglycan through a phosphodiester bond between the polymer and a C-6 hydroxyl of muramic acid in the peptidoglycan.

bacterium, Propionibacterium, Actinomyces, and Bifidobacterium.

Mycolic acid. Still another cell wall polymer is formed by bacteria belonging to the genus *Mycobacterium*, which include the causative agents of tuberculosis and leprosy. They have cell walls rich in waxy lipids called mycolic acids. When these bacteria are appropriately stained, the dye (basic fuschin) is not removed by dilute hydrochloric acid in ethanol (acid alcohol) because of the presence of the mycolic acid. They are therefore called *acid-fast* bacteria.

In summary, gram-positive cell walls have diverse types of neutral and acidic polysaccharides, glycolipids, lipids, and other compounds either free in the wall or covalently bound to the peptidoglycan. *The functions of most of these polymers is largely unknown,* although some may act as adhesins and/or presumably affect the permeability characteristics of the cell wall.

Gram-negative walls

The gram-negative cell wall is structurally and chemically complex. It consists of an outer membrane composed of lipopolysaccharide, phospholipid and protein, and an underlying peptidoglycan layer (Fig. 1.11). Between the outer and inner membrane (the cell membrane) is a compartment called the *periplasm*, wherein the peptidoglycan lies.

1. Lipopolysaccharide structure and function

The lipopolysaccharide (LPS) consists of three regions: *lipid A, core,* and a *repeating*

17

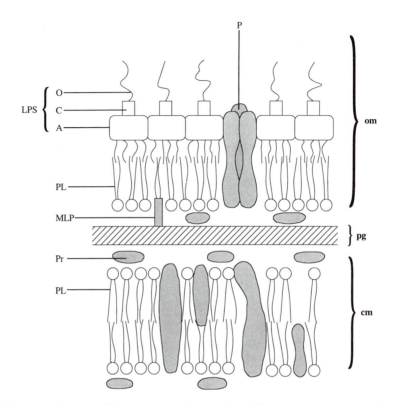

Fig. 1.11 Schematic drawing of the gram-negative envelope. The outer membrane consists of lipopolysaccharide, phospholipid, and proteins, most of which are porins. Underneath the outer membrane is the peptidoglycan layer which is noncovalently bonded to the outer membrane via murein lipoproteins, themselves covalently attached to the peptidoglycan. The cell membrane is composed of phospholipid and protein. The area between the outer membrane and the cell membrane is called the periplasm. The wavy lines are fatty acid residues which anchor the phospholipids and lipid A into the membrane. LPS, lipopolysaccharide; O, oligosaccharide; C, core; A, lipid A; P, porin; PL, phospholipid; MLP, murein lipoprotein; pg, peptidoglycan; Pr, protein; om, outer membrane; cm, cell membrane.

oligosaccharide, sometimes called *o-antigen* or *somatic antigen.* Its chemical structure and synthesis is described in Section 10.2. In the Enterobacteriacea (e.g., *E. coli*), the lipopolysaccharide is confined to the outer leaflet of the outer membrane and is arranged so that the lipid A portion is embedded in the membrane as part of the lipid layer, and the core and oligosaccharide extend into the medium (Fig. 1.11). Mutants that lack the oligosaccharide experience loss of virulence and it is believed that the LPS can increase pathogenicity. Loss of most of the core and oligosaccharide in *E. coli* and related bacteria is associated with increased sensitivity to hydrophobic compounds (e.g., antibiotics, bile salts, and hydrophobic dyes such as eosin and methylene blue). This is because the LPS

provides a permeability barrier to hydrophobic compounds. It is advantageous for the enteric bacteria to have such a permeability barrier because they live in the presence of bile salts in the intestine. In fact, the basis for selective media for gram-negative bacteria is the resistance of these bacteria to bile salts and/or hydrophobic dyes, which are included in the media. The bile salts and dyes inhibit the growth of gram-positive bacteria, but not gram-negative bacteria because of the LPS. An example of such a selective medium is eosin–methylene blue (EMB) agar, which is used for the isolation of gram-negative bacteria because it inhibits the growth of gram-positive bacteria.

The low permeability of the outer membrane to hydrophobic compounds is apparently due to the fact that the phospholipids are confined primarily to the inner leaflet of the outer envelope and the LPS is in the outer leaflet. The LPS presents a permeability barrier to hydrophobic substances. One suggestion as to how this might occur is that, since lipid A contains only saturated fatty acids, the LPS presents a somewhat rigid matrix, and this, plus the tendency of the large LPS molecules to engage in lateral noncovalent interactions, makes it difficult for hydrophobic molecules to penetrate between the LPS molecules to the phospholipid layer. The fact that mutants that lack a major region of the oligosaccharide and core are more permeable to hydrophobic compounds is due to the fact that there is more phospholipid in the outer leaflet of these mutants, and is also due to the likelihood that there is less lateral interaction between the LPS molecules. The asymmetric distribution of phospholipid to the inner leaflet of the outer envelope seems to be an adaptive evolutionary response of the enteric bacteria to hydrophobic toxic substances in the intestine of animals. Accordingly, not all gram-negative bacteria have an outer envelope with an asymmetric distribution of lipopolysaccharide and phospholipid.

Lipoproteins

In addition to lipopolysaccharide and phospholipid, a major component of the outer membrane is protein, of which there are several different kinds. One of the proteins is called the *murein lipoprotein.* This is a small protein with lipid attached to the amino terminal end (Fig. 1.12). The lipid end of the molecule extends into and binds hydrophobically with the lipids in the outer envelope. The protein end of some of the molecules is covalently bound to the peptidoglycan, thus anchoring the outer envelope to the peptidoglycan. In *E. coli,* about one-third of the murein lipoprotein is bound to the peptidoglycan. Mutants unable to synthesize the murein lipoprotein have unstable outer envelopes that bleb off into the medium at the cell poles and septation sites. Therefore, the murein lipoprotein may play a structural role in keeping the outer membrane attached to the cell surface. There are a small number of other outer membrane or cell membrane lipoproteins.[50] These were discovered by chemically cross-linking the peptidoglycan in whole cells to closely associated proteins with a bifunctional cross-linking reagent.[51] There is no evidence that these additional lipoproteins are covalently bonded to the peptidoglycan and their functions remain to be elucidated.

Porins and other proteins

The major proteins in the outer envelope are called porins. The porins form small nonspecific hydrophilic channels through the

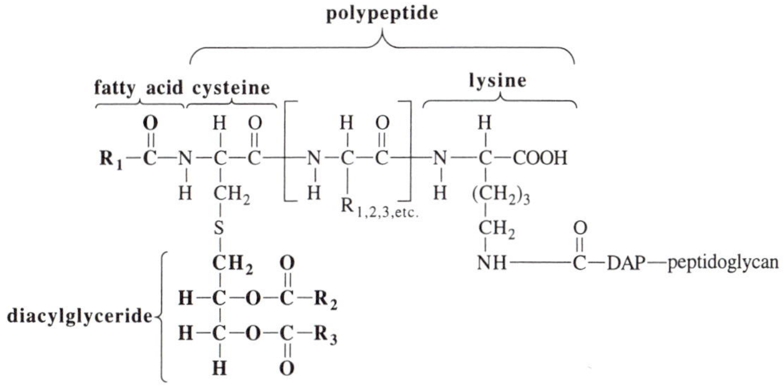

Fig. 1.12 Murein lipoprotein. Attached to the amino terminal cysteine is a diacylglyceride in thioether linkage and a fatty acid in amide linkage. The lipid portion extends into the outer envelope and binds hydrophobically with the fatty acids in the phospholipids and lipopolysaccharide. The carboxy terminal amino acid is lysine, which can be attached via an amide bond to the carboxyl group of diaminopimelic acid (DAP) in the peptidoglycan. The murein lipoprotein therefore holds the outer envelope to the peptidoglycan.

outer envelope allowing the diffusion of neutral and charged solutes of molecular weights less than 600 D. The channels are necessary to allow passage of small molecules into and out of the cell. E. coli has three major porins: OmpF, OmpC, and PhoE. Each porin makes a separate channel. Thus, there are OmpF, OmpC, and PhoE channels. The OmpC channel is approximately 7% smaller than the OmpF channel and is expected to make the outer envelope less permeable to larger molecules. OmpF and OmpC are both present under all growth conditions, although the ratio of the smaller OmpC to the larger OmpF increases in high osmolarity media and high temperature. The increased amounts of OmpC relative to OmpF presumably also occur in the intestine, which has a higher osmolarity and temperature than lakes and streams where E. coli is also found. This may be an advantage to the enterics because the smaller OmpC channel should present a diffusion barrier to toxic substances in the intestine, whereas the larger OmpF channel should be an advantage in more dilute environments outside of the body. The protein PhoE is produced only under conditions of inorganic phosphate limitation. This is because PhoE is a channel for phosphate (and other anions) whose synthesis under limiting phosphate conditions reflects the need to bring more phosphate into the cell. Porins appear to be widespread among gram-negative bacteria, although they are not all identical. As mentioned earlier, E. coli regulates the amounts of the various porins according to growth conditions. This is discussed in Chapter 17.

Since the porins exclude molecules with molecular weights larger than 600 D, one would expect to find other proteins in the outer membrane that facilitate the translocation of larger solutes across the outer membrane. This is the case. E. coli has an outer membrane protein called the LamB protein that forms channels for maltose and maltodextrins. Other proteins in the outer membrane of E. coli facilitate the transport of vitamin B_{12} (BtuB), nucleosides (Tsx), and several other solutes.

Archaeal cell walls

Archaeal cell walls are not all alike and are very different from bacterial cell walls. For example, *no archaeal cell wall contains peptidoglycan.* Archaeal cell walls may be either pseudopeptidoglycan, polysaccharide, or protein (the S layer). Pseudopeptidoglycan (also called pseudomurein) resembles peptidoglycan in consisting of glycan chains cross-linked by peptides (Fig. 1.13). However, the resemblance stops there. In pseudopeptidoglycan, although one of the sugars is N-acetylglucosamine as in peptidoglycan, the other is N-acetyltalosaminuronic acid instead of N-acetylmuramic acid. Furthermore, the

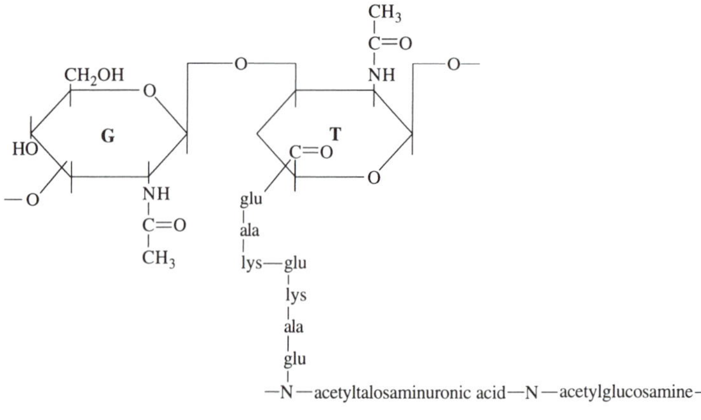

Fig. 1.13. Pseudopeptidoglycan in archaea. Pseudopeptidoglycan resembles peptidoglycan in being a cross-linked glycopeptide. It differs from peptidoglycan in the following ways: (1) N-acetyltalosaminuronic acid replaces N-acetylmuramic acid; (2) the glycosidic linkage is β1-3 instead of β1-4; (3) there are no D-amino acids. G, N-acetylglucosamine; T, N-acetyltalosaminuronic acid.

sugars are linked by a β,1-3 glycosidic linkage rather than a β,1-4, and the amino acids in the peptides are L-amino acids rather than D-amino acids. (The latter are present in peptidoglycan; compare Fig. 1.13 with Fig. 10.2.) Thus, pseudomurein is very different from murein.

1.2.4 Periplasm

Gram-negative bacteria have a separate compartment between the cell membrane and the outer membrane called the *periplasm*[52] (Fig.1.14). It can be seen in electron micrographs of thin-sections of cells as a space but should be considered an aqueous compartment containing protein, oligosaccharide, salts, and peptidoglycan. It seems that the peptidoglycan and oligosaccharides may exist in a hydrated state forming a periplasmic gel.[53] The periplasm should be considered a cellular compartment with specialized activities. These activities include oxidation–reduction reactions (Chapters 4 and 11), osmotic regulation (Chapter 14), solute transport (Chapter 15), protein secretion (Chapter 16), and hydrolytic activities.

Periplasmic components

The periplasm is chemically complex and carries out diverse functions. The following list of components and their functions reflects the importance of the periplasm. The list should be considered as a partial inventory emphasizing those periplasmic functions about which most is known.

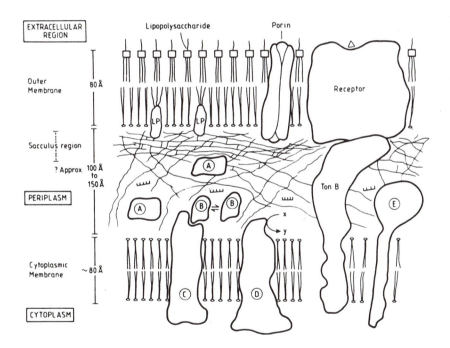

Fig. 1.14. A model of the periplasm in *E. coli*. The outer region of the periplasm is thought to consist of cross-linked peptidoglycan attached to the outer envelope via lipoprotein (LP) covalently bound to the peptidoglycan. The inner region of the periplasm (approaching the cell membrane) is believed to consist of fewer cross-linked peptidoglycan chains and oligosaccharides that are hydrated and form a gel. The gel phase is thought to contain periplasmic proteins (e.g., A and B). For example, A might be a periplasmic enzyme and B could be a solute-binding protein that interacts with a membrane transporter (C). D and E are integral membrane proteins, perhaps enzymes. The outer envelope is depicted as consisting of lipopolysaccharide, porins, and specific solute transporters (receptors) which require a second protein (TonB) for uptake. (From Ferguson, S. J. 1991. The periplasm. pp. 311–339. In: Prokaryotic Structure and Function, A New Perspective. S. Mohan, C. Dow, and J. A. Coles (eds.). Cambridge University Press, Cambridge.)

1. Oligosaccharides

The oligosaccharides in the periplasm are thought by some to be involved in osmotic regulation of the periplasm because their amounts decrease when the cells are grown in media of high osmolarity. This is a complex subject and is discussed more fully in Section 14.2.3.

2. Solute binding proteins

There are also solute binding proteins in the periplasm that assist in solute transport. The solute-binding proteins bind to solutes (e.g., sugars and amino acids that have entered the periplasm through the outer envelope) and deliver the solutes to specific transporters (carriers) in the membrane. This is an important means of bringing nutrients into the cell and is discussed in Section 15.3.3.

3. Cytochromes

Some of the enzymes in the periplasm are cytochromes c that oxidize carbon compounds or inorganic compounds, and deliver the electrons to the electron transport chain in the cell membrane. These oxidations are called periplasmic oxidations. There are other oxidoreductases in the periplasm as well, but the various cytochromes c are very common. Periplasmic oxidations are important for energy metabolism in many different gram-negative bacteria and are discussed in Chapters 4 and 11.

4. Hydrolytic enzymes

There are also hydrolytic enzymes in the periplasm that degrade nutrients to smaller molecules that can be transported across the cell membrane by specific transporters. For example, the enzyme amylase is a periplasmic enzyme that degrades oligosaccharides to simple sugars. Another example is alkaline phosphatase, which removes phosphate from simple organic phosphate monoesters. The inorganic phosphate is then carried into the cell via specific inorganic phosphate transporters.

5. Detoxifying agents

Some periplasmic enzymes are detoxifying agents. For example, the enzyme to degrade penicillin (β-lactamase) is a periplasmic protein.

6. TonB protein

Figure 1.14 also shows an interesting periplasmic protein in *E. coli* called the TonB protein whose mechanism of action is not understood. It is known that TonB is required for the uptake of several solutes that do not diffuse through the porins; rather, they require specific transport systems (also called receptors) in the outer envelope. All of these solutes have molecular weights larger than 600 D, which is the upper limit for molecules that enter via the porins. Examples of solutes with specific outer membrane receptors that require TonB for uptake are iron siderophores[54] and vitamin B_{12}. The interesting thing is that these solutes are brought into the periplasm against a large concentration gradient, sometimes 1,000 times higher than the concentration outside the cell. Since the source of metabolic energy is either in the cytoplasm (ATP) or the cell membrane (electrochemical potential), it is not understood how passage through the outer envelope is energized. In a way that is not understood, the TonB protein is thought to couple the energy in the cytoplasm or the cell membrane to the uptake of certain solutes through the outer envelope and into the periplasm.[55]

1.2.5 Cell membrane

We now come to what is certainly the most functionally complex of the cell structures (i.e., the cell membrane). The cell membrane is responsible for a broad range of physiological activities including solute transport, electron transport, photosynthetic electron transport, the establishment of electrochemical gradients, motility, ATP synthesis, biosynthesis of lipids, biosynthesis of cell wall polymers, secretion of proteins, intercellular signaling, and responses to environmental signals. To refer to the cell membrane simply as a lipoprotein bilayer does not do justice to the machinery embedded in the lipid matrix. It is clearly a complex mosaic of parts, whose structure and interactions at the molecular level are not well understood. As expected, the protein composition of cell membranes is complex. There can be more than 100 different proteins. Many of the proteins are clustered in functional aggregates (e.g., the proton translocating ATPase, the flagellar motor, electron transport complexes, and certain of the solute transporters). At the molecular level, the membrane is certainly a complex and busy place. What follows is a general description of the membrane, without reference to its microheterogeneity.

Bacterial cell membranes

Bacterial cell membranes consist primarily of phospholipids and protein in a fluid mosaic structure in which the phospholipids form a bilayer (Fig. 1.15). The structure is said to be fluid because there is extensive lateral mobility of bulk proteins and phospholipids. Although this is the case, it should be realized that certain protein aggregates (e.g., complex solute transporters and electron transport aggregates) remain as aggregates within which the proteins interact to catalyze sequential reactions.

The phospholipids are fatty acids esterified to two of the hydroxyl groups of phosphoglycerides (Fig. 1.16). The structure and synthesis of phospholipids is described in detail in Section 9.1.2. The third hydroxyl group in

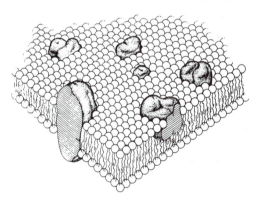

Fig. 1.15. Model of the cell membrane showing bimolecular lipid leaflet and embedded proteins. The phospholipid molecules are shown interacting with each other via their hydrophobic (apolar) "tails." The hydrophilic (polar) "heads" of the phospholipids face the outside of the membrane where they interact with proteins and ions. Proteins can span the membrane or be partially embedded. (From Singer, S. J. and G. L. Nicolson. 1972. The fluid mosaic model of the structure of cell membranes. *Science* 175:720–731. Copyright 1972 by the AAS.)

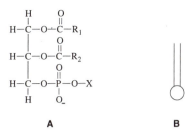

Fig. 1.16 Phospholipids have both a polar and a nonpolar end. A, phospholipid with two fatty acids (R) esterified to glycerol. The phosphate is substituted by X which determines the type of phospholipid. In bacteria, X is usually serine, ethanolamine, a derivative of glycerol, or a carbohydrate derivative. See Section 9.1.3 for a more complete description of bacterial phospholipids. B, a schematic drawing of a phospholipid showing the polar (circle) and nonpolar (straight lines) regions.

the glycerol backbone of the phospholipid is covalently bound to a substituted phosphate group, which makes one end of the molecule very polar due to a negative charge on the ionized phosphate group. Because the phospholipids are polar at one end and nonpolar at the other end (the end with the fatty acids), they are said to be *amphipathic*, and will spontaneously aggregate with their nonpolar fatty acid regions interacting with each other by hydrophobic bonding, while their polar phosphorylated regions face the aqueous phase where ionic interactions with cations, water, and polar groups on proteins occur. Phospholipids accomplish this by spontaneously forming lipid bilayers in water solutions or in cell membranes.

There are two classes of proteins in membranes, *integral* and *peripheral*. Integral proteins are embedded in the membrane and bound to the fatty acids of the phospholipids via hydrophobic bonding. They can be removed only with detergents or solvents. Peripheral proteins are attached to membrane surfaces to the phospholipids by ionic interactions, *and can be removed by washing the membrane with salt solutions*. The lipids and proteins diffuse laterally in the plane of the membrane but do not rotate across membrane. The insertion of the proteins into the membrane during membrane synthesis is discussed in Section 16.2.

The phospholipid bilayer acts as a permeability barrier to virtually all water-soluble molecules so that most solutes diffuse or are carried across the membrane through or on special protein transporters that bridge the phospholipid bilayer. This is discussed in the context of solute transport in Chapter 15. However, the lipid bilayer is permeable to water molecules, gases, and small hydrophobic molecules. An important consequence of the lipid matrix is that ions do not freely diffuse across the membrane unless they are carried on or through protein transporters. Because of this, the membrane is capable of holding a charge that is due to the unequal transmembrane distribution of ions. This is discussed in Chapter 3.

Archaeal cell membranes

1. The lipids

Archaeal membrane[56,57] lipids differ from those found in bacterial membranes. The archaeal lipids consist of *isopranoid alcohols,* either 20 or 40 carbons long, *ether-linked* to one glycerol to form monoglycerol diethers or to two glycerols to form diglycerol tetraethers. These are illustrated in Fig. 1.17. Their synthesis is described in Section 9.1.3. (Recall that bacterial glycerides are fatty acids esterified to glycerol. Refer to Fig. 1.15 for a comparison.) The C_{20} alcohol is a fully saturated hydrocarbon called *phytanol*. The C_{40} molecules are two phytanols linked together head to head in the diglycerol tetraether lipids. Thus, the lipids are either phytanyl glycerol diethers or diphytanyl diglycerol tetraethers. The diethers and tetraethers occur in varying ratios depending upon the bacterium. For example, there may be from 5 to 25 different lipids in any one bacterium. This is really quite a diverse mix and can be contrasted to the four or five different phospholipids found in a typical bacterium. The diversity of the archaeal lipids is due to the different polar head groups that exist, as well as to the mix of core lipid to which the head groups are attached (Fig. 1.17). Although the polar head group is responsible for the polarity of most phospolipids, there is some polarity at one end of the archaeal lipids without a polar head group

because of the free hydroxyl group on the glycerol. (Recall that hydroxyl groups are capable of forming hydrogen bonds with water and proteins.) It is usually stated that the ether linkages are an advantage over ester linkages in the acidic and thermophilic environments in which archaea live because of the greater stability of the ether linkage to hydrolytic cleavage. It is clear, however, that ether-linked lipids are not necessary for growth at high temperatures. A bacterium called *Thermotoga* does not have ether-linked lipids, yet grows in geothermally heated marine sediments alongside the sulfur-dependent thermophilic archaea. This emphasizes that the correlation between ether-linked lipids and habitat is not precise.

2. The proteins

There is little information regarding archaeal membrane proteins. An exception is bacteriorhodopsin and halorhodopsin in *Halobacterium*, whose conformational array in the cell membrane is dependent upon interaction with polar membrane lipids.[58] The functions of these two proteins are discussed in Sections 3.8.4 and 3.9.

3. The membrane

The thermoacidophilic archaea and some methanogens have tetraether glycerolipids. These lipids have a polar head group at both ends and span the membrane forming a *lipid monolayer* (Fig. 1.18). This is the only known example of a membrane where there is no midplane region. Since there is no midplane region, the lipid monolayer is more resistant to levels of heat that would disrupt the hydrophobic bonds holding the two lipids in the lipid bilayer together. The increased resistance to heat of the lipid monolayer may confer an advantage to organisms living at high temperature. However, it cannot be claimed that ether lipids or tetraether lipids are a *specific* adaptation to high temperatures, although they may be advantageous in these environments. This is because some mesophilic methanogens have tetraether lipids, whereas two extremely thermophilic archaeons, *Methanopyrus kandleri* and *Thermococcus celer*, do not have tetraether lipids.

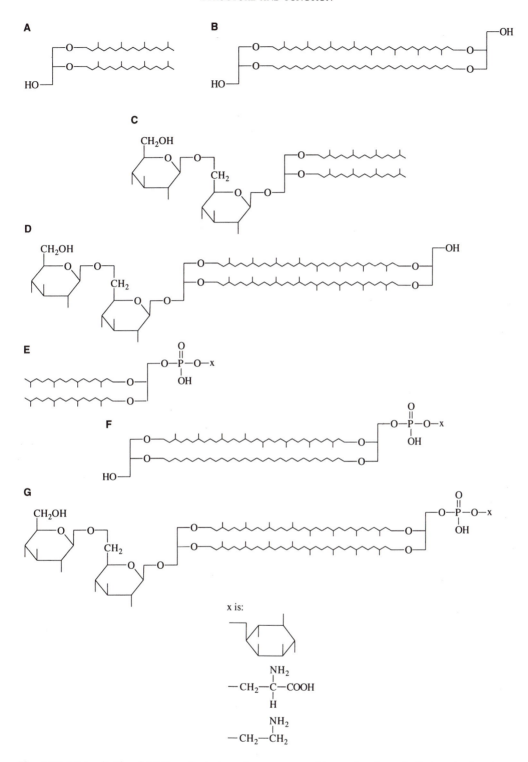

Fig. 1.17. Major lipids of *Methanobacterium thermoautotrophicum*. A. glycerol diether (archaeol); B, diglycerol tetraether (caldarchaeol); C, a glycolipid (gentiobiosyl archaeol); D, gentiobiosyl caldarchaeol; E, a phospholipid (archaetidyl-X), where X can be inositol, serine, or ethanoamine; F, a phospholipid (caldarchaetidyl-X); G, a phosphoglycolipid (gentiobiosyl caldarchatidyl-X). From Koga, Y., M. Nishihara, H. Morii, and M. Akagawa-Matsushita. 1993. Ether polar lipids of methanogenic bacteria: structures, comparative aspects, and biosynthesis. *Microbiol. Rev.* 57:164– 182.)

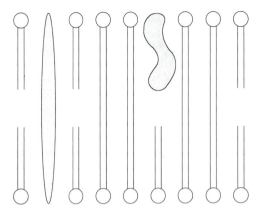

Fig. 1.18. A lipid layer with membrane proteins in archaebacteria membranes. The glycerol diethers form a lipid bilayer and the tetraethers form a monolayer. Some archaebacteria (e.g., the extreme halophiles) contain only the diethers. Most of the sulfur-dependent thermophiles have primarily the tetraethers, with only trace amounts of the diethers. Many methanogens have significant amounts of both the di- and tetraethers. The shaded areas represent protein.

1.2.6 Cytoplasm

The cytoplasm is defined as everything enclosed by the cell membrane. Cytoplasm is a viscous material containing a heavy concentration of protein (100–300 mg/ml),[59] salts, and metabolites. In addition, there are large aggregates of protein complexes designed for specific metabolic functions, various inclusions, and highly condensed DNA. Intracytoplasmic membranes are also present in many prokaryotes. The soluble part of the cytoplasm is called the *cytosol*.

Intracytoplasmic membranes
Many prokaryotes have intracytoplasmic membranes that have specialized physiological functions.[60] Intracytoplasmic membranes are often connected to the cell membrane and are generally believed to be derived from invaginations of chemically modified areas of the cell membrane. However, in some cases connections to the cell membrane are not seen and it is unknown whether the intracytoplasmic membranes are derived from an invagination of the cell membrane or are synthesized independently

of the cell membrane (e.g., the thylakoids of cyanobacteria). The following lists examples of a few prokaryotes with intracytoplasmic membranes and their physiological roles.

1. Methanotrophs
Bacteria that grow on methane as their sole source of carbon (methanotrophs) possess intracytoplasmic membranes that may function in methane oxidation. Methane oxidation is discussed in Section 12.2.1.

2. Nitrogen fixers
Bacteria utilizing nitrogen gas as a source of nitrogen use an oxygen-sensitive enzyme called *nitrogenase* to reduce the nitrogen to ammonia, which is subsequently incorporated into cell material. Many of these organisms have extensive intracytoplasmic membranes. One such nitrogen-fixing bacterium is *Azotobacter vinelandii* whose intracytoplasmic membranes increase with the degree of aeration of the culture. Since respiratory activity is localized in the membranes, it is probable that an important role for *Azotobacter* intracytoplasmic membranes is to increase the cellular respiratory activity in order to provide more ATP for nitrogen fixation and to remove oxygen from the vicinity of the nitrogenase. Nitrogen fixation is discussed in Section 11.3.

3. Nitrifiers
Intracellular membranes are also found in nitrifying bacteria (i.e., those bacteria that oxidize ammonia and nitrite as the sole source of electrons: *Nitrosomonas, Nitrobacter, Nitrococcus*). Several of the enzymes that catalyze ammonia and nitrite oxidation are in the membranes. This is discussed in Section 11.4.

4. Phototrophs
In bacteria that use light as a source of energy for growth (phototrophs), the intracytoplasmic membranes are the sites for the photosynthetic apparatus. The membrane structure varies and can be flat membranes, vesicles, flat sacs (thylakoids in cyanobacteria), and tubular invaginations of the cell membrane (photosynthetic bacteria). Photosynthesis and photosynthetic membranes are discussed in Chapter 5.

Inclusion bodies, multienzyme aggregates, and granules

Some bacteria contain specialized organelles in the cytoplasm. These differ from eukaryotic organelles in not being surrounded by a lipid bilayer–protein membrane, although they do have a membrane or coat. They are referred to by some researchers as inclusion bodies rather than organelles. In addition, there are numerous aggregates and multienzyme complexes of large size in all bacteria.

1. Gas vesicles

Aquatic bacteria such as cyanobacteria, certain photosynthetic bacteria, and some nonphotosynthetic bacteria have *gas vesicles* surrounded by a protein coat. Gas vesicles are hollow spindle-shaped structures about 100 nm long filled with gas in equilibrium with the gases dissolved in the cytoplasm. The gas vesicles allow the bacteria to float in lakes and ponds at depths that support growth because of favorable light, temperature, or nutrients. Many of these bacteria and cyanobacteria are plentiful in stratified freshwater lakes, but they are not as abundant in isothermally mixed waters. Other bacteria containing gas vesicles live in hypersaline waters (*Halobacterium*) and a few marine species of cyanobacteria belonging to the genus *Trichodesmium* have gas vesicles. When gas vesicles are collapsed by experimentally subjecting cells to high hydrostatic pressure or turgor pressure, the cells are no longer buoyant and sink. Collapsed vesicles do not recover and the cells acquire gas-filled vesicles only by *de novo* synthesis of new vesicles. Thus, during synthesis of the vesicles, water is excluded, presumably because of the hydrophobic nature of the inner protein surface. The green sulfur bacterium *Pelodictyon phaeoclathratiforme* forms gas vesicles only at low light intensities.[61] Perhaps this allows the bacteria to float at depths where the light is optimal for photosynthesis.

2. Carboxysomes

Bacteria that obligately grow on CO_2 as their sole or major source of carbon (strict autotrophs) sometimes have large (100 nm) polyhedral inclusions called *carboxysomes*.[62] These inclusions have been observed in nitrifying bacteria, sulfur oxidizers, and cyanobacteria. The distribution appears to be species specific (e.g., not all sulfur oxidizers have carboxysomes). Ribulose-bisphosphate carboxylase (RuBP carboxylase), the enzyme in the Calvin cycle that incorporates CO_2 into organic carbon, is stored in carboxysomes. The enzyme is discussed in Section 12.1.1. The physiological role for carboxysomes is not clear since many autotrophs do not have them. Perhaps they are simply storage sites for RuBP carboxylase.

3. Chlorosomes

Green sulfur photosynthetic bacteria (e.g., *Chlorobium*) have ellipsoid inclusions, *chlorosomes* (formerly called chlorobium vesicles), that lie immediately underneath the cytoplasmic membrane. The chlorosomes are surrounded by a nonunit membrane of galactolipid, with perhaps some protein. At one time it was believed that such vesicles were found only in the green sulfur photosynthetic bacteria. However, similar vesicles have been found in the green photosynthetic gliding bacteria, *Chloroflexus*. The major light-harvesting photopigments are located in the chlorosomes, whereas the photosynthetic reaction centers are in the cell membrane. This means that during photosynthesis in these organisms, light is absorbed by pigments in the chlorosomes and energy is transmitted to the reaction centers in the cell membrane where photosynthesis takes place. In those photosynthetic bacteria that do not have chlorosomes, the light-harvesting pigments surround the reaction centers in the cell membrane. The structure and function of chlorosomes are discussed in Section 5.6.

4. Granules and globules

Bacteria often contain cytoplasmic granules whose content varies with the bacterium. Many bacteria store a lipoidal substance called *poly-β-hydroxybutyric acid* (PHB) as a carbon and energy reserve. Other bacteria may store *glycogen* for the same purpose. Other granules found in bacterial cells can include *polyphosphate* and, in some sulfur-oxidizing bacteria, *elemental sulfur globules*.

5. Ribosomes

Ribosomes are the sites of protein synthesis.

They are small ribonucleoprotein particles, approximately 22 nm by about 30 nm or about the size of the smallest viruses, and consist of over 50 different proteins and three different types of RNA (23S, 16S, and 5S). Bacterial ribosomes sediment in a centrifugal field at a characteristic velocity of 70 Svedberg units (i.e., they are 70S), as opposed to eucaryotic cytosolic ribosomes that are 80S. Bacterial ribosomes are very similar regardless of the bacterium. However, there are some differences from archaeal ribosomes, which are also 70S. These differences are described in Section 1.1.1.

6. Nucleoid

The site of DNA and RNA synthesis is the nucleoid, an amorphous mass of DNA unbounded by a membrane, lying approximately in the center of the cell. Faster-growing bacteria may contain more than one nucleoid, but each nucleoid has but one chromosome, and all the chromosomes are identical. The DNA is very tightly coiled. If the DNA from *E. coli* were stretched out, it would be 500 times longer than the cell.

7. Cytosol

The cytosol is the liquid portion of the cytoplasm. This can be isolated in diluted form as the supernatant fraction obtained after centrifuging broken cell extracts at $105,000 \times g$ for 1–2 hours, which should sediment membranes, the DNA, the ribosomes, very large protein aggregates, and other intracellular inclusions. In the cytosol are found the enzymes that catalyze a major portion of the biochemical reactions in the cell, such as the enzymes of the central pathways for the metabolism of carbohydrates (glycolysis, the pentose phosphate pathway, and the Entner–Doudoroff pathway) as well as the central pathways for organic acid metabolism (citric acid cycle and glyoxylate pathways), and enzymes for other pathways such as the biosynthesis and degradation of amino acids, lipids, and nucleotides.

The concentration of proteins in the cytosol is very high, making it viscous, and it is expected that there are extensive protein-protein interactions among the enzymes, even those that do not exist in tight

complexes. However, if such interactions exist they must be weak because most of the enzymes that catalyze metabolic pathways in the cytosol (e.g., the glycolytic enzymes that catalyze the degradation of glucose to pyruvic acid or lactic acid), cannot be isolated as complexes. It generally has been assumed that they exist either as independent proteins or in loose associations that are easily disrupted during cell breakage and accompanying dilution of the proteins. (It should be realized that the concentration of proteins in broken cell extracts is orders of magnitude lower than in the aqueous portion of the unbroken cell because of the addition of buffers during the washing and suspension of the cells prior to breakage.) However, pathways such as glycolysis have numerous branch points and share intermediates with other pathways. It therefore is necessary that the intermediates leave the surface of the enzyme and join the soluble pool. For this reason, strict channeling of intermediates is not desirable.

It should not be thought that the cytoplasm is a random mixture of solutes and proteins. There are many examples of enzymes in the same pathway forming stable multienzyme complexes, reflecting strong intermolecular bonding.[63] For example, *pyruvate dehydrogenase* from *E. coli* is a complex of three different enzymes each present in multiple copies (50 proteins total) that oxidizes pyruvic acid to acetyl-CoA and CO_2. The size of the pyruvate dehydrogenase complex is $4.6–4.8 \times 10^6$ D. Contrast this with the size of a 70S ribosome (another multienzyme complex) that is about 2.7×10^6 D. Another enzyme complex that catalyzes a consecutive series of biochemical reactions is the α-*ketoglutarate dehydrogenase* complex. It consists of three different enzymes present in multiple copies (i.e., 48 proteins), 2.5×10^6 D (in *E. coli*), and oxidizes α-ketoglutarate to succinyl-CoA and CO_2. Another example is *fatty acid synthase* in yeast that consists of seven different enzymes, 2.4×10^6 D, which synthesizes fatty acids from acetyl-SCoA.[64] One of the advantages to multienzyme complexes is that it facilitates the channeling of metabolites, thus increasing the efficiency of catalysis. For example, in these stable enzyme complexes, the biochemical intermediates in the pathway are transferred directly from one enzyme to the next without entering the bulk phase. Thus, there is no dilution of intermediates and no reliance on random diffusion to reach a second enzyme.

1.3 Summary

There are two evolutionary lines of prokaryotes, the Bacteria and the Archaea. Archaea are similar to bacteria, but they differ in certain fundamental aspects of structure and biochemistry. These include differences in ribosomes, cell wall chemistry, membrane lipids, and coenzymes. In addition, certain archaea have metabolic pathways (e.g., methanogenesis) not found in bacteria.

Surrounding most bacteria are fimbriae and extracellular material called glycocalyx. The fimbriae are protein fibrils. The glycocalyx encompasses all extracellular polymers, polysaccharide or protein, including capsules and slime layers. The fimbriae and glycocalyx anchor the cell to specific animal and plant cell surfaces, as well as to inanimate objects, and in several cases to each other.

Some bacteria possess a fibril called a sex pilus that attaches the cell to a mating partner and retracts to draw the cells into intimate contact. When the cells are in contact, DNA is transferred unilaterally from the donor cell (i.e., the one with the sex pilus) to the recipient.

Protruding from many bacteria are flagella filaments that aid the bacterium in movement. At the base of the flagellum is a basal body that is embedded in the cell membrane. Part of the basal body is a rotary motor that runs on a current of protons.

There are two types of bacterial cell walls. One kind has a thick peptidoglycan layer to which there are covalently bonded polysaccharides, teichoic acids, and teichuronic acids. Bacteria with such walls stain gram-positive. A second type of wall has a thin peptidoglycan layer and an outer envelope consisting of lipopolysaccharide, phospholipid, and protein. Cells with such walls do not stain gram-positive and are called gram-negative. There is also a separate compartment in gram-negative bacteria called a periplasm, between the outer envelope and the cell membrane. The periplasm is the location of numerous proteins and enzymes, including proteins required for solute transport and enzymes that function in the oxidation and degradation of nutrients in the periplasm. Archaea have very different cell walls. No archaeon has peptidoglycan, although some have a similar polymer called pseudopeptidoglycan or pseudomurein.

The cell membranes of the bacteria are all similar. They consist primarily of phosphoglycerides in a bilayer, and protein. The phosphoglycerides are primarily fatty acids esterified to glycerol-phosphate. Archaeal cell membranes have lipids that are long-chain alcohols ether-linked to glycerol. Archaeal membranes can be a bilayer or a monolayer, or a mixture of the two.

The cytoplasm of prokaryotes is a viscous solution of protein. Salts, sugars, amino acids, and other metabolites are dissolved in the proteinaceous cytoplasm. In many bacteria, intracytoplasmic membranes are present. In many cases these membranes are invaginations of the cell membrane. They are sites for electron transport and specialized biochemical activities. Numerous inclusion bodies are also present. These include ribosomes, which are the sites of protein synthesis, large aggregates of enzymes that catalyze short metabolic pathways, gas vesicles, and carbon and energy reserves (e.g., glycogen particles). The DNA is also present in the cytoplasm; it is very tightly packed and referred to as a nucleoid. It is clear that although the cytoplasm of prokaryotes is not compartmentalized into membrane-bound organelles as is the eukaryotic cytoplasm, it nonetheless represents a very complex pattern.

The central metabolic pathways that are described in the ensuing chapters (i.e., glycolysis, the Entner–Doudoroff pathway, pentose phosphate pathway, and the citric acid cycle) all take place in the liquid part of the cytoplasm (i.e., the cytosol). Also found in the cytosol are most of the other pathways, including the enzymatic reactions for the synthesis and degradation of amino acids, fatty acids, purines, and pyrimidines.

The cell membrane is the site of numerous other metabolic pathways, including phospholipid biosynthesis, protein secretion, solute transport, electron transport, cell wall biosynthesis, the generation of electrochemical ion gradients, and ATP synthesis. The prokaryote cell can therefore be considered as having three metabolic domains: cytosol, particulate (ribosomes, enzyme aggregates, etc.), and membrane.

Study Questions

1. What chemical and structural differences distinguish the Archaea from the Bacteria?

2. What are the differences between the gram-positive and gram-negative cell walls?

3. What structures enable bacteria to adhere to surfaces? What is known about the molecules that mediate the adhesion (i.e., their chemistry, location, and receptors)?

4. Contrast the functions and cellular location of peptidoglycan, phospholipid, and lipopolysaccharide. What is it about the chemical structure of these three classes of compounds that is suitable for their functions and/or cellular location?

5. What are porins and what do they do? Are they necessary for gram-positive bacteria? Explain.

6. What are the physiological and enzymatic functions associated with the periplasm?

7. Summarize the cell compartmentalization of metabolic activities in prokaryotes.

8. What are some multienzyme complexes? What is the advantage to these?

9. Which metabolic pathways or activities are found in the cytosol and which in the membranes?

REFERENCES AND NOTES

1. Prokaryotes are organisms without a membrane-bound nucleus. It is also spelled procaryote. 'Karyo' (caryo) means nucleus. It is derived from the Greek word *karuon*, which means kernel or nut.

2. Woese, C. R. 1987. Bacterial evolution. *Microbiol. Rev.* 51:221–271.

3. Pace, N. R., D. A. Stahl, D. J. Lane, and G. J. Olsen. 1985. Analyzing natural microbial populations by rRNA sequences. *ASM News* 51:4–12.

4. The evolutionary relationships among all living organisms have been deduced by comparing the ribosomal RNAs of modern organisms. The structures of the ribosomal RNA molecules from different living organisms are sufficiently conserved in certain regions that it is possible to align conserved sequences and secondary structures in order to compare the differences in base sequences between RNA molecules in homologous regions. What is analyzed are the number of positions that differ between pairs of sequences as well as other parameters (e.g., which positions vary and the number of changes that have been made in going from one sequence to another). The number of nucleotide differences between homologous sequences is used to calculate the evolutionary distance between the organisms and to construct a phylogenetic tree.

5. Woese, C. R., O. Kandler, and M. L. Wheelis. 1990. Towards a natural system of organisms: Proposal for the domains Archaea, Bacteria, and Eucarya. *Proc. Natl Acad. Sci. USA* 87:4576–4579.

6. Stetter, K. O. 1989. Extremely thermophilic chemolithoautotrophic archaebacteria, pp. 167–176. In: *Autotrophic Bacteria*. H. G. Schlegel and B. Bowien (eds.). Springer-Verlag, Berlin.

7. Koga, Y., M. Nishihara, H. Morii, and M. Akagawa-Matsushita. 1993. Ether polar lipids of methanogenic bacteria: structures, comparative aspects, and biosynthesis. *Microbiol. Rev.* 57:164–182.

8. Joys, T. M. 1988. The flagellar filament protein. *Can. J. Microbiol.* 34:452–458.

9. Wilson, D. R., and T. J. Beveridge. 1993. Bacterial flagellar filaments and their component flagellins. *Can. J. Microbiol.* 39:451–472.

10. A great many bacteria move by gliding motility on solid surfaces. Organelles for gliding locomotion in bacteria have not been discovered.

11. Some rod-shaped or curved cells have flagella that protrude from one or both of the cell poles. A bacterium with a single, polar flagellum is said to be *monopolar*. Bacteria with a bundle of flagella at a single pole are *lophotrichous*. Bacteria with flagella at both poles are said to be *bipolar*. They may have either single or bundles of flagella at the poles. *Amphitrichous* refers to bundles of flagella at both poles. Some bacteria (e.g., spirochaetes) have *subpolar* flagella, where the flagella are inserted near but not exactly at the cell poles, and some curved bacteria (e.g., *Vibrio*) have a single, *medial* flagellum. If the flagella are arranged laterally all around the cell, then the condition is known as *peritrichous* (e.g., *Escherichia* and *Salmonella*). Peritrichous flagella coalesce into a trailing bundle during swimming.

12. DePamphilis, M. L., and J. Adler. 1971. Purification of intact flagella from *Escherichia coli* and *Bacillus subtilis*. *J. Bacteriol.* 105:384–395.

13. Aizawa, Shin-Ichi, G. E. Dean, C. J. Jones, R. M. Macnab, and S. Yamaguchi. 1985. Purification and characterization of the flagellar hook-basal body complex of *Salmonella typhimurium J. Bacteriol.* 161:836-849.

14. Khan, S., M. Dapice, and T. S. Reese. 1988. Effects of mot gene expression on the structure of the flagellar motor. *J. Mol. Biol.* 202:575-584.

15. Blair, D. F., and H. C. Berg. 1990. The MotA protein of *E. coli* is a proton-conducting component of the flagellar motor. *Cell* 60:439-449.

16. Pleier, E, and R. Schmitt. 1989. Identification and sequence analysis of two related flagellin genes in *Rhizobium meliloti. J. Bacteriol.* 171:1467-1475.

17. Iino, T. 1969. Polarity of flagellar growth in *Salmonella. J. Gen. Microbiol.* 56:227-239.

18. Emerson, S. U., K. Tokuyasu, and M. I. Simon. 1970. Bacterial flagella: polarity of elongation. *Science* 169:190-192.

19. Fuerst, J. A., and J. W. Perry. 1988. Demonstration of lipopolysaccharide on sheathed flagella of *Vibrio cholerae* 0:1 by protein A-gold immunoelectron microscopy. *J. Bacteriol.* 170:1488-1494.

20. Weissborn, A., H. M. Steinman, and L. Shapiro. 1982. Characterization of the proteins of the *Caulobacter crescentus* flagellar filament. *J. Biol. Chem.* 257:2066-2074.

21. Pleier, E., and R. Schmitt. 1989. Identification and sequence analysis of two related flagellin genes in *Rhizobium meliloti. J. Bacteriol.* 171:1467-1475.

22. Driks, A., Bryan, R., Shapiro, L., and D. J. DeRosier. 1989. The organization of the *Caulobacter crescentus* flagellar filament. *J. Mol. Biol.* 206:627-636.

23. Parales, J., Jr., and E. P. Greenberg. 1991. N-Terminal amino acid sequences and amino acid compositions of the *Spirochaeta aurantia* flagellar filament polypeptides. *J. Bacteriol.* 173:1357-1359.

24. Pearce, W. A., and T. M. Buchanan. 1980. Structure and cell membrane-binding properties of bacterial fimbriae, pp. 289-344. In: *Bacterial Adherence*. E. H. Beachey (ed.). Chapman and Hall.

25. Ottow, J. C. G. 1975. Ecology, physiology, and genetics of fimbriae and pili. *Ann. Rev. Microbiol.* 29:79-108.

26. Jones, G. W., and R. E. Isaacson. 1983. Proteinaceous bacterial adhesins and their receptors. *Crit. Rev. Microbiol.* 10:229-260.

27. Hultgren, S. J., S. Abraham, M. Caparon, P. Falk, J. W. St. Geme III, and S. Normark. 1993. Pilus and nonpilus bacterial adhesins: assembly and function in cell recognition. *Cell.* 73:887-901.

28. Yanagawa, R., and K. Otsuki. 1970. Some properties of the pili of *Corynebacterium renale. J. Bacteriol.* 101:1063-1069.

29. Cisar, J. O., and A. E. Vatter. 1979. Surface fibrils (fimbriae) of *Actinomyces viscosus Infect. Immun.* 24:523-531.

30. *C. renale* is the causative agent of bovine pyelonephritis and cystitis, and *A. viscosus* is part of the normal flora of the human mouth where it adheres to other bacteria in plaque and to teeth.

31. The fimbriae presumably attach to the erythrocytes because the erythrocytes carry cell surface molecules resembling the adhesin receptors found on the natural host cells.

32. Eisenstein, B. I. 1987. Fimbriae, pp. 84-90. In: Escherichia coli *and* Salmonella typhimurium: *Cellular and Molecular Biology*, Volume 1. Neidhardt, F. C. (ed.). American Society for Microbiology, Washington, DC.

33. Hultgren, S. J., S. Normark, and S. N. Abraham. 1991. Chaperone-assisted assembly and molecular architecture of adhesive pili. Annu. Rev. Microbiol. 45:383-415.

34. Hultgren, S. J., S. Abraham, M. Caparon, P. Falk, J. W. St. Geme, III, and S. Normark. 1993. Pilus and nonpilus bacterial adhesins: assembly and function in cell recognition. Cell 73:887-901.

35. Pathogenic strains of *Escherichia coli* cause gastrointestinal and urinary tract infections, *Neisseria gonorrhoeae* causes gonorrhoeae, *Bacteroides nodosus* causes bovine foot rot, *Moraxella bovis* causes bovine keratoconjunctivitis, *Vibrio cholerae* causes cholera, and *Pseudomonas aeruginosa* causes infections of the urinary tract and wounds. The adherence of these bacteria to their host tissue is the first stage in pathogenesis.

36. In addition to the bacterial chromosome, most bacteria carry smaller extrachromosomal DNA molecules called *plasmids*. There are many types of plasmids. They all carry genes for self-replication. Conjugative plasmids also carry genes for DNA transfer into recipient cells (transmissible plasmids). Other genes that may be carried by plasmids include genes for antibiotic resistance (resistance transfer factors or R-factors), and genes for the catabolism of certain nutrients. For example, *Pseudomonas* carries plasmids for the degradation of camphor, octane, salicylate, and napththalene. Some plasmids confer virulence because the plasmids carry genes for toxins or other virulence factors, such as fimbriae. The F plasmid in *E. coli* carries genes for self-transmission, including genes for the F pilus, and is also capable of integrating into the chromosome and promoting the transfer of chromosomal DNA.

37. Dunny, G. M. 1991. Mating interactions in Gram-positive bacteria p. 9-33. In: *Microbial Cell Interactions*. M. Dworkin (ed.). American Society for Micrbiology, Washington, DC.

38. An excellent reference for microbial cell–cell interactions is the book edited by Dworkin: *Microbial Cell–Cell Interactions*, 1991. M. Dworkin

(ed.). American Society for Microbiology, Washington, DC.

39. Costerton, J. W., T. J. Marrie, and K.-J. Cheng. 1985. Phenomena of bacterial adhesion, p. 3–43. In: *Bacterial Adhesion: Mechanisms and Physiological Significance*. D. C. Savage and M. Fletcher (eds). Plenum Press, New York.

40. Sleytr, U. B., and P. Messner. 1983. Crystalline surface layers on bacteria. *Annu. Rev. Microbiol.* 37:311–339.

41. Costerton, J. W., R. T. Irvin, and K.-J Cheng. 1981. The bacterial glycocalyx in nature and disease. *Annu. Rev. Microbiol.* 35:299–324.

42. Kolenbrander, P. E., Ganeshkumar, N., Cassels, F. J., and C. V. Hughes. 1993. Coaggregation: specific adherence among human oral plaque bacteria. *FASEB* 7:406–409.

43. Nikaido, H., and M. Vaara. 1987. Outer membrane, pp. 7–22. In: Escherichia coli *and* Salmonella typhimurium: *Cellular and Molecular Biology* Vol 1. F. C. Neidhardt (ed.). ASM Press, Washington, DC.

44. Weidel, W., and H. Pelzer. 1964. Bagshaped macromolecules – a new outlook on bacterial cell walls. *Adv. Enzymol.* 26:193–232.

45. Koch, A. L. 1983. The surface stress theory of microbial morphogenesis. *Adv. Microb. Physiol.* 24:301–366.

46. Koch, A. L., and S. Woeste. 1992. Elasticity of the sacculus of *Escherichia coli. J. Bacteriol.* 174:4811–4819.

47. Kellenber, E. 1990. The "Bayer bridges" confronted with results from improved electron microscopy methods. *Mol. Microbiol.* 4:697–705.

48. *S. pyogenes* is an important pathogen that causes most streptococcal infections in humans, including "strep" throat. The M protein is a cell surface adhesin that is thought to bind to the LTA that in turn binds to host cell surfaces.

49. Sutcliffe, I. C. and N. Shaw. 1991. Atypical lipoteichoic acids of Gram-positive bacteria. *J. Bacteriol.* 173:7065–7069.

50. Leduc, M. K. Ishidate, N. Shakibai, and L. Rothfield. 1992. Interactions of *Escherichia coli* membrane lipoproteins with the murein sacculus. *J. Bacteriol.* 174:7982–7988.

51. Cross-linking reagents can be useful for determining which molecules are physically associated. These are bifunctional reagents (i.e., they have a chemical group at both ends of the molecule that can form a covalent bond to functional groups on proteins or other molecules such as peptidoglycan) thus cross-linking two molecules that are within the length of the cross-linking reagent. For example the reagent dithio-bis-succinimidylpropionate (DSP) forms covalent cross-links between amino groups that are less than 12 Å (1.2 nm) apart. Therefore, DSP can cross-link proteins that are physically associated with the peptidoglycan. After treatment with DSP, the peptidoglycan and its cross-linked proteins can be isolated. DSP has two arms held together by a disulfide bond that can be cleaved with mercaptoethanol. It is called a cleavable cross-linking reagent. The cross-linked proteins that were held to the peptidoglycan by the DSP are released after mercaptoethanol treatment and can be analyzed by gel electrophoresis. Thus, the proteins cross-linked to the peptidoglycan can be identified. In addition to various lipoproteins, including the murein lipoprotein, an outer membrane protein, OmpA, can also be cross-linked to the peptidoglycan, indicating close association. The cross-linking of OmpA to the peptidoglycan reflects the fact that it extends through the outer envelope.

52. Ferguson, S. J. 1992. The periplasm. In: *Procaryotic Structure and Function: A New Perspective*. S. Mohan, C. Dow, and J. A. Cole (eds.). Society for General Microbiology, Symposium 47, Cambridge University Press.

53. Hobot, J. A., E. Carleman, W. Villiger, and E. Kellenberger. 1984. Periplasmic gel: new concept resulting from the reinvestigation of bacterial cell envelope ultrastructure by new methods. *J. Bacteriol.* 160:143–152.

54. Bacteria secrete specific iron chelators called siderophores for scavenging iron from the environment. Special transport systems bind the iron siderophores and bring the iron into the cell.

55. Postle, K. 1990. TonB and the Gram-negative dilemma. *Molec. Microbiol.* 4:2019–2025.

56. De Rosa, M., A. Gambacorta, and A. Gliozzi. 1986. Structure, biosynthesis, and physiochemical properties of archaebacterial lipids. *Microbiol. Rev.* 50:70–80.

57. Sternberg, B., C. LHostis, C. A. Whiteway, and A. Watts. 1992. The essential role of specific *Halobacterium halobium* polar lipids in 2D-array formation of bacteriorhodopsin. *Biochem. Biophys. Acta.* 1108:21–30.

58. Sternberg, B., C. Hostis, C. A. Whiteway, and A. Watts. 1992. The essential role of specific *Halobacterium halobium* polar lipids in 2D-array formation of bacteriorhodopsin. *Biochem. Biophys. Acta* 1108:21–30.

59. Westerhoff, H. V., and G. R. Welch. 1992. Enzyme organization and the direction of metabolic flow: Physiochemical considerations, pp.361–390. In: *Current Topics in Cellular Regulation*, Volume 33. E. R. Stadtman and P. B. Chock (eds) Academic Press, Inc., New York.

60. Reviewed by Drews, G. 1991. Intracytoplasmic membranes in bacterial cells, pp. 249–274. In: *Prokaryotic Structure and Function, A new*

Perspective. S. Mohan, D. Dow, and J. A. Coles (eds). Cambridge University Press.

61. Overmann, J., S. Lehmann, and N. Pfennig. 1991. Gas vesicle formation and buoyancy regulation in *Pelodictyon phaeoclathratiforme* (Green sulfur bacteria). *Arch. Microbiol.* **157**:29–37.

62. Codd, G. A. 1988. Carboxysomes and ribulose bisphosphate carboxylase/oxygenase, pp. 115–164. In: *Advances in Microbial Physiology*, Volume 29. A. H. Rose and D. W. Tempest (eds.). Academic Press, New York.

63. Molecular interactions among enzymes is reviewed by Srivastava, D. K. and S. A. Bernhard. 1986. Enzyme–enzyme interactions and the regulation of metabolic reaction pathways, pp.1–68. In: *Current Topics in Cellular Regulation*, Volume 28. B. L. Horecker and E. R. Stadtman (eds.). Academic Press, Inc., New York.

64. In bacteria fatty acid synthesis is catalyzed by similar enzymatic reactions as found in yeast and mammals but the enzymes cannot be isolated as a complex.

2

Growth

The study of growth of microbial populations is at the heart of microbial cell physiology because population growth characteristics reflect underlying physiological events in the individual cells. In fact, these cellular events are frequently manifested to the investigator only by changes in the growth of populations. Therefore, it is important to understand the changes in growth that microbial populations undergo and to be able to measure population growth accurately. Additionally, the investigator must be able to manipulate population growth (e.g., by placing the culture in balanced growth or continuous growth) in order to investigate certain aspects of cell physiology (e.g., the interdependencies between the rates of synthesis of the individual classes of macromolecules, such as DNA, RNA, and protein). This chapter describes methods to measure growth, an analysis of exponential growth kinetics, growth in batch culture and in continuous culture, and some important aspects of growth physiology under certain nutritional conditions.

2.1 Measurement of growth

2.1.1 Turbidity

Growth is defined as an increase in mass. The most direct measurement of growth is therefore a dry weight measurement. However, one can also measure any growth parameter

(e.g., cell number, turbidity, protein) provided it increases proportionally to the mass of the culture. Routinely, bacterial growth is measured using turbidity measurements because of the simplicity of the procedure. Within limits, the amount of light scattered by a bacterial cell is proportional to its mass. Therefore, light scattering is proportional to the total mass. (The relationship between total mass and light scattering deviates from linearity at very high cell densities.) If the average mass per cell remains a constant, then one can also use light scattering to measure changes in cell number. What is actually measured is the *fraction of incident light transmitted* through the culture (i.e., not the scattered light). The more dense the culture, the less light is transmitted. This relationship is given by the *Beer–Lambert law* (eq. 2.1). Thus if I_o is the incident light and I is the transmitted light, then the Beer–Lambert law states that in a population whose cell density is x, the fraction of light that is transmitted (I/I_o) will decrease as the logarithm of x to the base 10.

eq. 2.1 $I/I_o = 10^{-xl}$ where l is the light path in centimeters.

The situation can be drawn schematically as shown below. (Fig. 2.1).

If one takes the log to the base 10 of both sides of eq. 2.1, then $\log (I/I_o) = -xl$, where I/I_o is the fraction of incident light that is

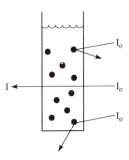

Fig. 2.1 Illustration of light scattering. The incident light is I_o and the transmitted light is I. The spectrophotometer usually provides the logarithm of the reciprocal of the fraction of transmitted light (i.e., $\log(I_o/I)$ which is called the optical density or turbidity.

transmitted. Note that the fraction of light that is transmitted decreases as a logarithmic function (base 10) (i.e., exponentially with the density of the culture). Bacteriologists, however, prefer to measure something that *increases* with cell density. They therefore use the reciprocal of $\log(I/I_o)$ which is $\log(I_o/I)$. The $\log(I_o/I)$ is called *turbidity*, or *absorbency* or *optical density*, and has the symbol O.D. (for optical density) or A (for absorbency). Thus turbidity or O.D. is directly proportional to cell mass/cell in the culture (or cell density if the mass/cell remains constant) and is written as follows:

eq. 2.2 $OD = A = xl$

The turbidity is measured with a colorimeter or spectrophotometer. In practice, one constructs a standard curve by measuring the turbidity of several different cell suspensions

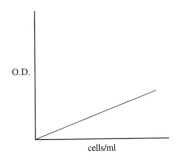

Fig. 2.2 Standard curve of optical density (O.D.) versus cells/ml. The line deviates from linearity at very high cell densities.

where the cell number is counted independently, and generates a straight line as shown in Fig. 2.2. The line deviates from linearity at high cell densities.

Whenever turbidity measurements are made on an unknown sample, a standard curve must be constructed to determine the cell density. Turbidity measurements are the simplest way to measure growth. Methods to determine the cell density by direct cell counts are described later.

2.1.2 Total cell counts

If the mass per cell is constant, one can measure growth by sampling the culture over a period of time and counting the cells. There are two ways to do this: total cell counts and viable cell counts. For *total cell counts*, one can use a counting chamber. This is simply a glass slide divided into tiny square wells of known area and depth. A drop of culture is placed on the slide and, when a cover slip is applied, each square well in the grid holds a known volume of liquid. The number of cells is counted microscopically and the value is converted to cells per milliliter using a conversion factor which is based on the total number of square wells counted and their volume. Although this is a simple procedure and one used widely, there are two limitations:

1. They do not distinguish between live and dead cells.

2. They cannot be performed on populations whose cell density is too low to count microscopically (less than 10^6 cells/ml).

A more sophisticated method is to use *electronic cell counting*. For electronic counting, the bacteria, suspended in a saline solution, are placed in a chamber with an electrode separated by a second chamber filled with the same saline solution and also provided with an electrode. A microscopic pore separates the two chambers. The bacterial suspension is pumped through the pore into the second chamber. Whenever a bacterium passes through the pore, the electrical conductivity of the circuit decreases because the electrical conductivity of a bacterial cell is less than that of the saline solution. This results in a voltage pulse that is counted electronically. The electronic counter has an advantage in that one can also measure the *size* of the bacterium. This is because the size of the pulse is proportional to the size of the bacterium. Thus, one can obtain both a total cell count and a size distribution.

2.1.3 Viable cell counts

A *viable cell count* is one in which the cells are serially diluted and then deposited on a solid growth medium. Each viable cell grows into a colony and the colonies are counted. Viable cell counts are routinely performed in microbiology laboratories. However, it is important to recognize that such counts may underestimate the number of viable cells for the following reasons:

1. A viable cell count will underestimate the number of live cells if the cells are clumped, since each clump of cells will give rise to a single colony.

2. Some bacteria plate with poor efficiency (i.e., single cells give rise to colonies with a low frequency.)

2.1.4 Dry weight and protein

The most direct way to measure growth is to quantify the *dry weight* of cells in a culture. The cells are harvested by either centrifugation or filtration, dried to a constant weight, and carefully weighed. Since protein increases parallel with growth, one can also measure growth by doing *protein measurements* of the cells. The cells are either harvested by centrifugation or filtration, or (more usually) precipitated first with acid or alcohol, and then recovered by centrifugation or filtration. A simple colorimetric test for protein is then performed.

2.2 Growth physiology

Growth physiology is an extensive topic. It includes the regulation of rates of synthesis of

macromolecules, the regulation of timing of DNA synthesis and cell division, adaptive physiological responses to nutrient availability, including changes in gene expression, homeostasis adaptations to the external environment, and the coupling of the rates of biosynthetic pathways to the rates of utilization of products for growth (metabolic regulation). From this point of view, this entire text is concerned with the physiology of growth and we therefore return to the topic in subsequent chapters. What follows here is an introduction to growth physiology as it pertains to the synthesis of macromolecules during steady-state growth, and very general adaptations of cells to nutrient depletion. We begin with a discussion of the sequence of growth phases through which a population of bacteria progresses when inoculated into a flask of fresh media.

2.2.1 Phases of population growth

When one measures the growth of populations of bacteria grown in batch culture, a progression through a series of phases can be observed (Fig. 2.3). The first phase is frequently a *lag phase*, where no net growth occurs (i.e., no increase in cell mass). This is followed by a phase of *exponential growth* where cell mass increases exponentially with time. Following exponential growth, the culture enters a phase of no net growth called the *stationary phase*. After a stationary period, cell death occurs in a final stage called the *death phase*. Notice in Fig. 2.3 that, prior to the exponential phase of growth, the cells increase in mass (i.e., they grow larger), before cell division occurs (solid line). At the end of exponential growth, the cells continue to divide after growth has ceased and become smaller (dotted line). It would seem to be an advantage to a population of bacteria to continue to divide after growth has ceased in the population so as to produce more cells for distribution to new sites where conditions for continued growth might be better. One consequence of the uncoupling of growth from cell division during lag and late log phase is that the size of the cell varies during

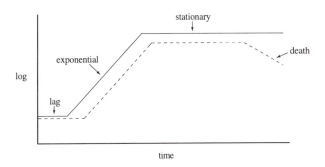

Fig. 2.3 Growth kinetics in batch culture. The solid line is mass, the dotted line is viable cell number. Note that if we define growth as an increase in mass, then only the solid line accurately reflects the growth of the culture. The dotted line reflects growth only when it is parallel to the solid line.

growth in batch culture. A practical consequence for the investigator of the non-coincidence of growth and cell division during stages of batch growth is that cell counts are not always a valid measurement of growth.

Lag phase

When cells in the stationary phase of growth are transferred to fresh media, a lag phase often occurs. Basically, the lag phase is due in this situation to the time required for the physiological adaptation of stationary phase cells in preparation for growth. Usually the longer the cells are kept in stationary phase, the longer is the lag phase when they are transferred to fresh media. Some of the physiological changes that occur in the stationary phase are described in Section 2.2.2. There are several possible reasons for the lag phase. The lag phase can be due to the time required for recovery of cells from toxic products of metabolism that may accumulate in the external medium, such as acids, bases, alcohols, or solvents. Sometimes, new enzymes or coenzymes must be synthesized before growth resumes if the fresh medium is different from the inoculum medium and requires a change in the enzyme composition of the cells. If significant cell death occurs in the stationary phase, then an apparent lag phase will be measured because the inoculum includes dead cells that contribute to the turbidity. The lag phase can be avoided if the inoculum is taken from the exponential phase of growth and transferred to fresh medium of the same composition.

Stationary phase

The reason why cells stop growing and enter stationary phase can vary. It may be because of the exhaustion of nutrients, limitation of oxygen, or the accumulation of toxic products (e.g., alcohols, solvents, bases, or acids). The accumulation of toxic products is frequently a problem for fermenting cells because most of the nutrient is not converted to cell material but excreted as waste products. The excretion of fermentation end products is discussed in Chapter 13.

Death

Death can result from several factors. Common causes of death include the depletion of cellular energy and the activity of autolytic (self-destructive) enzymes. Some bacteria begin to die within hours of entering the stationary phase. However, many bacteria remain viable for longer periods. For example, some bacteria sporulate or form cysts when exponential growth ceases. The spores and cysts are resting cells that remain viable and germinate in fresh media. As discussed below, even nonsporulating bacteria can adapt to nutrient depletion and remain viable for long periods in stationary phase.

2.2.2 Adaptive responses to starvation

In the natural environment, bacteria frequently are faced with starvation conditions and enter intermittent periods of no growth or very slow growth. In some environments, the generation time may be many days or even months because the nutrient levels are so dilute. Lately, increased attention is being given to physiological changes that occur in bacteria that enter the stationary phase when they are experimentally subjected to starvation.[1,2] These bacteria undergo physiological changes that result in metabolically less active cells that are more resistant to environmental hazards. This has been known for a long time for those bacteria that sporulate or form desiccation-resistant cysts upon nutrient deprivation. The spores and cysts represent metabolically inactive or less active stages of the life cycle of the organism. When nutrients become available once more, the spores and cysts germinate into vegetative cells that grow and divide. More recently, it has been discovered that even bacteria such as *E. coli*, *Salmonella*, and *Vibrio* that do not form spores, undergo adaptive changes when faced with nutrient deprivation. These changes also result in resistant, metabolically less active cells. However, not all of the effects can be rationalized in terms of survivability and their physiological role is not yet known. Some of the changes that occur in cells that are starved are described below.

Changes in cell size

As discussed earlier, cells that enter stationary phase because of starvation generally become smaller due to the fact that they keep dividing after growth has ceased. This results in the production of more cells that may be advantageous for dispersing the population. In some cases, there may be several cell divisions in the absence of growth so that the cells size differs radically from the growing cell. Some bacteria decrease in length from several micrometers, e.g., 5–$10 \, \mu m$, to approximately 1–$2 \, \mu m$ or even less.

Changes in cell walls and in surface properties

When certain marine bacteria are starved, the cell surface becomes hydrophobic and the cells are more adhesive. *Vibrio* synthesizes surface fibrils and forms cell aggregates when starved for a long time.

Changes in metabolic activity and chemical composition

For many bacteria there is a significant increase in the turnover (metabolic breakdown and resynthesis) of protein and RNA when they are starved. Presumably, in starved cells, the protein and ribosomal RNA can serve as an energy source to maintain viability and crucial cell functions. The latter would include solute transport systems, an energized membrane, and ATP pool levels. There are also specific biochemical changes that occur when cells enter stationary phase. In *E. coli*, all of the unsaturated fatty acids in the membrane phospholipids become converted to the cyclopropyl derivatives by the methylation of the double bonds. The advantage to the cyclopropane fatty acids is not known, since mutants unable to synthesize cyclopropane fatty acids appear not to be at a survival disadvantage when faced with environmental stresses such as starvation and high or low oxygen tension. Bacteria may also synthesize as many as 30–50 new proteins under conditions of carbon starvation. Many of the proteins serve specific functions related to the nutrient that is in low concentration. For example, phosphate starvation induces the synthesis of PhoE porin, which is an outer membrane channel for anions, including phosphate (Sections 1.2.3 and 17.6). This helps the bacterium to bring in more phosphate. Another example is the nitrogen fixation genes that are induced when the cells are starved for nitrogen (Sections 11.3 and 17.5). In addition to the proteins of known function, there are also many proteins made under all conditions of starvation for which there is presently no known function.

Changes in metabolic rate and in resistance to environmental stress

When bacteria enter stationary phase, their overall metabolic rate slows. The cells also become more resistant to environmental stresses such as heat, osmotic stress, and certain chemicals (e.g., hydrogen peroxide). Some of these changes are due to the synthesis of the KatF protein as discussed next.

The KatF gene

The response to starvation in *E. coli* is mediated by the product of the *katF* gene, which is required for the synthesis of many of the proteins induced by carbon starvation. *KatF* mutants do not have the resistance properties of the wild type, including stationary phase resistance to heat, osmotic stress, or to H_2O_2. They also lose viability faster than wild type when starved under certain conditions. The predicted amino acid sequence of the *katF* gene, based upon nucleotide sequence data, suggests that it may be a sigma factor. Proteins whose synthesis depends upon KatF include a catalase made during stationary phase (HPII), and which presumably accounts for resistance to H_2O_2, an exonuclease III, and an acid phosphatase.

The stringent response

There is controversy surrounding the question of what regulates the rate of synthesis of ribosomes, although it is clear that ribosome synthesis is coupled to growth rates (Section 2.2.3). Faster growing cells have more ribosomes per unit cell mass. Probably, several factors are involved in regulating rates of ribosome synthesis. One hypothesis, called the *stringent response*,[3] will be discussed here. The stringent response in *E. coli* includes a temporary inhibition in the synthesis of ribosomal RNA and transfer RNA when the cells are shifted to a medium where they are starved for a carbon source or an amino acid. It works in the following way. When the carbon source or the amino acid supply decreases so as to slow the growth rate, the "extra" ribosomes that are "idling" or "stalled" because of an insufficiency of aminoacylated tRNA synthesize a signaling molecule called guanosine tetraphosphate, (ppGpp). The ppGpp inhibits the transcription of ribosomal RNA (and transfer RNA). This leads to an inhibition of ribosomal protein synthesis because ribosomal proteins in the absence of rRNA inhibit their own translation. They do this because the mRNAs for the ribosomal proteins have binding sites similar to the binding sites for ribosomal proteins on rRNA. When rRNA becomes limiting, the ribosomal proteins bind to the mRNA instead of the rRNA and prevent their own translation. Hence, ribosome synthesis is curtailed and the cell lowers the content of ribosomes in accordance with the lower growth rate. When all of the ribosomes are working full-time, the ppGpp levels fall and ribosomal RNA synthesis resumes. It must be emphasized that the control of ribosomal RNA synthesis is complex and ppGpp may be only one factor in the regulation. The synthesis of ppGpp is due to the product of the *relA* gene.[4]

When bacteria are shifted from a rich to a minimal media, a variety of other physiological responses occur that are also part of the stringent response. In addition to inhibition of ribosomal RNA synthesis, the stringent response includes:

1. A large decrease in the rate of synthesis of protein.

2. A temporary cessation in the initiation of new rounds of DNA replication.

3. An increase in the biosynthesis of amino acids.

4. A decrease in the rates of synthesis of phospholipids, nucleotides, peptidoglycan, and carbohydrates.

The stringent response can therefore be viewed as an adaptation to conditions that limit growth. It occurs as a consequence of any nutrient or energy limitation (e.g., amino acid starvation, nitrogen limitation, or a shift to a poorer carbon or energy source). Little is known concerning how ppGpp might affect physiological events other than stable RNA synthesis.

2.2.3 Macromolecular composition as a function of growth rate

Much has been learned about growth physiology by observing changes in macromolecular composition of cells whose growth rates have been altered by changing the nutrient composition of the medium, or by manipulation of the dilution rates of cells growing in continuous culture (Section 2.4).

Figure 2.4 illustrates the changes in the ratios of cell mass, RNA, protein, and DNA

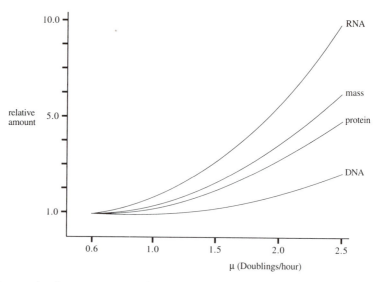

Fig. 2.4 Macromolecular composition of *E. coli* grown at different growth rates. The values are amounts per cell and have been normalized to values at a doubling time of 0.6 doublings/h. (From Neidhardt, F. C., Ingraham, J. L., and M. Schaechter. 1990. *Physiology of the Bacterial Cell*. Sinauer Associates, Inc. Sunderland, MA.)

of *E. coli* grown at increasingly faster growth rates due to differences in the nutrient composition of the growth media. On the ordinate are plotted relative amounts of macromolecules *per cell* normalized to cells undergoing 0.6 doublings per hour. On the abscissa are plotted the growth rate in units of doublings per hour. Notice that the mass of the cell as well as the relative concentrations of the macromolecules increase as exponential functions of the growth rate. However, the relative rates of increase are not equal so that the macromolecular composition of the cells is different at the different growth rates. For example, faster growing cells are enriched for RNA with respect to the other cell components. This reflects a higher proportion of ribosomes in faster growing cells. The reason for this is that, over a wide range of rapid growth rates, a ribosome polymerizes amino acids at approximately a constant rate. When a cell increases or decreases its growth rate, and therefore the number of proteins it must make per unit time, it adjusts the number of ribosomes rather than make each ribosome work substantially slower or faster. Figure 2.4 also shows that faster growing cells have more DNA per cell. The increased DNA per cell is rationalized in the following way. In *E. coli*, the minimum time required between the initiation of DNA replication and the onset of cell division for rapidly growing cells (i.e., those with doubling times between 20 and 60 min) is about 60 min.[5,6] If the generation time is shorter than 60 min, then DNA replication must begin in a previous cell cycle, which increases the average amount of DNA per cell. This is illustrated in Fig. 2.4 which shows that cells with a generation time more than 60 min have more DNA.

Why faster growing cells are larger

As seen in Fig. 2.4, cells that are grown at a faster growth rate have a greater mass (i.e., they are larger). For example, *E. coli* when growing at a doubling rate of 2.5 times an hour is approximately six times larger in mass than when growing at 0.6 doublings per hour. Why is that? It is due to a combination of the following three circumstances: (1) Cell division occurs approximately 60 min after the time of initiation of DNA replication. This is because it takes about 40 min to replicate the DNA and 20 min post-DNA replication before cell division occurs. For example, a cell that divides every 90 min begins replication of its DNA 30 min into the cell cycle, and a cell that divides every 70 min begins replication of its DNA 10 min into the cell cycle. (2) Therefore, faster growing cells must initiate DNA synthesis earlier in the cell cycle. (3) The initiation of DNA replication requires a minimum cell mass, called the critical cell mass (M_i). Points 1, 2, and 3 mean that the critical cell mass must be reached earlier in the cell cycle for faster growing cells and therefore faster growing cells must be larger, on average, than slower growing cells. These conclusions were reached by Donachie several years ago.[7,8] Actually, when the doubling times are faster than 60 min, then DNA replication must begin in the previous cell cycle, which means that faster growing cells may have DNA with multiple origins of replication (N_i). Donachie pointed out that bacteria initiate DNA synthesis only when they reach a critical mass in the cell cycle with respect to the number of DNA replication origins. In other words, the ratio M_i/N_i is a constant, where N_i is the number of replication origins. It was hypothesized that some cellular substance (e.g., a protein) is synthesized in proportion to the cell mass, and must reach a critical amount per replication origin in order for DNA replication to be initiated.

2.2.4 Diauxic growth

Many bacteria, including *E. coli*, will grow preferentially on glucose when presented with mixtures of glucose and other carbon sources. Preferential growth on one carbon source before growth on a second carbon source is called *diauxic growth*. The bacteria first grow exponentially on glucose, then enter a lag period, and finally grow exponentially on the second carbon source. An example of diauxic growth with glucose and lactose is shown in Fig. 2.5.

It is now known that glucose represses the synthesis of enzymes required to grow on certain alternative carbon sources such as lactose. Glucose also inhibits the uptake of certain other sugars into the cell. The result is that growth on the second carbon source does not proceed until the glucose is exhausted from the medium. During the lag period following the exponential growth on glucose, the bacteria synthesize the enzymes required to grow on the second carbon source. The repression of genes by glucose is called *catabolite repression* or *glucose repression*, and is widespread among bacteria. The inhibition of uptake of alternative sugars is called *inducer exclusion*. Both of these phenomena are explained for gram-negative

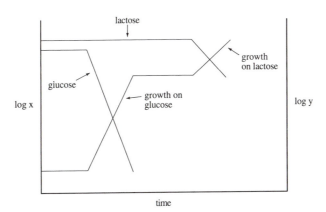

Fig. 2.5. Diauxic growth on glucose and lactose. *x*, bacterial mass. *y*, glucose or lactose concentration in the medium. There are two phases of growth: initially the cells grow on the glucose and, later, when the glucose is sufficiently depleted, a lag phase occurs followed by growth on lactose. During the lag phase the cells induce the synthesis of enzymes necessary for growth on lactose. Glucose prevents the induction of the genes necessary for growth on lactose. The inhibition by glucose is called catabolite repression and is discussed in Section 15.5.3.

bacteria in Section 15.3.4. A rationale for glucose repression is that glucose is one of the most common carbon sources in the environment and bacteria frequently grow more rapidly on glucose than on other carbon sources, especially other sugars. Therefore, bacteria in certain ecological habitats that preferentially utilize glucose may be at a competitive advantage with respect to the other cells that depend upon glucose for carbon and energy. Additionally, the enzymes required to metabolize glucose are constitutive in most bacteria (i.e., present under all growth conditions) and the cell is prepared to grow on glucose at any time.

However, not all bacteria preferentially grow on glucose, nor are the glucose catabolic enzymes necessarily constitutive. Several obligately aerobic bacteria will grow first on organic acids when given a mixture of glucose and the organic acid. This is also called catabolite repression except that it is the organic acid that is the repressor of glucose utilization. An example is the nitrogen fixing bacteria belonging to the genus *Rhizobium* that live symbiotically in root nodules of leguminous plants. These bacteria grow most rapidly in laboratory cultures on C_4 carboxylic acids such as succinate, malate, and fumarate. When presented with a mixture of succinate and glucose, *Rhizobium* shows diauxic growth by growing on the succinate first and the glucose second. This is because, in these bacteria, succinate represses key enzymes of the Entner–Doudoroff and Embden–Meyerhof–Parnas pathways that degrade glucose.[9] The mechanism of repression is not understood but is probably not the same as glucose repression.

2.2.5 Growth yields

When a single nutrient (e.g., glucose) is the sole source of carbon and energy, and when its quantity limits the production of bacteria, it is possible to define a *growth yield constant, Y*. The growth yield constant is the amount of dry weight of cells produced per weight of nutrient used (i.e., $Y = $ wt cells made/wt nutrient used). It has no units since the weights cancel out. [The molar growth yield constant is Y_m, which is the dry weight of cells

produced (in grams) per *mole* of substrate used.] For example, the $Y_{glucose}$ for aerobically growing cells is about 0.5, which means that about 50% of the sugar is converted to cell material and 50% oxidized to CO_2. For certain sugars and bacteria, the efficiency of conversion to cell material can be much lower (e.g., 20%). These differences are thought to relate to the amount of ATP generated from unit weight of the carbon source. The more ATP that is made, the greater the growth yields. This makes sense if you consider that growth is, after all, an increase in dry weight, and it requires a certain number of ATP molecules to synthesize each cell component. A value of 10.5 g of cells per mol of ATP called the Y_{ATP} (or molar growth yield) has been determined for fermenting bacterial cultures growing on glucose as the energy source. The experiments are performed in media in which all of the precursors to the macromolecules (e.g., amino acids, purines, pyrimidines) are supplied so that the glucose serves only as the energy source, and essentially all of the glucose carbon is accounted for as fermentation end products. Knowing the amount of ATP produced in the fermentation pathways, it is possible to calculate the Y_{ATP} from the $Y_{glucose}$. For example, 22 g of *Streptococcus faecalis* cells are produced per mol of glucose fermented. This organism ferments a mole of glucose to fermentation end products using the Embden–Meyerhof–Parnas pathway, which produces 2 mol of ATP (Section 8.1). Thus, the Y_{ATP} is 22/2 or 11. On the other hand, *Zymomonas mobilis* produces only 8.6 g of cells per mol of glucose fermented. These organisms use a different pathway for glucose fermentation, the Entner–Doudoroff pathway, which yields only 1 ATP per mol of glucose fermented (Section 8.5). Thus, the Y_{ATP} is 8.6. As mentioned, comparison of several different fermentations by bacteria and yeast has produced an average Y_{ATP} of 10.5. Although this value can vary, knowledge of the growth yields and the Y_{ATP} has allowed some deductions as to which fermentation pathway might be operating. Also, an unexpectedly high growth yield in fermenting bacteria can point to unrecognized sources of metabolic energy (Section 3.8).

2.3 Growth kinetics

2.3.1 The equation for exponential growth

During exponential growth, the mass in the culture doubles each generation. Equation 2.3 is the equation for exponential growth.

eq. 2.3. $x = x_0 2^Y$

In eq. 2.3, x is anything that doubles each generation, x_0 is the starting value, and Y is the number of generations. x can be cell number, mass, or some cell component (e.g., protein or DNA). Taking $\log_{10}$ of both sides:[10]

eq. 2.4. $\log x = \log x_0 + 0.301Y$

Let us define g as the generation time (i.e., the time per generation). Therefore, $g = t/Y$ and thus $Y = t/g$. t = time elapsed.

Therefore, eq. 2.4 becomes:

eq. 2.5 $\log x = \log x_0 + (0.301/g)(t)$

You will recognize this as an equation for a straight line. When x is plotted against t on semi-log paper (which is to base 10), the slope is $0.301/g$ and the intercept is x_0 (see Fig. 2.6).

The generation time, g

As stated, the generation time (doubling time) is the time that it takes for the population of cells to double. It is an important parameter of growth and probably the one most widely used by microbiologists. It is important that

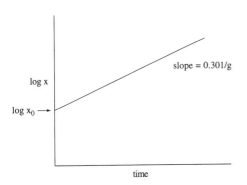

Fig. 2.6 Exponential growth plotted on a semi-log scale. g, generation time (time per generation).

the student learns how to find the generation time. The simplest way is graphically. In practice, one determines the generation time from inspection of a plot of x versus t on semi-log paper (Figs 2.3 or 2.5), and simply reads off the time it takes for x to double. It can also be done using eq. 2.5 if x, x_0, and t are known accurately for the exponential phase of growth.

The growth rate constant, k

The growth rate constant, k, is a measure of the *instantaneous* growth rate and has the units of reciprocal time. It is also a widely used parameter of growth. It is not to be confused with the generation time, g, which is the *average* time it takes for the population of cells to double. The distinction is made clear below. Recall that since each bacterial cell gives rise to two, the rate at which the population is growing at any instant, i.e., the instantaneous rate of growth, (dx/dt), must be equal to the number of cells at that time (x) times a growth rate constant (k). Thus:

eq. 2.6 $dx/dt = kx$, or $dx/x = k\,dt$

Integrate: $\ln x = kt + \ln x_0$ or,

eq. 2.7 $x = x_0 e^{kt}$

Taking $\log_{10}$ of both sides of eq. 2.7 converts the expression from the natural log ($\log_e$) to log base 10 so that x can be plotted against t on semi-log paper:

eq. 2.8 $\log x = 0.4342(kt) + \log x_0$

or $\log x = kt/2.303 + \log x_0$

Note that this gives the same line as in Fig. 2.6. However, the slope is equal to $k/2.303$. Since the slope in Fig. 2.6 is equal to $0.301/g$, it follows that $k/2.303$ must be equal to $0.301/g$. Thus $k = 0.693/g$ (or $kg = 0.693$). Therefore, k and g vary as the reciprocal of one another with the proportionality constant of 0.693. The slope of the curve in Fig. 2.6 can give you either k (the instantaneous growth rate) or g (the doubling time for the population).[11]

Using the growth equations

The growth equations can be used for finding g and k. Another, and more routine, circumstance is when one must choose the proper size inoculum when growing a culture to be used at a later time. A convenient equation to remember for this calculation is eq. 2.3 or eq. 2.5. Suppose you have a culture growing exponentially at a density of 10^8 cells/ml and you would like to subculture it so that 16 h later the density of the new culture will also be 10^8 cells/ml. The generation time (g) is 2 h. What should x_o be? One way to do this problem is to estimate the number of generations (Y) and then to use eq. 2.3 or eq. 2.5. For example, since $Y = t/g$, the number of generation is 16/2, or 8. Using eq. 2.3, $10^8 = x_o 2^8$ or $x_o = 10^8/2^8 = 3.9 \times 10^5$ cells/ml. This means that the initial cell density in the growth flask should be 3.9×10^5 cells/ml. But the cell density of the inoculum is 10^8/ml. This means that the inoculum must be diluted $10^8/3.9 \times 10^5$ or 2.6×10^2 times. If you wanted to grow 1 liter of cells, the inoculum size would have to be 3.8 ml.[12] Most growth problems that one might have to do routinely can be done with the following equations:

$$x = x_o 2^Y$$

$$Y = t/g$$

$$g(k) = 0.693$$

2.3.2 The relationship between growth rate constant (k) and nutrient concentration

In the natural environment, the concentrations of nutrients is so low that the growth rates are limited by the rates of nutrient uptake or the rate at which a previously stored nutrient is used. At very low nutrient concentrations, the growth rate can be shown to be a function of the nutrient concentration (Fig. 2.7). The curve approximates saturation kinetics and is probably due to the saturation of a transporter in the membrane that brings the nutrient into the cell. The curve can be compared to the kinetics of solute uptake on a transporter described in Section 15.2. One

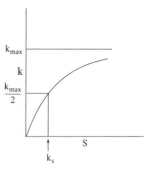

Fig. 2.7. Variation of growth rate as a function of substrate concentration. k_{max}, maximal growth rate constant; k, specific growth rate constant; k_s, nutrient concentration that gives $1/2\,k_{max}$; S, nutrient concentration. The concentrations where the growth rate is proportional to the nutrient concentrations are very low, in the micromolar range.

can rationalize the kinetics by assuming that, at very low concentrations of nutrient, the growth rate is proportional to the percentage of transporter that has bound the nutrient. At high nutrient concentrations, the rate of nutrient entry approaches its maximal value because all of the transporter has bound the nutrient. Therefore, the growth rate becomes independent of the nutrient concentration and becomes limited by some other factor. The curve in Fig. 2.7 can be described by the following equation, where k is the specific growth rate constant, k_{max} is the maximal growth rate constant, S is the nutrient concentration, and K_s is the nutrient concentration that gives one-half k_{max},

eq. 2.9 $\quad k = k_{max}S/(K_s + S)$

These are the same kinetics that describe the saturation of an enzyme by its substrate (i.e. Michealis–Menten kinetics) (Section 6.3.1). Equation 2.9 can be used to calculate growth yields during continuous growth (Section 2.4).

2.4 Steady-state growth and continuous growth

A culture is said to be in steady-state growth (balanced growth) when all of its components double at each division and maintain a constant ratio with respect to one another. Steady-state growth is usually achieved when a culture is maintained in exponential growth by subculturing (i.e., it is not allowed to enter stationary phase) or during continuous growth.

2.4.1 The chemostat

Continuous growth takes place in a device called a chemostat (Fig. 2.8). A reservoir feeds fresh medium continuously into a growth chamber at a flow rate (F) set by the operator. In the chemostat, the concentration of the limiting nutrient in the reservoir is kept very low so that growth is limited by the availability of the nutrient. When a drop of fresh medium enters the growth chamber, the growth-limiting nutrient is immediately used up and the cells cannot continue to grow until the next drop of media enters. Because of this, the growth rate is manipulated by adjusting the rate at which fresh medium enters the growth chamber.

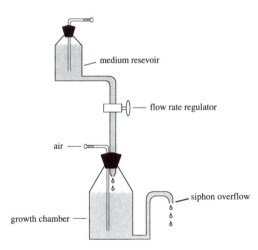

Fig. 2.8. A chemostat. A chemostat works because the growth of the cells in the growth chamber is limited by the rate at which a particular limiting nutrient (e.g., glucose) is supplied. That is to say, the concentration of the limiting nutrient in the reservoir is sufficiently low so that there is never an excess in the growth chamber. Each moment when fresh medium comes into the growth chamber the incoming limiting nutrient is rapidly utilized and growth cannot continue until fresh nutrient enters.

The dilution rate, D

The dilution rate is equal to F/V, where F is the flow rate and V is the volume of medium in the growth chamber. For example, if the flow rate (F) is 10 ml/h and the volume (V) in the growth chamber is 1 liter, then the dilution rate (D) is $10\,ml\,hr^{-1}/1{,}000\,ml$ or $0.01\,h^{-1}$. Notice that the units of D are reciprocal time. If one multiplies D by the number of cells, x, in the growth chamber, then the product Dx represents the rate of loss of cells from the outflow, and has the units of cells/time.

Relationship of dilution rate, D, to growth rate constant, k

In the steady state, $k=D$. To show this, consider that the rate of growth is $dx/dt=kx$. When a steady state is reached in the growth chamber, the rate of formation of new cells (kx) equals the rate of loss of cells from the outflow (Dx). That is to say, $kx=Dx$ and therefore, $k=D$. Therefore, in a chemostat, the growth rate of a culture can be changed merely by changing the flow rate, F, which determines the dilution rate, D.

Dependence of cell yield, x, on concentration of limiting nutrient in reservoir, S_R

The actual concentration of cells in the growth chamber is manipulated by changing the concentration of limiting nutrient in the reservoir. Let S_r be the concentration of limiting nutrient in the reservoir and S be the concentration of the nutrient in the growth chamber. The difference, $S_r - S$, must be equal to the amount of nutrient used up by the growing bacteria. If we define the growth yield constant, Y, as being equal to the mass of cells (x) produced per amount of nutrient used up ($S_r - S$), then $Y=x/(S_R-S)$ or rearranging,

eq. 2.10 $x=Y(S_R-S)$

S can be calculated from eq. 2.9 by substituting D for k. S is much smaller than S_R and can be conveniently ignored in most calculations.

Varying the cell density, x, and growth rate constant, k, independently of one another

Equation 2.10 predicts that if one were to increase the concentration of rate-limiting nutrient in the reservoir (S_R), then the cell density in the growth chamber will increase. When S_R increases, more nutrient enters the growth chamber per unit time. Initially, S must increase. Since the growth rate is limited by the concentration of S (Fig. 2.7), the cells will respond initially by growing faster and the cell density will increase. A new steady state will be reached in which the increased number of cells use up all of the available nutrient as it enters. Thus, even though S_R is increased, the dilution rate still controls the growth rate. What has been accomplished, therefore, when S_R is increased is a higher steady-state value of x, according to eq. 2.10, but no change in the growth rate constant (k). There are two conclusions to be drawn:

1. The only way to change the steady state-growth rate constant is to change the dilution rate (D), because this changes the rate of supply of S.

2. The only way to change the steady state-growth yield is to change the concentration of limiting nutrient in the reservoir, S_R.

2.5 Summary

Growth is defined as an increase in mass and can conveniently be measured turbidometrically. Other methods can be used provided that they measure something that parallels mass increase. These methods include measuring the rate of increase in cell number (viable or total), or specific macromolecules (e.g., protein, DNA, RNA).

Growth in batch culture can progress through a lag and log to stationary phase where net growth of the population (measured as mass or its equivalent) has ceased. Eventually, the viable cells may decrease in number and this is referred to as the death phase. The availability of nutrients, the need to synthesize specific enzymes to metabolize newly encountered nutrients, and the accumulation of toxic end products in the medium, can explain the different stages of the growth curve. In addition, specific physiological changes of an adaptive nature occur when cells enter the stationary phase because of nutrient depletion.

Growth during the log phase can be described by a simple exponential equation depicting a first-order autocatalytic process. That is to say, the mass doubles at each generation. From this equation, one can derive a generation time and an instantaneous growth rate constant. These constants are used to characterize growth under different physiological situations and to predict growth yields at specific times for experimental purposes.

Steady-state (continuous) growth is defined as the growth of a population of cells during which all the components of the cell double at each division.

When cells are grown at different growth rates because of nutritional alterations or chemostat growth, then the macromolecular composition changes. Faster growing cells are larger, have proportionally more ribosomes and have more DNA per cell. This can be rationalized in terms of an approximate constancy of ribosome efficiency in protein synthesis, and an almost constant period between the initiation of DNA replication and cell division.

Diauxic growth is characterized by two phases of population growth separated by a stationary phase when the cells are incubated with certain pairs of carbon and energy sources. For example, consider diauxic growth with glucose. The cells grow on the glucose first because it represses the expression of the genes necessary to grow on the second carbon source and because it prevents the uptake of other sugars. The repression is called catabolite repression or glucose repression. Prevention of sugar uptake is called inducer exclusion. There are at least two possible rationales for this. Glucose is the most widely used carbon and energy source, and cells are in a better situation to outgrow their neighbors if they use the glucose first. In doing so, they lower the supply of glucose to other cells. Furthermore, many bacteria always express the genes to metabolize glucose (i.e., they are constitutive) but the genes to metabolize other carbon sources are often not expressed unless the carbon source is present in the medium. This lowers the energy burden of carrying genetic information. However, glucose is certainly not a universal catabolite repressor among the bacteria. For example, several obligately aerobic bacteria such as *Rhizobium* are known to grow preferentially on organic acids such as succinate, malate, and fumarate, when given a mixture of glucose and one of these acids. This makes physiological sense for these organisms because growth on the C_4 carboxylic acids is faster than growth on glucose.

Cells can be grown in continuous culture using a chemostat. Two advantages to growing cells this way are: (1) they can be maintained in balanced growth, and (2) the growth rate constant can be easily changed simply by altering the flow rate. The growth yields can be separately manipulated by changing the concentration of limiting nutrient in the reservoir. Growth of continuous cultures is possible when the growth rate of the culture is limited by the supply of a nutrient that is continuously fed into the growth chamber.

When bacteria are starved of a carbon and/or energy source, many biosynthetic reactions slow down, including ribosome synthesis. This has been called the "stringent response," and is correlated with increased synthesis of guanosine tetra- and pentaphosphates.

Study Questions

1. What is the generation time (g) of a culture with a growth rate constant (k) of $0.01 \, min^{-1}$?

   ans. 69.3 min

2. Assume you wanted to grow a culture to 10^8 cells/ml in 3 hrs. The generation time is 30 min. What should be the starting cell density (x_o)?

   ans. 1.6×10^6

3. Assume a culture whose cell density is 10^8/ml. The generation time is 30 min. If you started with a 10^{-2} dilution, how many hours would you grow the culture?

   ans. 3.3 h

4. Assume you have a stock culture at 5×10^9/ml and you wish to inoculate 1 liter of fresh medium so that in 15 h the cell density will be 2×10^8/ml. Assume a generation time of 3.5 h. What should be the dilution? What size inoculum should be used?

   ans. 1/500 or 2×10^{-3}; 2 ml

5. Assume the yield coefficient for glucose (Y_g) is 0.5 g cells per g of glucose consumed. In a glucose-limited chemostat, what should be the concentration of glucose in the reservoir so that the mass of cells in the growth chamber (x) would be 0.1 mg/ml? For this problem, ignore S in eq. 2.10 because it is much smaller than S_R.

   ans. 0.2 mg/ml

6. In a 500 ml chemostat, what should be the flow rate (F) in minutes for a generation time (g) of 6 h?

   ans. 0.95 ml/min

7. Suppose you were operating a chemostat where S is the limiting nutrient. Assume that D is $0.2 \, h^{-1}$, K_s is 1×10^{-6} M, and k_{max} is $0.4 \, h^{-1}$. (a) What is the concentration of S in the growth chamber? (b) If the cell density were 0.25 mg/ml, what would be the concentration of S in the reservoir if Y_S is 0.5? (c) What is the concentration of S in the growth chamber when the cells are growing one-half as fast (i.e., $D = 0.1 \, h^{-1}$)?

   ans. (a) 1×10^{-6} M; (b) 0.5 mg/ml; (c) 3.3×10^{-7} M

REFERENCES AND NOTES

1. Siegele, D. A. and R. Kolter. 1992. Life after lag. *J. Bacteriol.* **174**:345–348.

2. Matin, A. 1991. The molecular basis of carbon-starvation-induced general resistance in *Escherichia coli*. *Molec. Microbiol.* **5**:3–10.

3. Cashel, M. and K. R. Rudd. 1987. The stringent response, p. 1410–1438. In: Escherichia coli *and* Salmonella typhimurium: *Cellular and Molecular*

Biology, Vol. 2. F. C. Neidhardt *et al.* (eds.). ASM Press. Washington, DC.

4. There are two mechanisms for the synthesis of ppGpp. As mentioned, the ppGpp can be synthesized on ribosomes that are "stalled" because of a restriction in the supply of aminoacylated tRNA. This occurs during amino acid starvation and requires the product of the *rel*A gene. The *rel*A gene was discovered in a search for mutants that failed to respond to amino acid starvation with the stringent response. These mutants were termed "relaxed", hence the name of the gene. The RelA protein (also called (p)ppGPP synthetase I) is a ribosome-associated protein that synthesizes either ppGpp or pppGpp by transferring a pyrophosphoryl group from ATP to either GDP or GTP. The synthetase is activated in starved (stalled) ribosomes. The guanosine polyphosphates are also synthesized in response to carbon starvation by a different enzyme called (p)ppGpp synthetase II.

5. Cooper, S. and C. R. Helmstetter. 1968. Chromosome replication and the division cycle of *Escherichia coli* B/r. *J. Mol. Biol.* **31**:519–539.

6. The length of time between the onset of DNA replication and cell division is the sum of two time periods [i.e., the time it takes for the chromosome to replicate (the C period) and the interval between the end of a round of DNA replication and cell division (the D period)]. The minimum C and D periods are 40 min and 20 min, respectively, in rapidly growing *E. coli*. Hence 60 min is the minimum time required between the onset of a round of DNA replication and cell division. These times gradually lengthen as doubling times increase.

7. Donachie, W. D. 1968. Relationhsip between cell size and time of initiation of DNA replication. *Nature.* **219**:1077–1079.

8. Donachie assumed that each cell grew exponentially and doubled in mass during the cell cycle. Knowing the size of the cell at the time of cell division, he was able to calculate the size at any time during the cell cycle. With this information and the knowledge that DNA initiation occurred 60 min before cell division, he was able to compute the mass of the cell at the time of initiation of DNA initiation. This mass, divided by the number of replication origins, was a constant.

9. Chandra, N. M. and P. K. Chakrabartty. 1993. Succinate-mediated catabolite repression of enzymes of glucose metabolism in root-nodule bacteria. *Curr. Microbiol.* **26**:247–251.

10. Suppose you want to convert $\log_2$ to $\log_{10}$. First write the exponential equation (e.g., $X = X_o 2^Y$). Now take $\log_{10}$ of both sides of the equation (i.e., $\log_{10} X = \log_{10} X_o + \log_{10} 2(Y)$). Since $\log_{10} 2 = 0.301$, the equation becomes $\log X = \log X_o + 0.301 Y$. Note that $\log_{10}$ is usually written simply as log.

11. Some investigators use the symbol μ for the instantaneous growth rate constant and k for the reciprocal of the generation time (i.e., $k = 1/g$).

12. The dilution is 2.6×10^2. Assume the volume of inoculum is X. Therefore, $2.6 \times 10^2 (X)$ must equal the final volume, which is $1000\,\text{ml} + X$. Solving for X gives 3.8 ml.

3

Membrane Bioenergetics: The Proton Potential

A major revolution in our conception of membrane bioenergetics[1,2] has taken place in the last 30 years as a result of the theoretical ideas of Peter Mitchell, referred to as the chemiosmotic theory.[3-5] Briefly, the chemiosmotic theory states that energy transducing membranes (i.e., bacterial cell membranes, mitochondrial, and chloroplast membranes) pump protons across the membrane, thereby generating an electrochemical gradient of protons across the membrane (the proton potential) that can be used to do useful work when the protons return across the membrane to the lower potential. In other words, bacterial, chloroplast, and mitochondrial membranes are energized by proton currents. Of course, the return of the protons across the membrane must be through special proton conductors that couple the translocation of protons to do useful cellular work. These proton conductors are transmembrane proteins. *Some membrane proton conductors are solute transporters, others synthesize ATP, and others are motors that drive flagellar rotation.* The proton potential provides the energy for other membrane activities besides ATP synthesis, solute transport, and flagellar motility (e.g., reversed electron transport and gliding motility). Because the chemiosmotic theory is central to energy metabolism, it lies at the foundation of all bacterial physiology. As explained in this chapter, the chemiosmotic theory brings together principles of

physics and thermodynamics in explaining membrane bioenergetics. The student should study the principles of the chemiosmotic theory for a deeper understanding of how the bacterial cell uses ion gradients to couple energy-yielding (exergonic) reactions to energy-requiring (endergonic) reactions. This chapter explains the principles of the theory.

3.1 The chemiosmotic theory

According to the chemiosmotic theory, protons are translocated out of the cell by exergonic (energy-producing) driving reactions which are usually biochemical reactions (e.g., respiration, photosynthesis, or ATP hydrolysis). Some of the translocated protons leave behind negative counterions (e.g., hydroxyl ions) thus establishing a membrane potential, outside positive. Protons may also accumulate electroneutrally in the extracellular bulk phase establishing a proton concentration gradient, high on the outside (outside acid). When the protons return to the inside, moving down the concentration gradient and towards the negative pole of the membrane potential, work can be done. Figure 3.1 illustrates the proton circuit in the bacterial cell membrane. From the perspective described here, the cell membrane is similar to a battery except that the current is one of protons rather than electrons. In Fig. 3.1, the

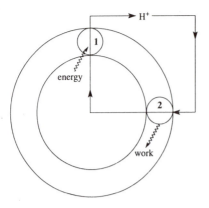

Fig. 3.1 The proton current. There is a proton circuit traversing the bacterial cell membrane. Protons are translocated to the cell surface, driven there by either chemical or light energy through a proton pump (1) and return through special proton transporters (2) that do work. The accumulation of protons on the outside surface of the membrane establishes a membrane potential, outside positive. A pH gradient can also be established, outside acid. In several gram-negative bacteria oxidizing certain inorganic compounds (lithotrophs), or single-carbon compounds such as methanol, protons that are released into the periplasm via periplasmic oxidations contribute to the proton current (Chapter 11). In some cases, periplasmic oxidations are the sole provider of protons for inward flux.

"battery" is charged by reactions 1 that translocate protons, and does work via reactions 2. Reactions 1 include redox reactions that occur during electron transport (Fig. 3.2, 1) and an ATP-driven proton pump (the ATP synthase) (Fig. 3.2, 7). The work that is done by the protons that enter the cell includes the extrusion of sodium ions (Fig. 3.2, 3), solute transport (Fig. 3.2 4), flagellar rotation (Fig. 3.2, 6), and the synthesis of ATP via the ATP synthase (Fig. 3.2, 7 and Section 3.6.2). As Mitchell emphasized, it is important that the membrane has a low permeability to protons so that the major route of proton re-entry is via the energy-transducing proton transporters rather than by general leakage. This, of course, would be expected of a lipid bilayer that is relatively nonpermeable to protons. Some bacteria couple respiration or the decarboxylation of carboxylic acids to the extrusion of sodium ions (Fig. 3.2, 2). (See Sections 3.7.1 and 3.8.2.) The re-entry of

sodium ions can also be coupled to the performance of work (e.g., solute uptake) (Fig. 3.2, 5). Once established, the membrane potential can energize the secondary flow of other ions. For example, the influx of potassium ions can be in response to a membrane potential, inside negative, created by proton extrusion. Mitochondrial and chloroplast membranes are also energized by proton gradients. Therefore, this is a widespread phenomenon, not restricted to the bacteria.

3.2 Proton electrochemical energy

When bacteria translocate protons across the membrane to the outside surface, energy is conserved in the proton gradient that is established. The energy in the proton gradient is both electrical and chemical. The *electrical* energy has to do with the fact that a positive charge, (i.e., the proton) has been moved to one side of the membrane creating a charge separation, and therefore a membrane potential. When the proton moves back into the cell towards the negatively charged surface of the membrane, the membrane potential is dissipated, (i.e., energy has been given up and work can be done). The energy dissipated when the proton moves to the inside of the cell is equal to the energy required to translocate the proton to the outside.

Stated more precisely, energy is required to move a charge *against* the electric field (i.e., to the side of the same charge). This energy is stored in the electric field. The energy that is stored in the electric field is called *electrical energy*. Conversely, the electric field gives up energy when a charge moves *with* the electric field, i.e., to the opposite pole, and work can be done. The amount of energy is the same, but opposite in sign.

The same description applies to chemical energy. Energy is required to move the proton against its concentration gradient. This energy is stored in the concentration gradient. The energy that is stored in a concentration gradient is called *chemical energy*. When the proton returns to the lower concentration, then the energy in the concentration gradient is dissipated and work can be done.

The sum of the change in electrical and

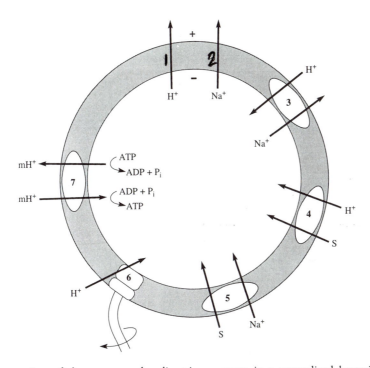

Fig. 3.2 An overview of the proton and sodium ion currents in a generalized bacterial cell. Driving reactions (metabolic reactions) deliver energy to create proton (1) and sodium ion (2) electrochemical gradients, high on the outside. The major driving reactions encompassed by (1) are the redox reactions that occur during electron transport. The establishment of sodium ion potentials coupled to metabolic reactions such as respiration are not widespread. When the ions return to the lower electrochemical potentials on the inside, work can be done. Built into the membrane are various transporters (porters) that translocate protons and sodium ions back into the cell, completing the circuit, and in the process doing work. There are three classes of porters: (a) *antiporters* that carry two solutes in opposite directions, (b) *symporters* that carry two solutes (S) in the same direction, and (c) *uniporters* that carry only a single solute. Illustrated are: (3), a Na^+/H^+ antiporter, which is the major mechanism for extruding Na^+ in bacteria, and also functions to bring protons into the cell for pH homeostasis in alkaliliphilic bacteria. In most bacteria the Na^+/H^+ antiporter creates the sodium potential necessary for the Na^+/solute symporter because a primary Na^+ pump is not present. The antiporter uses the proton electrochemical potential as a source of energy; (4), a H^+/solute symporter that uses the proton potential to accumulate solutes; (5), a Na^+/solute symporter that uses the sodium electrochemical potential to accumulate solutes; (6), a flagellar motor that turns at the expense of the proton electrochemical potential; (7), an ATP synthase that synthesizes ATP at the expense of the proton electrochemical potential. The ATP synthase is reversible and can create a proton electrochemical potential during ATP hydrolysis.

chemical energies is called *electrochemical energy*. The symbol for electrochemical energy is $\Delta\tilde{\mu}$ which is equal to $\tilde{\mu}_{in} - \tilde{\mu}_{out}$. For the proton, it would be $\Delta\tilde{\mu}_{H^+}$. The units of electrochemical energy are joules per mole, or calories per mole. (One calorie is equal to 4.18 joules.) The electrochemical energy is discussed in more detail in the following sections.

3.2.1 The electrochemical energy of protons

The protonmotive force
The electrochemical work that is performed when an ion crosses a membrane is a function of both the membrane potential, $\Delta\Psi$, and the difference in concentration between the solutions separated by the membrane. For ex-

ample, for one mole of protons:

� eq. 3.1. $\Delta\tilde{\mu}_{H^+} = F\Delta\Psi + RT\ln[H^+]_{in}/[H^+]_{out}$ J

This can be expressed as an electrical potential in mV:

✱ eq. 3.2 $\Delta\tilde{\mu}_{H^+}/F = \Delta\Psi - 60\,\Delta pH$ mV (at 30°C)

Usually, $\Delta\tilde{\mu}_{H^+}/F$ is called the protonmotive force and is denoted as Δp. Equation 3.2 is derived below. [See eq. 3.3–3.10.] Bacteria have an average Δp of approximately -140 to -200 mV. The values for respiring bacteria tend to be a little higher than those for fermenting bacteria.[6]

Units

It is important to distinguish between volts (V), electron volts (eV), and joules (J). Potential differences (e.g., ΔE or Δp) are expressed as volts or millivolts. These refer to the forces that can act on moving charges, whether they are ions or electrons. When one specifies the amount of charge that moves over the potential difference, then one is describing work, and the units are either joules or electron volts. One joule is the energy required to raise a charge of one coulomb (C) through a potential difference of one volt (i.e., J = C × V). (The older literature used calorie units instead of joules. One calorie = 4.184 J.) When calculating the change in joules when a mole of monovalent ions or electrons travels over a voltage gradient, one multiplies volts by the faraday (F), since the faraday is the number of coulombs of charge per mole of electrons or monovalent ions. One electron volt is the increase in energy of a single electron or monovalent ion when raised through a potential difference of one volt. The charge on the electron or monovalent ion is approximately 1.6×10^{-19} C. Therefore, one electron volt is equal to 1.6×10^{-19} J.

The electrical component of the $\Delta\mu_{H^+}$ *proton electrochemical energy*

A membrane potential, $\Delta\Psi$, exists across the cell membrane where $\Delta\Psi = \Psi_{in} - \Psi_{out}$. By convention, the $\Delta\Psi$ is negative when the inner membrane surface is negative. The units of $\Delta\Psi$ are volts. The work done on or by the

electric field when charges traverse the membrane potential is equal to the total charges carried by the ions or electrons multiplied by the $\Delta\Psi$. If a single electron or monovalent anion such as the proton moves across the membrane, then the work done is $\Delta\Psi$ electron volts. A single proton (or any monovalent ion, or electron) carries 1.6×10^{-19} C of charge. If we multiply this by Avogadros number, then we arrive at the charge carried by a mole of protons, which is 96,500 C, or the faraday (F). This means that the amount of work that is done per mole of protons that traverses the $\Delta\Psi$ is:

✱eq. 3.3 $\Delta G = F\Delta\Psi$ J   *for work done*

This is often expressed as electrical potential by dividing both sides of the equation by the faraday:

✱ eq. 3.4 $\Delta G/F = \Delta\Psi$ V   *for e⁻ potential*

The chemical component of the $\Delta\tilde{\mu}_{H^+}$

Of course, if a concentration gradient of protons exists, then we must add the chemical energy to the electrical energy in eq. 3.3 in order to obtain the expression for the electrochemical energy, $\Delta\tilde{\mu}_{H^+}$. The chemical energy of the proton as a function of its concentration is:[7]

✱ eq. 3.5 $G = G_o + RT\ln[H^+]$ J

The free energy change accompanying the transfer of one mole of protons between a solution of protons outside $[H^+]_{out}$ of the cell and inside $[H^+]_{in}$ of the cell is the difference between the free energies of the two solutions,

✱ eq. 3.6 $\Delta G = RT\ln[H^+]_{in} - RT\ln[H^+]_{out}$ J

or   ✱ $\Delta G = RT\ln[H^+]_{in}/[H^+]_{out}$ J

Equation 3.6 refers to the free energy change when one mole of protons moves from one concentration to another where the concentration gradient does not change (i.e., as applies to a steady state). It does not refer to the total energy released when the concentration of protons comes to equilibrium. Equation 3.6 can be used for the movement of any solute over a concentration gradient, not simply protons.

Usually, eq. 3.6 is expressed in electrical units of potential (V). In order to do this, one substitutes $8.3144 \, J \, deg^{-1} \, mol^{-1}$ for R, 303 K (30°C) for T, converts ln to $\log_{10}$ by multiplying by 2.303, and divides by the faraday to convert joules to volts, thus deriving eq. 3.7.

eq. 3.7 $\Delta G/F = 0.06 \log[H^+]_{in}/[H^+]_{out}$ V

or $= 60 \log[H^+]_{in}/[H^+]_{out}$ mV

since $\log[H^+]_{in}/[H^+]_{out} = pH_{out} - pH_{in}$, eq. 3.7 can be written as:

eq. 3.8 $\Delta G/F = 60(pH_{out} - pH_{in})$
$= -60(pH_{in} - pH_{out})$
$\Delta G/F = -60\Delta pH$ mV

The proton potential, Δp

We are now ready to derive an expression for the protonmotive force. The total energy change accompanying the movement of one mole of protons through the membrane is the sum of the energy due to the membrane potential (eq. 3.3) and the energy due to the concentration gradient (eq. 3.6). This sum is called the electrochemical energy, $\Delta\mu_{H^+}$.

eq. 3.9 $\Delta\tilde{\mu}_{H^+}$
$= F\Delta\Psi + RT \ln[H^+]_{in}/[H^+]_{out}$ J

One can also express the electrochemical energy as a potential in volts or millivolts (protonmotive force, or Δp) by adding eq. 3.4 and eq. 3.8:

eq. 3.10 $\Delta p = \Delta\Psi - 60 \Delta pH$ mV at 30°C.

The same equation is used to express the electrochemical potential for any ion (e.g., Na^+):

eq. 3.11 $\Delta\tilde{\mu}_{Na^+}/F = \Delta\Psi - 60\Delta pNa$ mV,

where pNa is $-\log(Na^+)$.

By convention, the values of the potentials are always negative when the cell membrane is energized for that particular ion.

3.2.2 Generating the membrane potential

We now consider some biophysical aspects of generating the protonmotive force. First we will examine the formation of the $\Delta\Psi$, and then its relationship to the ΔpH.

Electrogenic flow

The movement of an uncompensated charge creates the membrane potential. When a moving charge creates a membrane potential, the movement is said to be *electrogenic*. For example, a membrane potential is generated when a proton is translocated through the membrane to the outer surface leaving behind a negative charge on the inner surface. The energy required is given by eq. 3.3. A membrane potential can also develop if a molecule is oxidized on the outer membrane surface releasing $H^+ + e^-$, and the e^- returns across the membrane to the inner surface while the H^+ remains behind. The work done is the same because the electron carries the same charge as the proton (i.e., 1.6×10^{-19} coulombs). We will see that bacteria use this method as well as proton translocation to create a membrane potential.

The size of the membrane potential depends upon the capacitance of the membrane

The membrane potential that develops when even small numbers of protons move across the membrane can be more than 100 mV. This can be understood by considering the membrane to be a capacitor. The membrane is a capacitor because the membrane lipids prevent the protons from leaking back into the cell rapidly, and so the membrane stores positively charged protons on one surface, just like a capacitor stores charges on one surface.

The relationship between the charge that accumulates on one face of a capacitor and the voltage across the capacitor is:

eq. 3.12 $\Delta V = Q/C$

where ΔQ is the charge (in coulombs), ΔV is the voltage, and C is a proportionality constant called the capacitance. The value of C for biological membranes is low, about 1 microfarad (μF) cm^{-2} of membrane. The equation can be rewritten with different symbols and used to predict the theoretical membrane potential:

✱ eq. 3.13 $\Delta \Psi$ (volts) $= en/C$

where e is the charge per proton (1.6×10^{-19} C), n is the number of protons, and C is the capacitance. Assuming a membrane area of about 3×10^{-8} cm^2 (for a spherical cell the size of a typical bacterium), then $C =$ about 3×10^{-14} F. Therefore, only 40,000 protons translocated to the cell surface is sufficient to generate a membrane potential of -200 mV.[8,9] (By convention, $\Delta \Psi$ is said to be negative when the inside potential is negative.) The membrane potential that actually develops varies in magnitude from approximately -60 mV to about -200 mV depending upon the bacterium and the growth conditions.[10] The membrane potential that develops when a relatively small number of protons are translocated limits electrogenic proton pumping and the size of the membrane potential. (We will return to this point later when we discuss the generation of a ΔpH by proton translocation.)

Relationship of $\Delta \Psi$ to ΔpH

When a proton is translocated across the membrane, a $\Delta \Psi$ and a ΔpH cannot be created simultaneously. This is summarized in Fig. 3.3. Let us first discuss creating a ΔpH (i.e., accumulating protons in the external bulk phase). One should remember that the only way the bulk external medium can become acidified during proton translocation is if electrical neutrality is conserved (i.e., the protons in the bulk phase must have a negative counterion). This can happen if the proton is pumped out with an anion (i.e., H^+/R^-), or if a cation enters the cell from the external bulk phase in exchange for the proton (i.e., H^+ in exchange for R^+) (Fig. 3.3). However, these would be electroneutral events and a charge separation, hence a $\Delta \Psi$, would not develop. Therefore, in the absence

symport
antiport

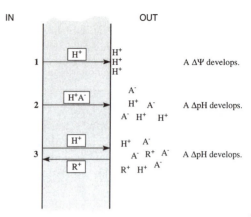

Fig. 3.3 A ΔpH in the bulk phase and a $\Delta \Psi$ cannot develop simultaneously by the movement of a proton. (1) Electrogenic movement of a proton. (2) Establishment of a ΔpH by the extrusion of both protons and counterions. (3) Establishment of a ΔpH by the exchange of protons for cations in the medium. Theoretically, a ΔpH could develop if the proton on the outer surface of the membrane exchanged with a cation from the bulk phase. But even under these circumstances the membrane potential (positive outside) would limit further efflux of protons.

of compensating ion flow, the protons that are pumped out of the cell remain on or very close to the membrane, and a $\Delta \Psi$ rather than a pH (measurable with a pH electrode) is created.

Of course, some of the protons could be electroneutrally released into the bulk phase to create a ΔpH and some might remain at the outer membrane surface to create a $\Delta \Psi$, but even under these circumstances, a large ΔpH cannot be generated in the face of a large $\Delta \Psi$. This is because the positive charge on the outside surface inhibits further efflux of protons. In fact, to demonstrate the formation of a ΔpH experimentally as a result of proton pumping, a large $\Delta \Psi$ must not be allowed to develop. This is done experimentally by making the membrane permeable either to a cation so that the incoming cations can compensate electrically for the outgoing protons, or to an anion that moves in the same direction as the proton. For example, in many experiments the K^+ ionophore valinomycin is added to make the membrane permeable to K^+ (Section 3.4). When this is done, K^+ exchanges for H^+, and a ΔpH can develop.

Although bacteria cannot make both a $\Delta\Psi$ and a ΔpH with the same proton, they can create a $\Delta\Psi$ during proton translocation, and then convert it to a ΔpH. Suppose a $\Delta\Psi$ is created because a few protons are translocated to the cell membrane outer surface. This cannot proceed for very long because a membrane potential develops quickly which limits further efflux of protons. However, a cation such as K^+ might enter the cell electrogenically. This would result in a lowering of the membrane potential because of the positive charge moving in. Now more protons can be translocated out of the cell. The protons can leave the outer membrane surface and accumulate in the bulk phase because they are paired with the anion that was formerly paired with the K^+. Thus, a membrane potential can form during proton translocation and be converted into a ΔpH by the influx of K^+. This is an important way for bacteria to maintain a ΔpH as described in Section 14.1.3.

3.3 The contributions of the $\Delta\Psi$ and the ΔpH to the overall Δp in neutrophiles, acidophiles, and alkaliphiles

Partly for the reasons stated in Section 3.2.2, the contributions of the $\Delta\Psi$ and the ΔpH to the Δp are never equal. Additionally, the relative contributions of the ΔpH and the $\Delta\Psi$ to the Δp vary, depending upon the pH of the environment where the bacteria naturally grow. Below are summarized the relative contributions of the $\Delta\Psi$ and the ΔpH to the Δp in neutrophiles, acidophiles, and alkaliphiles.[11] Notice that in acidophiles and alkaliphiles, the $\Delta\Psi$ (acidophiles) or the ΔpH (alkaliphiles) has the wrong sign and actually detracts from the Δp.

Neutrophilic bacteria

For neutrophilic bacteria (i.e., those that grow with a pH optimum near neutrality), the $\Delta\Psi$ contributes approximately 70–80% to the Δp, with the ΔpH contributing only 20–30%. This is reasonable when one considers that the intracellular pH is near neutrality and therefore the ΔpH cannot be very large.

Acidophilic bacteria

For acidophilic bacteria (i.e. those that grow between pH 1 and pH 4, and not at neutral pH), the $\Delta\Psi$ is positive rather than negative (below an external pH of 3, which is where they are usually found growing in nature), and thus lowers the Δp. Under these conditions, the force in the Δp is due entirely to the ΔpH. Let us examine an example of this situation. Thiobacillus ferroxidans is an aerobic acidophile that can grow at pH 2.0 (Section 11.4.1). Because it has an intracellular pH of 6.5, the ΔpH is 4.5, which is very large (remember, $\Delta pH = pH_{in} - \Delta H_{out}$). However, the aerobic acidophiles have a positive $\Delta\Psi$ when growing at low pH, and for T. ferroxidans the $\Delta\Psi$ is $+10\,mV$.[12,13] The contribution of the ΔpH to the Δp is $-60\Delta pH$ or $-270\,mV$. Since $\Delta p = \Delta\Psi -60\Delta pH$, the actual Δp would be $-260\,mV$. In this case, the $\Delta\Psi$ lowered the Δp by 10 mV. As discussed in Section 14.1.3, the inverted membrane potential is necessary for maintaining the large ΔpH in the acidophiles.

Alkaliphilic bacteria

An opposite situation holds for the aerobic alkaliphilic bacteria (i.e., those that grow above pH 9, often with optima between pH 10 and pH 12). For these organisms, the ΔpH is one to two units negative (because the cytoplasmic pH is less than 9.6) and consequently lowers the Δp, by 60–120 mV.[14] Therefore, in the alkaliphiles, the potential of the Δp may come entirely from the $\Delta\Psi$. In fact, as explained in Section 3.10, because of the large negative ΔpH, the calculated Δp in these organisms is so low that it raises conceptual problems regarding whether there is sufficient energy to synthesize ATP.

3.4 Ionophores

Before we continue with the discussion of the protonmotive force, we must explain ionophores and their use because reference will be made to these important molecules later. Ionophores are important research tools for investigating membrane bioenergetics and their use has contributed to an understanding of the role of electrochemical ion gradients in membrane energetics. As mentioned earlier,

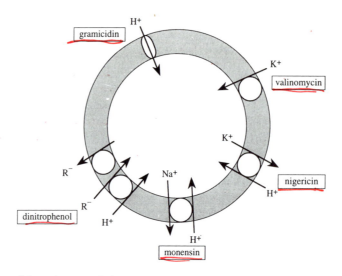

Fig. 3.4 Examples of ionophores and the transport processes that they catalyze. All of the reactions are reversible. *Valinomycin*: Transports K^+ and Rb^+. Valinomycin transports only one cation at a time. Since K^+ is positively charged, valinomycin carries out electrogenic transport (i.e., creates a membrane potential). If valinomycin is added to cells with high intracellular concentrations of K^+, then the K^+ will rush out of the cell ahead of counterions, and create a transient membrane potential, outside positive, predicted by the Nernst equation. In the presence of excess extracellular K^+ and valinomycin, the K^+ will rapidly diffuse into the cell collapsing the existing potential. *Nigericin*: Carries out an electroneutral exchange of K^+ for H^+. When nigericin is added to cells, one can expect a collapse of the pH gradient as the internal K^+ exchanges for the external H^+, but the membrane potential should not decrease. The combination of nigericin and valinomycin will collapse both the ΔpH and the membrane potential. *Monensin*: Carries out an electroneutral exchange of Na^+ or K^+ for H^+. There is a slight preference for Na^+. *Gramicidin*: Carries out electrogenic transport of $H^+ > Rb^+$, K^+, Na^+. Gramicidin differs from the other ionophores in that it forms polypeptide channels in the membrane. That is to say, it is not a diffusible carrier. Since the addition of gramicidin results in the equilibration of protons across the cell membrane, it will collapse the $\Delta\Psi$ and the ΔpH (i.e., the Δp). *Dinitrophenol*: This is an anion (R^-). It carries out electroneutral influx of H^+ and R^- into the cell, and returns to the outside without H^+ (i.e., R^-). It will therefore collapse both the ΔpH and the $\Delta\Psi$. This is the classic uncoupler of oxidative phosphorylation. *Carbonyl cyanide-p-trifluoromethylhydrazone (FCCP)* (not shown): This is a lipophilic weak acid that exists as the nonprotonated anion ($FCCP^-$) and the protonated form (FCCPH), both of which can travel through the membrane. Protons are carried into the cell in the form of FCCPH. Inside the cell the FCCPH ionizes to $FCCP^-$ which exits in response to the membrane potential, outside positive. The result is a collapse of both the ΔpH and the $\Delta\Psi$.

membranes are poorly permeable to ions and this is why the membrane can maintain ion gradients. Ionophores perturb these ion gradients. Most ionophores are organic compounds that form lipid-soluble complexes with cations (e.g., K^+, Na^+, H^+) and rapidly equilibrate these across the cell membrane (Fig. 3.4). The incorporation of an ionophore into the membrane is equivalent to short-circuiting an electrical device with a copper wire. Another way of saying this is that the ionophore causes the electrochemical potential difference of the ion to approach zero. Since some ionophores are specific for certain ions, it is sometimes possible to identify the ion current that is performing the work with the appropriate use of ionophores. For example, if ATP synthesis is prevented by a proton ionophore, then this implies that a current of protons carries the energy for ATP synthesis. One can also preferentially collapse the $\Delta\Psi$ or ΔpH with judicious use of ionophores, and perhaps gain information regarding the driving force for the ion current.

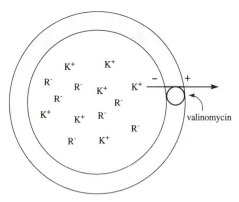

Fig. 3.5 Valinomycin-stimulated K^+ efflux can impose a temporary membrane potential. When valinomycin is added to cells or membrane vesicles loaded with K^+, the valinomycin dissolves in the membrane and carries the K^+ out of the cell. The efflux of K^+ creates a diffusion potential since the negative counterions lag behind the K^+. The membrane potential is predicted by the Nernst equation, $\Delta\Psi = -60 \log(K^+)_{in}/(K^+)_{out}$. The membrane potential is transient because it is neutralized by the movement of counterions.

For example, nigericin, which catalyzes an electroneutral exchange of K^+ for H^+ will dissipate the pH gradient but not the $\Delta\Psi$.[15] Valinomycin plus K^+, on the other hand, will dissipate the $\Delta\Psi$ because it electrogenically carries K^+ into the cell, thus setting up a potential opposite to the membrane potential. It is also possible to *create* a $\Delta\Psi$ using valinomycin. The addition of valinomycin to starved cells or vesicles loaded with K^+ will induce a temporary $\Delta\Psi$ predicted by the Nernst equation, $\Delta\Psi = -60 \log[S_i]/[S]_{out}$ mV. (This is the concentration gradient expressed as mV at 30°C.) What happens is that the K^+ moves from the high internal concentration to the lower external concentration in the presence of valinomycin (Fig. 3.5). Because the K^+ moves faster than its counterion, a temporary diffusion potential, outside positive is created. The diffusion potential is temporary because of the movement of the counterions. The use of ionophores has helped researchers to determine which ions are carrying the primary current that is doing the work, and to investigate the relative importance of the membrane potential and ion concentration gradients in providing the energy for specific membrane functions.

3.4.1 The effect of uncouplers on respiration

Uncouplers are ionophores that have the following effects:

1. They collapse the Δp and thereby inhibit ATP synthesis coupled to electron transport.

2. They stimulate respiration.

Why should uncouplers stimulate respiration? The flow of electrons through electron carriers in the membrane is *obligatorily* coupled to a flow of protons in a closed circuit (Section 4.5). This occurs at the coupling sites discussed in Section 3.7.1. Protons are translocated to the outer surface of the membrane and then re-enter via the ATP synthase. If re-entry is blocked by inhibitors of the ATP synthase [e.g., dicyclohexycarbodiimide (DCCD)] or slowed by depletion of ADP, then respiration is slowed. One possible explanation for the slowing of respiration is that, as the protons are translocated to the outside surface, the Δp rises and approaches the ΔG of the oxidation–reduction reactions. This might slow respiration since the oxidation–reduction reactions are reversible. One can view this as the Δp producing a "back-pressure." The re-entry of the protons via the ATP synthase can be viewed, in this context, as placing a limit on the rise of the Δp and thereby promoting respiration. Uncouplers might stimulate respiration because they collapse the Δp. In the presence of uncouplers such as dinitrophenol, H^+ rapidly enters the cell on the uncoupler rather than through the ATPase.

3.5 Measurement of the Δp

Measurements of the size of the Δp are a necessary part of analyzing the role that the Δp plays in the overall physiology of the cell. The two components of the Δp, the $\Delta\Psi$ and the ΔpH, are measured separately.

3.5.1 Measurement of $\Delta\Psi$

How is the membrane potential measured? Bacteria are too small for the insertion of electrodes, therefore the membrane potential is measured indirectly. Suppose a membrane potential exists and an ion was allowed to come to its electrochemical equilibrium in response to the potential difference. Then, at equilibrium, the electrochemical energy of the ion is zero and, using the equation for electrochemical energy (for a monovalent ion):

eq. 3.14 $\Delta\tilde{\mu}/F = 0$
$$= \Delta\Psi + 60 \log_{10}[S]_{in}/[S]_{out} \text{ mV}$$

Solving for $\Delta\Psi$, *NERNST*

eq. 3.15 $\Delta\Psi = -60 \log_{10}[S]_{in}/[S]_{out}$
mV at 30°C

Equation 3.15 is one form of the Nernst equation. It states that the measurement of the intracellular and extracellular concentrations of a permeant ion at equilibrium allows one to calculate the membrane potential. The bacterial cell membrane is relatively nonpermeable to ions. Therefore, to measure the membrane potential one must use either an ion plus an appropriate ionophore (e.g., K^+ and valinomycin) or a lipophilic ion (i.e., one that can dissolve in the lipid membrane and pass freely into the cell). When the inside is negative with respect to the outside, a cation (R^+) is chosen because it accumulates inside the cells or vesicles (Fig. 3.6). When the inside is positive, an anion (R^-) is used. It is important to use a small concentration of the ion to prevent the collapse of the membrane potential by the influx of the ion.

Cationic or anionic fluorescent dyes have also been used to measure the $\Delta\Psi$. The dyes partition between the cells and the medium in response to the membrane potential, and the fluorescence is quenched. The fluorescent dye will monitor relative changes in membrane potential (Fig. 3.7). When using a fluorescent dye to measure the absolute membrane potential, it is necessary to produce a standard curve. This involves measuring the fluorescence quenching in a sample where the membrane potential is known (i.e., has been measured by independent means such as the distribution of a permeant ion) (Fig. 3.8).

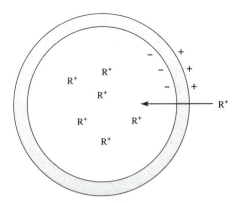

Fig. 3.6 Measurement of membrane potential by the accumulation of a lipophilic cation. The cation accumulates in response to the membrane potential until the internal concentraton reaches a point where efflux equals influx, (i.e., equilibrium has been reached). At this time the electrochemical energy of the cation is zero and $\Delta\Psi = -60 \log[(R^+)_{in}/(R^+)_{out}]$ mV at 30°C. Any permeant ion can be be used as long as it diffuses passively, is used in small amounts, is not metabolized, and accumulates freely in the bulk phase on either side of the membrane. Permeant ions that have been used include lipophilic cations such as tetraphenylphosphonium (Ph_4P^+).

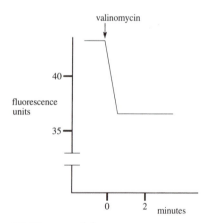

Fig. 3.7 The use of fluorescence to measure the membrane potential. An example of what one might expect if valinomycin were added to bacteria in buffer containing low concentrations of potassium ion in the presence of a fluorescent cationic lipophilic probe. The potassium inside would rush out creating a diffusion potential predicted by the Nernst equation. The cationic probe would enter in response to the membrane potential and the fluorescence would be quenched. It has been suggested that the quenching of fluorescence is due to the formation of dye aggregates with reduced fluorescence inside the cell.

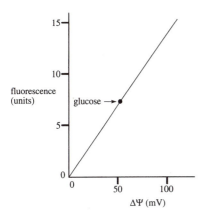

Fig. 3.8 The relationship between the membrane potential and fluorescence change of 1,1'-dihexyl-2,2'-oxacarbocyanine (CC_6). On the ordinate are plotted the changes in fluorescence (continuous line) corresponding to different membrane potentials imposed by the addition of valinomycin to *Streptococcus lactis* cells. The membrane potentials were calculated using the Nernst equation from potassium concentration ratios (IN/-OUT) in parallel experiments where the intracellular K^+ concentrations were about 400 mM and the extracellular concentrations were varied. Also shown is the membrane potential caused by the addition of glucose (rather than valinomycin) to the cells. (Adapted from Maloney, P. C., E. R. Kashket, and T. H. Wilson. 1975. Methods for studying transport in bacteria, pp. 1–49. In: *Methods in Membrane Biology*, Vol. 5, Korn, E. D. (ed.). Plenum Press, New York.)

3.5.2 Measuring the ΔpH

A common way to measure the ΔpH is to measure the distribution of a weak acid or weak base at equilibrium between the inside and outside of the cell.[16,17] The assumption is that the uncharged molecule freely diffuses across the membrane but that the ionized molecule cannot. Inside the cell, the acid becomes deprotonated or the base becomes protonated, the extent of which depends upon the intracellular pH. For example:

$$AH \longrightarrow A^- + H^+$$

or

$$B + H^+ \longrightarrow BH^+$$

On addition of a weak acid (or base) to a cell suspension, the charged molecule accumulates in the cell. One uses a weak acid (e.g., acetic acid) when $pH_{in} > pH_{out}$ because it will ionize more extensively at the higher pH, and a weak base (e.g., methylamine) when $pH_{in} < pH_{out}$ because it will become more protonated at the lower pH. At equilibrium, for a weak acid:

$$K_a = [H^+]_{in}[A^-]_{in}/[AH]_{in}$$
$$= [H^+]_{out}[A^-]_{out}/[AH]_{out}$$

where, K_a is the dissociation constant of the acid, HA, and

$$[AH]_{in} = [AH]_{out}$$

If pH_{in} and pH_{out} are at least 2 units higher than the pK, then most of the acid is ionized on both sides of the membrane and [AH] can be neglected. Therefore:

eq. 3.16 $pH_{in} - pH_{out} = \Delta pH$
$$= \log_{10}[A^-]_{in}/[A^-]_{out}$$

Thus, the $\log_{10}$ of the ratio of concentrations inside to outside of the weak acid is equal to the ΔpH. In practice, one uses radioactive acids or bases as probes and measures the amount of radioactivity taken up by the cells. A more complex equation must be used when the concentration of the unionized acid (AH) cannot be ignored.[17] Cytoplasmic pH is sometimes measured by ^{31}P-nuclear magnetic resonance (^{31}P-NMR) of phosphate whose spectrum is pH dependent.[18]

3.6 Use of the Δp to do work

The Δp provides the energy for several membrane functions, including solute transport and ATP synthesis discussed in this section. (See Chapter 15 for a more complete discussion of solute transport.)

3.6.1 Use of the Δp to drive solute uptake

As an example of how the Δp can be used for doing work, consider symport of an uncharged solute, S, with protons (Fig. 3.2, 4). The total driving force on solute S at 30°C is:

eq. 3.17 $y\Delta p + 60 \log[S]_{in}/[S]_{out}$ mV

where y is the ratio of H^+/S

At equilibrium, the sum of the forces equals 0, and therefore:

eq. 3.18 $y\Delta p = -60 \log[S]_{in}/[S]_{out}$

For $y = 1$ and $\Delta p = -0.180$ v,

$3 = \log[S]_{in}/[S]_{out}$

and,

$[S]_{in}/[S]_{out} = 10^3$

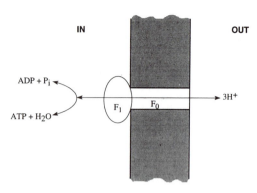

Fig. 3.9 ATP synthase. F_o is a proton channel that spans the membrane. F_1 is the catalytic subunit on the inner membrane surface that catalyzes the reversible hydrolysis of ATP. Under physiological conditions the ATP synthase reaction is poised to proceed in either direction. When Δp levels decrease relative to ATP, they can be restored by ATP hydrolysis. When the ATP levels decrease relative to Δp, then more ATP can be made.

Therefore, a Δp of -180 mV could maintain a 10^3 concentration gradient of S_{in}/S_{out} if the ratio of H^+/S were one. If the ratio of H^+/S were two, then a concentration gradient of 10^6 could be maintained. It can be seen that very large concentration gradients can be maintained using the Δp.

3.6.2 Use of the Δp to drive ATP synthesis

Built into the cell membranes of prokaryotes and mitochondrial and chloroplast membranes is an enzyme complex that couples the translocation of protons down a proton potential gradient to the phosphorylation of ADP to make ATP. It is called the proton-translocating ATP synthase, or simply ATP synthase. As discussed in Section 3.7.2, the ATP synthase is reversible and can pump protons out of the cell, generating a Δp.

Description of the ATP synthase
The ATP synthase consists of two regions (i.e., the F_o and F_1) (Fig. 3.9).[19] The F_o region spans the membrane and serves as a proton channel through which the protons are pumped. The F_1 region is located on the inner surface of the membrane and is the catalytic subunit responsible for the synthesis and

hydrolysis of ATP. The F_1 unit is also called the "coupling factor." The F_1 subunit from *E. coli* is made from five different polypeptides with the following stoichiometry, α_3, β_3, γ, δ, and ε. The β polypeptides each contain a single catalytic site for ATP synthesis and hydrolysis, although the sites are not equivalent at any one time. The F_o portion has three different polypeptides, which in *E. coli* are a_1, b_2, and c_{10}, all of which occur in multiples. All in all, *E.coli* uses eight different types of polypetides to construct a 22-polypeptide machine that acts as a *reversible* pump driven by the proton potential or by ATP hydrolysis.

Evidence that either a ΔΨ or a ΔpH can drive the synthesis of ATP via proton influx through the ATP synthase
It is possible to impose a $\Delta\Psi$ or a ΔpH on cells or membrane vesicles and demonstrate that the influx of protons through the ATP synthase results in the synthesis of ATP. A $\Delta\Psi$ can be imposed by using valinomycin and creating a potassium ion diffusion potential. (The use of valinomycin to impose a membrane potential is described in Section 3.4.) A ΔpH can be created by adding acid to the medium in which cells or membrane vesicles are suspended. These critical experiments

were done with the lactic acid bacterium *Streptococcus lactis*.[20,21] Washed cells of *Streptococcus lactis* containing a high intracellular concentration of K^+ (300–400 mM) were suspended in a medium containing a low concentration of K^+ (0.3–0.4 mM). The initial Δp was close to zero. Valinomycin was then added. The valinomycin caused the rapid efflux of K^+ causing a temporary potassium diffusion potential, outside positive, predicted by the Nernst equation. The cells responded to the valinomycin by making ATP (Fig. 3.10). This indicates that a $\Delta\Psi$ can drive ATP synthesis. A similar experiment was done but this time in the presence of N,N'-dicyclohexylcarbodiimide (DCCD) which is an inhibitor of the

ATP synthase. The DCCD inhibited ATP synthesis *and* proton influx (Fig. 3.11). This indicates that a major route of proton influx is through the ATP synthase and that the ATP synthase is responsible for ATP synthesis. Notice that proton entry continued even in the presence of DCCD. It was assumed that proton entry into DCCD-treated cells resulted from passive inflow or leakage by routes other than the ATP synthase.

To demonstrate that a ΔpH can drive ATP synthesis, washed cells of *S. lactis* were suspended in media to which a small volume of sulfuric acid was added to lower the external pH to about 3.5. The internal pH was initially about 7.6, giving a ΔpH of 4.1 and a driving force ($-60\,pH$) of about

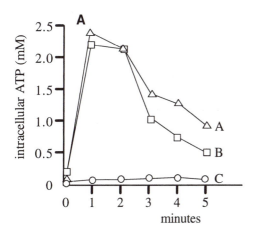
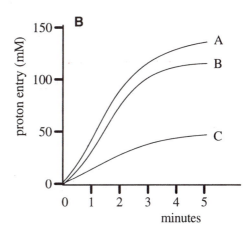

Fig. 3.10 Proton entry and ATP synthesis in response to an imposed membrane potential. A dense suspension of *Streptococcus lactis* cells was incubated in buffer A, B, or C. Average internal potassium was about 360 mM. In buffer B the external K^+ was 0.4 mM. In buffers A and C, KCl added to a final concentration of 3 mM. The external pH was set at pH 5 (sample A) or pH 6 (samples B and C), and proton entry was measured as the amount of acid added to maintain the external pH at the initial value, using a pH stat. Valinomycin was added at 0 time and intracellular ATP levels (A) and proton entry (B) were measured. Upon addition of valinomycin the cells immediately made ATP during proton influx (curves A and B). For sample A, the Δp was 200 mV, of which 125 mV was due to the measured membrane potential and 75 mV was due to the measured ΔpH. For sample B, the Δp was also about 200 mV despite the smaller contribution from the ΔpH (15 mV). This was because the external K^+ concentration in sample B was 0.4 mM instead of 3 mM, thus raising the membrane potential. Notice that the amounts of ATP made and proton influx in samples A and B were approximately the same, despite the differences in the membrane potential and the ΔpH. This suggests that proton influx and ATP synthesis depend upon the Δp rather than on the individual values of the $\Delta\Psi$ or the ΔpH. Sample C had a Δp of only 140 mV (membrane potential, 125 mV; ΔpH, 15 mV) and made no ATP. This suggests a threshold value for Δp for ATP synthesis. Presumably, below the threshold value of Δp, the ATP synthase pumps protons out of the cell at the expense of ATP. (From Maloney, P. C. 1977. Obligatory coupling between proton entry and the synthesis of adenosine 5'-triphosphate in *Streptococcus lactis*. *J. Bacteriol.* **132**:564–575.)

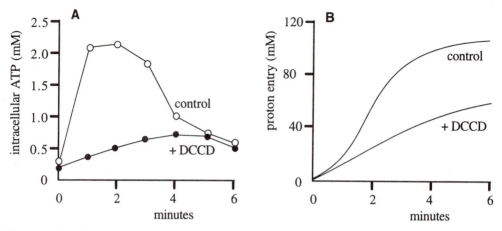

Fig. 3.11 The effect of DCCD on ATP synthesis (a) and proton entry (b). The conditions are similar to those for Fig. 3.10. Cells were suspended in buffer A [(K^+_{in}/K^+_{out}) about 120] and valinomycin was added at 0 time. Some samples were incubated with DCCD, which is an inhibitor of the ATP synthase. DCCD blocked both ATP synthesis and proton entry, suggesting that a major route of influx of protons is through the ATP synthase, and that this flow results in ATP synthesis. (From Maloney, P. C. 1977. Obligatory coupling between proton entry and the synthesis of adenosine 5'-triphosphate in *Streptococcus lactis. J. Bacteriol.* **132**:564–575.)

−246 mV.[22] The cells responded to the imposed ΔpH by making ATP (Fig. 3.12B) during proton influx (Fig. 3.12A). In these experiments, it was necessary to add valinomycin to the samples in order to stimulate proton influx and consequent ATP synthesis. This is because, in the absence of valinomycin, the entering protons established a membrane potential, inside positive, that impeded further net influx of protons. Valinomycin allowed K^+ to exit the cell in exchange for the proton, thus electrically compensating for

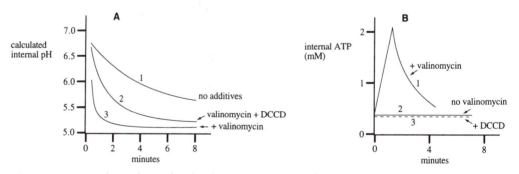

Fig. 3.12 ATP synthesis driven by the ΔpH. *Streptococcus lactis* cells were placed in buffer with various additions. At 0 time the external pH was lowered from pH 8 to pH 3.5 with sulfuric acid. (A) The internal pH dropped very slowly (curve 1) unless valinomycin was added (curve 3). This can be explained by assuming that in the absence of valinomycin the entering H^+ imposed a membrane potential, inside positive, that limited the uptake of protons. Valinomycin allowed the exit of internal K^+ thus diminishing the buildup of an internally positive potential. The addition of DCCD markedly inhibited proton uptake in the first minute, suggesting that most of the protons that entered rapidly were entering via the ATP synthase (curve 2). (B) The addition of the sulfuric acid at 0 time resulted in an immediate synthesis of ATP (curve 1). No such increase in ATP was observed if valinomycin was omitted (curve 2), presumably because net proton influx was slow in the absence of valinomycin. The ATP synthase inhibitor DCCD also prevented ATP synthesis (curve 3). (Adapted from Maloney, P. C. and F. C. Hansen, III. 1982. Stoichiometry of proton movements coupled to ATP synthesis driven by a pH gradient in *Streptococcus lactis. J. Memb. Biol.* **66**:63–75.)

the influx of the protons. In this case the exiting K^+ did not produce a membrane potential because excess K^+ (0.1 M) was added to the medium to lower the K^+_{in}/K^+_{out}. Both ATP synthesis and the influx of protons were inhibited by the ATP synthase inhibitor DCCD, indicating that proton entry was primarily through the ATPase.

Model of the mechanism of ATP synthase

The mechanism of ATP synthesis/hydrolysis by the ATP synthase is complex. A model that is favored by most investigators is called the "binding change mechanism." It is based upon the finding that purified F_1 will synthesize tightly bound ATP in the absence of a Δp.[23] (The K_d for the high-affinity site is 10^{-12} M.) The equilibrium constant between enzyme-bound ATP and bound hydro-lysis products (ADP and P_i) is approximately one, using soluble F_1.[24] The tight binding of ATP and the equilibrium constant of approximately one suggests that the energy requirement for net ATP synthesis is for the release of ATP from the enzyme rather than for its synthesis. The model proposes that, when protons move down the electrochemical gradient through the F_oF_1 complex, a conformational change in F_1 occurs that results in the release of newly synthesized ATP from the high-affinity site. This is illustrated in Fig. 3.13.

3.7 Exergonic reactions that generate a Δp

Section 3.6 describes the influx of protons down a proton potential gradient coupled to

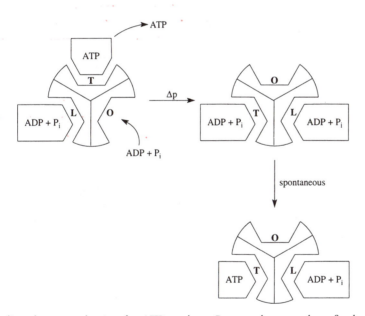

Fig. 3.13 Binding-change mechanism for ATP synthase. Because there are three β subunits encoded by a single gene and, because there is only one catalytic site per β subunit, a maximum of three catalytic sites are possible. However, the three catalytic sites are not functionally equivalent and are thought to cycle through conformational changes driven by the electrochemical proton gradient. As a result of the conformational changes, newly synthesized ATP is released from the catalytic site. The three conformational states are O (open, very low affinity for substrates), L (loose binding), and T (tight binding, the active catalytic site). ATP is made spontaneously at site T which converts to site O in the presence of a Δp and releases the ATP. The Δp also drives the conversion of the pre-existing site O to site L which binds ADP and P_i, as well as the conversion of the pre-existing site L to site T. In this model, the only two catalytic sites that bind substrates are L and T. (Adapted from Cross, L. R., Cunningham, D., and J. K. Tamura. 1984. Binding change mechanism for ATP synthesis by oxidative phosphorylation and photophosphorylation. *Curr. Top. Cell. Regul.* **24**:335–344.)

the performance of work (e.g., solute transport or ATP synthesis). These activities dissipate the Δp. Let us now consider driving forces that generate a Δp (i.e., move protons out of the cell towards a higher potential). The major driving reactions in most prokaryotes are the redox reactions in the cell membranes of respiring organisms (electron transport), ATP hydrolysis in fermenting organisms, and several other less frequently used driving reactions considered in Section 3.8. At this time, let us consider some general thermodyamic features common to all of the driving reactions that generate the proton gradient. The translocation of protons out of cells is an energy requiring process that can be written as:

$$yH^+_{\ in} \longrightarrow yH^+_{\ out}$$

$$\Delta G = yF\Delta p \quad J \quad or \quad y\Delta p \quad eV$$

That is to say, the amount of energy required to translocate y moles of protons out of the cell is $yF\Delta p$ J, or for y protons, $y\Delta p$ eV. (This is the same energy, but of opposite sign, that is released when the protons enter the cell.) Therefore, the ΔG of the driving reaction must be equal to or greater than $yF\Delta p$ J or $y\Delta p$ eV. If the reaction is near equilibrium, which is the case for the major driving reactions, then the energy available from the driving reaction is approximately equal to the energy available from the proton gradient and $\Delta G_{drivingreaction} = yF\Delta p$. This relationship allows one to calculate the Δp if y and $\Delta G_{drivingreaction}$ are known. The most common classes of driving reactions will be discussed first. They are oxidation–reduction reactions that occur during respiration and photosynthetic electron transport, and ATP hydrolysis. In all cases, we will write the ΔG for the driving reaction. Then we will equate the $\Delta G_{drivingreaction}$ with $yF\Delta p$. Finally, we will solve for Δp.

3.7.1 Oxidation–reduction reactions as driving reactions

Electrode potentials and energy changes during oxidation–reduction reactions
The tendency of a molecule (A) to accept an electron from another molecule is given by its electrode potential, E, also called the reduction potential. Under standard conditions (1 M for all solutes and gas at 1 atmosphere), the electrode potential has the symbol, E_o, and is related to the actual electrode potential, E_h, as follows: $E_h = E_o + RT \ln[ox]/[red]$. Usually standard potentials at pH 7 are quoted and the symbol is E'_o. You can see that it is important to know the concentrations of the oxidized and the reduced forms in order to know the actual electrode potential. Molecules with more negative E_h values are better donors (i.e., electrons spontaneously flow to molecules with more positive E_h values). When n electrons travel over a potential difference of E_h volts, work is done which is equal to $n\Delta E_h$ eV, where $\Delta E_h = E_{h,acceptor} - E_{h,donor}$. The reaction is endergonic (energy input) when the electron moves to the lower potential and exergonic (energy output) when it moves to the higher potential. Since the total charge carried by one mole of electrons is F coulombs, the work done per n mol of electrons is $\Delta G = -nF\Delta E_h$ J, where ΔG is the Gibbs-free energy. The negative sign is placed in the equation to show that energy is released when ΔE_h is positive.

Oxidation–reduction reactions can generate a Δp
An important method of generating a Δp is to couple proton translocation to oxidation–reduction reactions that occur during electron transport. The details of electron transport are discussed in Chapter 4 but here we will simply compute the Δp that can be created. Energy released from oxidation–reductions can generate a Δp because the oxidation–reduction reactions are coupled to proton translocation. The mechanism of coupling is described in Section 4.6 but we describe here only the energetic relationship between the ΔE_h and the Δp. Let us do a simple calculation to illustrate how to estimate the size of the Δp that can be generated by an oxidation–reduction reaction that is coupled to proton translocation. Consider what happens in mitochondria and in many bacteria.

A common oxidation–reduction reaction in the respiratory chain is the oxidation of reduced ubiquinone, UQH_2, by cytochrome c_1 (cyt.c_1) in an enzyme complex called the *bc_1 complex*. The oxidation–reduction is coupled to the translocation of protons across the membrane, hence the creation of a Δp. As stated earlier, the total energy change during an oxidation–reduction reaction is $\Delta G = -nF\Delta E_h$ joules, where n equals the number of moles of electrons. The total energy change during proton translocation is $yF\Delta p$ J, where y equals the number of moles of translocated protons. [One can convert to electrical potential (volts) by dividing by the faraday.] This is summarized below:

$$\Delta G \text{ (J)}$$

(1) $UQH_2 + 2\,\text{cyt.c}_1(\text{ox}) \longrightarrow UQ$
    $+ 2\,\text{cyt.c}_1(\text{red}) + 2H^+ \qquad -nF\Delta E_h$

(2) $yH^+_{\text{in}} \longrightarrow yH^+_{\text{out}} \qquad yF\Delta p$

Reactions (1) and (2) are coupled (i.e., one cannot proceed without the other). What is the size of the Δp that can be generated? These reactions are close to equilibrium and can proceed in either direction. Therefore, one can write that the total force available from the redox reaction is equal to the total force available from the proton potential:

eq. 3.19 $\Delta G/F = -n\Delta E_h = y\Delta p$.

Now we solve for Δp. Four protons ($y=4$) are translocated for every two electrons ($n=2$) that travel from reduced quinone to cytochrome c_1. The ΔE_h between quinone and cytochrome c_1 is approximately $+0.2$ V (200 mV). Substituting this value for ΔE_h, and using $n=2$ and $y=4$, we obtain a Δp of -0.1 V (-100 mV). That is to say, when two electrons travel down a ΔE_h of 200 mV and four protons are translocated, -100 mV are stored in the Δp. Another way of looking at this is that, when two electrons travel down

a redox gradient of 0.2 V, 0.4 eV are available for doing work (2×0.2 V). The 0.4 eV are used to raise each of four protons to a Δp of 0.1 V. Oxidation-reduction reactions in the respiratory chain that are coupled to proton translocation are called coupling sites, of which the bc_1 complex is an example. Coupling sites are discussed in more detail in Section 4.5.

Coupling of redox reactions during electron transport to proton translocation is the main process by which a Δp is created in respiring bacteria, in phototrophic prokaryotes, in chloroplasts, and in mitochondria.

Reversed electron transport

The fact that some of the redox reactions in the respiratory pathway are in equilibrium with the Δp has important physiological consequences. One of these is that it can be expected that the Δp can drive electron transport in reverse. That is to say, protons coming into the cell towards the lower proton potential can drive electrons to the more negative electrode potential. One test for the functioning of reversed electron transport is the inhibition of NAD^+ reduction by ionophores that collapse the Δp (Section 3.4). Reversed electron transport commonly occurs in bacteria that use inorganic compounds such as ammonia, nitrite, sulfur, and so on as a source of electrons to reduce NAD^+ (chemolithotrophs) because these electron donors are at a potential higher than NAD^+. The chemolithotrophs are discussed in Chapter 11.

Respiration coupled to sodium ion efflux

Although respiratory chains coupled to proton translocation appears to be the rule in most bacteria, a respiration linked Na^+ pump (a Na^+-dependent NADH-quinone reductase) has been reported in several halophilic marine bacteria that require high concentrations of Na^+ (0.5 M) for optimal growth.[25,26] The situation has been well studied with *Vibrio alginolyticus*, an alkalotolerant marine bacterium that uses a $\Delta\tilde{\mu}_{Na^+}$ for solute transport, flagella rotation, and ATP synthesis (at alkaline pH). *Vibrio alignolyticus* creates the $\Delta\tilde{\mu}_{Na^+}$ in two ways. At pH 6.5, a respiration-driven H^+ pump generates a $\Delta\tilde{\mu}_{H^+}$ which drives a Na^+/H^+ antiporter that creates the $\Delta\tilde{\mu}_{Na^+}$. The antiporter creates the sodium ion gradient by coupling the influx of protons (down the proton electrochemical gradient) with the efflux of sodium ions. However, at pH 8.5, the $\Delta\tilde{\mu}_{Na^+}$ is created directly by a respiration-driven Na^+ pump. In agreement with this conclusion, the generation of the membrane potential at alkaline pH (pH 8.5) is resistant to the proton ionophore, *m*-carbonylcyanide phenylhydrazone (CCCP), which short-circuits the proton current (Section 3.4), but is sensitive to CCCP at pH 6.5.[27] It has been suggested that switching to a Na^+-dependent respiration at alkaline pH may be energetically economical. The reasoning is that when the external pH is more alkaline than the cytoplasmic pH, the only part of the Δp that contributes energy to the antiporter is the $\Delta\Psi$, and therefore the antiporter must be electrogenic. That is to say, the H^+/Na^+ must be greater than one. If one assumes that the antiporter is electroneutral when the external pH is acidic, then the continued use of the antiporter at alkaline pH would necessitate increased pumping of protons out of the cell by the primary proton-linked respiration pumps. Rather than do this, the cells simply switch to a Na^+-dependent respiration pump to generate the $\Delta\tilde{\mu}_{Na^+}$. This argument assumes that the ratio Na^+/e^- and H^+/e^- are identical so that the energy economies of the respiration-linked cation pumps are the same. Several other bacteria have recently been found to generate sodium ion potentials by a primary process. For example, primary Na^+ pumping is also catalyzed by sodium-ion translocating decarboxylases in certain anaerobic bacteria described in Section 3.8.1. It must be pointed out, however, that the $\Delta\tilde{\mu}_{Na^+}$, although relied upon for solute transport and motility in many other Na^+-dependent bacteria, is usually created by Na^+/H^+ antiport rather than by a primary Na^+ pump. For example, the marine sulfate reducer *Desulfovibrio salexigens*, which uses a $\Delta\tilde{\mu}_{Na^+}$ for sulfate accumulation, generates

the $\Delta\tilde{\mu}_{Na^+}$ by electrogenic Na^+/H^+ antiport driven by the $\Delta\tilde{\mu}_{H^+}$, which is created by a respiration-linked proton pump.[28] Similarly, nonmarine aerobic alkaliphiles belonging to the genus *Bacillus*, which rely on the $\Delta\tilde{\mu}_{Na^+}$ for most solute transport and for flagella rotation, create the $\Delta\tilde{\mu}_{Na^+}$ using a Na^+/H^+ antiporter driven by a $\Delta\tilde{\mu}_{H^+}$ that is created by a respiration-linked proton pump (Section 3.10). Furthermore, in *D. Salexigens*, as well as in the alkaliphilic *Bacillus* species, the membrane-bound ATP synthase is H^+-linked rather than Na^+-linked.

3.7.2 ATP hydrolysis as a driving reaction for creating a Δp

Electron transport reactions are the major energy source for creating a Δp in respiring organisms, but not in fermenting bacteria.[29] A major energy source for the creation of the Δp in fermenting bacteria is ATP hydrolysis catalyzed by the membrane-bound proton-translocating ATP synthase, which yields considerable energy (Fig. 3.2, 7). Consider the coupled reactions summarized below:

(1) $ATP + H_2O \longrightarrow ADP + P_i \quad \Delta G_p$

(2) $yH^+_{in} \longrightarrow yH^+_{out} \qquad yF\Delta p$

When referring to the energy of ATP hydrolysis or synthesis using physiological concentrations of ADP, P_i, and ATP, the term ΔG_p (phosphorylation potential) is used instead of ΔG:

eq. 3.20 $\Delta G_p = \Delta G^{o'} + 2.303RT \times \log[ATP]/[ADP][P_i]$

The ATP synthase reaction is close to equilibrium and can operate in either direction. Therefore, the total force available from the proton potential equals the force available from ATP hydrolysis:

eq. 3.21 $\Delta G_p/F = y\Delta p \quad V$

The value of ΔG_p is about $-50,000$ J and the consensus value for y is 3. Therefore the hydrolysis of one ATP would generate a maximum Δp of -173 mV.

3.8 Other mechanisms for creating a $\Delta\Psi$ or a Δp

Redox reactions and ATP hydrolysis are the most common driving reactions for creating a proton potential. However, other mechanisms exist for generating proton potentials and even sodium ion potentials. These driving reactions are not as widespread among the prokaryotes as the others. Nevertheless, they are very important for certain groups of prokaryotes, especially anaerobic bacteria and halophilic archaea. In some instances, they may be the only source of ATP. Some of these driving reactions are considered next.

3.8.1 Sodium transport decarboxylases can create a sodium potential

Although chemical reactions directly linked to Na^+ translocation (primary transport of Na^+) are not widespread among the bacteria, as most primary transport is of the proton, they can be very important for certain bacteria[30-34]. An example of primary Na^+ transport is the Na^+ pump coupled to respiration in Vibrio alginolyticus that was discussed in Section 3.7.1. There is also the decarboxylation of organic acids coupled to sodium ion efflux. This occurs in anaerobic bacteria that generate a sodium gradient by coupling the decarboxylation of a carboxylic acid to the electrogenic efflux of sodium ions. The decarboxylases include methylmalonyl–CoA decarboxylase from Veillonella alcalescens and Propionigenium modestum, glutaconyl–CoA decarboxylase from Acidaminococcus fermentans, and oxaloacetate decarboxylase from Klebsiella pneumonia and Salmonella typhimurium. A description of these bacteria and their metabolism will explain the importance of the sodium translocating decarboxylases as a source of energy.

Propionigenium modestum is an anaerobe isolated from marine and freshwater mud, and from human saliva.[35] It grows only on carboxylic acids (i.e., succinate, fumarate, L-aspartate, L-malate, oxaloacetate, and pyruvate). The carboxylic acids are fermented to propionate and acetate, forming methylmalo-nyl–CoA as an intermediate. (See the description of the propionic acid fermentation in Section 13.7 for an explanation of these reactions.) The methylmalonyl–CoA is decarboxylated to propionyl–CoA and CO_2 coupled to the electrogenic translocation of 2 moles of sodium ions per mole of methylmalonyl–CoA decarboxylated.[36]

$$\text{Methylmalonyl–CoA} + 2Na^+_{in}$$
$$\longrightarrow \text{Propionyl–CoA} + CO_2 + 2Na^+_{out}$$

P. modestum differs from most known bacteria in that it has a Na^+-dependent ATP synthase rather than a H^+-dependent ATP synthase and thus relies on a Na^+ current to make ATP. The consequence of this is that, when it grows on succinate, the decarboxylation of methylmalonyl–CoA is its only source of ATP.

Another example is V. alcalescens . This is an anaerobic gram-negative bacterium that grows in the alimentary canal and mouth of humans and other animals. Like P. modestum, it is unable to ferment carbohydrates, but does ferment the carboxylic acids lactate, malate, and fumarate to propionate, acetate, H_2, and CO_2. During the fermentation, pyruvate and methylmalonyl–CoA are formed as intermediates, as in P. modestum.[37] Some of the pyruvate is oxidized to acetate, CO_2, and H_2 with the formation of one ATP via substrate level phosphorylation (Sections 7.3 and 13.6.1). Pyruvate can also be converted to methylmalonyl–CoA that is decarboxylated to propionyl–CoA coupled to the extrusion of sodium ions. The sodium-ion gradient that is created might be used as a source of energy for solute uptake via a process called Na^+-coupled symport, described in Chapter 15.

Acidaminococcus fermentans is an anaerobe that ferments the amino acid glutamate to acetate and butyrate. During the fermentation, glutaconyl–CoA is decarboxylated to crotonyl–CoA by a decarboxylase that is coupled to the translocation of sodium ions (eq. 3.21).[38]

$$\text{Glutaconyl–CoA} + yNa^+_{in}$$
$$\longrightarrow \text{Crotonyl–CoA} + CO_2 + yNa^+_{out}$$

The oxaloacetate decarboxylase in Klebsiella pneumoniae

We will examine the Na^+-dependent decarboxylation of oxaloacetate by oxaloacetate decarboxylase from the facultative anaerobe *K. pneumoniae* because this has been well studied. A substantial sodium potential develops because the decarboxylation of oxaloacetate is coupled to the electrogenic efflux of sodium ion. (The standard free energy for the decarboxylation reaction is approximately $-29,000$ J.) The enzyme from *K. pneumoniae* translocates two Na^+ out of the cell per oxaloacetate decarboxylated according to the following reaction:

$$2\ Na^+_{in} + oxaloacetate$$
$$\longrightarrow 2\ Na^+_{out} + pyruvate + CO_2$$

1. Evidence that oxaloacetate decarboxylase is a sodium pump

Inverted membrane vesicles (inside-out)[39] were prepared from *Klebsiella* and incubated with $^{22}Na^+$ and oxaloacetate. As seen in Fig. 3.14, oxaloacetate-dependent sodium ion influx took place. The uptake of sodium ion was prevented by avidin, an inhibitor of the oxaloacetate decarboxylase.

2. The structure of the oxaloacetate decarboxylase

Oxaloacetate decarboxylase consists of two parts, a peripheral catalytic portion attached to the inner surface of the membrane and an integral membrane portion that serves as a Na^+ channel.[40] The integral membrane protein also takes part in the decarboxylation step along with the peripheral membrane protein. How the decarboxylase translocates Na^+ to the cell surface is not understood.

3. Use of the sodium current

What does *Klebsiella* do with the energy conserved in the sodium potential? It uses the energy to transport the growth substrate, citrate, actively into the cell (Fig. 3.15).

3.8.2 Oxalate:formate exchange can create a Δp

Oxalobacter formigenes, an anaerobic bacterium that is part of the normal flora in

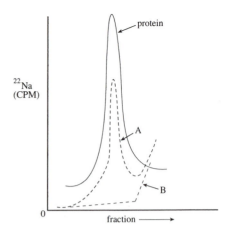

Fig. 3.14 Sodium ion influx into vesicles driven by oxaloacetate decarboxylation. Inside-out vesicles were prepared from *Klebsiella pneumoniae* and incubated with $^{22}Na^+$ in the presence (A) and absence (B) of oxaloacetate. After 1 min of incubation the vesicles were isolated by Sephadex chromatography which separates the vesicles from the $^{22}Na^+$ in the medium. The fractions from the Sephadex columns that contained vesicles were detected by absorbance at 280 nm (protein). Only when oxaloacetate was present did the vesicle fraction contain $^{22}Na^+$ (A). This demonstrates the dependency of sodium uptake upon oxaloacetate. In separate experiments it was demonstrated: (1) that oxaloacetate decarboxylase was present in the vesicles; (2) that the oxaloacetate was decarboxylated; and (3) that inhibition of the oxaloacetate decarboxylase by avidin prevented uptake of $^{22}Na^+$. (Adapted from Dimroth, P. 1980. A new sodium-transport system energized by the decarboxylation of oxaloacetate. *FEBS Lett.* **122**:234–236.)

mammalian intestines, uses dietary oxalic acid as its sole source of energy for growth. The organism has evolved a method for creating a proton potential at the expense of the free energy released from the decarboxylation of oxalic acid to formic acid and carbon dioxide.[41–43] What is especially interesting is that the enzyme is not in the membrane and therefore cannot act as an ion pump, yet a $\Delta\Psi$ is created. The reaction catalyzed by oxalate decarboxylase is:

$$-OOC\text{–}COO + H^+ \longrightarrow CO_2 + HCOO-$$
$$\text{oxalate} \qquad\qquad\qquad \text{formate}$$

For every mole of oxalate that enters the cell, one mole of formate leaves (Fig. 3.16). Since a dicarboxylic acid enters the cell in exchange

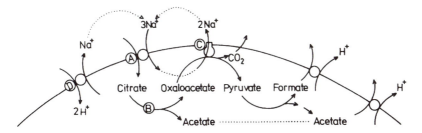

Fig. 3.15 Proposed ion currents in *Klebsiella pneumoniae*. (A) Citrate (three carboxyl groups) enters the cell electroneutrally via symport with three Na$^+$. (B) The citrate is cleaved to oxaloacetate and acetate by citrate lyase. (C) The oxaloacetate is decarboxylated to pyruvate and carbon dioxide with the extrusion of two sodium ions. (D) The third sodium ion (required for citrate uptake) is produced by a H$^+$/Na$^+$ antiporter. The pyruvate is split to acetate and formate with the production of ATP. It is suggested that the formate and acetate produce a Δp via coupled efflux with protons equivalent to that of lactate efflux in *Lactobacillus*. (From Dimroth, P. 1990. Energy transductions by an electrochemical gradient of sodium ions, pp. 114–127. In: *The Molecular Basis of Bacterial Metabolism*. Hauska and R. Thauer (eds.). Springer-Verlag, Berlin.)

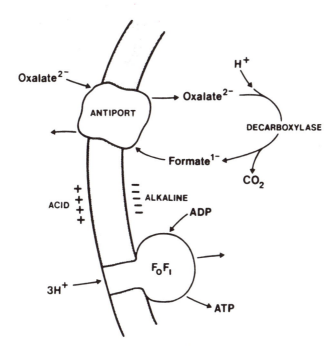

Fig. 3.16 The electrogenic oxalate:formate exchange and the synthesis of ATP in *Oxalobacter formigenes* using a proton current. Oxalate^{2-} enters via an oxalate^{2-}/formate^{1-} antiporter. The oxalate^{2-} is decarboxylated to formate^{1-} and CO_2 while a cytoplasmic proton is consumed. (The oxalate is first converted to oxalyl–CoA which is decarboxylated to formyl–CoA. The formyl–CoA transfers the CoA to incoming oxalate thus forming formate and oxalyl–CoA.) An electrogenic exchange of oxalate^{2-} for formate^{1-} creates a membrane potential, outside positive. It is suggested that the stoichiometry of the ATP synthase is 3H$^+$/ATP. Since the decarboxylation of one mole of oxalate results in the consumption of one mole of protons, the incoming current of protons during the synthesis of one mole of ATP is balanced by the decarboxylation of three moles of oxalate. In other words, a steady-state current of protons requires that 1/3 of a mole of ATP is made per mole of oxalate decarboxylated. (From Anantharam, V. M. J. Allison, and P. C. Maloney. 1989. Oxalate:formate exchange. *J. Biol. Chem.* **264:**7244–725.)

for a monocarboxylic acid, there is a net negative charge moving into the cell, thus creating a $\Delta\Psi$, inside negative. Also, a proton is consumed in the cytoplasm during the decarboxylation, which can contribute to a pH gradient. Recently the antiporter has been purified and shown to be a 38 kD hydrophobic polypeptide which catalyzes the exchange of oxalate and formate in reconstituted proteoliposomes.[44,45] (For a discussion of proteoliposomes, see Section 15.1.)

ATP synthesis

Assuming that the stoichiometry of the ATP synthase reaction is $3H^+/ATP$ and for oxalate decarboxylation it is $1H^+/oxalate$, then a steady proton current requires the decarboxylation of 3 mole of oxalate per mole of ATP synthesized. In other words, 1/3 mole of ATP can be synthesized per mole of oxalate decarboxylated. This is apparently the only means of ATP synthesis in *Oxalobacter formigenes*.

The decarboxylation of other acids may also create a Δp

In principle, the influx and decarboxylation of any dicarboxylic acid coupled to the efflux of the monocarboxylic acid can create a proton potential (Fig. 3.17). For example, a decarboxylation and electrogenic antiport was reported for the decarboxylation of malic acid to lactic acid during the malolactic fermentation in *Lactobacillus lactis*.[46]

3.8.3 End-product efflux as the driving reaction

Theoretically, it is possible to couple the excretion of fermentation end products down a concentration gradient to the translocation of protons out of the cell, thereby creating a Δp. This is the reverse of solute uptake by proton-coupled transport systems and is called the "energy recycling model."[47] In other words, the direction of solute transport

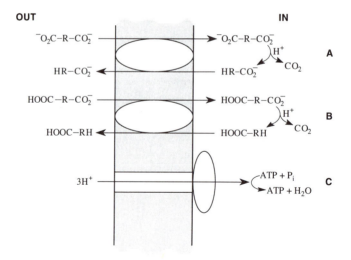

Fig. 3.17 How the decarboxylation of a dicarboxylic acid can be coupled to the development of a proton potential. (A and B) A dicarboxylic acid enters as a divalent anion (A) or a monovalent anion (B). In the cytoplasm a decarboxylase cleaves off CO_2 consuming a proton in the process and producing a monocarboxylic acid with one less negative charge than the original dicarboxylic acid. Exchange of the dicarboxylic acid and the monocarboxylic acid via the antiporter is electrogenic and produces a membrane potential, inside negative. The consumption of the proton during the decarboxylation creates a ΔpH. (C) Influx of protons via the ATP synthase completes the proton circuit and results in ATP synthesis. The energetics can be understood in terms of the decarboxylase removing the dicarboxylic acid from the inside, thus maintaining a concentration gradient that stimulates influx.

depends upon which is greater, the energy from the proton potential, Δp, which drives solute uptake creating an electrochemical solute gradient, $\Delta\tilde{\mu}/F$, or the electrochemical solute gradient which drives solute efflux, creating a Δp. Solute efflux coupled to proton translocation can spare ATP because, in fermenting bacteria, the hydrolysis of ATP (catalyzed by the ATP synthase) is used to pump protons to the outside to create the Δp. As the Δp rises, ATP hydrolysis is diminished.

Energetics

Consider solute/proton symport as reversible. Under these circumstances, the Δp can drive the uptake of a solute against a concentration gradient, or the efflux of a solute down a concentration gradient can create a Δp. The driving force for solute transport in symport with protons is the sum of the proton potential and the electrochemical potential of the solute:

eq. 3.22 Driving force (mV) $= y\Delta p + \Delta\tilde{\mu}/F$

where y is the number of protons in symport with the solute, S. The term $\Delta\tilde{\mu}_s/F$ represents the electrochemical potential of S, i.e.,

eq. 3.23 $\Delta\tilde{\mu}_s/F = m\Delta\Psi + 60\log[S]_{in}/[S]_{out}$ mV

where m is the charge of the solute. Therefore, the overall driving force (substituting for $\Delta\mu_s/F$ and Δp) is:

eq. 3.24 Driving force(mV)
$$= 60\log[S]_{in}/[S]_{out}$$
$$+ (y+m)\Delta\Psi - y60\,\Delta pH$$

Substituting $m = -1$ for lactate, eq. 3.24 becomes:

eq. 3.25 Driving force
$$= (y-1)/\Delta\Psi - y60\Delta pH$$
$$+ 60\log[S]_{in}/[S]_{out}$$

During growth, lactate transport is near equilbrium and the net driving force is close to zero. Thus by setting eq. 3.25 equal to zero, one can solve for y:

eq. 3.26 $y = (\Delta\Psi - 60\log[S]_{in}/[S]_{out})/\Delta p$

Note that eq. 3.25 states that when y, which is the number of moles of H^+ translocated per mole of lactate, is one, then the translocation is electroneutral and a $\Delta\Psi$ does not develop; only a small ΔpH develops due to the acidification of the external medium by the lactic acid. We will return to these points later when we discuss the physiological significance of energy conservation via end-product efflux. Let us now consider some data that support the hypothesis that lactate/H^+ symport and succinate/Na^+ symport can create a membrane potential in membrane vesicles derived from cells.

Symport of protons and sodium ions with fermentation end products

Coupled translocation of protons and lactate can be demonstrated in membrane vesicles prepared from lactic acid bacteria belonging to the genus *Streptococcus*.[48–51] Similarly, coupled translocation of sodium ions and succinate is catalyzed by vesicles prepared from a rumen bacterium belonging to the genus *Selenemonas*.[52] In both cases, a transporter exists in the membrane that simultaneously translocates the organic acid (R^-) with either protons or sodium ions out of the cell (symport) (Fig. 3.18).[53]

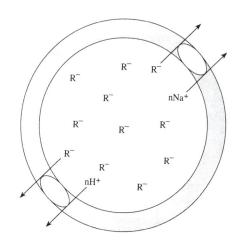

Fig. 3.18 End-product efflux in symport with protons or sodium ions. The high intracellular concentrations of R^- may drive the efflux of Na^+ or H^+ via symporters. If the ratio of protons or sodium ions to carboxyl (i.e., n/carboxyl) is >1 then the symport is electrogenic and a membrane potential develops.

Lactate efflux

L-Lactate-loaded membrane vesicles were prepared from *Streptococcus cremoris* in the following way: A concentrated suspension of cells was treated with lysozyme in buffer to degrade the cell walls. The resulting cell suspension was gently lysed (broken) by adding potassium sulfate. The cell membranes spontaneously resealed into empty vesicles. The vesicles were purified and then incubated for 1 h with 50 mM L-lactate which equilibrated across the membrane, thus loading the vesicles with L-lactate. Then the membrane vesicles loaded with L-lactate were incubated with the lipophilic cation, tetra-phenylphosphonium (Ph_4P^+), in solutions containing: buffer (Fig. 3.19), curve 1; buffer + 50 mM L-lactate, curve 2. Samples were filtered to separate the vesicles from the external medium, and the amount of Ph_4P^+ that accumulated was measured. The Ph_4P^+ accumulated inside the cells when the $(lactate)_{in}/(lactate)_{out}$ was high.

The accumulation of Ph_4P^+ implies that the efflux of lactate along its concentration gradient imposed a membrane potential on the vesicles.

The membrane potential was calculated using the Nernst equation and the Ph_4P^+ accumulation ratio, as explained in Section 3.5.1.

Succinate efflux

A similar experiment was done with vesicles from the rumen bacterium *Selenomonas ruminantium* loaded with succinate (Fig. 3.20).

The Ph_4P^+ was taken up by the vesicles when there was a concentration gradient of succinate, indicating that a membrane potential, inside negative, developed during succinate efflux (Fig. 3.20, curve 1).

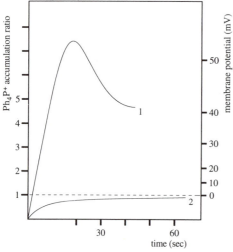

Fig. 3.19 Lactate efflux can produce a membrane potential. Lactate-loaded membrane vesicles from *Streptococcus cremoris* were incubated with the lipophilic cation Ph_4P^+ without lactate (curve 1) and with 50 mM lactate in the external medium (curve 2). In the absence of external lactate the lipophilic probe accumulated inside the vesicles, suggesting that a membrane potential developed as lactate left the cells. The addition of lactate to the external medium prevented the formation of the membrane potential, because the lactate concentration (IN/OUT) was lowered. Additional experiments showed that the electrogenic ion was the proton. (Adapted from Otto, R., R. G. Lageveen, H. Veldkamp, and W. N. Konings. 1982. Lactate efflux-induced electrical potential in membrane vesicles of *Streptococcus cremoris*. *J. Bacteriol.* **149**:733–738.)

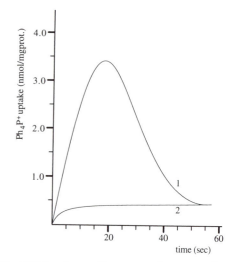

Fig. 3.20 Succinate efflux can produce a membrane potential. Succinate-loaded membrane vesicles from *Selenomonas ruminantium* were incubated with the lipophilic cation Ph_4P^+ without succinate (curve 1) or with succinate (curve 2) in the external medium. The uptake of Ph_4P^+ indicates that a membrane potential developed when the $(succinate)_{in}/(succinate)_{out}$ was high. (Adapted from Michel, T. A. and J. M. Macy. 1990. Generation of a membrane potential by sodium-dependent succinate efflux in *Selenomonas ruminantium*. *J. Bacteriol.* **172**:1430–1435.)

When the external buffer contained a high concentration of succinate, the membrane potential did not develop, indicating that the energy to establish the potential was derived from the efflux of succinate along its concentration gradient, $[succ]_{in}/[succ]_{out}$, (Fig. 3.20, curve 2). The molar growth yields of *S. ruminantium* are higher when succinate production is at a maximum, implying that there is more ATP available for biosynthesis as a result of the succinate efflux.

Physiological significace of end-product efflux as a source of cellular energy

In order to demonstrate the generation of a membrane potential due to end-product efflux, cells or membrane vesicles must be "loaded" with high concentrations of the end product by incubating de-energized cells or membrane vesicles for several hours before dilution into end product-free media. These are not physiological conditions, and a question has been raised as to the conditions under which the electrogenic extrusion of lactate coupled to protons is physiologically relevant. Assuming a given value of y, one can use eq. 3.26 to calculate the necessary lactate concentration gradient given experimentally determined values of $\Delta\Psi$ and ΔpH. When one substitutes $y = 2$ (for electrogenic lactate extrusion), a major problem appears. Substituting a measured external lactate concentration of 30 mM, a $\Delta\Psi$ of -100 mV, a $-60\Delta pH$ of -30 mV (at pH 7.0), and $y = 2$, the caculated internal lactate concentration would have to be 14 M in order for lactate secretion to occur.[54] This is a nonphysio-logical concentration. (Internal lactate concentrations reach concentrations of about 0.2 M.) In fact, the values of y were calculated by Brink *et al.*, using experimentally measured values of lactate concentrations, membrane potential, and ΔpH. Only when the external lactate concentrations were very low was lactate/proton efflux electrogenic (i.e., $y > 1$). For example, when the cells were grown at pH 6.34 and the lactate accumulated in the medium over time, the value of y decreased from about 1.44 to 0.9 when the external lactate concentrations increased from 8 mM to 38 mM. This means that only

growth conditions (low external lactate and high external pH so that the ΔpH is 0 or inverted), would one expect lactate efflux effectively to generate a Δp. This may occur in the natural habitat where growth of the producer may be stimulated by a population of bacteria that utilize lactate, thus keeping the external concentrations low.

3.8.4 Light absorbed by bacteriorhodopsin can drive the creation of a Δp

Certain archaea (i.e., the extremely halophilic archaea), have evolved a way to produce a Δp using light energy directly (i.e., without the intervention of oxidation–reduction reactions, and without chlorophyll).[55–58] [It has been reported that, under the proper conditions, they can be grown photohetero-trophically (i.e., on organic carbon with light as a source of energy) but these conclusions have been questioned.[59,60]] Halophilic archaea are heterotrophic organisms that carry out an ordinary aerobic respiration creating a Δp driven by oxidation–reduction reactions during electron transport.[61] The Δp is used to drive ATP synthesis via a membrane ATP synthase. However, conditions for respiration are not always optimal and, in the presence of light and low oxygen levels, the halophiles adapt by making photopigments (rhodop-sins), one of which (bacteriorhodopsin) functions as a proton pump that is energized directly by light energy.[62] Whereas photo-synthetic electron flow is an example of an indirect transformation of light energy into an electrochemical potential (via redox reactions), bacteriorhodopsin illustrates the direct transformation of light energy into an elec-trochemical potential. We will first consider data that demonstrate the light-dependent pumping of protons. This will be followed by a description of the proton pump, the photocycle, and a model for the mechan-ism of pumping protons. It should be added that bacteriorhodopsin is the best charac-terized ion pump and, for this reason, it will be examined in detail.

Evidence that halophiles can use light energy to drive a proton pump

This is shown in Fig. 3.21. A suspension of *Halobacterium halobium* was subjected to light of different intensities and for different periods of time, and the pH of the external medium was monitored. As illustrated in Fig. 3.21A, the light produced an efflux of protons from the cell. Higher light intensities produced a greater rate of proton efflux. (Compare I_c to I_a.) In Fig. 3.21B, the rate of proton efflux (ng/sec), as determined from the slopes in Fig. 3.21A, is plotted as a function of the light intensity (nEinstein/sec). (An Einstein is equal to a "mole" of photons.) From data such as these, a quantum yield (i.e., protons ejected per photon absorbed), was calculated to be 0.52 proton/photon absorbed.[63] The reported values for the *maximum* quantum yield for proton efflux are a little higher, (i.e., 0.6–7 proton/photon absorbed). The quantum yield for the photocycle (i.e., the fraction of bacteriorhodopsin molecules absorbing light that undergoes the photocycle described below) is 0.64 ± 0.04.[64,65] These values suggest that one proton is pumped per photocycle.

Bacteriorhodopsin is the proton pump

Built into the cell membrane of the halophilic archaebacteria is a pigment-protein called bacteriorhodopsin which is a pump responsible for the light-driven electrogenic efflux of protons. It consists of one large polypeptide (248 amino acids, 26,486 D) folded into seven α helices that form a transmembrane channel (Fig. 3.22). Located in the middle of the channel, and attached to the bacteriorhodopsin, is a pigment called retinal (a C_{20}-carotenoid) which is attached via a Schiff base to a lysine residue on the protein.[66] When the retinal absorbs light, the bacteriorhodopsin remarkably translocates protons out of the cell and a Δp is created.

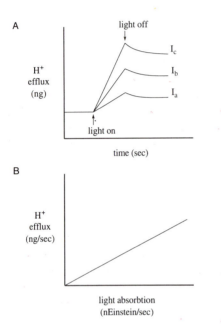

Fig. 3.21 Proton pumping by bacteriorhodopsin. Expected results if *Halobacterium* cells were illuminated by light. In order to measure proton outflow accurately, proton inflow through the ATP synthase must be blocked either with uncouplers or nigericin, which collapse the proton electrochemical potential, or an ATP synthase inhibitor such as DCCD. The extruded protons can be quantified with a pH meter. (A) Proton efflux measured as a function of time at increasing light intensities (nEinsteins/sec), $I_a < I_b < I_c$. The slope of each line is the rate of proton efflux. (B) The rate of proton efflux is plotted as a function of the light intensity. From these data one can calculate the quantum yield (i.e., the number of protons extruded per photon absorbed). (Adapted from data by Bogomolni, R. A., R. A. Baker, R. H. Lozier, and W. Stoeckenius. 1980. Action spectrum and quantum efficiency for proton pumping in *Halobacterium halobium. Biochemistry* 19:2152–2159.

The photocycle and a model for proton pumping

The photoevents occurring in halophiles can be followed spectroscopically because, when bacteriorhodopsin absorbs light, it loses its absorption peak at 568 nm (bleaches) while it is converted in the dark to a series of pigments that have absorption peaks at different wavelengths. This is called the photocycle. Also, site-specific mutagenesis of bacteriorhodopsin is being used to identify the amino acid side chains that transfer protons across the membrane.[67–70] The photocycle is shown in Fig. 3.23. The retinal is attached via a Schiff base to the epsilon amino group of lysine-216 (K_{216}). Before absorbing light, the retinal is protonated at the Schiff base and exists in the all trans (13-trans)

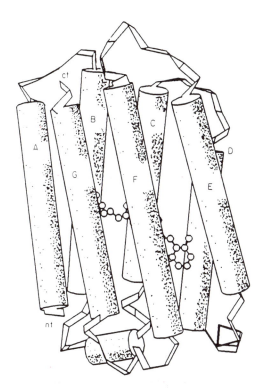

Fig. 3.22 Diagram of bacteriorhodopsin. The seven helices are shown as solid rods. The helices form a central channel where the retinal is attached to a lysine residue on helix G. (From Henderson, R., J. M. Baldwin, Ceska, T. A., Zemlin, F., Beckmann, E., and K. H. Downing. 1990. Model for the structure of bacteriorhodopsin based on high-resolution electron cryomicroscopy. *J. Mol. Biol.* **213**:899–929.)

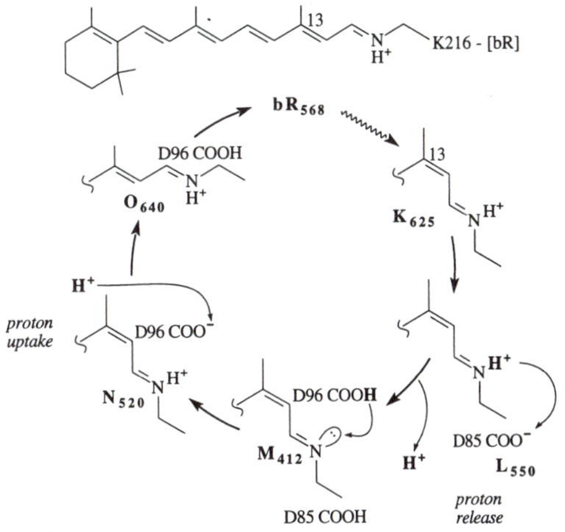

Fig. 3.23 The photochemical cycle of bacteriorhodopsin. Upon absorption of a photon of light, bR_{568} undergoes a trans to *cis* isomerization and is converted to a series of intermediates with different absorbance maxima. The Schiff base becomes deprotonated during the L to M transition and reprotonated during the M to N transition. A recent photocycle postulates two M states (i.e., L to M_1 to M_2 to N). Although not indicated, some of the steps are reversible.[70] (From Krebs, M. P. and H. Gobind Khorana. 1993. Mechanism of light-dependent proton translocation by bacteriorhodopsin. *J. Bacteriol.* **175**:1555–1560.)

configuration (bR_{568}). The subscript refers to the absorption maximum. Upon absorbing a photon of light, the retinal isomerizes to the 13-*cis* form (K_{625}). All subsequent steps do not require light and represent the de-energization of the bacteriorhodopsin via a series of intermediates that have different absorbance maxima to the original unexcited molecule. The transitions are very fast and occur in the nanosecond and millisecond range. These intermediates are, in the order of their appearance, K, L, M, N, and O. In converting from L to M, the Schiff base loses its proton to aspartate-85 but regains a proton from aspartate-96 in going from M to N. Aspartate-96 acquires a proton from the cytoplasm in going from N to O. Thus, a proton has moved from the cytoplasmic side through aspartate-96 to the Schiff base to

aspartate-85. From aspartate-85, the proton moves to the outside membrane surface.[72] Precisely how the proton travels through the bacteriorhodopsin channel from the cytoplasmic side to the Schiff base in the center of the channel, and from there to the external surface of the membrane, is not known. Probably the proton is passed from one amino acid side group that can be reversibly protonated to another. Two of these amino acids have been suggested from site-specific mutagenesis experiments to be aspartate-85 and aspartate-96. The protonatable residues would extend along the protein from the cytoplasmic surface to the outside. In the case of bacteriorhodopsin, where the three-dimensional structure is known, the protonatable residues are oriented towards the center of the channel. Any bound water might also participate in proton translocation. For example, there might be a chain of water and hydrogen-bonded protons connecting protonatable groups. Since the Schiff base gives up its proton to the extracellular side of the channel (transition L to M) and becomes protonated with a proton from the cytoplasm (transition M to N), it has been assumed that there is a switch that reorients the Schiff base so that it alternatively faces the cytoplasmic and extracellular sides of the channel. Logically, the switch would be at M. The mechanism for the switch is unknown, although it has been suggested to be a conformational change in the bacteriorhodopsin.[71,73] It is not obviously correlated with the retinal isomerizations because the retinal is in the *cis* configuration throughout most of the photocycle, including the protonation and deprotonation of the Schiff base. The transfer of protons along protonatable groups has been suggested for several proton pumps besides bacteriorhodopsin, including cytochrome oxidase and the proton-translocating ATP synthase.[74]

3.9 Halorhodopsin, a light-driven chloride pump

The halophiles have a second light-driven electrogenic ion pump, but one which does not energize the membrane. The second pump, called halorhodopsin, is structurally similar to bacteriorhodopsin and is used to accumulate Cl^- intracellularly in order to maintain osmotic stability. Recall that the halophiles live in salt water where the extracellular concentrations of NaCl can be 3–5 M. The osmotic balance is preserved by intracellular concentrations of KCl that match the extracellular Cl^- concentrations. (As discussed in Chapter 14, K^+ is important for osmotic homeostasis in eubacteria as well.) Since the membrane is negatively charged on the inside, energy must be used to bring the Cl^- into the cell and halorhodopsin accomplishes this purpose.[75] In the dark, the halophiles use another energy source, probably ATP, to accumulate chloride ions actively.[76]

3.10 The Δp and ATP synthesis in alkaliphiles

Alkaliphilic bacteria grow in habitats where the pH is very basic, usually around pH 10. These habitats include soda lakes, dilute alkaline springs, and desert soils where the alkalinity is usually due to sodium carbonate. A definition of *obligate* alkaliphiles are organisms that cannot grow at pH values of 8.5 or less, and usually have an optimum around 9.[77] These include *Bacillus pasteurii*, *B. firmus*, and *B. alcalophilus*. The Δp has been measured in obligate aerobic alkaliphiles that grow optimally at pH 10–12, and appears to be too low to drive the synthesis of ATP. The problem is that because the external pH is so basic, the internal pH is generally at least two pH units more acid. This gives the ΔpH a sign opposite to that found in the other bacteria, that is, negative. A ΔpH of two is equivalent to 60×2 or 120 mV. Thus, the Δp can be lowered by about 120 mV in these organisms. Typical membrane potentials for aerobic alkaliphiles are approximately -170 mV, outside positive. Therefore, the Δp values can be as low as $-170 + 120$ or -50 mV.[78] The ATPase in the alkaliphiles is a proton translocating enzyme as in most bacteria. Even if the ATPase translocated as many as four protons, this would generate only 0.2 eV, far short of the approximately 0.4 to 0.5 eV required to synthesize an ATP under physiological conditions. [The energy to synthesize an ATP (i.e., ΔG_p) is 40,000– 50,000 J. Dividing this number by the faraday gives 0.4–0.5 eV.] How can this dilemma be resolved? It is possible that the protons in the bulk extracellular phase may not be as important for the Δp as protons on the membrane or a few angstrom units away from the membrane. One suggestion is that proton pumps may be in frequent contact with the ATPases due to random collisons within the membrane, and that as soon as a proton is pumped out of the cell, it may re-enter via an adjoining ATPase without entering the pool of bulk protons.[79] This suggestion emphasizes the activity of protons at the face of the membrane rather than the ΔpH, which is due to the concentration of protons in the bulk phase, and raises questions about the details of the proton circuit. An important tenet of the chemiosmotic theory, that a delocalized Δp is used, may not apply in this situation. That is to say, the chemiosmotic theory postulates that proton currents couple any exergonic reaction with *any* endergonic reaction (i.e., proton circuits are delocalized over the entire membrane). In that way, proton extrusion during respiration can provide the energy for several different reactions, not simply the ATP synthase (Fig. 3.2). However, it may be that, in some alkaliphiles, there is direct transfer of protons extruded during respiration to the ATP synthase (i.e., *localized* proton circuits).

3.11 Summary

The energetics of bacterial cell membranes can be understood for most bacteria in terms of an electrochemical proton potential established by exergonic chemical reactions or light. The protons are raised from a low electrochemical potential on the inside of the cell to a high electrochemical potential on the outside of the cell. When the protons circulate back into the cell through appropriate carriers, work can be done (e.g., the synthesis of ATP via the membrane ATP synthase solute transport, and flagellar rotation).

The proton potential is due to a combination of a membrane potential ($\Delta\Psi$), outside positive, and a ΔpH, outside acid. Because of the low capacitance of the membrane, not very many protons need be extruded before a large membrane potential develops. The membrane potential seems to be the dominant component in the Δp for most bacteria, except for acidophiles which can have a reversed membrane potential. Other cations, especially sodium ions, can use the established membrane potential for doing work, principally solute accumulation. The sodium ions must be returned to the outside of the cell, and Na^+/H^+ antiporters serve this purpose in most bacteria. Thus, although the major ion circuit is a proton circuit, the sodium circuit is also important.

A Δp is created when an exergonic chemical reaction is coupled to the electrogenic flow of charge across the cell membrane and the liberation of protons on the outer membrane surface. Energy input of at least $yF\Delta p$ J is necessary in order to raise the electrochemical potential of y protons to Δp V. The three most widespread reactions that provide the energy to create the Δp are oxidation–reduction reactions during electron transport in membranes (respiration), oxidation–reduction reactions during electron transport stimulated by light absorption (photosynthesis), and ATP hydrolysis via the membrane ATP synthase. Respiration and ATP hydrolysis are reversible, and the Δp can drive reversed electron transport, as well as the synthesis of ATP. During reversed electron transport, protons enter the cells rather than leave the cells. The ATP synthase is an enzyme complex that reversibly hydrolyzes ATP and pumps protons out of the cell. When protons enter via the ATP synthase, ATP is made.

Light energy can also be used directly to create a Δp without the establishment of a redox potential. This occurs in the halophilic archaea. They create a Δp using a light-driven proton pump called bacteriorhodopsin, which forms a proton channel through the membrane. Bacteriorhodopsin is being studied as a model system to investigate the mechanism of ion pumping across membranes.

Fermenting bacteria have evolved additional ways to generate a Δp. Lactic acid bacteria can create a Δp via coupled efflux of protons and lactate (in addition to ATP hydrolysis) under certain growth conditions. Another anaerobic bacterium, *Oxalobacter*, creates a Δp by the oxidation of oxalic acid to formic acid coupled with the electrogenic exchange of oxalate for formate and the consumption of protons during the decarboxylation. Other examples similar to these will no doubt be discovered in the future.

Sodium potentials are also important in the prokaryotes, especially for solute transport. Although most sodium potentials are created secondarily from the proton potential via antiporters, there are some prokaryotes that couple a chemical reaction to the creation of a sodium potential. For example, some marine bacteria, as exemplified by *Vibrio alginolyticus*, couple respiration to the electrogenic translocation of sodium ions out of the cell at alkaline pH. These bacteria can couple the sodium potential to the membrane ATP synthase and therefore rely on a sodium current rather than a proton current when growing in basic solutions. When growing in slightly acidic conditions, they use a proton potential.

Several different fermenting bacteria can create a sodium potential by coupling the decarboxylation of organic acids with the translocation of sodium ions out of the cell, or by coupling the efflux of end products of fermentation with the translocation of sodium ions to the outside. In some bacteria (e.g. *Klebsiella pneumoniae*), the sodium potential is used to drive the influx of the growth substrate into the cell but not for the generation of ATP. ATP is synthesized by a substrate level phosphorylation during the conversion of pyruvate to formate and acetate. These bacteria may also generate a membrane potential by coupled efflux of the end products of citrate degradation (formate and acetate) with protons.

It should be emphasized that in fermenting bacteria, ATP is hydrolyzed via the ATP synthase to create the Δp that is necessary for membrane activities (e.g., solute transport and flagellar rotation). A decrease in the Δp should result in even more ATP hydrolysis. Thus, reactions that create a membrane potential (e.g., electrogenic efflux of sodium ions or protons in symport with end-products of fermentations), or during de-carboxylation reactions, are expected to conserve ATP.

Measurements of the $\Delta\Psi$ and the ΔpH are necessarily indirect because of the small size of the bacteria. The $\Delta\Psi$ is measured using cationic or anionic fluorescent dyes that equilibrate across the membrane in response to the potential. The distribution of the dyes is monitored by fluorescence quenching. A second way to measure the membrane potential is by the equilibration of a permeant ion which achieves electrochemical equilibrium with the membrane potential. The membrane potential is computed using the Nernst equation, and the intracellular and extracellular ion concentrations. The ΔpH is measured using a weak acid or weak base whose log ratio of concentrations inside the cell to outside the cell is a function of the ΔpH.

Study Questions

1. The E'_o for ubiquinone (ox)/ubiquinone (red) is $+100\,mV$ and for $NAD^+/NADH$ it is $-320\,mV$. What is the $\Delta E'_o$?

ans. $420\,mV$

2. In the electron transport chain, oxidation–reduction reactions with a ΔE_h of about $200\,mV$ appear to be coupled to the extrusion of protons. Assume that 100% of the oxidation–reduction energy is converted to the Δp, for a two electron transfer and the extrusion of two protons, what is the expected Δp?

ans. $-200\,mV$

3. What is the maximum Δp (i.e., 100% energy conversion) when the hydrolysis of one ATP is coupled to the extrusion of four protons? Three protons? Assume the free energy of hydrolysis is 50,000 J.

ans. $-130\,mV$, $-173\,mV$

4. Design an experimental approach that can show that the efflux of an organic acid along its concentration gradient is coupled to proton translocation and can generate a membrane potential. (You must not only be able to demonstrate the membrane potential but also show that the proton is the conducting charge.)

5. Explain how the decarboxylation of oxalic acid by *Oxalobacter* creates a ΔpH and a $\Delta\Psi$.

6. What is the reason for stating that light creates a Δp indirectly in photosynthesis but directly in the extreme halophiles?

7. Lactate efflux was in symport with protons whereas succinate efflux was in symport with sodium ions. Which ionophores might you use to distinguish which cations are involved?

8. A reasonable figure for the actual free energy of hydrolysis of ATP inside cells is $-50,000\,J\,mol^{-1}$. It is believed that the hydrolysis of ATP is coupled to the extrusion of three protons in many systems. If a Δp of $-150\,mV$ were generated, what would be the efficiency of utilization of ATP energy to create the Δp?

ans. 87%

9. Assume a reduction potential of $+400\,mV$ for an oxidant, and a potential of $-100\,mV$ for a reductant. How many joules of energy

are released when two moles of electrons flow from the reductant to the oxidant?

ans. 96,500 J

10. Assume that 45 kJ are required to synthesize one mole of ATP. What would be the required Δp, assuming that three moles of H^+ entered via the ATPase per mole of ATP made?

ans. -155 mV

11. Assume that the Δp is -225 mV. If the ΔpH at 30°C is 1.0, what is the membrane potential?

ans. -165 mV

12. How much energy in joules is required to move a mole of uncharged solute into the cell against a concentration gradient of 1000 at 30°C? If transport were driven by the Δp, what would be the minimal value of the Δp required if one H^+ were co-transported?

ans. 17,370 J, 180 mV

13. Assume membrane vesicles loaded with K^+ so that K^+_{in}/K^+_{out} is 1,000. The temperature is 30°C. Upon adding valinomycin, what is the predicted initial membrane potential in mV?

ans. -180 mV

14. What is the rationale for adding valinomycin and K^+ to an experimental system where the number of protons being pumped out of the cell is measured?

15. Briefly, what does the chemiosmotic theory state?

REFERENCES AND NOTES

1. Harold, F. M. 1986. *The Vital Force: A Study of Bioenergetics.* W. H. Freeman and Co., New York.

2. Nichols, D. G. and S. J. Ferguson. 1992. *Bioenergetics 2.* Academic Press, London.

3. Mitchell, P. 1961. Coupling of phosphorylation to electron and hydrogen transfer by a chemiosmotic type of mechanism. *Nature (London)* 191:144–148.

4. Mitchell, P. 1966. Chemiosmotic coupling in oxidative and photosynthetic phosphorylation. *Biol. Rev. Cambridge Philos. Soc.* 41:445–502.

5. Mitchell, P. 1979. Compartmentation and communication in living systems. Ligand conduction: A general catalytic principle in chemical, osmotic and chemiosmotic reaction systems. *Eur. J. Biochem.* 95:1–20.

6. Kashket, E. R. 1985. The proton motive force in bacteria: A critical assessment of methods. *Ann. Rev. Micro.* 39:219–242.

7. The actual equation is $G = G_o + RT \ln a$, where a = activity. The activity is a product of the molal concentration (c) and the activity coefficient (γ) for the particular compound, $a = \gamma c$. In practice, concentrations are usually used instead of activities, and the concentrations are in molar units instead of molal. The symbol T is the absolute temperature in degrees kelvin ($273 + °C$). The symbol R is the ideal gas constant, and G_o is the standard free energy (i.e., when the concentration of all reactants is 1 M). When one uses 8.3144 J $deg^{-1} mol^{-1}$ as the units of R, then the free energy (G) is given in J $mole^{-1}$.

8. Cecchini, G., and A. L. Koch. 1975. Effect of uncouplers on "downhill" β-galactoside transport in energy-depleted cells of *Escherichia coli*. *J. Bacteriol.* 123:187–195.

9. Gould, J. M., and W. A. Cramer. 1977. Relationship between oxygen-induced proton efflux and membrane energization in cells of *Escherichia coli*. *J. Biol. Chem.* 252:5875–5882.

10. Kashket, E.R. 1985. The proton motive force in bacteria: a critical assessment of methods. *Ann. Rev. Micro.* 39:219–242.

11. Padan, E., D. Zilberstein, and S. Schuldner. 1981. pH homeostasis in bacteria. *Biochim. Biophys. Acta* 650:151–166.

12. Reviewed in Cobley, J. G., and J. C. Cox. 1983. Energy conservation in acidophilic bacteria. *Microbiol. Rev.* 47:579–595.

13. The pumping of protons out of the cell or the electrogenic influx of electrons will create a membrane potential, positive outside. However, in the aerobic acidophilic bacteria [i.e., those bacteria that live in environments of extremely low pH (pH 1 to 4)] other events act to reverse the membrane potential. These bacteria have positive membrane potentials (i.e., inside positive with respect to outside at low pH). It is not clear why the aerobic acidophiles have a positive $\Delta \Psi$. One possibility is that they have an energy-dependent K^+ pump that brings K^+ into the cells at a rate sufficient to establish a net influx of positive charge, creating an inside positive membrane potential. This point is discussed further in Section 14.1.3.

14. Krulwich, T. A., and A. A. Guffanti. 1986. Regulation of internal pH in acidophilic and alkalophilic bacteria Vol. 125, pp. 352–365. In:

Methods in Enzymology. S. Fleischer and B. Fleischer (eds.). Academic Press, New York.

15. Actually, what happens in the presence of nigericin is that an equalization of the K^+ and H^+ gradients occurs.

16. Padan, E., D. Zilberstein, and S. Schuldiner. 1981. pH homeostasis in bacteria. *Biochim. Biophys. Acta* **650**:151–166.

17. Bakker, E. P. The role of alkali-cation transport in energy coupling of neutrophilic and acidophilic bacteria: an assessment of methods and concepts. 1990. *FEMS Microbiol. Rev.* **75**:319–334.

18. This is because the resonance frequency of inorganic phosphate or of the γ-phosphate of ATP in a high magnetic field is a function of the degree to which the phosphate is protonated. (Ferguson, S. J. and M. C. Sorgato. 1982. Proton electrochemical gradients and energy-transduction processes. *Annu. Rev. Biochem.* **51**:185–217.)

19. Cross, R. L. 1992. The reaction mechanism of F_oF_1-ATP synthases, pp. 317–330. In: *Molecular Mechanisms in Bioenergetics.* L. Ernster (ed.) Elsevier Science Publishers, Amsterdam.

20. Maloney, P. C. 1977. Obligatory coupling between proton entry and the synthesis of adenosine 5′-triphospahte in *Streptococcus lactis.* *J. Bacteriol.* **132**:564–575.

21. Maloney, P. C., and F.C. Hansen, III. 1982. Stoichiometry of proton movements coupled to ATP synthesis driven by a pH gradient in *Streptococcus lactis.* *J. Memb. Biol.* **66**:63–75.

22. The actual force was about 239 mV because the temperature was 20–21°C rather than 30°C.

23. Kandpal, R. P., Stempel, K. E., and P. B. Boyer. 1987. Characteristics of the formation of enzyme-bound ATP from medium inorganic phosphate by mitochondrial F_1 adenosinetriphosphatase in the presence of dimethyl sulfoxide. *Biochemistry* **26**:1512–1517.

24. Grubmeyer, C., Cross, R. L., and H. S. Penefsky. 1982. Mechanism of ATP hydrolysis by beef heart mitochondrial ATPase: Rate constants for elementary steps in catalysis at a single site. *J. Biol. Chem.* **257**:12092–12100.

25. Reviewed in Unemoto, T., H. Tokuda, and M. Hayashi. 1990. Primary sodium pumps and their significance in bacterial energetics, pp. 33–54. *The Bacteria*, Vol. XII. T. A. Krulwich (ed.). Academic Press, New York.

26. Reviewed in, Skulachev, V. P. 1992. Chemiosmotic systems and the basic principles of cell energetics, pp. 37–73. In: *Molecular Mechanisms in Bioenergetics.* Ernster, L. (ed.). New Comprehensive Biochemistry: Molecular Mechanisms in Bioenergetics, Vol. 23, Elsevier, Amsterdam.

27. Tokuda, H., and T. Unemoto. 1982. Characterization of the respiration-dependent Na^+ pump in the marine bacterium *Vibrio alginolyticus. J. Biol Chem.* **257**:10007–10014.

28. Kreke, B., and H. Cypionka. 1994. Role of sodium ions for sulfate transport and energy metabolism in *Desulfovibrio salexigens. Arch. Microbiol.* **161**:55–61.

29. Many non-fermenting anaerobic bacteria carry out electron transport using either organic compounds such as fumarate, or inorganic compounds such as nitrate as electron acceptors. Thus, electron flow in these bacteria can be coupled to proton efflux and the establishment of a Δp. Furthermore, even fermenting bacteria can carry out some fumarate respiration generating a Δp. However, the major source of energy for the Δp in most fermenting bacteria is ATP hydrolysis.

30. Dimroth, P. 1990. Energy transductions by an electrochemical gradient of sodium ions, pp. 114–127. In: *The Molecular Basis of Bacterial Metabolism.* G. Hauser and R. Thauer (eds.). Springer-Verlag, Berlin.

31. Dimroth, P. 1980. A new sodium-transport system energized by the decarboxylation of oxaloacetate. *FEBS Lett.* **122**: 234–236.

32. Dimroth, P., and A. Thomer. 1988. Dissociation of the sodium-ion-translocating oxaloacetate decarboxylase of *Klebsiella pneumoniae* and reconstitution of the active complex from the isolated subunits. *Eur. J. Biochem.* **175**: 175–180.

33. Hilpert, W., and P. Dimroth. 1983. Purification and characterization of a new sodium transport decarboxylase. Methylmalonyl–CoA decarboxylase from *Veillonella alcalescens. Eur. J. Biochem.* **132**:579–587.

34. Buckel, W., and R. Semmler. 1983. Purification, characterization and reconstitution of glutaconyl–CoA decarboxylase. *Eur. J. Biochem.* **136**:427–434.

35. Schink, B., and N. Pfennig. 1982. *Propionigenium modestum* gen. nov. sp. nov. A new strictly anaerobic nonsporing bacterium growing on succinate. *Arch. Microbiol.* **133**:209–216.

36. Hilpert, W., Schink, B., and P. Dimroth. 1984. Life by a new decarboxylation-dependent energy conservation mechanism with Na^+ as coupling ion. *EMBO J.* **3**:1665–1670.

37. De Vries, W., Theresia, R., Rietveld-Struijk, M, and A. H. Stouthamer. 1977. ATP formation associated with fumarate and nitrate reduction in growing cultures of *Veillonella alcalescens. Antonie van Leeuwenhoek* **43**:153–167.

38. Buckel, W., and R. Semmler. 1982. A biotin-dependent sodium pump: glutaconyl-CoA decarboxylase from *Acidaminococcus fermentans. FEBS Lett.* **148**:35–38.

39. Inside-out vesicles are prepared by sonicating whole cells, or shearing them with a French pressure cell. Right-side out vesicles are prepared by first removing the cell wall with lysozyme in a hypertonic medium, and then osmotically lysing the protoplasts or spheroplasts in hypotonic medium.

40. Dimroth, P., and A. Thomer. 1988. Dissociation of the sodium-ion-translocating oxaloacetate decarboxylase of *Klebsiella pneumoniae* and reconstitution of the active complex from the isolated subunits. *Eur. J. Biochem.* **175**:175–180.

41. Anantharam, V., M. J. Allison, and P. C. Maloney. 1989. Oxalate:formate exchange. *J. Biol. Chem.* **264**:7244–7250.

42. Baetz, A. L., and M. J. Allison. 1989. Purification and characterization of oxalyl–coenzyme A decarboxylase from *Oxalobacter formigenes*. *J. Bacteriol.* **171**:2605–2608.

43. Baetz, A. L., and M. J. Allison. 1990. Purification and characterization of formyl–coenzyme A transferase from *Oxalobacter formigenes*. *J. Bacteriol.* **172**:3537–3540.

44. Ruan, Z., V. Anantharam, I. T. Crawford, S. V. Ambudkar, S. Y. Rhee, M. J. Allison, and P. C. Maloney. 1992. Identification, purification, and reconstitution of OxlT, the oxalate:formate antiport protein of *Oxalobacter formigenes*. *J. Biol. Chem.* **267**:10537–10543.

45. To prepare proteoliposomes one disperses phospholipids (e.g., those isolated from *E. coli*), in water where they spontaneously aggregate to form spherical vesicles called liposomes consisting of concentric layers of phospholipid. The liposomes are then subjected to high frequency sound waves (sonic oscillation) which breaks them into smaller vesicles surrounded by a single phospholipid bilayer resembling the lipid bilayer found in natural membranes. Then purified protein (e.g., the OxlT antiporter) is mixed with the sonicated phospholipids in the presence of detergent and the suspension is diluted into buffer. The protein becomes incorporated into the phospholipid bilayer and membrane vesicles called proteoliposomes are formed. When the proteoliposomes are incubated with solute they catalyze uptake of the solute into the vesicles provided the appropriate carrier protein has been incorporated. In addition, one can "load" the proteoliposomes with solutes (e.g., oxalate) by including these in the dilution buffer.

46. Poolman, B., D. Molenaar, E. J. Smid, T. Ubbink, T. Abee, P. P. Renault, and W. N. Konings. 1991. Malolactic fermentation: Electrogenic malate uptake and malate/lactate antiport generate metabolic energy. *J. Bacteriol.* **173**:6030–6037.

47. Konings, W. N. 1985. Generation of metabolic energy by end-product efflux. *TIBS* August, 317–319.

48. Michels, J. P., J. Michel, J. Boonstra, and W. N. Konings. 1979. Generation of an electrochemical proton gradient in bacteria by the excretion of metabolic end-products. *FEMS Microbiol. Lett.* **5**:357–364.

49. Otto, R., R. G. Lageveen, H. Veldkamp, and W. N. Konings. 1982. Lactate efflux-induced electrical potential in membrane vesicles of *Streptococcus cremoris*. *J. Bacteriol.* **149**:733–738.

50. Brink, B. T., and W. N. Konings. 1982. Electrochemical proton gradient and lactate concentration gradient in *Streptococcus cremoris* cells grown in batch culture. *J. Bacteriol.* **152**:682–686.

51. Driessen, A. J. M., and W. N. Konings. 1990. Energetic problems of bacterial fermentations: Extrusion of metabolic end products, pp. 449–478. In: *Bacterial Energetics*. T. A. Krulwich (ed.). Academic Press, New York.

52. Michel, T. A., and J. M. Macy. 1990. Generation of a membrane potential by sodium dependent succinate efflux in *Selenomonas ruminantium*. *J. Bacteriol.* **172**:1430–1435.

53. The lactic and succinic acids are presumed to be in the ionized form because the intracellular pH is much larger than the pK_a values.

54. Brink, B. T., R. Otto, U. Hansen, and W. N. Konings. 1985. Energy recycling by lactate efflux in growing and nongrowing cells of *Streptococcus cremoris*. *J. Bacteriol.* **162**:383–390.

55. Oesterhelt, D., and J. Tittor. 1989. Two pumps, one principle: light-driven ion transport in halobacteria. *TIBS*. **14** :57–61.

56. Bogomolni, R. A., R. A. Baker, R. H. Lozier, and W. Stoeckenius. 1980. Action spectrum and quantum efficiency for proton pumping in *Halobacterium halobium*. *Biochemistry*. **19**:2152–2159.

57. Henderson, R., J. M. Baldwin, and T.A. Ceska. 1990. Model for the structure of bacteriorhodopsin based on high-resolution electron cryo-microscopy, *J. Mol. Biol.* **213**:899–929.

58. The extreme halophiles require unusually high external NaCl concentrations [at least 3–5 M (i.e., 17–28%)] in order to grow. They inhabit hypersaline environments such as the solar salt evaporation ponds near San Francisco and salt lakes (e.g., the Great Salt Lake and the Dead Sea). There are now six recognized genera, two of them being the well-known *Halobacterium* and *Halococcus*.

59. Oesterhelt, D. and G. Krippahl. 1983. Phototrophic growth of halobacteria and its use for isolation of photosynthetically-deficient mutants. *Ann. Microbiol. (Inst. Pastuer)* **134B**:137–150.

60. Gest, H. 1993. Photosynthetic and quasi-photosynthetic bacteria. *FEMS Microbiol. Lett.* **112**:1–6.

61. Some can carry out anaerobic respiration using nitrate as an electron acceptor.

62. Respiration can be severely limited under certain growth conditions since the oxygen content of hypersaline waters, the normal habitat of these organisms, is usually 20% or less than is found in normal sea water and, in unstirred ponds, oxygen becomes even more scarce. The halobacteria can derive energy from the fermentation of amino acids, but in the absence of a fermentable carbon source and respiration, light is the only source of energy.

63. Bogomolni, R. A., R. A. Baker, R. H. Lozier, and W. Stoeckenius. 1980. Action spectrum and quantum efficiency for proton pumping in *Halobacterium halobium*. *Biochemistry*. **19**:2152–2159.

64. Tittor, J., and D. Oesterhelt. 1990. The quantum yield of bacteriorhodopsin. *FEBS Lett.* **263**:269–273.

65. Govindjee, R., Balashov, S. P., and T. G. Ebrey. 1990. Quantum efficiency of the photochemical cycle of bacteriorhodopsin. *Biophys. J.* **58**:597–608.

66. A Schiff base is an imine and has the following structure: R—CH=N—R'. It is formed between a carbonyl group and a primary amine, (i.e., R—CHO + H_2N—R' → R – CH=N—R' + H_2O). In bacteriorhodopsin, the amino group is donated by lysine in the protein and the carbonyl is donated by the retinal.

67. Lanyi, J. K. 1992. Proton transfer and energy coupling in the bacteriorhodopsin photocycle. *J. Bioenerg. Biomemb*. **24**:169–179.

68. Oesterhelt, D., J. Tittor, and E. Bamberg. 1992. A unifying concept for ion translocation by retinal proteins. *J. Bioenerg. Biomemb*. **24**:181–191.

69. Fodor, S. P. A., J. B. Ames, R. Gebhard, E. M. M. van den Berg, W. Stoeckenius, J. Lugtenburg, and R. A. Mathies. 1988. Chromophore structure in bacteriorhodopsin's N intermediate: Implications for the proton-pumping mechanism. *Biochemistry* **27**:7097–7101.

70. Krebs, M. P., and H. G. Khorana. 1993. Mechanism of light-dependent proton translocation by bacteriorhodopsin. *J. Bacteriol.* **175**:1555–1560.

71. Lanyi, J. K. 1992. Proton transfer and energy coupling in the bacteriorhodpsin photocycle. *J. Bioenerg. Biomemb.* **24**:169–179.

72. Aspartate-85 remains protonated while a proton is released into the aqueous phase, and therefore it must be concluded that the immediate source of the released proton is a different amino-acid residue.

73. Fodor, S. P. A., J. B. Ames, R. Gebhard, E. M. M. van den Berg, W. Stoeckenius, J. Lugtenburg, and R. A. Mathies. 1988. Chromophore structure in bacteriorhodopsin's N intermediate: Implications for the proton-pumping mechanism. *Biochemistry* **27**:7097–7101.

74. Senior, A. E. 1990. The proton-translocating ATPase of *Escherichia coli*. *Ann. Rev. Biophys. Chem.* **19**:7–41.

75. Lanyi, J. K. 1990. Halorhodopsin, a light-driven electrogenic chloride-transport system. *Physiol. Rev.* **70**:319–330.

76. Duschl, A., and G. Wagner. 1986. Primary and secondary chloride transport in *Halobacterium halobium*. *J. Bacteriol.* **168**:548–552.

77. There are many non-obligate alkaliphilic bacteria whose optimal growth pH is 9 or greater, but which can grow at pH 7 or below. These are widely distributed among the bacterial genera, and include both eubacteria and archaebacteria. There are also *alkaliphilic tolerant* bacteria that can grow at pH values of 9 or more, but whose optimal growth pH is around neutrality.

78. Reviewed in Krulwich, T. A., and D. M. Ivey. 1990. Bioenergetics in extreme environments, pp. 417–447. In: *The Bacteria*, Vol. XII Krulwich, T. A. (ed.). Academic Press, New York

79. Krulwich, T. A., and A. A. Guffanti. 1989. Alkalophilic bacteria. *Annu. Rev. Microbiol.* **43**:435–463.

4

Electron Transport

The main method by which energy is generated for growth-related physiological processes such as biosynthesis and solute transport in *respiring* prokaryotes is by coupling the flow of electrons in membranes to the creation of an electrochemical proton gradient (Fig. 4.1).[1] (The student should review Section 3.2.2 for a discussion of the generation of membrane potentials, and Section 3.7.1 for a discussion of oxidation–reduction reactions coupled to the generation of a Δp.) The electrons flow from primary

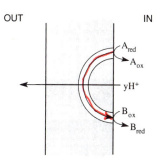

OUT   IN

A_{red}
A_{ox}

yH^+

B_{ox}
B_{red}

Fig. 4.1 Oxidation–reduction reactions in the cell membrane result in a proton potential. Electrons flow from A to B through a series of electron carriers in the membrane from a low potential towards a higher potential. The intermediate redox reactions between A and B are not shown. Certain redox reactions in the series are coupled to the translocation of protons across the cell membrane. These are called coupling sites. In this way a redox potential (ΔE) is converted into a proton potential (Δp).

electron donors to terminal electron acceptors through a series of *electron carrier proteins* and a class of lipids called *quinones*. One refers to electron flow via electron carriers in membranes as respiration. If the terminal electron acceptor is oxygen, then electron flow is called aerobic respiration. If it is not oxygen, then it is called anaerobic respiration. Proton translocation takes place at coupling sites described in Sections 3.7.1 and 4.5. The energy to make the proton potential is derived from the difference in electrode potentials (ΔE_h) of the electron carriers, as described in Section 3.7.1. In other words, prokaryotic cell membranes convert an electrode potential difference, ΔE_h, into a proton electrochemical potential difference (Δp). The proton potential is then used to drive solute transport, ATP synthesis, flagella rotation, and other membrane activities. In mitochondria, electron transport pathways are pretty much all the same. However, prokaryotes are diverse creatures and their electron transport pathways differ depending upon the primary donor and terminal acceptor. This chapter describes electron transport pathways and how they are coupled to the formation of a Δp.

4.1 Aerobic and anaerobic respiration

There is a steady current of electrons through electron carriers in prokaryotic cell membranes from low potential electron donors (the primary donors or *reductants*) to high potential electon acceptors (the terminal acceptor or *oxidant*). Electron acceptors can be oxygen or some other inorganic acceptor such as nitrate or sulfate. An example of an organic electron acceptor is fumarate. Thus, there is oxygen respiration, nitrate respiration, sulfate respiration, fumarate respiration, and so on.

4.2 The electron carriers

The electrons flow through a series of electron carriers. These are:

1. Flavoproteins
2. Quinones
3. Iron-sulfur proteins
4. Cytochromes

The quinones are lipids, whereas the other electron carriers are proteins, which exist in multiprotein enzyme complexes called *oxidoreductases*[2]. The electrons are not carried in the protein *per se*, but in a nonprotein molecule bound to the protein. The nonprotein portion that carries the electron is called a *prosthetic group*[3]. The prosthetic group in iron–sulfur proteins is a cluster of iron sulfide. These proteins are abbreviated as FeS or sometimes Fe/S. The prosthetic group in flavoproteins is a flavin. Flavoproteins are abbreviated Fp, FAD, or FMN. The prosthetic group in cytochromes is heme. The chemistry of the prosthetic groups is described in Section 4.2.1. Some of the prosthetic groups (flavins) carry hydrogen as well as electrons and they are referred to as hydrogen carriers. The quinones are also hydrogen carriers. Some of the prosthetic groups (FeS and heme) carry only electrons and they are referred to as electron carriers.

Each of the electron carriers has a different electrode potential, and the electrons are transferred sequentially to a carrier of a higher potential.

The standard potentials at pH 7 of the electron carriers and some electron donors and acceptors are listed in Table 4.1.

Table 4.1 Standard electrode potentials (pH 7)

Couple	E'_o (mV)
Fd_{ox}/Fd_{red} (spinach)	−432
CO_2/formate	−432
H^+/H_2	−410
Fd_{ox}/Fd_{red} (*Clostridium*)	−410
$NAD^+/NADH$	−320
FeS(ox/red) in mitoch	−305
Lipoic/dihydrolipoic	−290
S^o/H_2S	−270
$FAD/FADH_2$	−220
Acetaldehyde/ethanol	−197
$FMN/FMNH_2$	−190
Pyruvate/lactate	−185
OAA/malate	−170
Menaquinone(ox/red)	−74
Cyt b_{558}(ox/red)	−75 to −43
Fum/succ	+33
Ubiquinone(ox/red)	+100
Cyt b_{556}(ox/red)	+46 to +129
Cyt b_{562}(ox/red)	+125 to +260
Cyt d(ox/red)	+260 to +280
Cyt c(ox/red)	+250
FeS(ox/red) in mitochondria	+280
Cyt a(ox/red)	+290
Cyt c_{555}(ox/red)	+355
Cyt $_a3$(ox/red) in mitochondria	+385
NO_3^-/NO_2^-	+421
Fe^{3+}/Fe^{2+}	+771
O_2 (1 atm)/H_2O	+815

Key: Fd, ferredoxin; OAA, oxaloacetate; Fum, fumarate; succ, succinate.
Sources: Thauer, R. K., K. Jungermann, and K. Decker. 1977. Energy conservation in chemotrophic anaerobic bacteria. *Bacteriol Rev.* 41:100–180; Metzler, D. E. 1977. *Biochemistry: The Chemical Reactions of Living Cells.* Academic Press., New York.

4.2.1 Flavoproteins

Flavoproteins (Fp) are electron carriers that have as their prosthetic group an organic molecule called flavin. [Flavin is derived from the Latin word *flavius* which means yellow, in reference to the color of flavins. They are synthesized by cells from the vitamin riboflavin (vitamin B_2).] There are two flavins, flavin mononucleotide (FMN) and flavin adenine dinucleotide (FAD) (Fig. 4.2). Phosphorylation of riboflavin at the ribityl 5′-OH yields FMN, and adenylylation of FMN yields FAD. As Fig. 4.2 illustrates, when flavins are reduced, they carry 2H (two electrons and

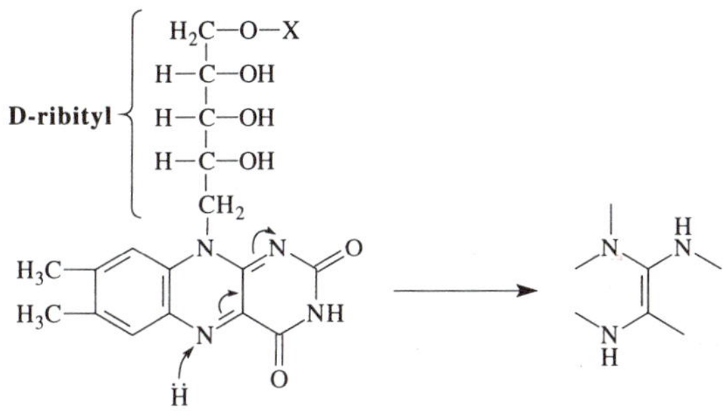

Fig. 4.2 Structures of riboflavin, FMN, and FAD (riboflavin: X=H; FMN: X=PO₃H₂; FAD: X=ADP). For the sake of convenience, the reduction reaction is drawn as proceeding via a hydride ion even though this need not be the actual mechanism in all flavin reductions.

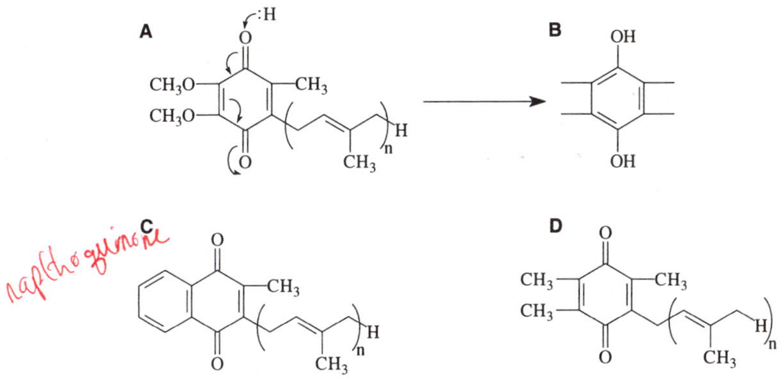

Fig. 4.3 The structure of quinones. (A) oxidized ubiquinone; (B) reduced ubiquinone; (C) oxidized menaquinone; (D) oxidized plastoquinone. The value of n can be 4–10 and is 8 for both quinones in *E. coli*. In *E. coli*, ubiquinone plays a major role in aerobic and nitrate respiration, whereas menaquinone is dominant during fumarate respiration. One reason for this is that ubiquinone has a potential (E'_o) of $+100$ mV compared to $+30$ mV for fumarate. It is therefore at too high a potential to deliver electrons to fumarate. Menaquinone has a low potential, -74 mV, and is thus able to deliver electrons to fumarate. Plastoquinone is used in chloroplast and cyanobacterial photosynthetic electron transport.

two hydrogens), one on each of two ring nitrogens. There are many different flavoproteins and they catalyze diverse oxidation–reduction reactions in the cytoplasm, not merely those of the electron transport chain in the membranes. Although all the flavoproteins have FMN or FAD as their prosthetic group, they differ in which oxidations they catalyze, as well as in their redox potentials. These differences are due to differences in the protein component of the enzyme, not in the flavin itself.

4.2.2 Quinones

Quinones are lipid electron carriers. Due to their hydrophobic lipid nature, some are believed to be highly mobile in the lipid phase of the membrane, and carry hydrogen and electrons to and from the complexes of protein electron carriers which are not mobile. Their structure and oxidation–reduction reactions are shown in Fig. 4.3. All quinones have hydrophobic isoprenoid side chains that contribute to their lipid solubility. The number of isoprene units varies but is typically six to ten. Bacteria make two types of quinones that function during respiration, ubiquinone (UQ), a quinone also found in mitochondria, and menaquinone (MQ or sometimes MK). Menaquinones, which are derivatives of vitamin K, differ from ubiquinones in being naphthoquinones where the additional benzene ring replaces the two methoxy groups present in ubiquinones (Fig.

4.3). They also have a much lower electrode potential than ubiquinones and are used predominantly during anaerobic respiration where the electron acceptor has a low potential (e.g., during fumarate respiration). A third type of quinone, plastoquinone, occurs in chloroplasts and cyanobacteria, and functions in photosynthetic electron transport. In plastoquinones, the two methoxy groups are replaced by methyl groups.

4.2.3 Iron-sulfur proteins

Iron–sulfur proteins contain nonheme iron and usually acid-labile sulfur (Fig. 4.4). The term acid-labile sulfur means that, when the pH is lowered to approximately one, H_2S is released from the protein. This is because there is sulfide attached to iron by bonds which are ruptured in acid. Generally, the proteins contain clusters in which iron and acid-labile sulfur are present in a ratio of 1:1. However, there may be more than one iron–sulfur cluster per protein. For example, in mitochondria, the enzyme complex that oxidizes NADH has at least four FeS clusters (see Fig. 4.9). The FeS clusters have different E_h values and the electron travels from one FeS cluster to the next towards the higher E_h. It appears that the electron may not be localized on any particular iron atom and the entire FeS cluster should be thought of as carrying one electron, regardless of the number of Fe atoms. These proteins also contain cysteine sulfur, which is not acid

labile, and which bonds the iron to the protein. There are several different types of iron–sulfur proteins and these catalyze numerous oxidation–reduction reactions in the cytoplasm as well as in the membranes.[4] The iron–sulfur proteins have characteristic electron spin resonance spectra (EPR) because of an unpaired electron in either the oxidized or reduced form of the FeS cluster in different FeS proteins.[5] The iron–sulfur proteins cover a very wide range of potentials from approximately -400 mV to $+350$ mV. They therefore can carry out oxidation–reduction reactions at both the low potential end and the high potential end of the electron transport chain, and indeed are found in several locations. An example of an FeS cluster is shown in Fig. 4.4. Note that each Fe is bound to two acid-labile S and two cysteine S. This would be called an Fe_2S_2 cluster.

4.2.4 Cytochromes

Cytochromes are electron carriers that have heme as the prosthetic group. Heme consists of four pyrrole rings attached to each other by methene bridges (Fig. 4.5). Because hemes have four pyrroles, they are called *tetrapyrroles*. Each of the pyrrole rings is substituted by a side chain. Substituted tetrapyrroles are called *porphyrins*. Therefore, hemes are also called porphyrins. (An unsubstituted tetrapyrrole is called a porphin.) Hemes are placed in different classes, described below, on the basis of the side chains attached to the pyrrole rings. In the center of each heme there is an iron atom which is bound to the nitrogen of the pyrrole rings. The iron is the electron carrier and is oxidized to ferric or reduced to ferrous ion during electron transport. Cytochromes are therefore one electron carriers. The E_h of the different cytochromes differ depending on the protein and the molecular interactions with surrounding molecules.

Classes of cytochromes

There are four classes of cytochromes (i.e., a, b, c, and d) which have hemes a, b, c, and d. As mentioned previously, the hemes can be distinguished according to the side groups

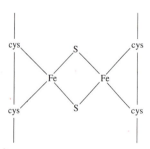

Fig. 4.4. FeS cluster. This is an Fe_2S_2 cluster. More than one cluster may be present per protein. The sulfur atoms held only by the iron are acid labile. The iron is bonded to the protein via sulfur in cysteine residues.

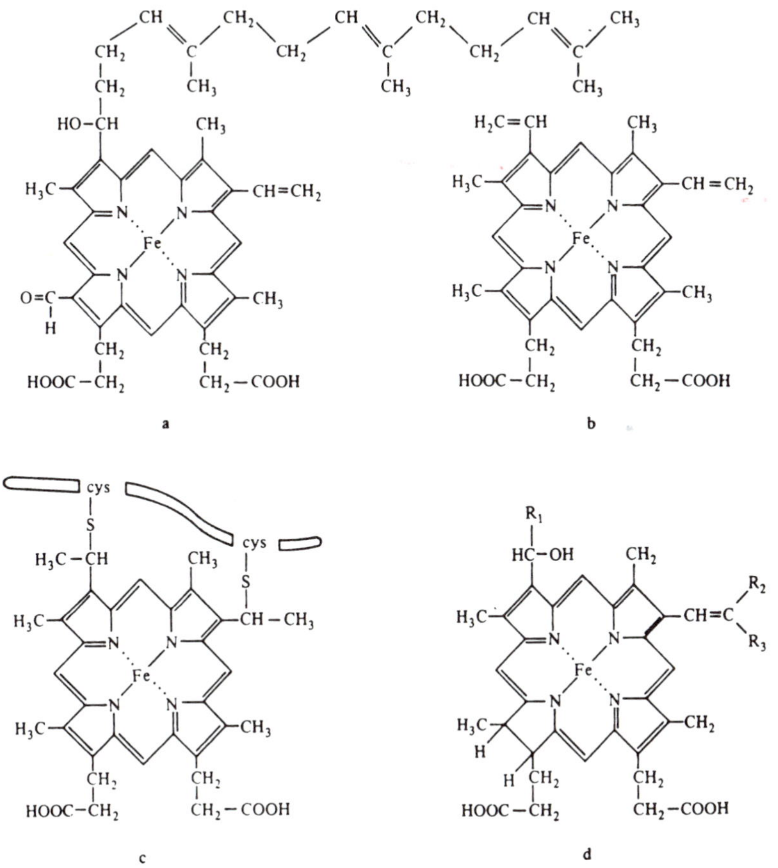

Fig. 4.5 The prosthetic groups of the different classes of cytochromes. The hemes vary according to their side groups. Heme c is covalently bound to the protein via a sulfur bridge to a cysteine residue on the protein. The structures of R_1, R_2, and R_3 are not known. (From Gottschalk, G. 1986. *Bacterial Metabolism*, Springer-Verlag, New York.)

that they possess. (Cytochrome o in bacteria is a b-type cytochrome.) Hemes can also be distinguished spectrophotometrically. When cytochromes are in the reduced state, they have characteristic light absorption bands in the visible range due to absorption by the heme. These are the α, the β, and the γ bands. The α bands absorb light between 500 and 600 nm, the β bands absorb at a lower wavelength, and the γ bands are in the blue region of the spectrum. The spectrum for a cytochrome c is shown in Fig. 4.6. Cytochromes are distinguished, in part, by the position of the maximum in the α band. For example, cyt. b_{556} and cyt. b_{558} differ because the former has a peak at 556 nm, and the latter a peak at 558 nm.

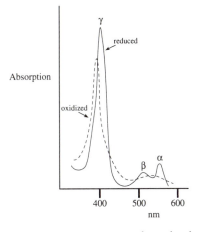

Fig. 4.6 Absorption spectra of oxidized and reduced cytochrome c. The α band in the reduced form is used to identify cytochromes.

Reduced minus oxidized spectra

It is very difficult to resolve the different peaks of individual cytochromes in whole cells because of light scattering and nonspecific absorption unless one employs difference spectroscopy. For difference spectroscopy, the cells are placed into two cuvettes in a split-beam spectrophotometer and monochromatic light from a single monochromator scan is split to pass through both cuvettes. In one cuvette the cytochromes are oxidized by adding an oxidant and, in the second cuvette, they are reduced by adding a reductant. The spectrophotometer subtracts the output of one cuvette from the other to give a reduced minus oxidized difference spectrum. In this way, nonspecific absorption and light scattering are eliminated from the spectrum, and the cytochromes in the preparation are identified.

Dual beam spectroscopy

To follow the kinetics of oxidation or reduction of a particular cytochrome, a dual-beam spectrophotometer is used. In a dual-beam spectrophotometer, there are two monochromators. Light from one monochromator is set at a wavelength at which absorbance will change during oxidation or reduction, and the second beam of light is at a nearby wavelength for which absorbance will not change. The light is sent alternatively from both monochromators through the sample cuvette and the difference in absor-

bance between the two wavelengths is automatically plotted as a function of time.

4.2.5 Standard electrode potentials of the electron carriers

Table 4.1 shows standard electrode potentials at pH 7 (E_o') of some electron donors, acceptors, and electron carriers. Notice that redox couples are generally written in the form, oxidized/reduced (ox/red). Many of the oxidation–reduction reactions in the electron transport chain can be reversed by the Δp as discussed in Section 3.7.1. This means that the ratio ox/red for several of the electron carriers (flavoproteins, cytochromes, quinones, FeS proteins) must be close to one. Thus, for these reactions, the E_h (actual potential at pH 7) of the redox couples are close to their midpoint potentials, E_m', which is the potential at pH 7 when the couple is 50% reduced (i.e., [ox] = [red]).

4.3 Organization of the electron carriers in mitochondria

The electron carriers are organized as an electron transport chain that transfers electrons from electron donors at a low electrode potential to electron acceptors at a higher electrode potential (Fig. 4.7). Electrons can enter at the level of flavoprotein, quinone, or

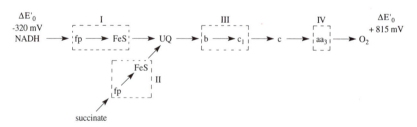

Fig. 4.7 Electron transport scheme in mitochondria. Electrons travel in the electron transport chain from a low to a high electrode potential. Complexes I–IV are bracketed by dotted lines. Complex I is NADH dehydrogenase, also called NADH–ubiquinone oxidoreductase. Complex II is succinate dehydrogenase, also called succinate–ubiquinone oxidoreductase. Complex III is the bc_1 complex, also called ubiquinol–cytochrome c oxidoreductase. Complex IV is the cytochrome aa_3 oxidase, also called cytochrome c oxidase. There are several FeS clusters in complexes I and II, and an FeS protein in complex III. Complex II also has a cytochrome b. fp, flavoprotein; FeS, iron–sulfur protein; UQ, ubiquinone; b, cytochrome b; c_1, cytochrome c_1; c, cytochrome c; aa_3, cytochrome aa_3.

cytochrome, depending upon the potential of the donor. The carriers are organized in the membrane as individual complexes. The complexes can be isolated from each other by appropriate separation techniques after mild detergent extraction which removes the lipids but does not destroy the protein–protein interactions. The separated complexes can be analyzed for their components, and also can be incorporated into proteoliposomes in order to study the oxidation–reduction reactions that they each catalyze, in addition to proton translocation. (Proteoliposomes are artificial constructs of purified lipids and proteins. They are described in Section 15.1.) Four complexes can be recognized in mitochondria. They are complex I (NADH–ubiquinone oxidoreductase), complex II (succinate dehydrogenase), complex III (ubiquinol–cytochrome c oxidoreductase, also called the bc_1 complex), and complex IV (cytochrome c oxidase, which is cytochrome aa_3). Each complex can have several proteins. The most intricate is complex I from mammalian mitochondria which has about 40 polypeptide subunits, at least four iron–mononucleotide (FMN), and one or two bound ubiquinones. Analogous complexes have been isolated from bacteria, but in some cases (e.g., NADH–ubiquinone oxidoreductase and the bc_1 complex) they have fewer protein components.[6–9] Note the pattern in the arrangement of the electron carriers; a dehydrogenase complex accepts electrons from a primary donor and transfers the electrons to a quinone. The quinone then transfers the electrons to an oxidase complex via intervening cytochromes. As described next, the same pattern exists in bacteria.

4.4 Organization of the electron carriers in bacteria

Bacterial electron transport chains vary among the different bacteria, and also according to the growth conditions. These variations will be discussed later (Section 4.7). First, the common features in bacterial electron transport schemes will be described and comparisons will be made with mitochondrial electron transport. As with the mitochondrial electron transport chain, the bacterial chains are organized into dehydrogenase and oxidase complexes connected by quinones (Fig. 4.8).

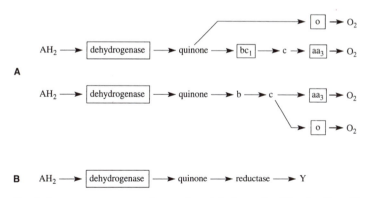

Fig. 4.8 Generalized electron transport pathways found in bacteria. The details will vary depending upon the bacterium and the growth conditions. (A) Aerobic respiration: a dehydrogenase complex removes electrons from an electron donor and transfers these to a quinone. The electrons are transferred to an oxidase complex via a branched pathway. Depending upon the bacterium, the pathway may branch at the quinone or at cytochrome. Many bacteria have bc_1, cytochrome c, and cytochrome aa_3 in one of the branches and, in this way, resemble mitochondria. Other bacteria do not have a bc_1 complex, and may or may not have cytochrome aa_3. (B) Anaerobic respiration: under anaerobic conditions the electrons are transferred to reductase complexes which are synthesized anaerobically. Several reductases exist, each one specific for the electron acceptor. Y represents either an inorganic electron acceptor other than oxygen (e.g., nitrate) or an organic electron acceptor (e.g., fumarate). More than one reductase can simultaneously exist in a bacterium (i.e., the anaerobic pathways can also be branched).

The quinones accept electrons from dehydrogenases and transfer these to oxidase complexes that reduce the terminal electron acceptor. Bacteria are capable of using electron acceptors other than oxygen (e.g., nitrate and fumarate), during anaerobic respiration. The enzyme complexes that reduce electron acceptors other than oxygen are called *reductases*, rather than oxidases. Some of the dehydrogenase complexes are NADH and succinate dehydrogenase complexes analogous to complex I and II in mitochondria. In addition to these dehydrogenases, there are several others that reflect the diversity of substrates oxidized by the bacteria. For example, there are H_2 dehydrogenases (called hydrogenases), formate dehydrogenase, lactate dehydrogenase, methanol dehyhdrogenase, methylamine dehydrogenase, and so on. Depending upon the source of electrons and electron acceptors, bacteria can synthesize and substitute one dehydrogenase complex for another, or reductase complexes for oxidase complexes. For example, when growing anaerobically, *E. coli* makes the reductase complexes instead of the oxidase complexes, represses the synthesis of some dehydrogenases and stimulates the synthesis of others. (This is discussed more fully in Chapter 17.) For this reason, the electron carrier complexes in bacteria are sometimes referred to as modules since they can be synthesized and "plugged" into the respiratory chain when needed.

4.4.1 Bacterial electron transport chains are branched

Two major differences between mitochondrial and bacterial electron transport chains are that (1) the routes to the terminal electron acceptor in the bacteria are branched, the branch point being at the quinone or cytochrome, and (2) many bacteria can alter their electron transport chains depending upon growth conditions (Fig. 4.8). Under aerobic conditions, there are often two or three branches leading to different oxidases. For example, a two-branched electron transport chain might contain a branch leading to cytochrome o oxidase and a branch leading

to cytochrome aa_3 oxidase (cytochrome c oxidase). Other bacteria (e.g., *E. coli*) have cytochrome o and d oxidase branches but do not have the cytochrome aa_3 branch. The ability to synthesize branched electron transport pathways to oxygen confers flexibility on the bacteria, because the branches differ not only in the Δp that can be generated (because they may differ in the number of coupling sites), but their terminal oxidases differ with respect to affinities for oxygen. Switching to an oxidase with a higher affinity for oxygen allows the cells to continue to respire even when oxygen tensions fall to very low values. This is important to insure the reoxidation of the reduced quinones. For example, in *E. coli*, cytochrome o has a low affinity for oxygen and is a coupling site, whereas cytochrome d has a higher affinity for oxygen than does cytochrome o but is not a coupling site. The adaptability of the bacteria with respect to their electron transport chains can be seen with bacteria that can grow either aerobically or anaerobically. Under anaerobic conditions, they do not make the oxidase complexes but instead synthesize reductases. For example, during anaerobic growth, *E. coli* synthesizes fumarate reductase, nitrate reductase, and trimethylamine-N-oxide (TMAO) reductase. (The regulation of synthesis of these reductases is discussed in Chapter 17.) The different reductases enable the bacteria to utilize alternative electron acceptors under anaerobic conditions. (Some facultative anaerobes will ferment when a terminal electron acceptor is unavailable. Fermentation is discussed in Chapter 13.)

4.5 Coupling sites

Sites in the electron transport pathway where redox reactions are coupled to proton translocation and a Δp is created are called *coupling sites*.[10] Each coupling site is also a site for ATP synthesis since the protons extruded re-enter via ATP synthase to make ATP. In mitochondria there are three coupling sites, and they are called sites 1, 2, and 3. These are shown in Fig. 4.7. Site 1 is the NADH dehydrogenase complex (complex I), site 2 is the bc_1 complex (complex III), and site 3 is the cytochrome aa_3 complex (complex IV). The succinate dehydrogenase complex (complex II) is not a coupling site. The ratio of protons translocated per two electrons varies, depending upon the complex. A consensus value of 10 protons is extruded per two electrons that travel from NADH to oxygen. The bc_1 complex translocates four protons per two electrons, and depending upon the reported value, complexes I and IV translocate from two to four protons per two electrons. (The consensus value for mitochondrial complex I is $4H^+/2e^-$.) During reversed electron flow, protons enter the cell through coupling sites 1 and 2, driven by the Δp, and the electrons move towards the lower redox potential. This creates a positive ΔE at the expense of the Δp. (See eq. 3.19.) Coupling site 3 is not physiologically reversable. Thus, water cannot serve as a source of electrons for NAD^+ reduction using reversed electron flow. However, during oxygenic photosynthesis, light energy can drive electrons from water to $NADP^+$. The mechanism of photoreduction of $NADP^+$ is different from reversed electron flow and is discussed in Chapter 5.

4.5.1 The identification of coupling sites

For an understanding of the physiology of energy metabolism during electron transport, it is necessary to study the mechanism of proton translocation, and for this the coupling sites must be identified and isolated. The coupling sites can be identified by the use of electron donors that feed electrons into the chain at different places and by measuring the amount of ATP made per $2\ e^-$ transfer through the respiratory chain. The number of ATPs made per $2\ e^-$ transfer to oxygen is called the *P/O ratio*. It is equal to the number of ATP molecules formed per atom of oxygen taken up. When an electron acceptor other than oxygen is used, then $P/2e^-$ is substituted for P/O. In mitochondria, the oxidation of NADH results in a P/O ratio of about three, indicating that three coupling sites occur between NADH and O_2. The use of succinate

as an electron donor results in a P/O ratio of approximately two. Since electrons from succinate enter at the ubiquinone level, this indicates that coupling site 1 occurs between NADH and ubiquinone (i.e., the NADH–ubiquinone oxidoreductase reaction) (Fig. 4.7). The other two coupling sites must occur between ubiquinone and oxygen. When electrons enter the respiratory chain after the bc_1 complex, the P/O ratio is reduced to one, indicating that the bc_1 complex is the second coupling site and that site 3 is cytochrome aa_3 oxidase. Site 3 can be demonstrated by bypassing the bc_1 complex. The bc_1 complex can be bypassed using an artificial electron donor to reduce cytochrome c [e.g., ascorbate and tetramethylphenylenediamine (TMPD)], thus channeling electrons from ascorbate to TMPD to cytochrome c to cytochrome aa_3. Alternatively, one can simply use reduced cytochrome c as an electron donor to directly reduce cytochrome aa_3. Each of these sites is characterized by a drop in midpoint potential of about $200\,mV$, which is sufficient for generating the Δp (Fig. 4.9). The size of the Δp that can be generated with respect to the ΔE is discussed in Section 3.7.

4.5.2 The actual number of ATPs that can be made per two electrons travelling through the coupling sites

The maximum amount of ATP that can be made per coupling site is equal to the ratio of protons extruded at the coupling site to protons that re-enter via the ATP synthase (Fig. 4.10). For example, if two protons are translocated at a coupling site and three protons enter through the ATP synthase, then 2/3 of an ATP can be made when two electrons pass through the coupling site. The ratio of protons translocated per two electrons travelling through all of the coupling sites to O_2 is called the H^+/O ratio. It can be measured by administering a pulse of a known amount of oxygen to an anaerobic suspension of mitochondria or bacteria and measuring the initial efflux of protons with a pH electrode as the small amount of oxygen is used up. The experiment requires that valinomycin plus K^+ or a permeant anion

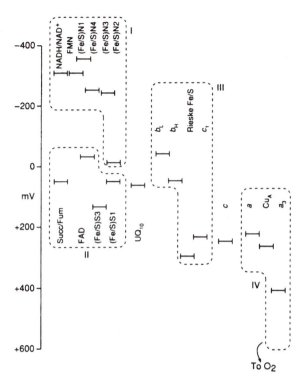

Fig. 4.9 Average mid-point potentials, E'_m, for components of the mitochondrial respiratory chain. The complexes are in dotted boxes. The actual potentials (E_h) for most of the components are not very different from the mid-point potentials. An exception is cytochrome aa_3 whose E_h is much more positive than its mid-point potential. There are three sites where there are changes in potential of 200 mV or more. These drive proton translocation. One site is within complex I, a second within complex III, and the third between complex IV and oxygen. (Adapted from Nicholls, D. G. and S. J. Ferguson. 1992. *Bioenergetics 2*. Academic Press, London.)

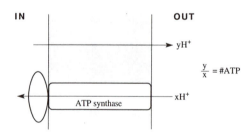

Fig. 4.10 The ratio of protons extruded to protons translocated through the ATPase determines the amount of ATP made.

such as thiocyanate, SCN^-, be in the medium to prevent a $\Delta\Psi$ from developing. (See Section 3.2.2 for a discussion of this point.) The reported values for H^+/O for NADH oxidation vary. However, there is a consensus that the true value is probably around 10. The ratio of protons entering via the ATP synthase to ATP made is called the H^+/ATP.

It can be measured using inverted sub-mitochondrial particles prepared by sonic oscillation. These particles have the synthase on the outside and will pump protons into the interior upon addition of ATP. Similar inverted vesicles can be made from bacteria by first enzymatically weakening or removing the cell wall and breaking the spheroplasts or protoplasts by passage through a French press at high pressures.[11] Values of H^+/ATP from two to four have been reported, and a consensus value of three can be used for calculations. For intact mitochondria, an additional H^+ is required to bring P_i electroneutrally from the cytosol into the mitochondrial matrix in symport with H^+, so H^+/ATP would be four. A value of 10 for H^+/O predicts a maximum P/O ratio of 2.5 (10/4) for mitochondria. This means that the often stated value of three for a P/O ratio for

NADH oxidation by mitochondria may be too high. The number of protons ejected per $2e^-$-travelling between succinate and oxygen in mitochondria is six. Therefore, the maximal P/O ratio for this segment of the electron transport chain may be 6/4 or 1.5. The P/O ratios in bacteria can be higher since a proton is not required to bring P_i into the cell. One significant aspect of branched aerobic respiratory chains in bacteria is that the number of coupling sites, and therefore the ratio of H^+/O, can differ in the branched chains. Thus, the different branches are not equally efficient in generating a Δp or making ATP. We will return to this point later.

Of course, an ATP can be made when three protons enter via the ATP synthase only if the Δp is sufficiently large. As an excercise, we can ask how large the $y\Delta p$ must be. Recall that $y\Delta p$ is the work that can be done in units of electron volts when y protons traverse a proton potential of Δp volts. If H^+/ATP is three, then the number of electron volts made available by proton influx through the ATP synthase is $3\Delta p$. How many electron volts are needed to synthesize an ATP? The free energy of formation of ATP at physiological concentrations of ATP, ADP, and P_i is the phosphorylation potential, ΔG_p, which is approximately 45,000–50,000 J.[12] Dividing by the faraday (96,500 C) expresses the energy required to synthesize an ATP in electron volts. For 45,000 J this is 0.466 eV. Therefore, $3\Delta p$ must be greater than or equal to 0.466 eV. Thus, the minimum Δp is 0.466/3 or -0.155 V. Values of Δp approximate to this are easily generated during electron transport. (However, see the discussion in Section 3.10 regarding the low Δp in alkaliphiles.)

4.6 How a proton potential might be created at the coupling sites: Q loops, Q cycles, and proton pumps

The previous discussion points out that proton translocation takes place at coupling sites when electrons travel "downhill" over a potential gradient of at least 200 mV (see Fig. 4.9.) However, the mechanism by which the redox reaction is actually coupled to proton translocation was not explained.

There are two ways in which this is thought to occur: (1) a Q loop or Q cycle, and (2) a proton pump. In the Q loop or Q cycle, reduced quinone carries hydrogen across the membrane and becomes oxidized, releasing protons on the external face of the membrane as the electrons return electrogenically via electron transport carriers to the inner membrane surface. Proton pumps are electron carrier proteins that couple electron transfer to the electrogenic translocation of protons through the membrane. These two fundamentally different mechanisms are described below.

4.6.1 The Q loop

The essential feature of the Q loop model is that the electron carriers alternate between those that carry both hydrogen and electrons (flavoproteins and quinones), and those that carry only electrons (iron–sulfur proteins and cytochromes). This is illustrated in Fig. 4.11. The electron carriers and their sequence are: flavoprotein (H carrier), FeS protein (e^- carrier), quinone (H carrier), and cytochromes (e^- carriers). The flavoprotein and FeS protein comprise the NADH dehydrogenase which can be a coupling site (Section 4.6.3), although the mechanism of proton translocation by the NADH dehydrogenase is not understood. When electrons are transferred from the FeS protein to quinone (Q) on the inner side of the membrane, two protons are acquired from the cytoplasm. According to the model, the reduced quinone (QH_2), called quinol, then diffuses to the outer membrane surface and becomes oxidized, releasing the two protons. The electrons then return via cytochromes to the inner membrane surface where they reduce oxygen. This would create a $\Delta\Psi$ because the protons are left on the outer surface of the membrane as the electrons move electrogenically to the inner surface. Thus, quinol oxidation is a second coupling site. As mentioned previously, the energy to create the membrane potential is derived from the ΔE_h between the oxidant and the reductant. One way to view this is that the energy from the ΔE_h "pushes" the electron to the negative membrane potential on the inside surface. During

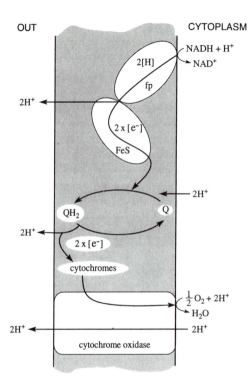

OUT   CYTOPLASM

NADH + H⁺

2[H]

→ NAD⁺

fp

2H⁺ ←

2 x [e⁻]

FeS

← 2H⁺

QH₂   Q

2H⁺ ←

2 x [e⁻]

cytochromes

½O₂ + 2H⁺

→ H₂O

2H⁺ ←   ← 2H⁺

cytochrome oxidase

Fig. 4.11 Proton translocation showing a Q loop and a proton pump. It is proposed that the electron carriers exist in an alternating sequence of hydrogen [H] and electron [e⁻] carriers. This would be flavoprotein (fp), iron-sulfur protein (FeS), quinone (Q), and cytochromes. Oxidation of the flavoprotein deposits two protons on the outer membrane surface. The electrons return to the inner membrane surface where a quinone is reduced, taking up two protons from the cytoplasm. The reduced quinone diffuses to the outer surface of the membrane where it is oxidized, depositing two more protons on the surface. The electrons return to the cytoplasmic surface via cytochromes where they reduce oxygen in a reaction that consumes protons. Some cytochrome oxidases function as proton pumps. During anaerobic respiration, the cytochrome oxidase is replaced by a reductase and the electrons reduce some other electron acceptor (e.g., nitrate or fumarate). It should be noted that electrons can also enter at the level of quinone (e.g., from succinate dehydrogenase.)

reduction of the terminal electron acceptor, cytoplasmic protons are consumed, thus contributing to the proton gradient. Note that the role of the quinone is to ferry the hydrogens across the membrane, presumably by diffusing from a reduction site on the

cytoplasmic side of the membrane to an oxidation site on the outer side.

4.6.2 The Q cycle

Although the linear Q loop as described above for the oxidation of quinol may accurately describe quinol oxidation in *E. coli* and some other bacteria, it is inconsistent with experimental observations of electron transport in mitochondria, chloroplasts, and many bacteria. For example, the Q loop predicts that the ratio of H^+ released per QH_2 oxidized is two, whereas the measured ratio in mitochondria and many bacteria is actually four. In order to account for the extra two protons, Peter Mitchell suggested a new pathway for the oxidation of quinol called the *Q cycle*.[13] The Q cycle operates in an enzyme complex called the bc_1 complex (complex III). The bc_1 complex from bacteria contains three polypeptides. These are cytochrome b with two b-type hemes, an iron–sulfur protein containing a single 2Fe-2S cluster (the Rieske protein), and cytochrome c_1 with one heme. The complex spans the membrane and has a site for binding reduced ubiquinone, UQH_2, on the outer surface of the membrane called site P (for positive), and a second site on the inner surface for binding UQ, site N (for negative) (Fig. 4.12). At site P, UQH_2 is oxidized to the semiquinone anion, UQ^- (Fig. 4.12A), and the two protons are released to the membrane surface. The electron travels to the FeS protein, and from there to cytochrome c_1 on its way to the terminal electron acceptor. The UQ^- is then oxidized to UQ by the removal of the second electron, which is transferred to b_{556}, also called b_L because of its relatively low E'_m. The electron is transferred across the membrane to b_{560}, also called b_H because of its relatively high E'_h. Heme b_H transfers the electron to UQ bound at site N, reducing it to the semiquinone anion, UQ^-. *Electron flow from the Q_P site to the Q_N site is transmembrane and creates a membrane potential.* A second UQH_2 is oxidized at the P site and the UQ_N^- is reduced to UQH_2, picking up two protons from the cytoplasm (Fig. 4.12B). The UQH_2 enters the quinone pool. Thus, for every two

UQH$_2$ that are oxidized releasing four protons, one UQH$_2$ is regenerated. The net result is the oxidation of one UQH$_2$ to UQ with the release of four protons. The situation can be summarized as the following:

P site: $2UQH_2 + 2$ cyt.c$_{ox}$ $\longrightarrow$ $2UQ$
$+2$ cyt.c$_{red} + 4H^+_{out} + 2e^-$

N site: $UQ + 2e^- + 2H^+_{in}$ $\longrightarrow$ UQH_2

$UQH_2 +$ cyt.c$_{ox} + 2H^+_{in}$ $\longrightarrow$ UQ
$+$ cyt.c$_{red} + 4H^+_{out}$

Bioenergetics of the Q cycle

Since the Q cycle translocates two protons per electron that flows to the terminal electron acceptor, it generates a larger proton current than the Q loop that translocates only one proton per electron. This can result in more ATP synthesis. Consider the situation where an ATP synthase requires the influx of three protons to make one ATP. If the transfer of an electron through the electron transport pathway resulted in the translocation of one proton, then 1/3 of an ATP could be made per electron. On the other hand, if electron transport resulted in the translocation of two protons per electron, then 2/3 of an ATP could be made per electron. In other words, the size of the proton current generated by respiration determines the upper value of the amount of ATP that can be made.

Distinguishing the Q loop from the Q cycle

When examining the physiology of electron transport in particular bacteria, it is important to learn whether a Q loop or a Q cycle is operating. There are several features of the Q cycle that help to distinguish it from the linear pathway of electron flow in the Q loop. These include:

1. The ratio of H$^+$ extruded in the Q cycle per e$^-$ is two rather than one.

2. Cytochrome b can be reduced by quinol at two sites, site P or site N (by reversal), whereas in a linear respiratory chain there would be only one site, because there is only one site for UQH$_2$ oxidation. These

sites can be distinguished using inhibitors, as described next.

Two pathways for the reduction of cytochrome b can be demonstrated with the use of inhibitors. Myxothiazol and stigmatellin block the oxidation of ubiquinol at site P (Fig. 4.12, reaction 1), whereas antimycin blocks the oxidation of ubiquinol at site N (Fig. 4.12, reaction 4). These inhibitions are shown in Fig. 4.13. Therefore, antimycin alone does not inhibit the reduction of cytochrome c$_1$ or cytochrome b in the presteady state, since the electrons are coming from site P. Also, myxothiazol and stigmatellin do not inhibit the reduction of cytochrome b since electrons can come from site N. Thus, the presence of both myxothiazol (or stigmatellin) and antimycin are necessary to block the reduction of cytochrome b.

The fact that one requires two different inhibitors of UQH$_2$ oxidation to block the reduction of cytochrome b in the presteady state indicates that there are two routes for cytochrome b reduction by UQH$_2$, and is evidence for a Q cycle.

4.6.3 Pumps

Proton pumps also exist. These catalyze the electrogenic translocation of protons across the membrane rather than electrons (Fig. 4.11). For example, proton extrusion accompanies the cytochrome aa$_3$ oxidase reaction when cytochrome c is oxidized by oxygen in mitochondria and some bacteria. This can be observed by feeding electrons into the respiratory chain at the level of cytochrome c, thus bypassing the quinone. The experimental procedure is to incubate the cells in lightly buffered anaerobic media with a reductant for cytochrome c and a permeant anion (e.g., SCN$^-$) or valinomycin plus K$^+$. Changes in pH are measured with a pH electrode. Upon the addition of a pulse of oxygen, given as an air-saturated salt solution, a sharp, transient acidification of the medium occurs, and the H$^+$/O can be calculated. Since the only enzymatic reaction is the cytochrome aa$_3$ oxidase reaction, it can be concluded that the cytochrome aa$_3$ oxidase pumps protons out

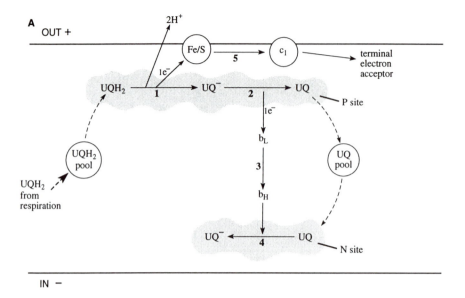

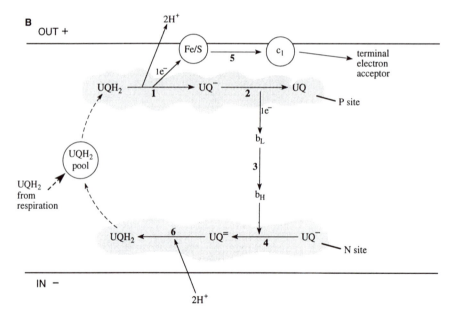

Fig. 4.12 The bc_1 complex. The bc_1 complex isolated from bacteria contains three polypeptides which are a cytochrome b containing two heme b groups, b_H and b_L, an iron–sulfur protein, Fe/S (the Rieske protein), and cytochrome c_1. It is widely distributed among the bacteria, including the photosynthetic bacteria. (Mitochondrial bc_1 complexes are similar but contain an additional 6–8 polypeptides without prosthetic groups.) The iron–sulfur protein and cytochrome c_1 are thought to be located on the outside (positive or p) surface of the membrane and the cytochrome b is believed to span the membrane acting as an electron conductor. On the outer surface there is a binding site in the bc_1 complex for ubiquinol, UQH_2 (the P site). On the inner surface (negative or n) there is a binding site in the bc_1 complex for ubiquinone, UQ (the N site). (A) Reduced ubiquinone binds to the P site and one electron is removed forming the semiquinone anion, UQ^-, (reaction 1). At this time two protons are released on the outer membrane surface. The electron that is removed is transferred to the iron–sulfur protein and from there to cytochrome c_1 (step 5). (Cytochrome c_1 transfers the electron to the terminal electron acceptor via a series of electron carriers in other reactions.) In reaction 2, the second electron is removed

of the cell during the redox reaction. When similar experiments were done with *Paracoccus denitrificans*, it was found that the *P. denitrificans* cytochrome aa$_3$ oxidase translocates protons with a stoichiometry of $1H^+/1e$.[14] Cytochrome oxidase pumping activity can also be demonstrated in proteoliposomes made with purified cytochrome aa$_3$. Because the proton pumps move a positive charge across the membrane leaving behind a negative charge, a membrane potential, outside positive, develops. The membrane potential should be the same as when an electron moves inward since the proton and the electron carry the same charge (i.e., 1.6×10^{-19} C). The mechanism of pumping is not known but probably requires conformational changes in cytochrome oxidase resulting from its redox activity. Conformational changes probably also occur in bacteriorhodopsin during proton pumping (Section 3.8.4).

The mitochondrial NADH dehydrogenase complex (NADH:ubiquinol oxidoreductase) translocates four protons per NADH oxidized.[15] However, the mechanism has not yet been elucidated. Two types of NADH:ubiquinol oxidoreductases exist in bacteria.[16–19] One of these, called NDH-1, is similar to the mitochondrial complex I in that it is a multi-subunit enzyme complex (approximately 14 polypeptide subunits) consisting of FMN and FeS clusters, and translocates protons across the membrane during NADH oxidation ($4H^+/2e^-$). The mechanism of proton translocation is not known. [The marine bacterium, *Vibrio alginolyticus* has a sodium-dependent NDH (Na-NADH).] A second NADH–ubiquinol oxidoreductase, called NDH-2, is also present in bacteria. NDH-2 differs from NDH-1 in consisting of a single polypeptide and FAD, and not being

an energy-coupling site. In *E. coli,* NDH-1 and NDH-2, are present simultaneously.[20] How *E. coli* regulates the partitioning of electrons between the two NAD dehydrogenases is not known but clearly has important energetic consequences.

4.7 Patterns of electron flow in individual bacterial species

Although the major principles of electron transport as outlined above apply to bacteria in general, several different patterns of electron flow exist in particular bacteria, often within the same bacterium grown under different conditions.[21] The patterns of electron flow reflect the different sources of electrons and electron acceptors that are used by the bacteria. For example, bacteria may synthesize two or three different oxidases in the presence of air and several reductases anaerobically. In addition, certain dehydrogenases are made only anaerobically because they are part of an anaerobic respiratory chain. Furthermore, whereas electron donors such as NADH and FADH$_2$, generated in the cytoplasm, are oxidized inside the cell, there are many instances of oxidations in the periplasm in gram-negative bacteria. Examples of substances which are oxidized in the periplasm include hydrogen gas, methane, methanol, methylamine, formate, perhaps ferrous ion, reduced inorganic sulfur, and elemental sulfur. In most of these instances periplasmic cytochromes c accept the electrons from the electron donor and transfer them to electron carriers in the membrane. The discussion that follows is not complete, but is meant to convey the diversity of electron transport systems found in bacteria. The electron transport chains for the respira-

---

Fig. 4.12 *continued from facing page*
from the semiquinone anion producing the fully oxidized quinone, UQ. The second electron is transferred transmembrane via cytochrome b to ubiquinone at site N (steps 3 and 4). Because the electron travels transmembrane, a membrane potential is created, outside positive. (B) A second reduced ubiquinone is oxidized at the P site, releasing two more protons, and the sequence of electron transfers is repeated. The UQ^- at site N becomes reduced to UQH$_2$, having acquired two protons from the cytoplasm. Note that for every two UQH$_2$ molecules that are oxidized, four protons are released, and one UQH$_2$ is regenerated. Therefore, four protons are released per one UQH$_2$ oxidized. Another way of saying this is that the ratio (H^+/e^-) is two.

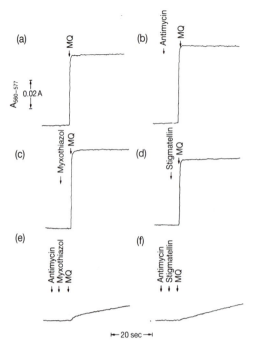

Fig. 4.13 Inhibitor studies can provide evidence for a Q cycle. One distinguishing feature of the Q cycle is that there are two sites for cytochrome b reduction (i.e., sites P and N). At site P quinol is oxidized by the iron–sulfur protein generating the semiquinone anion (Q^-). This oxidation is inhibited by stigmatellin or myxothiazol. The semiquinone anion, Q^-, reduces cytochrome b. Thus, stigmatellin or myxothiazol prevent the oxidation of quinol and the reduction of cytochrome b at site P. The reduced cytochrome b is reoxidized by Q or Q^- in an antimycin-sensitive step at site N. If site P is blocked (e.g., by myxothiazol or stigmatellin, or by removal of the iron-sulfur protein), then cytochrome b can be reduced by quinol by a reversal of the antimycin-sensitive step at site N. There are therefore two routes for the reduction of cytochrome b as opposed to a single route in a linear pathway between quinol and cytochrome b. This experiment was done with a purified bc$_1$ complex (isolated from the bacterium *Paracoccus denitrificans*) which was incorporated into liposomes. The reductant was menaquinol (MQ). Cytochrome b reduction is reflected by a rise in the trace which is a measure of the absorbance of reduced cytochrome b. The addition of antimycin, myxothiazol, or stigmatellin alone did not prevent the reduction of cytochrome b by MQ. However, the addition of both antimycin and stigmatellin, or antimycin and myxothiazol, blocked cytochrome b reduction. Trace (a), control without inhibitors. Traces (b), (c), (d), antimycin, myxothiazol, and stigmatellin, respectively. Traces

tory metabolism of ammonia, nitrite, inorganic sulfur, and iron are described in Chapter 11.

4.7.1 Escherichia coli

E. coli is a gram-negative heterotrophic facultative anaerobe. It can be grown aerobically using oxygen as an electron acceptor, anaerobically (e.g., using nitrate or fumarate as the electron acceptor) or anaerobically via fermentation (Chapter 13). The bacteria adapt to their surroundings, and the electron transport system that the cells assemble reflects the electron acceptor that is available.

Aerobic respiratory chains
When grown aerobically, *E. coli* makes two different cytochrome oxidase complexes, cytochrome bo complex (cytochrome o) and cytochrome bd complex (cytochrome d), resulting in a branched respiratory chain to oxygen (Fig. 4.14).[22] In these pathways, electrons flow from ubiquinol to the terminal oxidase complex and, therefore, the oxidases are also called ubiquinol oxidases. There are four polypeptides tightly associated in the cytochrome bo complex. These contain two Cu atoms and two hemes of the b type. The cytochrome bo complex from *E. coli* is a proton pump, with a stoichiometry of $1H^+/e^-$. Cytochrome bo complex is the predominant cytochrome oxidase when the oxygen levels are high. When the oxygen tensions are lowered, *E. coli* makes cytochrome oxidase bd. Cytochrome bd has two polypeptide chains, two hemes of the b type, and one heme d. The cytochrome bd oxidase complex has a higher affinity for oxygen (lower K_m) than does the cytochrome bo complex, suggesting a rationale for why it is the dominant oxidase under low oxygen

(e) and (f), antimycin with myxothiazol, and antimycin with stigmatellin. (From Yang, X. and B. L. Trumpower. 1988. Protonmotive Q cycle pathway of electron transfer and energy transduction in the three-subunit ubiquinol:cytochrome c oxidoreductase complex of *Paracoccus denitrificans*. *J. Biol. Chem.* 263:1196–11970.)

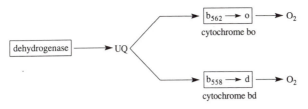

Fig. 4.14 Aerobic respiratory chain in *E. coli*. A variety of dehydrogenase complexes are possible (e.g., NADH dehydrogenase, succinate dehydrogenase, lactate dehydrogenase, and glycerol-3-phosphate dehydrogenase). The chain branches at the level of ubiquinone (UQ) to two alternate quinol oxidases, cytochrome bo and bd. The cytochrome bd complex has a higher affinity for oxygen and is synthesized under low oxygen tensions, where it becomes the major route to oxygen. Coupling sites (i.e., sites where protons are translocated to the outer surface) occur during NADH oxidation, catalyzed by NADH–ubiquinone oxidoreductase, also called NDH-1, and quinol oxidation, catalyzed by cytochrome bo complex or cytochrome bd complex. Additionally, cytochrome bo is a proton pump. *E. coli* has a second NADH dehydrogenase, called NADH-2, which does not translocate protons. Therefore, the number of protons translocated per NADH oxidized can vary from two (NADH-2 and bd complex) to eight (NADH-1 and bo complex).

tensions. Cytochrome bd is not a proton pump.

The bo pathway is more energetically efficient.[23] For every two electrons travelling from ubiquinol to oxygen via cytochrome bo, four protons are translocated (i.e., two in the Q loop and two by the cytochrome bo pump). Assuming a H^+/ATP of 3 (i.e., the ATP synthase translocates inwardly 3 protons per ATP made), then 4/3 or 1.33 ATPs can be made. Since cytochrome bd is not a proton pump, only 2/3 or 0.67 ATPs can be made per two electrons during quinol oxidation via this route.

Physiological significance of alternate electron routes that differ in the number of coupling sites

Since the NDH-1 dehydrogenase may translocate as many as 4 protons per $2\,e^-$, whereas the NDH-2 dehydrogenase translocates 0, the number of protons translocated per NADH oxidized can theoretically vary from 2 (NDH-2 and bd complex) to 8 (NDH-1 and bo complex). This means that the ATP yields per NADH oxidized can vary fourfold from 2/3 or 0.67 to 8/3 or 2.7. It also means that *E. coli* has great latitude in adjusting the Δp generated during respiration. Since a large Δp can drive reversed electron transport and thus slow down oxidation of NADH and quinol, it may be an advantage to be able to direct electrons along alternate routes that bypass coupling sites and translocate fewer protons.

This could ensure adequate rates of oxidation of NADH and quinol.

Anaerobic respiratory chains

In the absence of oxygen, *E. coli* can use either nitrate or fumarate as an electron acceptor (Fig. 4.15). This occurs only in the absence of oxygen because oxygen represses the synthesis of the nitrate and fumarate reductases. When presented with nitrate anaerobically, only the nitrate reductase is made because nitrate represses the synthesis of the fumarate reductase. (The regulation by oxygen and nitrate of the electron transport chain is discussed in Chapter 17.) Thus, there exists a hierarchy of electron acceptors for *E. coli* (i.e., oxygen > nitrate > fumarate). This may reflect the energy yields of the electron transport pathways, since oxygen has the highest electrode potential and fumarate the lowest.

4.7.2 Paracoccus denitrificans

P. denitrificans is a non-fermenting gram-negative facultative anaerobe that can obtain energy from either aerobic respiration or nitrate respiration.[24,25] The bacteria can grow heterotrophically on a wide variety of carbon sources, or autotrophically on H_2 and CO_2 under anaerobic conditions, using nitrate as the electron acceptor. It can be isolated from soil by anaerobic enrichment with media

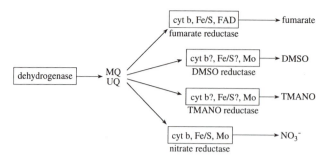

Fig. 4.15 Anaerobic respiratory chains in *E. coli*. When oxygen is absent, *E. coli* synthesizes any one of several membrane-bound reductase complexes depending upon the presence of the electron acceptors. Nitrate induces the synthesis of nitrate reductase and represses the synthesis of the other reductases. Menaquinone must be used to reduce some of the reductases (e.g., fumarate reductase), because it has a sufficiently low midpoint potential ($-74\,mV$). Ubiquinol can reduce nitrate reductase. Each reductase may be a complex of several proteins and prosthetic groups through which the electrons travel to the terminal electron acceptor. The transfer of electrons from the dehydrogenases to the reductases results in the establishment of a proton potential. If the dehydrogenase has site one activity, there can theoretically be two coupling sites: one at the dehydrogenase step, and one linked to quinol oxidation at the reductase step. cyt b, cytochrome b; Fe/S, nonheme iron–sulfur protein; FAD, flavoprotein with flavin adenine dinucleotide as the prosthetic group; Mo, molybdenum; TMANO, trimethylamine *N*-oxide; DMSO, dimethylsulfoxide; MQ, menaquinone; UQ, ubiquinone.

containing H_2 as the source of energy and electrons, Na_2CO_3 as the source of carbon, and nitrate as the electron acceptor. Electron transport in *P. denitrificans* receives a great deal of research attention because certain features closely resemble electron transport in mitochondria.

Aerobic pathway

P. denitrificans differs from *E. coli* in that, in addition to a cytochrome o pathway, it also has a bc$_1$ complex and a cytochrome aa$_3$ oxidase (cytochrome c oxidase) and, in this way, resembles mitochondria. The composition of the electron transport chain varies with the growth conditions; other oxidases can be present.[26] The aerobic pathway is thought by some to branch at ubiquinone (UQ) or perhaps at cytochrome c (Fig. 4.16).[27] It is not known to what extent the cytochrome o branch functions during aerobic growth. It has been suggested that the cytochrome o pathway serves as a Δp "release valve." The idea is that, under conditions where ATP consumption diminishes, electron flow through the two coupling sites (the bc$_1$ complex and cytochrome aa$_3$ complex) generates a Δp which can build to a point of retarding electron flow through the bc$_1$ complex.

(Recall that electron flow can be driven in the reverse direction by the Δp.) Under these conditions, the electrons would be shunted to cytochrome o. In this way, the rates of oxidation of NADH and ubiquinol could be maintained.

P. denitrificans has three coupling sites. Site 1 is the NADH dehydrogenase (NADH–ubiquinone oxidoreductase) and this translocates two to three protons per two electrons. Site 2 is the bc$_1$ complex which, due to the Q cycle, translocates four protons per two electrons. Site 3 is the cytochrome aa$_3$ oxidase which pumps two protons per two electrons across the membrane. As described later, *P. denitrificans* also oxidizes methanol and, in this case, the electrons enter at cytochrome c, thus bypassing the bc$_1$ site.

Anaerobic pathway

P. denitrificans can also grow anaerobically using nitrate as an electron acceptor, reducing it to nitrogen gas in a process called *denitrification* (Fig. 4.16).[28] During anaerobic growth on nitrate, the electron transport chain is very different than during aerobic growth. The cells contain cytochrome o, nitrate reductase, nitrite reductase, nitrous oxide reductase, and possibly nitric oxide re-

aerobic

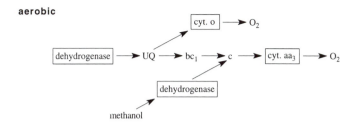

anaerobic

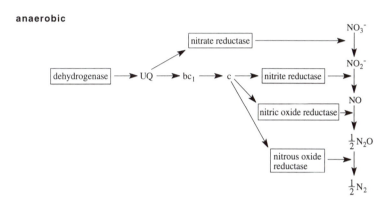

Fig. 4.16 A model for electron transport pathways in *Paracoccus denitrificans*. *Aerobic*: When using oxygen as an electron acceptor, *P. denitrificans* has an aerobic pathway, branched at ubiquinone, and leading either through a bc_1 complex to cytochrome aa_3 (cyt aa_3), or bypassing the bc_1 complex to cytochrome o (cyt o). Some researchers place the branch to cytochrome o at the level of cytochrome c. The ratio of cytochrome o to cytochrome aa_3 increases under low oxygen tensions. *Paracoccus* also uses methanol as an electron donor. The electrons from methanol are transferred to cytochrome c. *Anaerobic*: When the bacteria are grown anaerobically using nitrate as the electron acceptor, the cytochrome aa_3 levels are very low and the electrons travel from ubiquinone to nitrate reductase and also through the bc_1 complex to nitrite reductase, nitric oxide reductase, and nitrous oxide reductase. It has not been firmly established that nitric oxide is a free intermediate or that nitric oxide reductase exists. It is possible that nitrite reductase reduces nitrite to nitrous oxide.

ductase.[29,30] The levels of cytochrome aa_3 are very low. There are several different cytochromes c, some membrane bound and some in the periplasm.[27] The nitrate is reduced to nitrogen gas via a membrane-bound nitrate reductase, periplasmic nitrite reductase, and periplasmic nitrous oxide reductase. The reduction of two nitrites (NO_2^-) to nitrous oxide (N_2O) by nitrite reductase requires two 2-electron reductions per mole of N_2O. The electron transport pathway includes several branches to the individual reductases. Ubiquinol transfers electrons directly to nitrate reductase, bypassing the bc_1 complex and the c cytochromes. In agreement with the model,

electron flow to nitrate reductase is not sensitive to inhibitors of the bc_1 complex. However, electron flow from ubiquinol to the nitrite and nitrous oxide reductases is sensitive to bc_1 inhibitors, suggesting that the electrons travel through the bc_1 complex and cytochromes c. There are two coupling sites: site one is the NADH dehydrogenase and site two is the bc_1 complex.

Periplasmic oxidation of methanol
Many gram-negative bacteria oxidize substances in the periplasm and transfer the electrons to membrane-bound electron carriers, often via periplasmic cytochromes c. An

example is *P. denitrificans* which can grow aerobically on methanol (CH_3OH) by oxidizing it to formaldehyde (HCHO) and $2H^+$ with a periplasmic dehydrogenase, methanol dehydrogenase (Fig. 4.17).

$$CH_3OH \longrightarrow HCHO + 2H^+$$

(Growth on methanol is autotrophic since the formaldehyde is eventually oxidized to CO_2 which is assimilated via the Calvin cycle.) The electrons are transferred from the dehydrogenase to c-type cytochromes in the periplasm. The cytochromes c transfer the electrons to cytochrome aa_3 oxidase in the membrane, which reduces oxygen to water on the cytoplasmic surface. A Δp is established due to the inward flow of electrons and the outward pumping of protons by the cytochrome aa_3 oxidase, as well as the release of protons in the periplasm during methanol oxidation and consumption in the cytoplasm during oxygen reduction. Since the oxidation of methanol bypasses the bc_1 coupling site, the ATP yields are lower. In addition to

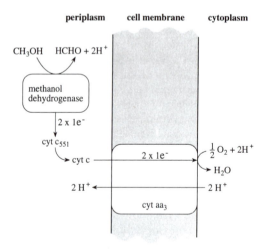

Fig. 4.17 Oxidation of methanol by *P. denitrificans*. Methanol is oxidized to formaldehyde by a periplasmic methanol dehydrogenase. The electrons are transferred to periplasmic cytochromes c and to a membrane-bound cytochrome aa_3 which is also a proton pump. A Δp is created due to the electrogenic influx of electrons and the electrogenic efflux of protons, accompanied by the release of protons in the periplasm (methanol oxidation) and uptake in the cytoplasm (oxygen reduction).

methanol, the bacteria can grow on methylamine ($CH_3NH_3^+$), which is oxidized by methylamine dehydrogenase to formaldehyde NH_4^+), and $2H^+$.

$$CH_3NH_3^+ + H_2O \rightarrow HCHO + NH_4^+ + 2H^+$$

Methylamine dehydrogenase is also located in the periplasm and donates electrons via cytochromes c to cytochrome aa_3, bypassing the bc_1 complex.

4.7.3 Electron transport chain in *Rhodobacter sphaeroides*

Rhodobacter sphaeroides is a purple photosynthetic bacterium that can be grown photoheterotrophically under anaerobic conditions or aerobically in the dark. When growing aerobically in the dark, the bacteria obtain energy from aerobic respiration. The respiratory chain resembles the mitochondrial and *P. denitrificans* respiratory chains, in that it consists of NADH–ubiquinone oxidoreductase (coupling site 1), a bc_1 complex (coupling site 2), and a cytochrome aa_3 complex (coupling site 3). As in *E. coli*, there are two NADH–ubiquinone oxidoreductases: NDH-1 and NDH-2.

4.7.4 Fumarate respiration in *Wolinella succinogenes*

Fumarate respiration occurs in a wide range of bacteria growing anaerobically.[31] This is probably because fumarate itself is formed from carbohydates and protein during growth. The following is a description of the electron transport pathway in *W. succinogenes*, a gram-negative anaerobe isolated from the rumen. *W. succinogens* can grow at the expense of H_2 or formate, both produced in the rumen by other bacteria. The electron transport pathway is shown in Fig. 4.18A. The active sites for both the hydrogenase and formate dehydrogenase are periplasmic, whereas the active site for the fumarate reductase is cytoplasmic. An examination of the topology of the components of the respiratory chain reveals how a Δp is generated.

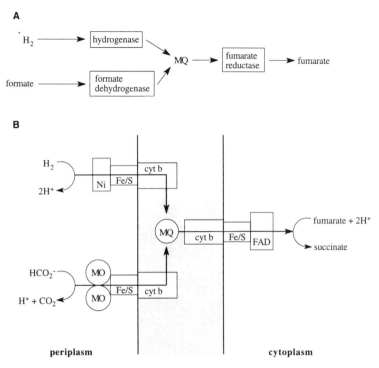

Fig. 4.18 A model for the electron transport system of *W. succinogenes*. (A) Electrons flow from H_2 and formate through menaquinone (MQ) to fumarate reductase. (B) Illustration showing that the catalytic portions of hydrogenase and formate dehydrogenase are periplasmic, whereas fumarate reductase reduces fumarate on the cytoplasmic side. Electrons flow electrogenically to fumarate. A Δp is created because of electrogenic influx of electrons together with the release of protons in the periplasm, and their consumption in the cytoplasm. (Modified from Kroger, A., V. Geisler, E. Lemma, F. Theis, and R. Lenger. 1992. Bacterial fumarate respiration. *Microbiol.* 158:311–314.)

Topology of the components of the electron transport pathway

The electron transport chain consists of a periplasmic enzyme that oxidizes the electron donor, a membrane-bound menaquinone (MQ) that serves as an intermediate electron carrier, and membrane-bound fumarate reductase which accepts the electrons from the menaquinone and reduces fumarate on the cytoplasmic side of the membrane (Fig. 4.18B). Both the hydrogenase and formate dehydrogenase are made of three polypeptide subunits, two facing the periplasm and one an integral membrane protein (cytochrome b). Note that cytochrome b serves not only as a conduit for electrons, but also binds the dehydrogenases into the membrane. The two periplasmic subunits of the hydrogenase are a Ni-containing protein subunit and an iron–sulfur protein. The two periplasmic subunits of the formate dehydrogenase are a Mo-containing protein subunit and an iron–sulfur protein. The fumarate reductase is a complex containing three subunits. One subunit of the fumarate reductase is a flavoprotein with FAD as the prosthetic group (subunit A). A second subunit has several FeS centers (subunit B). And the third subunit has two hemes of the b type (subunit C), which binds the fumarate reductase to the membrane. Fumarate reductase is similar in structure to succinate dehydrogenase isolated from several different sources, which catalyzes the oxidation of succinate to fumarate in the citric acid cycle.

Electron flow and the establishment of a Δp

Electron flow is from the dehydrogense or hydrogenase to cytochrome b to menaquinone to fumarate reductase. Two electrons are electrogenically transferred across the membrane to fumarate per H_2 or formate oxidized, leaving two H^+ on the outside, thus establishing a Δp. A value of 1.1 was obtained for the ratio of H^+/e^- during fumarate reduction when studying whole cells, suggesting perhaps that a mechanism of proton translocation through the membrane may also exist.[32] If one assumes that 1.1 is a correct number, and also assumes a stoichiometry for the ATP synthase of $3H^+/ATP$, then the theoretical maximum number of ATPs that can be formed from the transfer of two e^- to fumarate is 2.2/3, or 0.73. The actual number measured by experimentation was 0.56. Note that the quinone functions as an electron carrier between the cytochromes b, but does not take part in hydrogen translocation across the membrane as in a Q loop or Q cycle. This is because protons are released in the periplasm during the oxidation of the electron donor, and therefore there is no need to move protons from the cytoplasm to the periplasm. This is a common feature in periplasmic oxidations. In other bacteria, the electron donor and fumarate (the electron acceptor) are both on the cytoplasmic side of the membrane. For example, NADH, which is the electron donor for fumarate in many bacteria, is in the cytoplasm along with fumarate. In this situation, one would expect that the quinone becomes reduced on the cytoplasmic side and ferries the hydrogens across to the periplasmic side where they are released when the quinol is oxidized (Section 4.6.1).

4.8 Summary

All electron transport schemes can be viewed as consisting of membrane-bound dehydrogenase complexes, such as NADH dehydrogenase or succinate dehydrogenase. The dehydrogenases remove electrons from their substrates and transfer the electrons to quinones that, in turn, transfer the electrons to oxidase or reductase complexes. The latter complexes reduce the terminal electron acceptors. In contrast to mitochondria, which all have the same electron transport scheme, bacteria differ in the details of their electron transport pathways, although the broad outlines of all the electron transport schemes

are similar. In bacteria, the dehydrogenase, oxidase, and reductase complexes are sometimes referred to as modules because specific ones are synthesized under certain growth conditions and "plugged" into the respiratory pathway. For example, in facultative anaerobes such as E. coli, the oxidase modules are synthesized in an aerobic atmosphere and the reductase modules are synthesized under anaerobic conditions.

Other dehydrogenases besides NADH dehydrogenase and succinate dehydrogenase exist. These oxidize various electron donors (e.g., methanol, hydrogen, formate, H_2, glycerol, and so on), and are located in the periplasm or the cytoplasm. The coenzyme or prosthetic group for these soluble dehydrogenases vary (e.g., they may be NAD^+ or flavin). The electrons from the soluble dehydrogenase are transferred to one of the electron carriers (e.g., quinone or cytochrome).

An important difference between electron transport chains in bacteria and those in mitochondria is that the former are branched. Branching can occur at the level of quinone or cytochrome. The branches lead to different oxidases or reductases, depending upon whether the bacterium is growing aerobically or anaerobically. Many bacteria, including E. coli, transfer electrons from reduced quinone to cytochrome o (a type-b cytochrome), which is the major one used when oxygen levels are high, and to cytochrome d, which is used when oxygen becomes limiting. Other bacteria may have, in addition, or instead of, an electron transport pathway in which electrons travel from reduced quinone through a bc_1 complex, to cytochrome aa_3, which reduces oxygen. This is the same as the electron transport pathway found in mitochondria, although the carriers may not be identical. The alternative branches may differ in the number of coupling sites, and this could have important regulatory significance regarding the rates of oxidation of reduced electron carriers, as well as ATP yields.

Another difference from mitochondria is that bacteria can have either aerobic or anaerobic electron transport chains or, as is the case with facultative anaerobes such as E. coli, either can be present, depending upon the availability of oxgyen or alternative electron acceptors. A hierarchy of electron acceptors is used. For *E. coli*, oxygen is the preferred acceptor, followed by nitrate, and finally fumarate.

With respect to cytochrome c oxidase, there are two classes. Cytochrome aa_3 is the major class and has been reported in many bacteria, including *Paracoccus denitrificans*, *Nitrosomonas europaea*, *Pseudomonas AM1*, *Bacillus subtilis*, and *Rhodobacter sphaeroides*. A different cytochrome c oxidase, that consists of cytochrome o (b type) rather than cyt aa_3, has been reported for *Azotobacter vinelandii*, *Rhodobacter capsulata*, *Rhodobacter sphaeroides*, *Rhodobacter palustris*, and *Pseudomonas aeruginosa*.[33] Apparently, both classes of cytochrome c oxidase coexist in the same organism where the heme o containing oxidases serve as an alternate route to oxygen.

The main energetic purpose of the respiratory electron transport pathways is to convert a redox potential (ΔE_h) into a proton potential (Δp). This is done at coupling sites. A membrane potential is created by electrogenic influx of electrons, leaving the positively charged proton on the outside, or by electrogenic efflux of protons during proton pumping, leaving a negative charge on the inside. Influx of electrons occurs when oxidations take place on the periplasmic membrane surface or in the periplasm, and electrons move vectorially across the membrane to the cytoplasmic surface where reductions take place. There are two situations where this occurs. One is when the substrate (e.g., H_2, methanol, and so on) is oxidized by dehydrogenases in the periplasm, and the electrons move across the membrane to reduce the electron acceptor. The second situation is when quinones are reduced on the cytoplasmic side of the membrane, diffuse across the membrane, and are oxidized on the outside membrane surface. Quinone oxidation can be via a Q loop or a Q cycle.

A Q loop is an electron transport pathway in which a reduced quinone carries hydrogen to the outside surface of the cell membrane and releases $2H^+$ as a result of oxidation. The $2e^-$ return to the inner surface, via cytochromes, where they reduce an electron acceptor. The inward transfer of the electron is electrogenic (i.e., a membrane potential is created). In the Q loop, which is a linear pathway of electron flow, one H^+ is released per e^-. A more complicated pathway is called the Q cycle and results in two H^+ released per e^-. In the Q cycle, QH_2 gives up two H^+ and two e^-, but one of the electrons is recycled back to oxidized quinone. Thus, the ratio of H^+/e^- is two, rather than one. The Q cycle is more efficient since it results in more protons available for influx through the ATP synthase per electron, and thus can increase the yield of ATP.

A second method of coupling oxidation–reduction reactions to the establishment of a proton potential relies on some of the electron carriers acting as proton pumps. A membrane potential is created when protons are pumped electrogenically out of the cell. A well-established example is cytochrome aa_3 oxidase. However, there is evidence that other oxidases are also proton pumps, including cytochrome bo in *E. coli*. Possibly some NADH dehydrogenase complexes are also proton pumps.

Study Questions

1. What is it about the solubility properties and electrode potentials of quinones that make them suitable for their role in electron transport?

2. Design an experiment that can quantify the H^+/O for the cytochrome aa_3 reaction in proteoliposomes.

3. What are two features that distinguish the Q cycle from linear flow of electrons? How can you verify that the Q cycle is operating?

4. What is the relationship between H^+/O, H^+/ATP, and the number of ATPs that can be synthesized?

5. What is the relationship between the Δp and number of ATPs that can be synthesized?

6. Draw a schematic outline of three ways that a cell might create a membrane potential during respiration. You can include both periplasmic and cytoplasmic oxidations.

7. Assume a Δp of $-200\,mV$ and a H^+/ATP of 3, calculate what the ΔG_p must be in joules/mole if ΔG_p is in equilibrium with Δp. Assume that the temperature is 30°C, and that $\Delta G_o'$ is 37 kcal/mol. What will be the ratio of $ATP/[ADP][P_i]$?

8. What are the similarities and differences between bacterial respiratory pathways and the mitochondrial respiratory pathway?

9. Calculate the maximum number of ATPs that can be made per NADH oxidized by *E. coli* when using the NDH-1 and bo combination, and the NDH-2 and bd combination.

REFERENCES AND NOTES

1. As discussed in Chapter 3, some marine bacteria couple electron transport to the creation of a sodium ion gradient.

2. For example, the enzyme complex that oxidizes NADH and reduces quinone is an NADH–quinone oxidoreductase. It consists of a flavoprotein and iron–sulfur proteins. (Reduced quinone is called quinol.) An enzyme complex that oxidizes quinol and reduces cytochrome c is called a quinol–cytochrome c oxidoreductase. It consists of cytochromes and an iron–sulfur protein.

3. Here are some definitions that are useful to know. *Cofactors* are nonprotein molecules, either metals or organic, bound with varying degrees of affinity to enzymes and required for enzyme activity. *Coenzymes* are organic cofactors that shuttle back and forth between enzymes carrying electrons, hydrogen, or organic moieties (e.g., acyl groups). *Prosthetic* groups are cofactors (organic or inorganic) that are tightly bound to the protein and do not dissociate from the protein. Thus, one difference between a coenzyme and a prosthetic group is that the former shuttles between enzymes and the latter remains tightly bound to the enzyme. Coenzyme A (carrier of acyl groups) and NAD^+ (electron and hydrogen carrier) are examples of coenzymes, whereas FAD (electron and hydrogen carrier) and heme (electron carrier) are examples of prosthetic groups.

4. The first iron–sulfur protein identified was a soluble protein isolated from bacteria and called ferredoxin. A protein similar to ferredoxin with a low redox potential can be isolated from chloroplasts and mediates electron transfer to $NADP^+$ during noncyclic electron flow. Iron–sulfur proteins that have no other prosthetic groups are divided into four classes based upon the number of iron atoms per molecule and whether or not the sulfur is acid labile. Examples from the four different classes are: rubredoxin (no-acid labile sulfur), isolated from bacteria; high potential iron protein (HiPIP); isolated from photosynthetic bacteria; chloroplast ferredoxin; and various bacterial ferredoxins.

5. Electron spin resonance detects unpaired electrons such as the ones that exist in the FeS cluster. Monochromatic microwave radiation is absorbed by the unpaired electron when a magnetic field is applied. The size of the magnetic field required for radiation absorption depends upon the molecular environment of the unpaired electron. A spectrum is obtained by varying the magnetic field and keeping the frequency of the microwave radiation constant. The spectra differ for different FeS clusters. A spectroscopic constant, called the g value, is obtained which is characteristic of the FeS cluster in the protein.

6. Yagi, T., X. Xu, and A. Matsuno-Yagi. 1992. The energy-transducing NADH–quinone oxidoreductase (NDH-1) of *Paracoccus denitrificans*. *Biochim. Biophys. Acta* **1101**:181–183.

7. Finel, M. 1993. The proton-translocating NADH:Ubiquinone oxidoreductase: a discussion of selected topics. *J. Bioenerg. Biomemb.* **25**:357–366.

8. Ohnishi, T. 1993. NADH-quinone oxidoreductase, the most complex complex. *J. Bioenerg. Biomemb.* **25**:325–329.

9. Yagi, T., Yano, T, and A. Matsuno-Yagi. 1993. Characteristics of the energy-transducing NADH–quinone oxidoreductase of *Paracoccus denitrificans* as revealed by biochemical, biophysical, and molecular biological approaches. *J. Bioenerg. Biomemb.* **25**:339–345.

10. Reviewed in *J. Bioenerg. Biomemb.* **25**(4), 1993.

11. Right-side out vesicles are made by osmotic lysis of the spheroplasts or protoplasts.

12. The phosphorylation potential, ΔG_p, is dependent upon the standard free energy of formation of ATP and the actual ratios of ATP, ADP, and P_i, according to the following equation, $\Delta G_p = \Delta G_o' + RT\ln[ATP]/[ADP][P_i]$ J/mol. The free energy of formation is proportional to the equilibrium constant (i.e., $\Delta G_o' = -RT\ln K_{eq}'$.

13. Reviewed in Trumpower, B. L. 1990. Cytochrome bc1 complexes of microorganisms. *Microbiol. Rev.* **54**:101–129.

14. Verseveld, H. W. Van, K. Krab, and A. H. Stouthamer. 1981. Proton pump coupled to cytochrome c oxidase in *Paracoccus denitrificans*. *Biochem. Biophys. Acta* **635**:525–534.

15. Hinkle, P. C., M. a. Kumar, A. Resetar, and D. L. Harrs. 1991. Mechanistic stoichiometry of mitochondrial oxidative phosphorylation. *Biochemistry*. **30**:3576–3582.

16. Yagi, T., X. Xu, and A. Matsuno-Yagi. 1992. The energy-transducing NADH–quinone oxidoreductase (NDH-1) of *Paracoccus denitrificans*. *Biochim. Biophys. Acta* **1101**:181–183.

17. Finel, M. 1993. The proton-translocating NADH:ubiquinone oxidoreductase: A discussion of selected topics. *J. Bioenerg. Biomemb.* **25**:357–366.

18. Ohnishi, T. 1993. NADH–quinone oxidoreductase, the most complex complex. *J. Bioenerg. Biomemb.* **25**:325–329.

19. Yagi, T., Yano, T, and A. Matsuno-Yagi. 1993. Characteristics of the energy-transducing NADH–quinone oxidoreductase of *Paracoccus denitrificans* as revealed by biochemical, biophysical, and molecular biological approaches. *J. Bioenerg. Biomemb.* **25**:339–345.

20. Matsushita, K., T. Ohnishi, and H. R. Kaback. 1987. NADH-ubiquinone oxidoreductases of the *Escherichia coli* aerobic respiratory chain. *Biochemistry* **26**:7732–7737.

21. Sled, V. D., Freidrich, T., Leif, H., Weiss, H., Meinhardt, S. W., Fukumori, Y., Calhoun, M. W., Gennis, R. B., and T. Ohnishi. 1993. Bacterial NADH–quinone oxidoreductases: iron sulfur clusters and related problems. *J. Bioenerg. Biomemb.* **25**:347–356.

22. Poole, R. K. and W. J. Ingledew. 1987. Pathways of electrons to oxygen, pp. 170–200. In: Escherichia coli *and* Salmonella typhimurium, *Cellular and Molecular Biology,* Vol. 1. F. C. Neidhardt, J. L. Ingraham, K. B. Low, B. Magasanik, M. Schaechter and H. E. Umberger (eds.). ASM Press, Washington, DC.

23. Calhoun, M. W., K. L. Oden, R. B. Gennis, M. J. Teixeira de Mattos, and O. M. Neijssel. 1993. Energetic efficiency of *Escherichia coli*: Effects of mutations in components of the aerobic respiratory chain. *J. Bacteriol.* **175**:3020–3025.

24. Trumpower, B. L. 1990. Cytochrome bc1 complexes of microorganisms. *Microbiol. Rev.* **54**:101–129.

25. Verseveld, H. W., and A. H. Stouthamer. 1992. The genus *Paracoccus*, pp. 2321–2334. In: *The Prokaryotes*, Vol. III, 2nd ed. A. Balows, H. G. Truper, M. Dworkin, W. Harder, and K.-H. Schleifer (eds.). Springer-Verlag, Berlin.

26. For example, when the bacteria are grown under low oxygen tensions, cytochrome aa$_3$ is decreased, and cytochrome o, cytochrome cd (nitrite reductase with oxidase activity), and cytochrome a$_1$ are increased.

27. There are several cytochromes c (at least eight), some of which are in the membrane, and some in the periplasm.

28. Ferguson, S. J. 1987. Denitrification: a question of the control and organization of electron and ion transport. TIBS **12**:354–357.

29. Whether nitric oxide is a free intermediate and whether nitric oxide reductase exists as a separate enzyme is discussed by Ferguson. The evidence in favor of it being an intermediate is that small amounts are produced during the nitrite reductase reaction, and *P. denitrificans* can use exogenously added nitric oxide as a source of energy for solute uptake. The evidence against nitric oxide being an intermediate is that when $^{15}NO_2^-$ and ^{14}NO were added together to cells there was no detectable formation of ^{15}NO (or of $^{15,14}N_2$).

30. Nitrate reductase is a membrane-bound molybdenum protein with three subunits, α, β, γ. The γ subunit is a cytochrome b. It accepts the electrons from ubiquinol and transfers them to the α subunit which has both an FeS center and molybdenum. The α subunit reduces the nitrate. The molybenum is part of a cofactor (prosthetic group) called the molybdenum cofactor (Moco), which consists of molybdenum bound to a pterin. The Moco prosthetic group is in all molybdenum-containing proteins, except nitrogenase. [The molybdenum cofactor from nitrogenase contains non-heme iron and is called FeMo cofactor or FeMoco (Section 11.3.2).] Nitrite reductase is a periplasmic enzyme that reduces nitrate to either nitrous acid or to nitric oxide. There is some uncertainty as to whether nitric oxide is a free intermediate. Nitrite reductase from *P. denitrificans* has two identical subunits. These have hemes c and d$_1$. Nitrous oxide reductase is a periplasmic enzyme that reduces nitrous acid to nitrogen gas. It has two identical subunits containing Cu.

31. Reviewed in, Kroger, A., V. Geisler, E. Lemma, F. Theis, and R. Lenger. 1992. Bacterial fumarate respiration. *Microbiology* **158**: 311–314.

32. Reviewed in, Kroger, A., V. Geisler, E. Lemma, F. Theis, and R. Lenger. 1992. Bacterial fumarate respiration. *Microbiology* **158**: 311–314.

33. Anraku, Y. 1988. Bacterial electron transport chains, pp. 101–132. In: *Annual Review of Biochemistry*, Vol. 57. C. C. Richardson, P. D. Boyer, I. B. Dawid and and A. Meister. Annual Reviews, Inc., Palo Alto.

5

Photosynthesis

Photosynthesis is the conversion of light energy into chemical energy used for growth. Organisms that obtain most or all of their energy in this way are called phototrophic or photosynthetic. Depending upon the photosynthetic system, light energy can do one or both of the following: (1) In all photosynthetic systems, light energy can drive the phosphorylation of ADP to make ATP (called photophosphorylation). (2) In some photosynthetic systems, light can also drive the transfer of electrons from H_2O ($E_{m,7} = +820$ mV) to $NADP^+$ ($E_{m,7} = -320$ mV). The latter is referred to as the photoreduction of $NADP^+$. During photoreduction of $NADP^+$ the water becomes oxidized to oxygen gas and the $NADP^+$ becomes reduced to NADPH. The student will recognize that this is the reverse of the spontaneous direction in which electrons flow during aerobic respiration (Chapter 4). During both ATP and NADPH synthesis, electromagnetic energy is absorbed by photopigments in the photosynthetic membranes and converted into chemical energy. This chapter explains how this is done. At the heart of the process is the light-driven oxidation of chlorophyll or bacteriochlorophyll. This initiates electron transport that results in the generation of a Δp and subsequent synthesis of ATP and, in the case of chlorophyll, $NADP^+$ reduction. It is of interest that (bacterio)chlorophyll-based photosynthesis is widely distributed, being found in bacteria, green plants, and algae, but is not present in the known archaea. This chapter begins with backround information on the different groups of phototrophic prokaryotes, and then describes photosynthetic electron transport in all of the photosynthetic systems.

5.1 The phototrophic prokaryotes

The phototrophic prokaryotes are a diverse assemblage of organisms that share the common feature of being able to use light as a source of energy for growth. Their classification is based upon physiological differences, including whether they produce oxygen, and what they use as a source of electrons for biosynthesis. For example, there are both oxygenic phototrophs (produce oxygen) and anoxygenic phototrophs. Oxygenic phototrophs include the well-known *cyanobacteria* (formerly called blue–green algae) and members of the genera *Prochloron*, *Prochlorothrix*, and *Prochlorococcus*. The latter three genera consist of organisms phylogenetically related to the cyanobacteria but having chlorophyll b as a light-harvesting pigment instead of phycobilins. The anoxygenic phototrophs are the *purple photosynthetic bacteria*, the *green photosynthetic bacteria*, and the *heliobacteria*.[1] These photosynthesize only under anaerobic conditions. The purple and

Table 5.1 Phototrophic prokaryotes

Type	Pigments*	Electron donor	Carbon source	Aerobic dark growth
Oxygenic				
Cyanobacteria	Chl a, phycobilins	H_2O	CO_2	No
Prochloron	Chl a, Chl b	H_2O	CO_2	No
Prochlorococcus	Chl a, Chl b	H_2O	CO_2	No
Prochlorothrix	Chl a, Chl b	H_2O	CO_2	No
Anoxygenic				
Purple sulfur	Bchl a or Bchl b	$H_2S^\dagger$, S°, $S_2O_{3=}$, H_2, organic	CO_2, organic	Yes§
Purple nonsulfur	Bchl a or Bchl b	H_2, organic, H_2S(some)	CO_2, organic	Yes
Green sulfur	Bchl a and c, d, or e	$H_2S^\ddagger$, S°, $S_2O_{3=}$, H_2	CO2, organic	No
Green gliding	Bchl a and c or d	H_2S, H_2, organic	CO_2, organic	Yes
Heliobacteria	Bchl g	organic	Organic	No

*Carotenoids are usually present.
†Members of the genus *Chromatium* accumulate S° intracellularly. Members of the genus *Ectothiorhodospira* accumulate extracellular sulfur.
‡Accumulates S° extracellularly.
§In the natural habitat growth occurs primarily anaerobically in the light. However, several purple sulfur species (Chromatiaceae) can be grown continuously in the laboratory aerobically in the dark at low oxygen concentrations.
Source: Overmann, J., and N. Pfennig. 1992. Continuous chemotrophic growth and respiration of Chromoatiacèae species at low oxygen concentrations. *Arch. Microbiol.* **158**:59–67.

green photosynthetic bacteria are further subdivided according to whether they use sulfur as a source of electrons. Thus there are purple sulfur and nonsulfur photosynthetic bacteria and green sulfur and nonsulfur photosynthetic bacteria. The various phototrophic prokaryotes and some of their properties are summarized in Table 5.1.

5.1.1 Oxygenic phototrophs

The oxygenic prokaryotic phototrophs use H_2O as the electron donor for the photosynthetic reduction of $NADP^+$. Most of the species belong to the cyanobacteria which are widely distributed in nature, occurring in fresh and marine waters, and terrestrial habitats. They have only one type of chlorophyll (chlorophyll a) and light-harvesting pigments called phycobilins. Another kind of oxygenic microbial phototroph are the prochlorophytes, which are a diverse group of photosynthetic prokaryotes that are evolutionarily related to the cyanobacteria but differ from the latter in having chlorophyll b

rather than phycobilins as the light-harvesting pigment.[2] There are three genera among the prochlorophytes: *Prochloron*, *Prochlorothrix*, and *Prochlorococcus*. Members of the genus *Prochloron* are obligate symbionts of certain ascidians (sea squirts). *Prochlorothrix* is a flilamentous, free-living microorganism that lives in freshwater lakes. *Prochlorococcus* is a free-living marine microorganism.

5.1.2 Anoxygenic phototrophs

The other phototrophic prokaryotes do not produce oxygen (i.e., H_2O is not a source of electrons). Instead, they use organic compounds, inorganic sulfur compounds, or hydrogen gas as a source of electrons. *These organisms will grow phototrophically only anaerobically or when oxygen tensions are low.* There are four major groups of anoxygenic phototrophs (see Table 5.1 and note 1):

1. Purple photosynthetic bacteria, which includes both the purple sulfur and the purple nonsulfur bacteria;

2. Green sulfur photosynthetic bacteria;

3. Green nonsulfur photosynthetic bacteria, also called green gliding bacteria;

4. Heliobacteria.

Purple sulfur phototrophs

The purple sulfur photosynthetic bacteria grow photoautotrophically in anaerobic environments using hydrogen sulfide as the electron donor and CO_2 as the carbon source. Their natural habitats are freshwater lakes and ponds, or marine waters where the sulfide content is high due to sulfate-reducing bacteria. They oxidize sulfide to elemental sulfur, which accumulates as granules intracellularly in all known genera except in the genus *Ectothiorhodospira*, which deposits sulfur extracellularly. However, some can grow photoheterotrophically and several have been grown chemoautotrophically under low partial pressures of oxygen with reduced inorganic sulfur as an electron donor and energy source. There are many genera, including the well-studied *Chromatium*.

Purple nonsulfur phototrophs

The purple nonsulfur photosynthetic bacteria are extremely versatile with regard to sources of energy and carbon. Originally, it was thought that these bacteria were not able to utilize sulfide as a source of electrons, hence the name "nonsulfur." However, it turns out that the concentrations of sulfide used by the purple sulfur bacteria are toxic to the nonsulfur purples and, provided the concentrations are sufficiently low, some purple nonsulfur bacteria, (i.e., *Rhodopseudomonas* and *Rhodobacter*) can use sulfide and/or thiosulfate as a source of electrons during photoautotrophic growth. The purple nonsulfur phototrophs can grow photoautotrophically or photoheterotrophically in anaerobic environments. When growing photoautotrophically, they use H_2 as the electron donor and CO_2 as the source of carbon. Photoheterotrophic growth uses simple organic acids such as malate or succinate as the electron donor and source of carbon, and light as the source of energy. If these organisms are placed in the dark in the presence of oxygen, they will carry out an ordinary aerobic respiration and grow chemoheterotrophically (e.g., on succinate or malate). A few can even grow very slowly and fermentatively in the dark. Because of their physiological versatility, the purple nonsulfur photosynthetic bacteria have received much attention in research. They are found in lakes and ponds with a low sulfide content. Representative genera are *Rhodobacter* and *Rhodospirillum*.

Green sulfur phototrophs

The *green sulfur phototrophic bacteria* are strict anaerobic photoautotrophs that use H_2S, S^o, $S_2O_3^{2-}$ (thiosulfate), or H_2 as the electron donor and CO_2 as the source of carbon. The green sulfur bacteria coexist with the purple sulfur phototrophs in sulfide-rich anaerobic aquatic habitats, although they are found in a separate layer. Their light-harvesting pigments are located in special inclusion bodies called *chlorosomes*, which will be described later, and their reaction centers (that part of the photosynthetic membrane where the bacteriochlorophyll is oxidized) differ from those of the purple photosynthetic bacteria. There are several genera, including the well-studied *Chlorobium*.

Green nonsulfur phototrophs (green gliding bacteria)

The best-studied green nonsulfur photosynthetic bacterium is *Chloroflexus*, which is a thermophilic filamentous, gliding, green phototroph. *Chloroflexus* can be isolated from alkaline hot springs, whose pH can be as high as 10. Most isolates have a temperature optimum for growth from 52 to 60°C. It can be grown as a photoheterotroph, photoautotrophically with either H_2 or H_2S as the electron donor, or chemoheterotrophically in the dark in the presence of air. However, it really grows best as a photoheterotroph. *Chloroflexus* has chlorosomes, but its reaction center appears to differ from those in the green sulfur bacteria and more closely resembles the reaction centers in the purple bacteria.

Heliobacteria

The heliobacteria are recently discovered strictly anaerobic photoheterotrophs that differ from the other phototrophs in having bacterio-chlorophyll g as their reaction center chlorophyll, containing little carotenoids, and having no internal photosynthetic membranes or chlorosomes. The two known genera are *Heliobacterium* and *Heliobacillus*. The tetra-pyrrole portion of bacteriochlorophyll g is similar to chlorophyll a, and in fact isomerizes into chlorophyll a in the presence of air. Both 16S rRNA analyses and the lack of lipopoly-saccharide in the cell wall place these organisms with the gram-positive bacteria, and it has been suggested that they are phylogenetically related to the clostridia. Interestingly, endospores which resemble those produced by *Bacillus* or *Clostridium* are made by two species of *Heliobacterium*.

5.2 The purple photosynthetic bacteria

In all photosynthetic systems, light is absorbed by light-harvesting pigments (also called accessory pigments) and the energy is transferred to reaction centers where bacterio-chlorophyll is oxidized, creating a redox potential, ΔE. The light-harvesting pigments, the photosynthetic pigments in the reaction center, and the energy transfer to the reaction center are considered in Sections 5.6 and 5.7. The structure of the photosynthetic membrane systems in the prokaryotes as revealed by electron microscopy is described in Section 5.8. The generation of the ΔE and photosynthetic electron transport are described below.[3,4]

5.2.1 Photosynthetic electron transport

This section discusses the oxidation of bacteriochlorophyll, the generation of the ΔE, and electron transport. Photosynthetic electron flow in the purple photosynthetic bacteria is illustrated in Fig. 5.1A and summarized below. A detailed discussion of the reaction center is given in Section 5.2.2.

Step 1. A dimer of two bacteriochlorophyll molecules (P_{870}) in the reaction center absorbs light energy and one of its electrons becomes excited to a higher energy level (P_{870}^*). P_{870}^* has a very low reduction potential (E_m') in contrast to P_{870}.

Step 2. The electron is transferred to an acceptor molecule within the reaction center, thus oxidizing the P870 and reducing the acceptor molecule. The acceptor molecule is bacteriopheophytin (Bpheo), which is bacteriochlorophyll without magnesium. The light creates a $\Delta E_m'$ of about 1 V between P_{870}^+/P_{870}, which is around $+0.45$ V, and $Bpheo^+/Bpheo$ which is around -0.6 V.

Step 3. The electron is transferred from the bacteriopheophytin to a quinone called ubiquinone A (UQ_A). The E_m' for UQ_A/UQ_A^- is about -0.2 V.

Step 4. UQ_A transfers the electron to a ubiquinone in site B, called UQ_B. UQ_B becomes reduced to UQ_B^-. Steps 1–4 are repeated so that UQ_B accepts a second electron and becomes reduced to UQ_B^{2-}.

Step 5. UQ_B^{2-} picks up two protons from the cytoplasm and is released from the reaction center as UQH_2. The UQH_2 joins the quinone pool in the membrane.

Steps 6 and 7. UQH_2 transfers the electrons to a bc_1 complex and the bc_1 complex reduces cytochrome c_2. The bc_1 complex translocates protons to the cell surface thus generating a Δp. (See Section 4.5.3 for a discussion of the bc_1 complex and the Q cycle.)

Step 8. The cytochrome c_2 returns the electron to the oxidized bacteriochlorophyll molecule in the reaction center. Thus light energy causes a cyclic electric current in the membrane.

Steps 1–8 describe cyclic electron flow. We can follow the path of the electron starting with cytochrome c_2 (Fig. 5.1B). Notice that an electron moves "uphill" (to a lower electrode potential) from cytochrome c_2 to ubiquinone when "boosted" in the reaction center by the energy from a quantum of light. As pointed out above, in order to reduce UQ to UQH_2, two electrons and therefore two

A

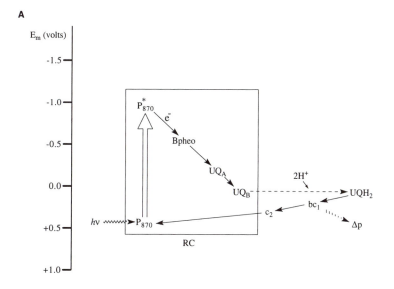

B

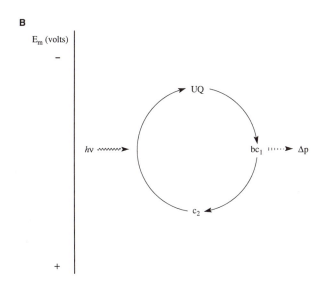

Fig. 5.1 Photosynthetic electron flow in purple photosynthetic bacteria. (A) Light energizes bacteriochlorophyll, here shown as P_{870} (bacteriochlorophyll a) in the reaction center (RC). (The number in the subscript refers to the major long wavelength absorption peak.) Some purple photosynthetic bacteria have bacteriochlorophyll b instead, which absorbs at 1020–1035 nm. The energized bacteriochlorophyll reduces bacteriopheophytin (Bpheo), which is bacteriochlorophyll without Mg^{2+}. The electron then travels through two ubiquinones, UQ_A and UQ_B, in the reaction center. (Some species use menaquinone.) After a second light reaction, UQ_B becomes reduced to UQH_2 and leaves the reaction center. The electron returns to the reaction center via a proton-translocating bc_1 complex and cytochrome c_2. ATP is synthesized via a membrane ATP synthase driven by the Δp (not shown). Midpoint potentials at pH 7 (approximate): $Bchl_{870}$ (P_{870}), +450 mV; Bpheo, −600 mV; UQ_A, −200 mV; UQ_B, +80 mV; c cytochromes, +380 mV. (B) A simplified version of A that emphasizes the cyclic flow of electrons. (Data from Mathis, P. 1990. Compared structure of plant and bacterial photosynthetic reaction centers. Evolutionary implications. *Biochem. Biophys. Acta* **1018**:163–167.)

light reactions are required. This can be written as follows:

eq. 5.1 $2\,\mathrm{cyt}\,c_2(\mathrm{red}) + 2H^+ + UQ$

$$\xrightarrow{\text{2 quanta of light}} 2\,\mathrm{cyt}\,c_2(\mathrm{ox}) + UQH_2$$

Because two electrons are required before reduced quinone leaves the reaction center, UQ_B is referred to as a *two-electron gate*. As illustrated in Fig. 5.1, the reduced quinone transfers the electrons to cytochrome c_2 via a bc_1 complex outside of the reaction center. The bc_1 complex translocates protons across the membrane via a Q cycle:

eq. 5.2 $UQH_2 + 2H^+{}_{in} + 2\,\mathrm{cyt}\,c_2(\mathrm{ox})$

$$\xrightarrow{bc_1} UQ + 4H^+{}_{out} + 2\,\mathrm{cyt}\,c_2(\mathrm{red})$$

The Δp that is generated is used to drive the synthesis of ATP via a membrane-bound ATP synthase:

eq. 5.3 $ADP + P_1 + 3H^+{}_{out}$

$$\xrightarrow{\text{ATP synthase}} ATP + H_2O + 3H^+{}_{in}$$

Thus, the net result of photosynthesis by purple photosynthetic bacteria is the synthesis of ATP. This is called *cyclic photophosphorylation*.

The actual amount of ATP made depends upon the number of protons translocated by the bc_1 complex and the number of protons that enter via the ATP synthase. For example, if four protons per two electron transfers are translocated out of the cell by the bc_1 complex and three protons re-enter via the ATP synthase, then the maximum number of ATP molecules made per two electrons is 4/3.

Figure 5.2 illustrates the topographical relationships between the reaction center, the bc_1 complex, and the ATP synthase in photosynthetic membranes and can be compared to Fig. 5.1, which illustrates the sequential steps in electron transfer. In the reaction center, electrons travel across the membrane from a periplasmic cytochrome c_2 to the quinone at site A. This creates a membane potential, which is negative inside.

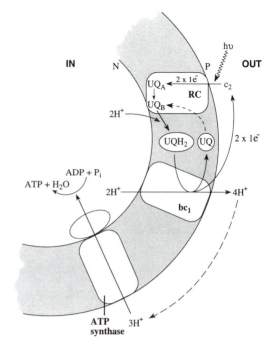

Fig. 5.2 Relationship between reaction center, bc_1 complex, and ATP synthase in photosynthetic membranes. The reaction center (RC) takes two protons from the cytoplasm and two electrons from a periplasmic cytochrome c to reduce ubiquinone (UQ) to UQH_2. The UQH_2 leaves the reaction center and diffuses to the bc_1 complex. The bc_1 complex oxidizes UQH_2 and translocates four protons to the periplasmic surface in a Q cycle, which is described in Section 4.5.3. The electrons travel back to cytochrome c_2 (cyclic flow). The translocated protons re-enter via the ATP synthase which makes ATP.

After a second electron is transferred, the quinone at site B accepts two protons from the cytoplasm and leaves the reaction center as UQH_2. The UQH_2 diffuses through the lipid matrix to the bc_1 complex where it is oxidized. The bc_1 complex translocates four protons to the outside per UQH_2 oxidized via a Q cycle, creating a Δp. The electrons are transferred from the bc_1 complex to a periplasmic cytochrome c_2 which returns them to the reaction center. The proton circuit is completed when the protons are returned to the cytoplasmic side via the ATP synthase accompanied by the synthesis of ATP. *Thus, light is used to drive an electron circuit, which drives a proton circuit, that drives ATP synthesis.* This successful method for harnessing light energy to do chemical work is used in all phototrophic organisms from bacteria to plants.

5.2.2 A more detailed examination of the reaction center and what happens there

The description given above for electron flow in the reaction center is incomplete in that it does not include all of the components and electron carriers, or the structure of the reaction center. A more complete description is given below.

Structure and composition of the reaction center

The reaction centers from *Rhodopseudomonas viridis* and *Rhodobacter sphaeroides* have been crystallized, and the structures determined from high-resolution X-ray diffraction studies.[5,6] The reaction centers are similar. The following proteins and pigments are found in the reaction center from *R. sphaeroides*:

1. *Reaction center protein.* The reaction center protein has 11 membrane-spanning alpha helices and a globular portion on the cytoplasmic side of the membrane. *It consists of three polypeptides: H, L, and M.* The reaction center protein serves as a scaffolding to which the bacteriochlorophyll and bacteriopheophytin are attached, and to which the quinones and non-heme iron are bound.

2. *Four bacteriochlorophyll molecules (Bchl).* There are four bacteriochlorophyll molecules in the reaction center. Two of the bacteriochlorophyll molecules exist as a dimer $(Bchl)_2$, and two are monomers ($Bchl_A$ and $Bchl_B$).

3. *Two molecules of bacteriopheophytin ($Bpheo_A$ and $Bpheo_B$), which is Bchl without Mg^{2+}.*

4. *Two molecules of ubiquinone (UQ_A and UQ_B).*

5. *One molecule of nonheme ferrous iron (Fe).*

6. *One carotenoid molecule.* The function of the carotenoid is to protect the reaction center pigments from photodestruction.[7,8]

Electron transfer in the reaction center

The sequence of redox reactions in the reaction center summarized in Fig. 5.1 (boxed area) is shown in more detail in Fig. 5.3, along with the proposed time scale for the electron transfer events. As shown in Fig. 5.3, the arrangement of the pigment molecules and the quinones shows two-fold symmetry with a right and left half very similar. On the periplasmic side of the membrane, there sits a pair of bacteriochlorophyll molecules—$(Bchl)_2$ (Fig. 5.3). These bacteriochlorophyll molecules are P_{870} in Fig. 5.1. When the energy from a quantum of light is absorbed by $(Bchl)_2$, an electron becomes excited [i.e., $(Bchl)_2$ becomes $(Bchl)_2^*$] (Fig. 5.3 step 1). The redox potential of $(Bchl)_2^*$ is sufficiently low so that it reduces bacteriochlorophyll A ($Bchl_A$), forming $(Bchl)_2^+$ and $Bchl_A^-$ (Fig. 5.3 step 2). The electron then moves to bacteriopheophytin$_A$ ($Bpheo_A$), forming $Bpheo_A^-$ (Fig 5.3 step 3). (As reflected by the question mark in step 2 of Fig. 5.3, the exact route of the electron is not known. It has not been unequivocally demonstrated that the electron moves through the $Bchl_A$ monomer. Some researchers suggest that the electron travels from $Bchl_2^*$ to $Bchl_A$ and very quickly moves on to $Bpheo_A$, whereas other investigators postulate that the electron moves directly from $(Bchl^*)_2$ to $Bpheo_A$.) Then the electron moves to ubiquinone bound to site A on the cytoplasmic side of the reaction center (UQ_A) forming the semiquinone, UQ_A^- (Fig. 5.3 step 4). At this time, an electron is returned to $(Bchl)_2^+$ from reduced cytochrome c_2 which is periplasmic. Then, the electron is transferred from UQ_A^- to UQ_B, to form UQ_B^- (Fig. 5.3 step 6). Electron flow is therefore across the membrane from periplasmic c_2 to UQ_B located on the cytoplasmic side. This generates a membrane potential, outside positive. A second light reaction occurs and steps 1 through 6 in Fig. 5.3 are repeated so that UQ_B has two electrons (UQ_B^{2-}). Two protons are picked up from the cytoplasm to form UQH_2 which leaves the reaction center to

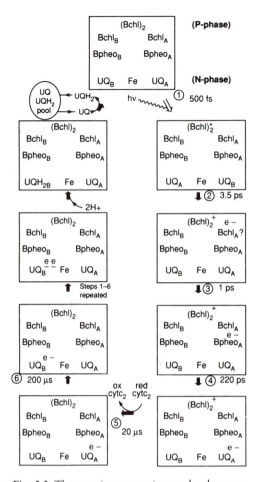

join the quinone pool in the membrane. (Although Fig. 5.3 indicates that two protons enter after both electrons arrive at UQ_B, it has been suggested that one proton enters after UQ_B receives the first electron.) Eventually, the protons carried by UQH_2 are released on the periplasmic side during oxidation of UQH_2 by the bc_1 complex. Thus, the reaction center and the bc_1 complex cooperate to translocate protons to the outside. As previously explained, because UQ_B cannot leave the reaction center until it has accepted two electrons, it is called a *two-electron gate*. The length of time that it takes the electron to travel from $(Bchl)_2$ to UQ_A is a little more than 200 psec. (One picosecond is one trillionth of a second, or 10^{-12} sec.) The rates of subsequent steps are slower (but still very fast) and measured in microseconds. (One microsecond is a millionth of a second.) Determining the pattern and timing of electron transfer in the reaction center requires the use of picosecond laser pulses and very rapid recording of the absorption spectra of the electron carriers. Interestingly, only one side of the reaction center (the A side) appears to be involved in electron transport. For example, there is photoreduction of only one of the bacteriopheophytin molecules. The reason why only one branch appears to function in electron transport is not known.

Fig. 5.3 The reaction center in purple photosynthetic bacteria. The reaction center, which spans the membrane, is represented by the boxed area. The absorption of light creates a transient membrane potential, positive (P-phase) on the periplasmic side and negative (N-phase) on the cytoplasmic side. Step 1: Absorption of a photon of light energizes a bacteriochlorophyl dimer $(Bchl)_2$. Step 2: The energized $(Bchl)_2$ reduces $Bchl_A$. The question mark refers to the fact that this has not been unequivocally demonstrated. Step 3: Bacteriopheophytin A $(Bpheo_A)$ is reduced. Step 4: The electron is transferred to a quinone (UQ_A). Step 5: The oxidized $(Bchl)_2$ is reduced via cytochrome c_2. Step 6: The electron moves from UQ_A to UQ_B. Steps 1–6 are repeated forming UQ_B^{2-}. Two protons are acquired from the cytoplasm to produce UQH_2 which enters the reduced quinone pool in the membrane and returns the electrons to oxidized cytochrome c_2 via the bc_1 complex. It may be that one proton is acquired by UQ_B^- and the second proton is acquired when the second electron arrives. (From Nicholls, D. G. and S. J. Ferguson. 1992. *Bioenergetics 2*. Academic Press, London.)

The contribution to the Δp by the reaction center

When reaction centers absorb light energy, a $\Delta \Psi$ and a proton gradient are created. The reason for the $\Delta \Psi$ is that the electrons move electrogenically from cytochrome c_2 to UQ_B across the membrane from outside to inside (Fig. 5.2). (This will produce a $\Delta \Psi$, outside positive, because the inward movement of a negative charge is equivalent to the outward movement of a positive charge.) The creation of the membrane potential is detected by a shift in the absorption spectra of membrane carotenoids.[9] The value of this technique is that very rapid changes in membrane potential can be monitored. The $\Delta \Psi$ produced by the reaction center is delocalized and increases the $\Delta \Psi$ made in the bc_1 complex

(Section 5.2.). Additionally, two protons are taken from the cytoplasm to reduce UQ_B to UQH_2. Eventually, the two protons will be released on the outside during the oxidation of UQH_2 by the bc_1 complex. Thus, the reaction center contributes to both the $\Delta\Psi$ and the ΔpH components of the Δp, as well as creating a ΔE between UQ/UQH_2 and $c_{2,ox}/c_{2,red}$.

5.2.3 Source of electrons for growth

In order to grow, all organisms must reduce $NAD(P)^+$ to $NAD(P)H$. That is because NADH and NADPH are the electron donors for almost all of the biosynthetic reactions in the cell, including the reduction of carbon dioxide to carbohydrate. When the electron donor is of a higher potential than the $NAD^+/NADH$ couple, then energy is required to reduce NAD^+. For example, this is the case for the purple photosynthetic bacteria that use certain inorganic sulfur compounds or succinate as a source of electrons. The purple photosynthetic bacteria drive electron transport in reverse using the Δp created by light energy (Fig. 5.4). During reversed electron flow, ubiquinol reduces NAD^+ via the NADH–ubiquinone oxidoreductase. (See Section 3.7.1 for a discussion of reversed electron transport.) Figure 5.4 illustrates the situation where electrons from the electron donor do not pass through the reaction center but simply enter the electron transport chain and travel directly to NAD^+ via ubiquinone. That is to say, there is no noncyclic electron transport. Many investigators hold this view. However, another scenario has been suggested, which is useful to consider because it illustrates the role that an increase in the Δp can play in slowing the rate of electron transfer. It should be recalled that ubiquinone can also accept electrons from bacteriopheophytin during cyclic electron transport, and reduce the bc_1 complex which generates a Δp (Fig. 5.1). It has been suggested that as the Δp grows larger, it might exert "backpressure" on the oxidation of ubiquinol by the bc_1 complex, thus slowing down the oxidation of ubiquinol via this route and making it available for NAD^+ reduction via reversed electron transport.[10] To the extent that this might occur, the electron donor (succinate or inorganic sulfur compounds) would replenish electrons to the bacteriochlorophyll via either ubiquinone or cytochrome c.

5.3 The green sulfur bacteria

5.3.1 Electron transport

The green sulfur photosynthetic bacteria have reaction centers that are distinguished from the reaction centers of the purple photosynthetic bacteria by:

1. Reducing $NAD(P)^+$ instead of quinone;

2. Possessing iron–sulfur centers as reaction center intermediate electron carriers.

However, the principles underlying the transformation of electrochemical energy into a ΔE and a membrane potential are the same as for the purple photosynthetic bacteria.[11,12] Light energizes a bacteriochlorophyll a molecule (P_{840}) which reduces a primary acceptor (A_o), establishing a redox potential difference greater than 1 V (Fig. 5.5). The primary electron acceptor (A_o) in the green sulfur bacteria has recently been reported to be an isomer of chlorophyll a called bacteriochlorophyll 663.[13] It might be added that the

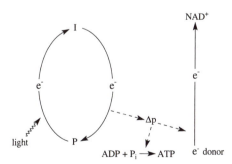

Fig. 5.4 Relationship between cyclic electron flow and reversed electron transport in the purple photosynthetic bacteria. The electron is driven by light from bacteriochlorophyll (P) to bacteriopheophytin (I), and then returns via electron carriers to bacteriochlorophyll with production of a Δp. The Δp is used to drive ATP synthesis as well as reversed electron transport.

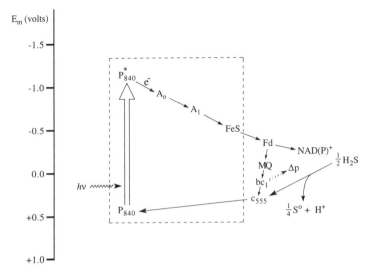

Fig. 5.5 A model for electron flow in the green sulfur bacteria. Both cyclic and noncyclic flow of electrons are possible. Reaction center bacteriochlorophyll a (P_{840}) becomes energized and reduces A_o, which is bacteriochlorophyll 663. The electron then travels to A_1, a quinone-like molecule, and then through two or three iron–sulfur centers (FeS). The first reduced product outside of the reaction center is the iron–sulfur protein, ferredoxin (Fd). The ferredoxin reduces $NAD(P)^+$ in the noncyclic pathway. Cyclic flow occurs when the electron reduces menaquinone (MQ) instead of $NAD(P)^+$, and returns to the reaction center via a bc_1 complex and cytochrome c_{555}. A Δp is created in the bc_1 complex. In the noncyclic pathway the electron donor is a reduced inorganic sulfur compound, here shown as hydrogen sulfide. Elemental sulfur or thiosulfate can also be used. The inorganic sulfur is oxidized by cytochrome c_{555}, which feeds electrons into the reaction center. It is possible that the electrons from the sulfide may instead enter at the menaquinone level, thus going through the bc_1 complex.

primary acceptor of heliobacteria is also a chlorophyll a derivative, hydroxychlorophyll a, reflecting the similarities known to exist between the reaction centers of the green sulfur bacteria, heliobacteria, and photosystem I of chloroplasts and cyanobacteria.[14] The electron flows from A_o to a quinone-like acceptor called A_1. The electron is then transferred from A_1 through two or three iron–sulfur centers to ferredoxin. There can be both cyclic and noncyclic electron flow. In cyclic flow, the electron returns to the reaction center via menaquinone (MQ) and a bc_1 complex, creating a Δp. In noncyclic electron flow, inorganic sulfur donates electrons which travel through the reaction center to NAD^+. According to the scheme in Fig. 5.5, electrons from inorganic sulfur enter at the level of cytochrome c and the bc_1 complex is bypassed. This is a widely held view based upon available data using isolated oxidoreductases. From an energetic point of view,

it is wasteful because a coupling site is bypassed, even though the E'_m values of some of the sulfur couples are sufficiently low to reduce menaquinone. For example, the E'_m for sulfur/sulfide ($n=2$) is -0.27, for sulfite/sulfide ($n=6$) it is -0.11 V, and for sulfate/sulfite ($n=2$) it is -0.54 V. They are all at a potential low enough to reduce menaquinone, which has an E'_m of -0.074 V. However, the point of entry of the electron is still an unresolved issue.[15]

Summary of photosynthesis by green sulfur bacteria (Chlorobiaceae)

$H_2S + NAD^+ + ADP + P_i$

light and bchl
$$\longrightarrow S^\circ + NADH + H^+ + ATP$$

(These organisms can also use elemental sulfur, oxidizing it to sulfate.)

5.4 Cyanobacteria and chloroplasts

Photosynthesis in cyanobacteria and chloroplasts differs from photosynthesis discussed thus far in three important respects:

1. H_2O is the electron donor and oxygen is evolved;

2. There are two light reactions in series, hence two different reaction centers;

3. Electron flow is primarily noncyclic, producing both ATP and NADPH. (Noncyclic flow also occurs in the green sulfur bacteria as described in Section 5.3.1.)

5.4.1 Two light reactions

As we shall see, chloroplasts and cyanobacteria have essentially combined the light reactions of purple photosynthetic bacteria and green sulfur photosynthetic bacteria in series, so that two light reactions energize a single electron that energizes ATP synthesis and reduces $NADP^+$. The initial evidence for two light reactions came from early studies of photosynthesis performed with algae by Emerson and his colleagues.[16,17] They observed that the efficiency of photosynthesis, measured as the moles of oxygen evolved per einstein absorbed (i.e., the *quantum yield*) is high over all of the wavelengths absorbed by chlorophyll and the light-harvesting pigments, but drops off sharply at 685 nm despite the fact that chlorophyll continues to absorb light between 680 and 700 nm (Fig. 5.6). This became known as the "red drop" effect because 700 nm light is red. One can restore the efficiency of photosynthesis of 700 nm light by supplementing it with light at shorter wavelengths (e.g., 600 nm light). The explanation for this is that there are two reaction centers, one of which is energized by light at a wavelength of around 700 nm, called reaction center I, and one of which is energized by lower wavelengths of light, called reaction center II. Reaction centers I and II operate in series and therefore both must be energized to maintain the electron flow from water to $NADP^+$. The lower wavelengths of light can energize both reaction centers, but 700 nm light can energize only reaction center I. It is for this reason that 700 nm light is effective only when given in combination with supplemental doses of shorter wavelengths.

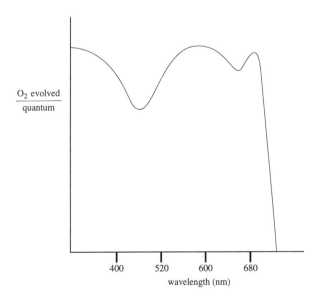

$\dfrac{O_2 \text{ evolved}}{\text{quantum}}$

wavelength (nm)

400   520   600   680

Fig. 5.6 Quantum yield of photosynthesis. When the rate of O_2 evolution per quantum absorbed is plotted against wavelength, the rate drops off sharply above 680 nm. (After Stryer, L. 1988. *Biochemistry.* W. H. Freeman and Co., New York.)

5.4.2 Electron transport

A schematic drawing of the overall pattern of electron flow in cyanobacteria and chloroplasts is shown in Fig. 5.7. These systems have two reaction centers, RC I and RC II, connected by a short electron transport chain that includes the analog to the bc_1 complex (i.e., the b_6f complex). Let us begin with reaction center II, which functions like the reaction center in the purple photosynthetic bacteria. Chlorophyll a, with a major absorption peak at 680 nm (i.e., P_{680}—probably a dimer), becomes energized and reduces an acceptor molecule, which is pheophytin (pheo). Having lost an electron, the P_{680}^+/P_{680} couple is of a sufficiently high redox potential (estimated at about $+1.1$ V) to replace the lost electron with one from water, thus oxidizing $1/2 H_2O$ to $1/4 O_2$. The energized electron travels to chlorophyll pheophytin which has an E'_m of about -0.6 V. The pheophytin reduces plastoquinone (PQ) which is structurally similar to ubiquinone (Fig. 4.3). A two-electron gate, similar to the two-electron gate in the reaction center of the purple photosynthetic bacteria, operates during quinone reduction. Electrons leave the reaction center in PQH_2 and are transferred through a b_6f complex (structurally and functionally similar to the bc_1 complex, except that cytochrome f replaces cytochrome c_1 and there are some differences in the cytochrome b) to a copper-containing protein called plastocyanin (Pc). A Δp is created by the b_6f complex, but there is some controversy concerning whether the b_6f complex catalyzes a Q cycle.[18] The plastocyanin reduces P_{700}^+ in reaction center I, which was previously oxidized by a light reaction described next.

Reaction center I is similar to the reaction center in the green photosynthetic bacteria. A photon of light energizes P_{700}, which is chlorophyll a with a major absorption peak at 700 nm. The P_{700}^* reduces a chlorophyll a molecule, called A_o. For this initial redox reaction, the energized electron travels from an E'_m of about $+0.5$ V (P_{700}^+/P_{700}) to an E'_m of about -1.0 V (A_o/A_o^-). Note that this is a far more negative potential than that generated in reaction center II. From A_o the electrons travel to A_1, which is a phylloqui-

none. Phylloquinones have a structure similar to menaquinone, but with only one double bond in the isoprenoid chain. (See Fig. 4.3 for the structure of menaquinone.) The electron is transferred from the quinone through several iron–sulfur centers (FeS) which reduce the iron–sulfur protein, ferredoxin (Fd) that is outside of the reaction center. Ferredoxin in turn reduces $NADP^+$. Thus, the two light reactions in series energize electron flow from H_2O to $NADP^+$, which is over a net potential difference of about 1.1 V. This is called noncyclic electron flow because the electron never returns to the reaction center. However, cyclic electron flow is possible. There is a branch point at the ferredoxin step and it is possible for the electron to cycle back to reaction center I via the b_6f complex, augmenting the Δp, rather than reducing $NADP^+$. This may be a way to increase the relative amounts of ATP to NADPH. The Calvin cycle, which is the pathway for reducing CO_2 to carbohydrate in oxygenic phototrophs, requires three ATPs per two NADPHs in order to reduce one CO_2 to the level of carbohydrate.

Photosynthesis by green plants, algae, cyanobacteria

$H_2O + NADP^+ + ADP + P_i$

$$\xrightarrow[\text{light and chl}]{} 1/2\,O_2 + NADPH + H^+ + ATP$$

5.5 Efficiency of photosynthesis

The efficiencies of photosynthesis based upon input light energy and products of photosynthesis are calculated below as an exercise. The calculated efficiencies are only approximations based upon assumptions regarding ATP yields, actual redox potentials, standard free energies, and so on.

5.5.1 ATP synthesis

Basically, photosynthesis is work done by energized electrons. The work is the phosphorylation of ADP and the reduction of $NAD(P)^+$. Each electron is energized by a photon (quantum) of light, and each mole of electrons by an einstein (6.023×10^{23} quanta) of light. It is instructive to ask how much of this energy is conserved in ATP and NADPH. Let us consider the synthesis of ATP. Assume that each energized electron results in the translocation of two protons (e.g., during the Q cycle in the bc_1 complex) but that three protons must re-enter through the ATP synthase to make one ATP. As mentioned before, a value of $3H^+/ATP$ is reasonable in light of experimental data. Thus, each energized electron (or each photon) results in the synthesis of 2/3 of an ATP. How much energy is required to synthesize 2/3 of an ATP? The ΔG_p for the phosphorylation of one mole of ADP to make ATP is about 45 kJ. Therefore, 2/3 of a mole of ATP should require approximately $+45$ (2/3) kJ$=30$ kJ. An einstein of 870 nm light, which corresponds to the absorption maximum of bacteriochlorophyll a (found in purple phototrophs), has 138 kJ of energy. Therefore, the efficiency is (30/138)(100) or about 22%. One can also calculate the efficiency using electron volts instead of joules. The energy in a photon of light at 870 nm is 1.43 eV. The synthesis of 2/3 of a mole of ATP requires 30,000 J. Dividing this number by the faraday gives the energy in electron volts (i.e., 0.31 eV).

5.5.2 ATP and NADPH synthesis

What about photosystems that reduce $NADP^+$ as well as make ATP (i.e., photosystems I and II)? The E_m' for O_2/H_2O is $+0.82$ V and for $NADP^+/NADPH$ it is -0.32 V. Therefore, the energized electron must have $0.82-(-0.32)$ or 1.14 eV to move from water to $NADP^+$. (The answer would not be very different if E_h values were used instead of E_m'.) If 2/3 of an ATP are made, then a total of $0.31+1.14=1.45$ eV would be required to make 2/3 of an ATP and 1/2 of an NADPH. A photon of light at wavelength 680 nm (the major long-wave absorption peak of chlorophyll a in reaction center II) has 1.82 eV. Two photons, or the equivalent of about 3.6 eV are used. Therefore, approximately 40% of the light energy is conserved as ATP and NADPH.

5.5.3 Carbohydrate synthesis and oxygen production

One can also estimate the approximate efficiency by considering the number of light quanta required to produce oxygen.

$$6CO_2 + 12H_2O \longrightarrow C_6H_{12}O_6 + 6H_2O + 6O_2$$

$$\Delta G_o' = 2870 \text{ kJ}$$

Therefore, per mole of O_2 produced, the standard free energy requirement at pH 7 is 2870/6 or 478 kJ. The number of einsteins of light required to produce one mole of O_2 is eight (Fig. 5.7). An einstein of 680 nm light carries 176 kJ of energy. Therefore, 176×8 or 1408 kJ of light energy are used to produce one mole of O_2. The efficiency is thus (478/1408)(100) or 34%.

5.6 Photosynthetic pigments

5.6.1 Light-harvesting pigments

The photosynthetic pigments are divided into two categories. They are *reaction center pigments* (primarily chlorophylls) and *light-harvesting pigments* (carotenoids, phycobilins, chlorophylls). The light-harvesting pigments are sometimes called accessory pigments or antennae pigments. By far, most of the photosynthetic pigments are light-harvesting pigments. The light-harvesting pigments are critical to photosynthesis because they absorb light of different wavelengths and funnel the energy to the reaction center. Whole cell absorption spectra of a purple photoysynthetic bacterium and a cyanobacterium showing the absorption wavelengths of the pigments are shown in Fig. 5.8, and the various pigments and their absorption peaks are listed in Table 5.2. The nature of the light-harvesting pigments will vary with the type of organism, but some important generalizations about them can be made:

1. They absorb light at wavelengths different from reaction center chlorophyll and

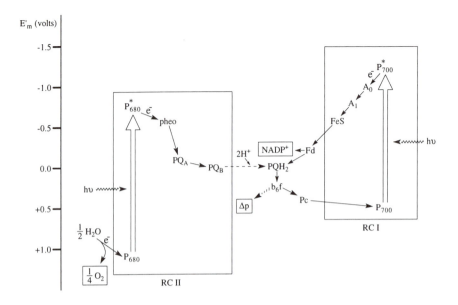

Fig. 5.7 Photosynthesis in cyanobacteria and chloroplasts. Light stimulates electron flow from water through reaction center II (RC II) to reaction center I (RC I) to $NADP^+$. Cyclic flow is possible using RC I from ferredoxin through the b_6f complex. P_{680}, chlorophyll a with a major absorption peak at 680 nm; pheo, chlorophyll pheophytin; PQ, plastoquinone; b_6f, cytochrome b_6f complex; Pc, plastocyanin; P_{700}, chlorophyll a with a major absorption peak at 700 nm; A_o, chlorophyll a; A1, phylloquinone; FeS, one of several iron–sulfur centers; Fd, ferredoxin.

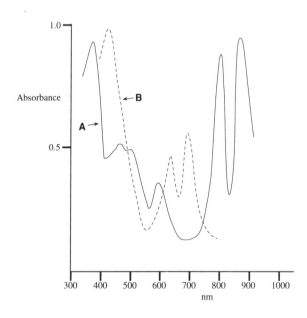

Fig. 5.8 Absorption spectra of a purple photosynthetic bacterium, *Rhodopseudomonas palustris*, (A) and a cyanobacterium (B). (A) Bacteriochlorophyll a peaks at 360, 600, 805, and 870 nm. Carotenoid peaks at 450–525 nm. (B) Chlorophyll a peaks at 440 and 680 nm. The phycocyanin peak is around 620 nm. (Adapted from Brock, T. D. and M. T. Madigan. 1988. *Biology of Microorganisms.* Reprinted by permission of Prentice Hall, Englewood Cliffs, NJ.)

Table 5.2 *In vivo* long wavelength absorption maxima of photosynthetic pigments of phototrophic prokaryotes

Pigment	Wavelength (nm)
Chlorophyll	
a	680
b	675
Bacteriochlorophyll	
a	800–810, 850–910
b	835–850, 1020–1035
c	745–760
d	725–745
e	715–725
g	670, 788
Carotenoid	
Chlorobactene	458
Isorenieratene	517
Lycopene, rhodopin	463, 490, 524
Okenone	521
Rhodopinal	497, 529
Spheroidene	450, 482, 514,
Spirilloxanthin	486, 515, 552,
β carotene	433, 483,
γ carotene	433, 483,
Phycobilins	
Phycocyanin	620–650
Phycoerythrin	560–566

Source: Stolz, J. F. 1991. *Structure of Phototrophic Prokaryotes.* CRC Press, Boston.

therefore extend the range of wavelengths over which photosynthesis is possible (Table 5.2).

2. In the purple photosynthetic bacteria, they are embedded in the membrane as pigment–protein complexes that are in close physical association with the reaction center.

3. In the green sulfur bacteria and Chloroflexus, they exist in chlorosomes, which are separate "organelles," also called inclusion bodies, attached to the inner surface of the cell membrane.

4. In the cyanobacteria and eukaryotic red algae, the light-harvesting pigments that transfer energy to reaction center II are localized in granules called phycobilisomes that are attached to the photosynthetic membranes.

Light-harvesting pigments of the purple photosynthetic bacteria

The light-harvesting complexes of the purple photosynthetic bacteria are *bacteriochlorphyll–protein–carotenoid complexes* localized in the cell membrane in close association with the reaction centers. The pigment–protein complexes absorb light in the range of 800–1000 nm (i.e., in the near infrared range). Upon treatment of the photosynthetic membranes with a mild detergent [e.g., lauryl dimethyl-amine-N-oxide (LDAO)], the light-harvesting complexes can be separated from the reaction centers and analyzed. In *Rhodobacter sphaeroides,* there are two light-harvesting complexes (LH 1 and LH 2), each containing two polypeptides (α and β) to which the pigments are attached (Fig. 5.9). Analysis of the hydrophobic domains of the polypeptides suggests that they span the membrane. It should be emphasized that, in order for effective energy transfer to take place between pigment molecules, they must be positioned correctly with respect to one another (Section 5.7). Organization on protein molecules probably accomplishes this. When light is absorbed by the light harvesting pigments, the energy is quickly transferred to the reaction center (Fig. 5.9).

Light-harvesting pigments in the green sulfur bacteria

The major light-harvesting pigment of the green sulfur bacteria is either bacteriochlorophyll c, d, or e, depending upon the species (Fig. 5.14). In addition, there is bacteriochlorophyll a and carotenoids. In contrast to the purple bacteria, most of the light-harvesting pigments are not in the cell membrane but rather in *chlorosomes.* The chlorosomes are interesting inclusion bodies, which are about $150 \times 70 \times 30$ nm and attached to the inner membrane surface by a baseplate which is attached to the reaction centers (Fig. 5.10). The chlorosomes contain all of the bacteriochlorophylls c, d, or e, much of the carotenoid, and some of the bacteriochlorophyll a. Within the chlorosomes are rod elements of about 10 nm in diameter in *Chlorobium*, which run lengthwise through the chlorosome. The rods are composed of a

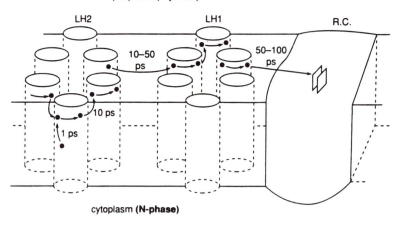

Fig. 5.9 Light-harvesting complexes, LH 1 and LH 2, in the purple nonsulfur photosynthetic bacterium, *R. sphaeroides*. The arrow shows transfer of energy to reaction center (RC). Each cylinder represents a polypeptide unit. The small filled circles symbolize bound pigments. Excitation energy begins in the LH 2 units and then travels to the LH 1 units. The LH 1 units transfer the excitation energy to the bacteriochlorophyll dimer in the reaction center. ps, picoseconds. (From Nicholls, D. G. and S. J. Ferguson. 1992. *Bioenergetics 2*. Academic Press, London.)

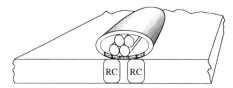

Fig. 5.10 Structure of the chlorosome. The chlorosome is surrounded by a galactolipid layer which contains some protein and is attached to the reaction centers in the cell membrane via a baseplate (shaded area). There is bacteriochlorophyll a in the baseplate and in the membrane. One model for the inside of the chlorosomes postulates rod-shaped proteins to which is attached the major light-harvesting pigment, bacteriochlorophyll c, d, or e. An alternative model is that the bacteriochlorophyll exists as an aggregate not attached to protein. It is suggested that light is absorbed by bacteriochlorophyll in the interior of the chlorosome and the energy is transmitted to bacteriochlorophyll a in the baseplate, then to bacteriochlorophyll a attached to the reaction centers, and from there to reaction center bacteriochlorophyll a.

polypeptide to which the bacteriochlorophyll c, d, or e is thought to be attached. However, some investigators have suggested that the bacteriochlorophyll exists as self-aggregated clusters in the chlorosomes rather

rather than attached to protein. Bacteriochlorophyll a exists as a protein–pigment complex in the baseplate. In addition, there is bacteriochlorophyll a in the membrane which is bound to the reaction centers. Most of the light is absorbed primarily by the bacteriochlorophyll c, d, or e, in the chlorosome, and the energy is transferred to bacteriochlorophyll a in the baseplate. The bacteriochlorophyll in the baseplate then transfers the energy to membrane bacteriochlorophyll a, which transfers the energy to reaction center bacteriochlorophyll a.

Light-harvesting pigments in the cyanobacteria

The photosynthetic membranes of cyanobacteria are actually intracellular membrane-bound sacs called thylakoids that have both reaction centers as well as light-harvesting complexes called *phycobilisomes*.[19,20] The phycobilisomes are numerous granules covering the thylakoids and attached to reaction center II (Fig. 5.11). They can be removed from the photosynthetic membranes with nonionic detergent (Triton X-100) for purification and analysis. The phycobilisomes consist of proteins called *phycobiliproteins* that absorb light in the range of 450–660 nm

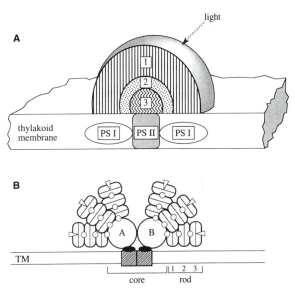

Fig. 5.11 Structure of phycobilisome. (A) Hemispherical phycobilisome common in red algae. They are approximately $48\,nm \times 32\,nm \times 32\,nm$. The major pigment is phycoerythrin. Light energy absorbed by the pigments in the outer region is transferred through an inner core to the reaction center. 1, phycoerythrin; 2, phycocyanin; 3, allophycocyanin. (B) Phycobilisomes from cyano-bacteria are generally fan-shaped with dimensions about $60\,nm$ wide, $30\,nm$ high, and $10\,nm$ thick. The major pigment is phycocyanin. A detailed model of a phycobilisome from the cyanobacterium *Synechococcus*. The outermost region of the phycobilisome consists of six cylindrical rods of phycocyanin. Each rod is made from three hexamers of phycocyanin, 1, 2, and 3. (*Synechococcus* does not have phycoerythrin.) The innermost region of the phycobilisome (the core) consists of two closely associated cylinders (A and B) of allophycocyanin. Each cylinder of allophycocyanin consists of two hexamers (not shown). The cross-hatched rectangle within the thylakoid membrane (TM) represents the reaction center and associated chlorophylls. The solid filled ovals between the core and the thylakoid mem-brane represent high-molecular weight polypeptides thought to anchor the phycobilisome to the thylakoid membrane. There is a linker polypeptide associated with each phycocyanin hexamer that is drawn as an open circle between the hexamers. The open triangle at the terminus of each rod represents a polypeptide that may terminate the rod substructure. (Adapted from Gantt, E. 1986. Phycobilisomes, pp. 260–268, In: *Photosynthesis III: Photosynthetic Membranes and Light Harvest-ing Systems*. L. A. Staehelin and C. J. Arntzen (eds.). Springer-Verlag, Berlin; and Grossman, A. R., M. R. Schaefer, G. G. Chiang, and J. L. Collier. 1993. The phycobilisome, a light-harvest-ing complex responsive to environmental conditions. *Microbiol. Rev.* 57:725–749.)

due to open-chain tetrapyrroles. The tetra-pyrroles are covalently bound to the protein via thioether linkages to a cysteine residue in the phycobiliprotein (Fig. 5.12). Examination of the absorption spectra of chlorophylls and carotenoids reveals that the absorbance is relatively poor in the 500–600 nm range, which is where the phycobiliproteins absorb. There are three classes of phycobiliproteins: *phycoerythrin, phycocyanin,* and *allophyco-cyanin*. Allophycocyanin is usually at a much lower concentration than phycoerythrin and phycocyanin and, as described below, is considered to be an energy funnel to the

reaction center. Although most species of cyanobacteria contain both phycoerythrin (red) and phycocanin (blue), the phycocya-nins generally predominate. The phycobili-proteins are arranged in layers around each other with phycoerythrin on the outside, phycocyanin in the middle, and allophycocya-nin in the inside, closest to the reaction center. A simplified drawing is shown in Fig. 5.11A. Figure 5.11B represents a detailed model at the molecular level of the phycobilisomes from the cyanobacterium *Synechococcus*, which does not contain phycoerythrin. These are fan-shaped (hemidiscoidal) phycobili-

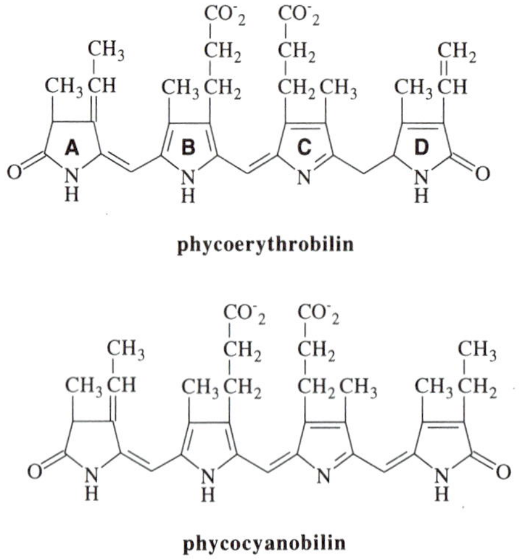

phycoerythrobilin

phycocyanobilin

Fig. 5.12 Structures of phycoerythrobilin and phycocyanobilin. The protein is covalently bonded via a thioether linkage between a cysteine residue in the phycobiliprotein and C2 of CH_3–CH= in ring A.

somes that consist of a dicylindrical core (A and B) containing allophycocyanin, and six cylindrical phycocyanin rods radiating from the core. (In other cyanobacteria, the core may consist of three cylinders.) If phycoerythrin cylinders were present, they would be an extension of the phycocyanin rods. The shorter wavelengths of light are absorbed by the phycoerythrin, the longer wavelengths by phycocyanin, and the still longer wavelengths (650 nm) by the allophycocyanin. These pigments are thus well positioned to transfer energy via inductive resonance to the reaction center. (See Section 5.7 for a discussion of energy transfer.)

A second kind of light-harvesting complex is located in the thylakoids and transmits energy to the closely associated reaction center I. It consists of chlorophyll a, carotenoids, and protein.

Light-harvesting pigments in algae
All of the algae contain chlorophyll a and the carotenoid, carotene. In addition, there are more than 60 other carotenoids found among algae, some of which are shown in Fig. 5.13.

1. The *green algae* (Chlorophyta) contain chlorophyll b and xanthophylls (modified carotenes) called lutein.

2. The *brown algae* (Chromophyta) have chlorophyll c_1 and c_2, and xanthophylls called fucoxanthin and peridinin.

3. The *red algae* (Rhodophyta) have phycobilins (phycoerythrin) (Fig. 5.12), chlorophylls, carotenes, and xanthophylls.

The pigment–protein complexes can be divided into two classes. One class is called the inner antenna β-carotene–Chl a–protein complex which exists as part of reaction centers I and II. The second class is called the outer antenna pigment–protein class that forms light-harvesting complex I (LHC I) associated with reaction center I, and light-harvesting complex II (LHC II) associated with reaction center II. The inner antenna complex that is part of the reaction centers is the same in all algae and higher plants, but the composition and structure of LHC I and LHC II varies. The main function of LHC I and II is to harvest light and transmit the energy to the pigment complex in the reaction center.

5.6.2 *Structures of the chlorophylls, bacteriochorophylls, and carotenoids*

Chlorophylls and bacteriochlorophylls
The chlorophylls are substituted tetrapyrroles related to heme (Fig. 4.5 and Fig. 5.14). The differences are that chlorophylls have Mg^{2+} in the center of the tetrapyrrole rather than $Fe^{2+(3+)}$, and a fifth ring (ring V) is present. The different substitutions on the pyrrole rings and the number of double bonds in ring II distinguish the chlorophylls from each other.

Carotenoids
Carotenoids are long *isoprenoids*, usually C_{40} tetraterpenoids made of eight isoprene units (Fig. I5.13).[21] Carotenoids have a system of conjugated double bonds which accounts for their absorption spectrum in the visible range at wavelengths poorly absorbed by chlorophylls, and makes them valuable as light-harvesting pigments. Light energy absorbed by carotenoids is transferred to the reaction center chlorophyll (Section 5.7). Carotenoid

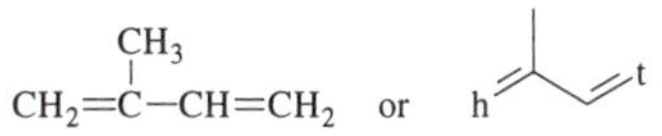

isoprene

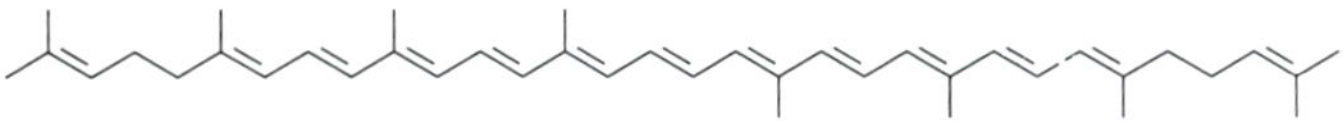

lycopene

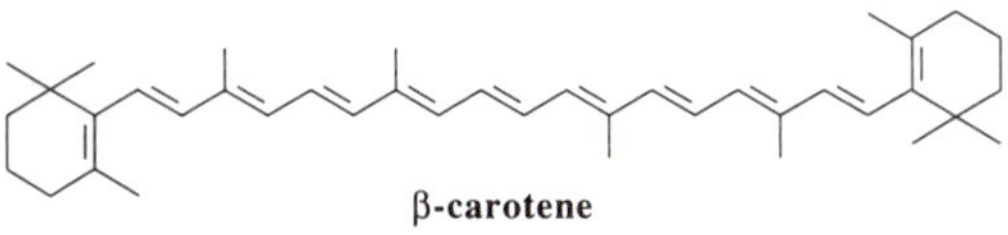

β-carotene

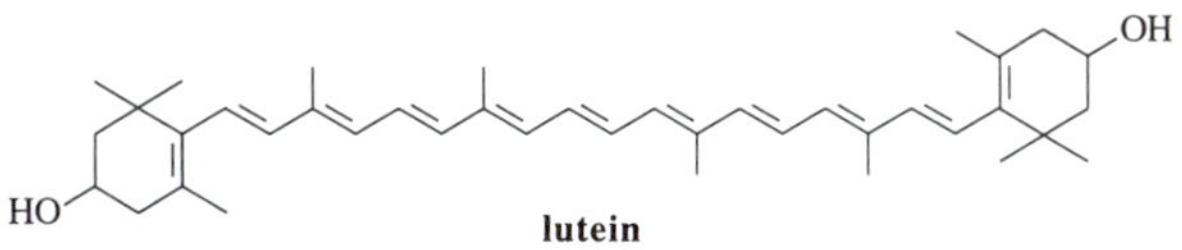

lutein

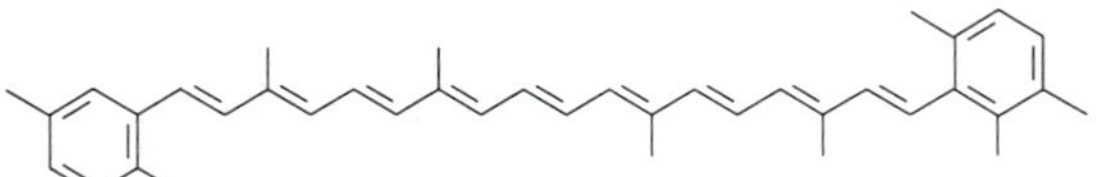

isorenieratene

Fig. 5.13 Structure of carotenoids. An isoprene molecule is shown at the top. One end can be called the head (h) and the other the tail (t). All carotenoids are made of isoprene subunits with a "tail to tail" connection in the middle. Carotenoids in which the ends have circularized into rings are called carotenes. Modified carotenes (e.g., by hydroxylation) are called xanthophylls. Lycopene is the carotenoid responsible for the red color in tomatoes. *β*-Carotene is part of the inner antenna complex in all plants and algae. Lutein is a xanthophyll found in algae. Isorenieratene is a carotenoid found in green sulfur bacteria.

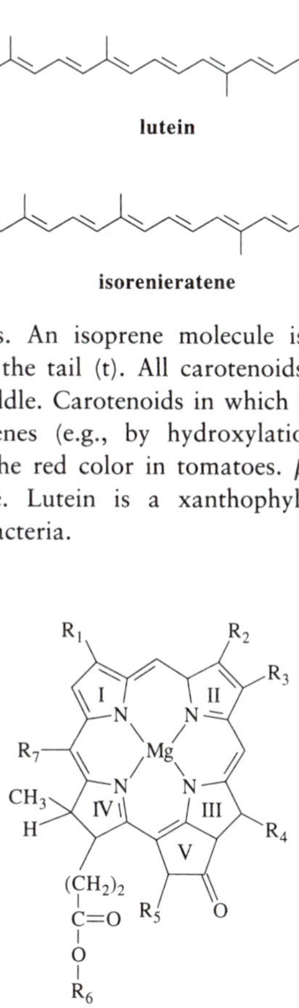

R	bchl a	bchl b	bchl c	bchl d	bchl e	bchl g	chl a
1	$COCH_3$	$COCH_3$	$CHOHCH_3$	$CHOHCH_3$	$CHOHCH_3$	$CH{=}CH_2$	$CH{=}CH_2$
2	CH_3	CH_3	CH_3	CH_3	CHO	CH_3	CH_3
3	CH_2CH_3	$=CHCH_3$	CH_2CH_3	CH_2CH_3	CH_2CH_3	$=CHCH_3$	CH_2CH_3
4	CH_3	CH_3	CH_2CH_3	CH_2CH_3	CH_2CH_3	CH_3	CH_3
5	$COOCH_3$	$COOCH_3$	H	H	H	$COOCH_3$	$COOCH_3$
6	phytyl	phytyl	farnesyl	farnesyl	farnesyl	farnesyl	phytyl
7	H	H	CH_3	H	CH_3	H	H

Fig. 5.14 Structure of bacteriochlorophyll. There is no double bond in ring II of bacteriochlorophyll a, b, and g between C3 and C4. (Nitrogen is position 1.)

pigments with six-membered rings at both ends of the molecule are called carotenes (e.g., β-carotene) (Fig. 5.14). Modification of carotenes (e.g., by hydroxylation) produces xanthophylls, which comprise a major portion of the light-harvesting pigments in algae. Carotenoids perform a function in addition to serving as a light-harvesting pigment. As described in Section 5.2.2, they protect against photooxidation.

5.7 The transfer of energy from the light-harvesting pigments to the reaction center

5.7.1 Mechanism of energy transfer

When light is absorbed in one region of the pigment system, excitation energy is transferred to other regions and eventually to the reaction center. The mechanism which probably accounts for the transfer of excitation energy from one pigment complex to another throughout the pigment system is called *inductive resonance transfer*.

Inductive resonance transfer
In an unexcited molecule, all the electrons occupy molecular orbitals having the lowest available energy (the ground state). When a photon of light is absorbed, an electron becomes energized and occupies a higher

energy orbital which had been empty. This usually occurs without a change in the spin quantum number (Fig. 5.15). The electron is now in the excited singlet state. The excitation can be transferred from one pigment complex to another via *inductive resonance*. During resonance transfer, electrons do not travel between complexes, but energy is transferred. Essentially what happens is that when the electron in the excited molecule drops back to its ground state, the energy is transferred to a molecule close by, resulting in an electron in that molecule being raised to the excited singlet state. This can occur if there is an overlap between the fluorescence spectrum of the donor molecule and the absorption spectrum of the acceptor, and if the molecules are in the proper orientation to each other. That is why the steric relationships of the light-harvesting pigments to each other and to the reaction center is of critical importance. It must be emphasized that transfer by this mechanism does not occur as a result of emission and reabsorption of light.

Delocalized excitons
A second method of excitation transfer occurs over very short distances (i.e., less than 2 nm). When the excited electron is raised to the singlet state, it is said to leave a positive "hole" in the ground state (Fig. 5.15). The combination of the excited electron and the

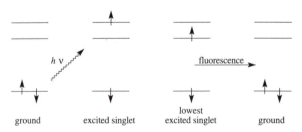

Fig. 5.15 Light raises an electron to a higher energy level in an unoccupied orbital. The electron is then said to be in the excited singlet state. As the diagram shows, the electron may drop down to a lower excited singlet state. The electron may return to the ground state releasing energy as light (fluorescence) or heat. Under appropriate conditions, the energy made available when the electron returns to the ground state is not released but instead is used to raise an electron in a nearby pigment to an excited singlet state, thereby transferring the energy by a process called inductive resonance. In the reaction center, electrons raised to a higher energy orbital are transferred to a nearby acceptor molecule, thereby creating a redox gradient.

positive hole is called an *exciton*. If two or more similar molecules are very close to each other ($< 2\,nm$), the exciton migrates over the molecular orbitals belonging to both molecules (i.e., the exciton becomes *delocalized*). Thus, the excitation energy is actually *shared* by the group of interacting molecules. However, delocalized excitons in the pigment complexes extend only over very short distances and cannot connect the outer regions of the light-harvesting complex with the reaction center. On the other hand, the sharing of excitation energy by delocalized excitons is more rapid than inductive resonance transfer, and is therefore expected to be an important factor for molecules at small intermolecular distances.

5.7.2 Evidence that energy absorbed by the light-harvesting pigments is transferred to the reaction center

When isolated chloroplasts are irradiated, fluorescence from reaction center chlorophyll occurs but not from the light-harvesting pigments. This means that the energy absorbed by the light-harvesting pigments is efficiently transferred to reaction center chlorophyll (or bacteriochlorophyll). Within the reaction center, an electron in a chlorophyll molecule becomes excited to a higher energy level and reduces the primary electron acceptor. It has been suggested that the initial redox reaction is initiated in reaction center chlorophyll by an electron in the lowest excited singlet state.[22]

5.8 The structure of photosynthetic membranes in bacteria

The anoxygenic phototrophic bacteria have a variety of types of photosynthetic membrane structures, depending upon the organism (Fig. 5.16). For example, the purple photosynthetic bacteria have numerous intracellular photosynthetic membranes where the light-harvesting pigments and the reaction centers are closely associated. The green bacteria separate most of the light-harvesting pigments into chlorosomes.

5.9 Summary

When the energy from a photon of light reaches reaction center chlorophyll, an electron in the chlorophyll becomes energized and is transferred to a primary acceptor molecule within the reaction center. This leaves the chlorophyll in an oxidized form. The fate of the electron lost by the primary donor differentiates two different types of reaction centers. One type of reaction center reduces quinone. Quinone-reducing reaction centers are the reaction centers in purple photosynthetic bacteria and reaction center II in chloroplasts and cyanobacteria. In the photosynthetic bacteria, the quinone is ubiquinone. In chloroplasts and cyanobacteria, it is plastoquinone. A second type of reaction center reduces an FeS protein. The latter reaction centers are reaction center I in chloroplasts and cyanobacteria, and the reaction center in green sulfur bacteria and heliobacteria.

In the purple photosynthetic bacteria, the reduced quinone transfers the electrons to a bc_1 complex. The bc_1 complex reduces cytochrome c_2 which returns the electron to the reaction center. This is called cyclic electron flow. A Δp is generated by the bc_1 complex.

In cyanobacteria and chloroplasts, the reduced quinone leaves reaction center II and reduces the b_6f complex, which is similar to the bc_1 complex. The b_6f complex reduces plastocyanin which transfers the electron to reaction center I. A Δp is established by the b_6f complex. Having lost an electron, reaction center II accepts an electron from water, thus producing oxygen as a byproduct of the light reactions in the cyanobacteria and in chloroplasts. Reaction center I is also energized by light and transfers the electron to ferredoxin. The ferredoxin reduces $NADP^+$. The combination of reaction centers I and II is called noncyclic electron flow. In noncyclic flow, both ATP and NADPH are synthesized. Cyclic flow is also possible as the electron returns to reaction center I from reduced ferredoxin.

The green sulfur photosynthetic bacteria have a reaction center, similar to reaction center I, that reduces ferredoxin and NAD^+. The electron donors are inorganic sulfur

GREEN BACTERIA (CHLOROSOMES)

SIMPLE PHOTOSYNTHETIC BACTERIA
(NO CHLOROSOMES OR INTERNAL MEMBRANES)

PURPLE BACTERIA (INTRACYTOPLASMIC MEMBRANES)

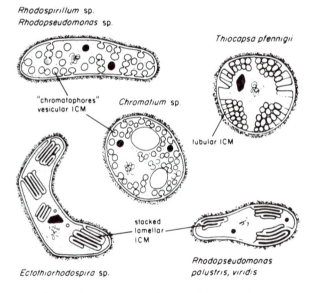

Fig. 5.16 Photosynthetic membranes in anoxygenic phototrophic bacteria. The green bacteria (*Chlorobium, Chloroflexus,* and *Prosthecochloris*) contain chlorosomes attached to the cytoplasmic surface of the cell membrane. Most of the purple photosynthetic bacteria have extensive invaginations of the cell membrane which can take the form of stacked lamellae (*Ectothiorhodospira* sp., *Rhodopseudomonas palustris, viridis*), vesicles (*Rhodospirillum* sp., *Rhodopseudomonas* sp.) or tubules (*Thiocapsa pfennigii*). A few photosynthetic bacteria do not have internal membranes or chlorosomes (*Rhodospirillum tenue, Heliobacterium chlorum*). (From S. G. Sprague and A. R. Varga. 1986. Membrane architecture of anoxygenic photosynthetic bacteria pp. 603–619. In: L. A. Staehelin, and C. J. Arntzen (eds.). *Photosynthesis III: Photosynthetic Membranes and Light Harvesting Systems.* Springer-Verlag, Berlin.)

compounds (e.g., sulfide and elemental sulfur). The electron can return to the reaction center via a quinone and a bc_1 complex in cyclic flow rather than reduce NAD^+.

Most of the light energy used in photosynthesis is not absorbed by reaction centers directly but is instead absorbed by light-harvesting pigments. These are organized either within the membrane and closely positioned to the reaction centers, in special inclusions called chlorosomes (attached to the membrane) or in granules called phycobilisomes (attached to the membrane). There is considerable variability in the composition and structure of the light-harvesting complexes. However, they all consist of pigment--protein complexes that transfer energy to the reaction center. The mechanism of energy transfer between the pigment molecules demands that they be positioned very closely to each other in a precise orientation. The proteins in the light-harvesting complexes and in the reaction centers fulfil this function.

Study Questions

1. If one measures oxygen production per quantum of absorbed light, the efficiency falls when using monochromatic light of 700 nm. The efficiency can be restored if the 700 nm light is supplemented with low intensity light of shorter wavelength (e.g., 600 nm). Explain these results in terms of the light reactions of oxygenic photosynthesis.

2. Describe the similarities and differences between reaction centers I and II.

3. When bacteriochlorophyll absorbs a photon of 870 nm light (about 1.4 eV of energy), an electron travels out of the reaction center, through a bc_1 complex, and back to the reaction center (cyclic flow) with the extrusion of two protons to the outer membrane surface at the site of the bc_1 complex. Protons return to the cytoplasm through the ATP synthase with a H^+/ATP of three. Assuming that the ΔG_p is 45 kJ, what is the efficiency of photophosphorylation?

4. For all anoxygenic photosynthetic bacteria except the green sulfur bacteria, electrons from the reductant do not pass through the reaction center. This means that light does not stimulate electron flow from the reductant to NAD^+. But many of the reductants have reduction potentials more positive than that of the $NAD^+/NADH$ couple. What energizes electron flow to NAD^+? Devise an experiment using ionophores that would support your conclusion.

5. Energy from light absorbed by the accessory pigments is transferred to the reaction center. What is the evidence for that?

6. Explain how the absorption of light by photosynthetic membranes creates a ΔE. Explain how the ΔE is converted into a membrane potential and a proton gradient. Explain how the Δp is converted into ATP.

7. Why are two light reactions required for oxygenic photosynthesis but only one light reaction for anoxygenic photosynthesis?

REFERENCES AND NOTES

1. These groupings do not reflect the complex taxonomy of the photosynthetic bacteria as revealed by ribosomal RNA sequencing. There are five distinct evolutionary lines of photosynthetic prokaryotes (Fig. 1.1 and Table 1.1). The purple photosynthetic bacteria are a heterogeneous assemblage that are part of the evolutionary line of prokaryotes known as the *purple bacteria*. The purple bacteria are subdivided into four subdivisions (i.e., the alpha, beta, gamma, and delta groups). Photosynthetic bacteria, along with nonphotosynthetic bacteria, are in the alpha, beta, and gamma subdivisions, whereas the delta subdivision contains only nonphotosynthetic bacteria. The purple nonsulfur photosynthetic bacteria are in the alpha and beta subdivision whereas the purple sulfur bacteria are in the gamma subdivision. In the alpha group are *Rhodospirillum*, *Rhodopseudomonas*, *Rhodobacter*, *Rhodomicrobium*, and *Rhodopila*, all of which are nonsulfur purple bacteria. Another nonsulfur purple bacterium, *Rhodocyclus*, is in the beta group. The gamma group contains purple sulfur bacteria, including *Chromatium* and *Thiospirillum*. The *green sulfur bacteria*, including *Chlorobium* and *Chloroherpeton* are only distantly related to the purple bacteria and are in a distinctly separate evolutionary line. The *green nonsulfur bacteria* (the *Chloroflexus* group) are in a separate evolutionary line phylogenetically distinct from the purple photosynthetic bacteria and the green sulfur bacteria. The *cyanobacteria* occupy a fourth evolutionary line. The photosynthetic *heliobacteria*, which consist of *Heliobacterium*, *Heliospirillum*, and *Heliobacillus*, are grouped in the evolutionary line that consists of the gram-positive bacteria.

2. Bullerjahn, G. S., and A. F. Post. 1993. The prochlorophytes: Are they more than just chlorophyll a/b-containing cyanobacteria? *Crit. Rev. Microbiol.* 19:43–59.

3. Dutton, P. L. 1986. Energy transduction in anoxygenic photosynthesis, p. 197–237. In L. A. Staehelin and C. J. Arntzen (eds.). *Photosynthesis III: Photosynthetic Membranes and Light Harvesting Systems*. Springer-Verlag, Berlin.

4. Mathis, P. 1990. Compared structure of plant and bacterial photosynthetic reaction centers. Evolutionary implications. *Biochim. Biophys. Acta.* **1018**:163–167.

5. Deisenhofer, J., H. Michel, and R. Huber. 1985. The structural basis of photosynthetic light reactions in bacteria. *TBS* **10**:243–248.

6. Komiya, H., T. O. Yeates, D. C. Rees, J. P. Allen, and G. Feher. 1988. Structure of the reaction center from *Rhodobacter sphaeroides* R-26 and 2.4.1: symmetry relations and sequence comparisons between different species. *Proc. Natl. Acad. Sci.* USA 9016.

7. Cogdell, R. J. 1978. Carotenoids in photosynthesis. *Phil. Trans. R. Soc. Lond. B.* **284**:569–579.

8. During photosynthesis under high light intensities an electron in bacteriochlorophyll can go from the excited singlet state to the lower energy triplet state (Section 5.7). In the triplet state, the spin of the excited electron has changed, so that there are now two unpaired electrons as opposed to paired electrons in the singlet state. (Phosphorescence occurs from the triplet state.) When triplet state bacteriochlorophyll reacts with oxgyen, the oxygen becomes energized to its first excited state, singlet oxygen, $^1O_2$. Singlet oxygen is very reactive and combines with cell molecules such as unsaturated fatty acids, amino acids, and purines, thus causing oxidative damage to various cell components, including lipids, enzymes, and nucleic acids. The carotenoid in the reaction center quenches the triplet state in bacteriochlorophyll. Thus, carotenoids prevent photooxidation under high light intensities. When the carotenoid absorbs the energy from the triplet bacteriochlorophyll, the carotenoid itself becomes energized to the triplet state (triplet–triplet energy transfer). However, the energy in triplet state carotenoid is dissipated harmlessly as heat to the medium. Because the triplet state energy of carotenoids is below the singlet state energy of oxygen, carotenoids will also quench singlet oxygen.

9. Carotenoids are isoprenoid membrane pigments. They are useful in detecting changes in membrane potential. The energy levels of electrons in the conjugated double bond system of the carotenoids is altered by the electric field of the membrane potential. When the membrane potential changes, the carotenoid undergoes a rapid shift in its absorption spectrum peaks. This can be followed using split beam spectroscopy similar to that discussed for examining spectral changes of cytochromes during oxidation–reduction reactions (Section 4.2.4).

10. Dutton, P. L. 1986. Energy transduction in anoxygenic photosynthesis, pp. 197–237. In L. A. Staehelin and C. J. Arntzen (eds.). *Photosynthesis III: Photosynthetic Membranes and Light Harvesting Systems.* Springer-Verlag, Berlin.

11. Nitschke, W., U. Feiler, and A. W. Rutherford. 1990. Photosynthetic reaction center of green sulfur bacteria studied by EPR. *Biochemistry.* **29**:3834–3842.

12. Miller, M., X. Liu, S. W. Snyder, M. C. Thurnauer, and J. Biggins. 1992. Photosynthetic electron-transfer reactions in the green sulfur bacterium *Chlorobium vibrioforme*: Evidence for the functional involvement of iron–sulfur redox centers on the acceptor side of the reaction center. *Biochemistry.* **31**:4354–436.

13. Meent, van de, E. J., M. Kobayashi, C. Erkelens, P. A. van Veelen, S. C. M. Otte, K. Inoue, T. Watanabe, and J. Amesz. 1992. The nature of the primary electron acceptor in green sulfur bacteria. *Biochim. Biophys. Acta* **1102**:371–378.

14. The reaction centers of the green sulfur bacteria, heliobacteria, and photosystem I of chloroplasts differ from reaction center II and the reaction center in purple photosynthetic bacteria in producing low potential reduced FeS proteins, rather than reducing quinone.

15. Reviewed in Dutton, 1986. Energy transduction in anoxygenic photosynthesis, pp. 197–237, In: *Photosynthesis III: Photosynthetic Membranes and Light Harvesting Systems.* Staehelin, L. A. and C. J. Arntzen (eds). Springer-Verlag, Berlin

16. Emerson, R. and C. M. Lewis. 1943. The dependence of the quantum yield of Chlorella photosynthesis on wave length of light. *Am. J. Botany* **30**:165–178.

17. Emerson, R., Chalmers, R., and Cederstrand, C. 1957. Some factors influencing the long wave limit of photosynthesis. *Proc. Natl. Acad. Sci. USA* **43**:133–143.

18. Knaff, D. B. 1993. The cytochrome bc_1 complexes of photosynthetic purple bacteria. *Photosynthesis Res.* **35**:117–133.

19. Grossman, A. R., M. R. Schaefer, G. G. Chiang, and J. L. Collier. 1993. The phycobilisome, a light-harvesting complex responsive to environmental conditions. *Microbiol. Rev.* **57**:725–749.

20. Gantt, E. 1986. Phycobilisomes, pp. 260-268. In: *Photosynthesis III: Photosynthetic Membranes and Light Harvesting Systems.* Staehelin, L. A. and C. J. Arntzen (eds.). Springer-Verlag, Berlin.

21. Isoprenoids are an important class of molecule. Other isoprenoids are the side chains of quinones, the phytol chain in chlorophyll, and the retinal chromophore in bacteriorhodopsin, halorhodopsin, and rhodopsin.

22. Diner, B.A. 1986. Photosystems I and II: Structure, proteins, and cofactors, pp. 422–436. In: *Photosynthesis III.* L.A. Staehelin and C.J. Arntzen (eds.). Springer-Verlag, New York.

6

The Regulation of Metabolic Pathways

Bacteria catalyze a very large number of chemical reactions (approximately 1000–2000) that are organized into interconnecting metabolic pathways. It is of the utmost importance that there is coordination between the pathways to avoid inefficiency, if not chaos. *Pathways are regulated by adjusting the rate of one or more regulatory enzymes that govern the overall rate of the pathway.* The rates at which regulatory enzymes operate are modified in two ways. One method is by the noncovalent binding to the enzyme of certain biochemical intermediates of the pathways (called *allosteric effectors*). The effectors either stimulate or inhibit the regulatory enzyme and, in this way, signal to the regulatory enzyme whether the pathway is producing optimal amounts of intermediates (Sections 6.1.1 and 6.1.2). Regulatory enzymes are also called *allosteric* enzymes because, in addition to having binding sites for their substrates, they also have separate binding sites for the effector molecules. (*Allo* is a Greek term meaning "other." It refers to the fact that allosteric enzymes have a second, "other," site besides the substrate site.) A second method of altering enzyme activity is by the *covalent modification* of the enzyme (Section 6.4). Covalent modifications in-clude the attachment and removal of chemical groups such as phosphate and nucleotides.

6.1 Patterns of regulation of metabolic pathways

In addition to learning the mechanisms of enzyme regulation (allosteric or covalent modification), the student should be aware of the *patterns* of regulation. What is meant by "patterns of regulation" will be made clear below. Patterns of regulation will vary, *even within the same pathway in different organisms.* Nevertheless, some common patterns of metabolic regulation have evolved. These are described next.

6.1.1 Feedback inhibition by an end product of the pathway

For biosynthetic pathways, the end product is usually a negative allosteric effector for a branch point enzyme. For example, in Fig. 6.1, F and J are negative effectors for the regulatory enzymes 1 and 2, respectively. Such control is called *end-product inhibition* or *feedback inhibition* by an end product. Its role is to maintain a steady state where the end product is utilized as rapidly as it is synthesized. When the rates of utilization of the end product increase, then the concentration of the end product decreases. The decrease in concentration relieves the inhibition and the regulatory enzyme speeds up,

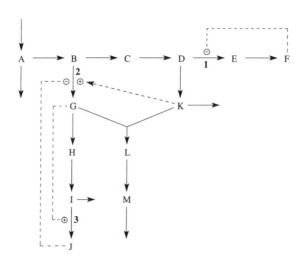

Fig. 6.1 Branched metabolic pathways. The pathways provide precursors to other pathways at the branch points. Both positive and negative regulation occurs. F and J are negative effectors that prevent their own overproduction by inhibiting the regulatory enzymes, 1 and 2. K is a positive effector that stimulates the production of G (in a different pathway) which is needed to react with K to form L. G is a precursor that activates a later reaction (enzyme 3) in its own pathway.

resulting in more end-product synthesis to meet the demands dictated by more rapid utilization (e.g., during rapid growth). There are three recognized patterns of feedback inhibition in biosynthetic pathways. They are *simple*, *cumulative*, and *concerted* (Fig. 6.2). Simple feedback inhibition describes the situation where the regulatory enzyme is inhibited by a single end product (Fig. 6.2A). It is encountered in linear biosynthetic pathways. Cumulative and concerted inhibition refers to situations where more than one end product inhibits the enzyme (Fig. 6.2B). They are seen in branched pathways. Concerted inhibition refers to a situation where both end products must bind to the regulatory

enzyme simultaneously to achieve any inhibition (Fig. 6.2B). In cumulative inhibition the enzyme is not completely inhibited by any single end product. For example, one end product might inhibit the enzyme by 25% and a second might inhibit the enzyme by 45%. Both end products together might inhibit the enzyme by 60% (not 70%) (Fig. 6.2B). Thus the inhibition is cumulative but not necessarily additive. The necessity for cumulative or concerted inhibition is that branched pathways may share a common regulatory enzyme prior to the point that leads to the separate branches of the pathway. Under these circumstances, it is important that a single end product does not shut down

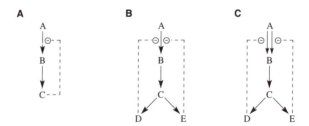

Fig. 6.2 Patterns of feedback inhibition. (A) Simple feedback inhibition in an unbranched pathway. (B) concerted or cumulative inhibition. In concerted inhibition, there is no inhibition unless both D and E bind. In cumulative inhibition D and E each exert partial inhibition but the combination is less than additive. (C) simple inhibition in a branched pathway using isoenzymes.

the activity of the common enzyme and, therefore, the synthesis of all the end products. However, sometimes branched biosynthetic pathways are regulated by simple feedback inhibition. This may occur when a reaction shared by the various branches uses enzymes that have the same catalytic sites but different effector sites (isoenzymes) (Fig. 6.2C). (Thus, isoenzymes catalyze the same reaction but are responsive to different end products.) It can be seen that different regulatory patterns (cumulative, concerted, isoenzymes) can produce similar results (i.e., control of the rate of the pathway by the levels of one or more of the end products).

It should not be concluded that feedback inhibition applies only to biosynthetic pathways. Catabolic pathways, including those for the breakdown of sugars and carboxylic acids, may also be subject to feedback inhibition by an end product.

6.1.2 Positive regulation

A metabolic pathway can also be positively regulated, sometimes by an intermediate in a second pathway. A situation where this occurs is shown in Fig. 6.1. One pathway produces an intermediate, G, which combines with K in a second pathway to produce L. If insufficient G is formed, then K accumulates and stimulates the enzyme that produces G. An example of this is the activation of PEP carboxylase by acetyl-CoA described in Section 8.8.1. Another pattern of positive regulation is sometimes called "precursor activation." The latter refers to a situation where a precursor intermediate stimulates a regulatory enzyme "downstream" in the same pathway. This ensures that the rate of the downstream reactions matches that of the upstream reactions. In Fig. 6.1, G is a precursor that activates enzyme 3, which converts I to J. An example of precursor activation is the activation of pyruvate kinase by fructose-1,6-bisphosphate described in Section 8.1.2. We will encounter other examples of both positive and negative regulation by multiple effectors when we discuss the regulation of the enzymes of central metabolism in Chapter 8.

6.1.3 Regulatory enzymes catalyze irreversible reactions at branch points

There are certain generalizations that one can make about regulatory enzymes and the reactions that they catalyze. *The reactions catalyzed by regulatory enzymes are usually at a metabolic branch point.* For example, in Fig. 6.1, the intermediates B, D, and I are at branch points. Also, *regulatory enzymes often catalyze reactions that are physiologically irreversible* (i.e., *they are far from equilibrium*). Thus they are poised to accelerate in one direction when stimulated.

6.2 Kinetics of regulatory and non-regulatory enzymes

To understand how effector molecules alter the activities of regulatory enzymes, one must understand enzyme kinetics and the enzyme kinetic constants, K_m and V_{max}. This is best done by beginning with the kinetics of nonregulatory enzymes.

6.2.1 Nonregulatory enzymes

A plot of the rate of formation of product (or disappearance of substrate, S) as a function of substrate concentration for most enzyme-catalyzed reactions generates a hyperbolic curve similar to the one shown in Fig. 6.3. As the substrate concentration [S] is increased, the rate of the reaction approaches a maximum, V_{max}, because the enzyme becomes saturated with substrate. For each enzyme there is a substate concentration that gives $1/2 V_{max}$. This is called the *Michaelis–Menten constant*, or K_m. The equation that describes the kinetics of enzyme activity is called the Michealis–Menten equation (eq. 6.1).

eq. 6.1 $v = (V_{max}S)/(K_m + S)$

When the substrate concentration, S, is very small compared to K_m, then the initial velocity, v, is proportional to S (actually to $V_{max}S/K_m$). However, when S becomes much

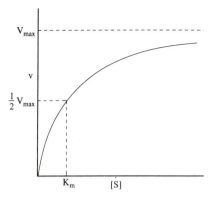

Fig. 6.3 Michealis–Menten kinetics. When substrate concentrations [S] are plotted against initial velocity (v), a hyperbolic curve is obtained that approaches a maximum rate (V_{max}). The substrate concentration that yields V_{max} is a constant for each enzyme, and is called the Michealis–Menten constant or K_m.

larger than K_m, then S cancels out and v approaches V_{max}. There are two units for V_{max} (i.e., specific activity and turnover number). The units of specific activity are micromoles of substrate converted per minute per milligram of protein, and the units for turnover number are micromoles of substrate converted per minute per micromole of enzyme.

Derivation of the Michealis–Menten equation

In 1913, L. Michaelis and M. L. Menten formulated a theory of enzyme action that explained the kinetics shown in Fig. 6.3. The following derivation of the Michealis–Menten equation was developed later by Briggs and Haldane. Consider the situation where a substrate (S) is converted to a product in an enzyme catalyzed reaction. During the reaction, S combines with the enzyme (E) at the substrate site on the enzyme to form an enzyme–substrate complex [(ES)], which then breaks down to enzyme and product. For the Briggs and Haldane derivation, it is necessary to assume a steady state [i.e., the enzyme is present in catalytic amounts (S ≫ total enzyme)]. Under these conditions, the rate of formation and breakdown of ES are equal, which results in a steady state level of ES.

The situation can be summarized as follows:

$$S + E(\text{free}) \; \underset{k_2}{\overset{k_1}{\rightleftarrows}} \; (ES) \; \underset{k_4}{\overset{k_3}{\rightleftarrows}} \; E + P$$

Notice that there are four rate constants, k_1, k_2, k_3, and k_4. If one were to measure the initial velocity (v), then the rate of the reaction (e.g., the rate of formation of P or disappearance of S) would be proportional to $k_3(ES)$. In this treatment we are ignoring the formation of (ES) from E and P, and k_4. This would be small, in any event, if the initial rates were measured before P accumulated. If all of the enzyme were bound to S, then all of E would be in the form of (ES), and the rate would be the maximum rate. These relationships are shown below (eqs 6.2 and 6.3).

eq. 6.2 $v = dP/dt = k_3(ES)$

and substituting total enzyme, E_t, for E, when all of the enzyme is saturated with substrate (eq. 6.3):

eq. 6.3 $V_{max} = k_3(E_tS)$

(For enzymes that may have complex reaction pathways that depend upon more rate constants than simply k_3, the symbol k_{cat} is used for the maximal catalytic rate (i.e., $V_{max} = k_{cat}(E_tS)$).[1]

The instantaneous rate of formation of (ES) is given by eq. 6.4:

eq. 6.4 $d(ES)/dt = k_1(E_t - ES)(S)$

The concentration of free enzyme (E) is $[E_t - (ES)]$. S represents the substrate concentration.

Equation 6.4 describes the instantaneous rate of a bimolecular reaction between the substrate whose concentration is (S), and the free enzyme whose concentration is $[E_t - ES]$. Now we must consider the rate of breakdown of ES. This is given by eq. 6.5:

eq. 6.5 $-d(ES)/dt = k_2(ES) + k_3(ES)$

Let us assume the reaction reaches a steady

state at the time of measurement. This is ensured by using substrate concentrations that are in large excess over the total amount of enzyme. In the steady state, the rate of formation of ES is equal to its rate of breakdown, therefore:

eq. 6.6 $k_1(E_t - ES)(S) = k_2(ES) + k_3(ES)$

Rearranging the above equation gives eq. 6.7:

eq. 6.7 $(S)(E_t - ES)/(ES) = [k_2 + k_3]/k_1 = K_m$

The constant, K_m, is called the Michealis–Menten constant. Note that if k_3 is small compared to k_2 (i.e., if the rate of product formation is small with respect to the rate that the substrate dissociates from the enzyme), then the K_m is approximately equal to k_2/k_1 [i.e., the dissociation constant for the enzyme (K_D)]. This is true for some enzyme substrate combinations. However, the student should recognize that it is not always true, and it is a mistake to assume (as is often done) that $1/K_m$ is a direct measure of the affinity of the substrate for the enzyme. It is best to refer to the K_m as being equal to the substrate concentration that gives 1/2 maximal velocity, as described in Fig. 6.3.

Now we can solve for (ES):

eq. 6.8 $(ES) = (E_t)(S)/(K_m + S)$,

where (E_t) is total enzyme

Since the initial rate (v) is proportional to $k_3(ES)$, we can write eq. 6.9:

eq. 6.9 $v = k_3(E_t)(S)/(K_m + S)$

$V_{max} = k_3(ES)$ when $(ES) = E_t$, therefore, $V_{max} = k_3(E_t)$. Equation 6.9 then becomes eq. 6.10:

eq. 6.10 $v = (V_{max}S)/(K_m + S)$

Equation 6.10 is the Michealis–Menten equation that describes the kinetics in Fig. 6.3. Typical K_m values range from $10^{-4}\,M$ (100 μM) and lower.

The K_m is the substrate concentration that gives 1/2 V_{max}

If one substitutes 1/2 V_{max} for v in the Michealis–Menten equation and solves for S, then S is equal to K_m. That is to say, the K_m is equal to the substrate concentration that gives 1/2 V_{max}. This is shown in eq. 6.11:

eq. 6.11 $V_{max}/2 = V_{max}(S)/(K_m + S)$

Divide both sides by V_{max}:

eq. 6.12 $1/2 = S/(K_m + S)$

Rearrange:

eq. 6.13 $K_m + S = 2S$

and:

eq. 6.14 $K_m = S$

Because the V_m and therefore the K_m can only be approximated from the plot shown in Fig. 6.3, eq. 6.10 is frequently written as the reciprocal:

eq. 6.15 $1/v = 1/V_{max} + (K_m/V_{max})(1/S)$

$1/v$ is plotted against $1/[S]$ to give a *double reciprocal* or *Lineweaver–Burk* plot (Fig. 6.4). One can find K_m/V_{max} from the slope, or $-1/K_m$ from the intercept on the y axis, and $1/V_{max}$ from the intercept on the x axis.

6.2.2 Regulatory enzymes

Regulatory enzymes may not follow simple Michaelis–Menten kinetics. Instead, they typically show sigmoidal kinetics (Fig. 6.5). An explanation for substrate-dependent sigmoidal kinetics is that the binding of one substrate molecule increases the affinity of the enzyme for a second substrate molecule, or increases the rate of formation of product from sites already occupied. This is called *positive cooperativity*. Positive cooperativity makes sense for a regulatory enzyme. It makes enzyme catalysis very sensitive to small changes in substrate level concentrations when the substrate concentrations are very

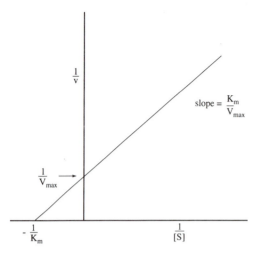

Fig. 6.4. Lineweaver–Burk plot.

absence of effector molecules. Some of the effectors raise the K_m and some lower the K_m. An effector that raises the K_m will decrease the velocity of the reaction, and one that lowers the K_m will increase the velocity of the reaction, provided that the enzyme is not saturated with substrate. This is shown in Fig. 6.5. Notice that the negative effector (curve 3) increases the sigmoidicity of the curve and also increases the K_m. The positive effector (curve 1) decreases the sigmoidicity of the curve and lowers the K_m. Effectors can also change the V_{max}. However, in Fig. 6.5, the V_{max} is not changed.

small compared to the K_m. (Notice in Fig. 6.5 how the enzyme rate rapidly changes with small changes of substrate at critical concentrations.) Allosteric effectors (positive and negative) can also bind cooperatively to enzymes, making the enzyme more sensitive to small changes in effector concentration. Even though sigmoidal kinetics are not explained by the Michealis–Menten equation, one can still measure the K_m and V_{max} because these are defined operationally. When one measures these "constants" for regulatory enzymes, one finds that they are subject to change, depending upon the presence or

6.3 Conformational changes in regulatory enzymes

When the effector binds to the allosteric site (the effector site), the protein undergoes a conformational change and this changes its kinetic constants. Many of the regulatory enzymes are *oligomeric* or *multimeric* i.e., they have multiple subunits. However, for illustrative purposes we will first consider a monomeric polypeptide (Fig. 6.6). Assume the polypeptide has three binding sites: one for the substrate (substrate site), a second for a positive effector, and a third for a negative effector. Further assume that the enzyme exists in two states, A and B. When the enzyme is in conformation A, it binds the substrate and also binds the positive effector which locks the enzyme in state A. You might say that state A has a low K_m. State B has a high K_m and also binds the positive effector poorly. But state B does bind the negative effector which locks it into state B. Thus, the positive effector lowers the K_m and the negative effector raises the K_m. In the cell, there is an interplay of positive and negative effectors that adjust the ratio of active and less active enzyme.

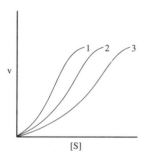

Fig. 6.5 Sigmoidal kinetics of regulatory enzymes. Initial velocities (v) are plotted against substrate concentrations [S] in the presence of a positive effector (curve 1); no effector (curve 2); or a negative effector (curve 3). The positive effector decreases the K_m as well as decreasing the sigmoidicity of the curve. The negative effector increases the K_m and the sigmoidicity of the curve. In other cases the effectors may change the V_{max}.

Regulatory subunits

Many regulatory enzymes are multimeric. Some consist of both regulatory and catalytic subunits. The catalytic subunits bear the active sites and bind the substrate. The regulatory subunits bind the effectors. When

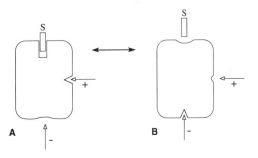

Fig. 6.6 Conformational changes in a regulatory enzyme. A monomeric enzyme is shown with three binding sites: one for the substrate (S), one for the positive effector (+), and one for the negative effector (−). (A) When the enzyme binds the positive effector, a conformational change occurs that lowers the K_m. (B). When the enzyme binds the negative effector there is an increase in the K_m and loss of affinity for the positive effector. In the presence of a sufficient concentration of positive effector, essentially all of the enzyme will be in conformation A and the enzyme will be turned "on." On the other hand, if sufficient negative effector is present, then the enzyme will be mostly in conformation B and turned "off." The fraction of enzyme in the "on" or "off" conformation will depend upon the relative concentrations of substrate, positive effector, and negative effector. Multimeric enzymes are regulated in a similar fashion except that the effector binding sites can be on a subunit (regulatory subunit) separate from the substrate binding site (the catalytic subunit). Conformational changes in one subunit induce conformational changes in the attached subunit.

the regulatory subunits bind the effectors, the polypeptide undergoes a conformational change induced by the binding. The re-

gulatory subunit then either inhibits or activates the catalytic subunit resulting in changes in the kinetic constants.

6.4 Regulation by covalent modification

Although most regulatory proteins appear to be regulated by conformational changes induced by the binding of allosteric effectors as described above, there are many important instances of regulation by covalent modification of the protein in both prokaryotes and eukaryotes. The enzyme may also be regulated by allosteric interactions. Covalent modification occurs by the reversible attachment of chemical groups such as acetyl groups, phosphate groups, methyl groups, adenyl groups, and uridyl groups. The covalent attachment of a chemical group can activate or inhibit the protein. Table 6.1 summarizes some examples of covalent modification of proteins, several of which are discussed in later chapters.

6.5 Summary

Metabolism is regulated by key regulatory enzymes that help to keep metabolism in a steady state where concentrations of intermediates do not change. Generally, the regulatory enzymes catalyze physiologically irreversible reactions at metabolic branch points. The enzymes can be regulated by negative effectors, positive effectors, both

Table 6.1 Covalent modifications of bacterial enzymes and other proteins

Enzyme	Organism	Modification
Glutamine synthetase	E. coli and others	Adenylylation
Isocitrate lyase	E. coli and others	Phosphorylation
Isocitrate dehydrogenase	E. coli and others	Phosphorylation
Chemotaxis proteins	E. coli and others	Methylation
P$_{II}$	E. coli and others	Uridylylation
Ribosomal protein L7	E. coli and others	Acetylation
Citrate lyase	Rhodopseudomonas gelatinosa	Acetylation
Histidine protein kinase	Many bacteria	Phosphorylation
Phosphorylated response regulators	Many bacteria	Phosphorylation

Source: Neidhardt, F. C., J. L. Ingraham, and M. Schaechter. 1990. *Physiology of the Bacterial Cell*. Sinauer Associates, Inc., Sunderland, MA.

negative and positive effectors, and by covalent modification. In biosynthetic pathways, the negative effectors are the end products. This is called feedback or end-product inhibition. Positive regulation by a precursor intermediate can speed up a subsequent reaction in the same metabolic pathway to avoid buildup of the precursor. Also, an intermediate in one pathway may regulate a reaction in a second pathway positively, if the second pathway provides a precursor metabolite for the first pathway. Enzymes that are regulated by effector molecules that bind to effector sites are called allosteric enzymes and the effector molecules are called allosteric effectors.

Nonregulatory enzymes display simple saturation kinetics called Michealis–Menten kinetics. Two kinetic constants are the V_{max} and the K_m. The V_{max} is the velocity of the enzyme when it is saturated with substrate. The K_m is the substrate concentration yielding $1/2\ V_{max}$. Regulatory enzymes usually show sigmoidal kinetics. Nevertheless, a V_{max} and K_m can be measured. A positive effector can raise the V_{max} and lower the K_m, but not necessarily both. A negative effector does just the opposite. Regulatory enzymes are frequently multimeric, and consist of catalytic subunits and regulatory subunits. The allosteric effector binds to the regulatory subunit and effects a conformational change in the catalytic subunit that alters its kinetic constants.

Study Questions

1. Plot the following data and derive a K_m and V_{max}:

Time (min)	Substrate (M $\times\ 10^{-6}$)			
	16	24	48	144
		Product (arbitrary units)		
0.4	0.55	0.70	0.70	1.00
0.8	1.10	1.30	1.35	1.70
1.2	1.65	1.85	2.00	3.00
1.6	2.05	2.35	2.55	3.15
2.0	2.50	2.85	3.10	3.80

2. What is a rationale for having regulatory enzymes catalyze physiologically irreversible reactions at branch points?

3. What are three patterns of feedback inhibition for the regulation of a branched biosynthetic pathway?

4. Under what circumstances might you expect to see positive regulation of enzyme activity?

5. What is meant by positive cooperativity? What is the physiological advantage to it?

REFERENCES AND NOTES

1. The turnover number is equal to k_{cat} (k_3), and is related to the concentration of active sites (E_t) (i.e., $V_{max} = k_{cat}[E_t]$). The turnover number is measured with pure enzyme whose molecular weight is known.

7

Bioenergetics in the Cytosol

Two different kinds of energy drive all cellular reactions. One of these is electrochemical energy, which refers to ion gradients across the cell membrane. In bacteria, these are primarily proton gradients (Chapter 3). Electrochemical energy energizes solute transport, flagella rotation, ATP synthesis by the ATP-synthase, and other membrane activities. The second type of energy is chemical energy in the form of high-energy molecules (e.g., ATP) in the soluble part of the cell. High-energy molecules are important because they drive biosynthesis in the cytoplasm, including the synthesis of nucleic acids, proteins, lipids, and polysaccharides. Additionally, the uptake into the cell of certain solutes is driven by high-energy molecules rather than by electrochemical energy. This chapter introduces the various high-energy molecules used by the cell and explains *why* "high-energy" is applied to them, how they are made, and how they are used. The major high-energy molecules include ATP as well as other nucleotide derivatives, phosphoenolpyruvate, acyl phosphates, and acyl-CoA derivatives. The chapter begins with a discussion of the chemistry of the major high-energy molecules (Section 7.1), followed by an explanation of how high-energy molecules are used to drive biosynthetic reactions (Section 7.2), and how the high-energy molecules are synthesized (Section 7.3). The information in this chapter will be referred to in later chapters when the metabolic roles of the high-energy molecules are considered in more detail.

7.1 High-energy molecules and group transfer potential

High-energy molecules such as ATP have bonds that have a high free energy of hydrolysis that are sometimes depicted by a "squiggle" ($\sim$). These bonds are often called *"high-energy" bonds*, but as will be discussed below, this is a misnomer because the bonds have normal bond energy. The important point is that the chemical group attached to the "squiggle" is readily transferred to *acceptor molecules*. Therefore, the high-energy molecules are said to have a *high group transfer potential*. When the chemical groups are transferred, new linkages between molecules (e.g., ester linkages, amide linkages, glycosidic linkages, and ether linkages) are made. This results in the synthesis of the different small molecules in the cell (e.g., complex lipids, nucleotides, and so on) as well as polymers such as nucleic acids, proteins, polysaccharides, and fatty acids. We will first consider group transfer reactions in general, and then explain why certain molecules have a high group transfer potential.[1] Finally, some examples of how group transfer reactions are used in biosynthesis will be discussed.

7.1.1 Group transfer potential

A common chemical group that is transferred between molecules is the phosphoryl group, and phosphoryl group transfer will be used as an example of a group transfer reaction. Let us consider a generic phosphoryl group transfer reaction (shown in Fig. 7.1). We will examine both the chemical mechanism of transfer and the thermodynamics. Notice that the phosphorus in all phosphate groups carries a positive charge (i.e., the P=O bond is drawn as the semipolar P^+-O^- bond). This is because phosphorus forms double bonds poorly, and the electrons in the bond are shifted toward the electron-attracting oxygen. During the phosphoryl group transfer reaction, the phosphorus atom is attacked by a nucleophile (an electronegative atom seeking a positive center) shown in Fig. 7.1. The chemical group, Y, is displaced with its

also written

where

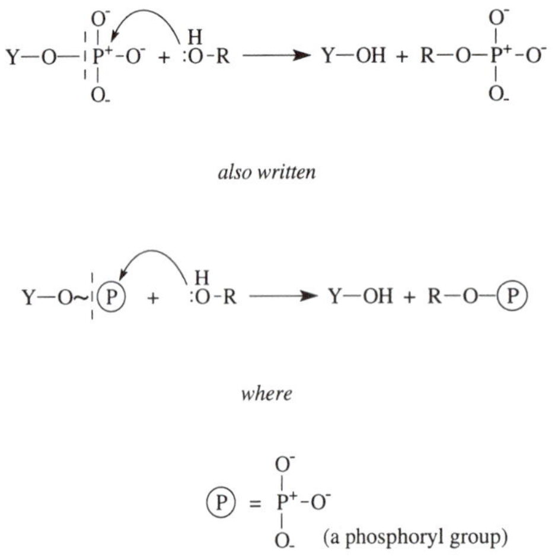

Fig. 7.1 Phosphoryl group transfer reaction. The phosphate group is shown as ionized. The phosphorus–oxygen bond exists as a semipolar bond because phosphorus forms double bonds poorly. The positively charged phosphorus is attacked by the electronegative oxygen in the hydroxyl. The leaving group is YOH. If ROH is water, then the reaction is an hydrolysis, and the product is inorganic phosphate and YOH. One can compare the tendency of different molecules to donate phosphoryl groups by comparing the free energy released when the acceptor is water (i.e., the free energy of hydrolysis). The group transfer potential is the negative of the free energy of hydrolysis.

bonding electrons, and the phosphoryl group is transferred to the hydroxyl, forming ROP. This is a general scheme for all group transfer reactions, not simply phosphoryl group transfers. That is to say, a nucleophile bonds to an electropositive center and displaces a leaving group with its bonding electrons. The reactions are called *nucleophilic displacements*. As we shall see later, various molecules such as ATP, acyl phosphates, and phosphoenolpyruvate have a high phosphoryl group transfer potential and undergo similar nucleophilic displacement reactions. But, how can one compare the phosphoryl group transfer potential of all of these molecules since the acceptors differ? A scale is used, where the standard nucleophile is the hydroxyl group of water, and the phosphoryl donors are all compared with respect to the tendency to donate the phosphoryl group to water. *The group transfer potential is thus defined as the negative of the standard free energy of hydrolysis at pH 7.* It is a quantitative assessment of the tendency of a molecule to donate the chemical group to a nucleophile. For example, suppose the standard free energy of hydrolysis of the phosphate ester bond in YOP is $-29,000$ J/mol. Then its phosphoryl group transfer potential is the negative of this number, or $+29,000$ J/mol. *Bonds that have a standard free energy of hydrolysis at pH 7 equal to or greater than $-29,000$ J are usually called "high energy" bonds.* The group transfer potential is not really a potential in an electrical sense, but a free energy change per mole of substrate hydrolyzed. However, the word "potential" is widely used in this context and the convention will be followed here. The molecules with high phosphoryl group transfer potentials that we will revisit in the ensuing chapters are listed in Table 7.1. Also listed is glucose-6-phosphate, which has a low phosphoryl group transfer potential.

Group transfer potentials are a convenient way of estimating the direction in which a reaction will proceed. For example, the phosphoryl group transfer potential of ATP at pH 7 is 35 kJ/mole and for glucose-6-phosphate, it is only 14 kJ/mol. This means that ATP is a more energetic donor of the phosphoryl group than glucose-6-phosphate.

Table 7.1 Group transfer potentials

Compound	G'_{ohyd} (kJ/mol)	Phosphoryl group transfer potential (kJ/mol)
$PEP + H_2O \longrightarrow pyruvate + P_i$	-62	$+62$
$1,3\text{-}BPGA + H_2O \longrightarrow 3\text{-}PGA + P_i$	-49	$+49$
$Acetyl\text{-}P + H_2O \longrightarrow acetate + P_i$	-47.7	$+47.7$
$ATP + H_2O \longrightarrow ADP + P_i$	-35	$+35$
$Glucose\text{-}6\text{-}P + H_2O \longrightarrow glucose + P_i$	-14	$+14$

It also means that ATP will transfer the phosphoryl group to glucose to form glucose-6-phosphate with the release of 35–14 or 21 kJ/mol under standard conditions, pH 7.

$$ATP + glucose \longrightarrow glucose\text{-}6\text{-}phosphate + ADP$$
$$\Delta G'_o = -21 \text{ kJ/mol}$$

This can also be seen by summing the hydrolysis reactions because their sum equals the transfer of the phosphoryl group from ATP to glucose. This can be done for thermodynamic calculations, even though glucose-6-phosphate is not synthesized by hydrolysis reactions as written below, because the overall energy change is independent of the path of the reaction.

$$ATP + H_2O \longrightarrow ADP + P_i$$
$$\Delta G'_o = -35 \text{ kJ/mol}$$
$$glucose + P_i \longrightarrow glucose\text{-}6\text{-}phosphate + H_2O$$
$$\Delta G'_o = +14 \text{ kJ/mol}$$

$$ATP + glucose \longrightarrow glucose\text{-}6\text{-}phosphate + ADP$$
$$\Delta G'_o = -21 \text{ kJ/mol}$$

The release of 21 kJ per mole means that the equilibrium lies far in the direction of glucose-6–phosphate. Because $\Delta G'_o = -RT \ln K'_{eq} = -5.80 \log_{10} K'_{eq} \text{ kJ}$ (at 30°C), a $\Delta G'_o$ of -21 kJ means that the equilibrium constant is 4.2×10^3. The actual free energy change, $\Delta G'$, is a function of the physiological concentrations of products and reactants and is equal to $\Delta G'_o + RT \ln(\text{glucose-6-phosphate})(ADP)/(ATP)(glucose)$. Reactions proceed only in the direction of a negative $\Delta G'$. This means that (glucose-6-phosphate) (ADP)/(ATP)(glucose) would have to be greater than 4.2×10^3 to change the sign of the $\Delta G'$ so that the reaction would proceed in the direction of ATP and glucose. This does not occur, and under physiological conditions the direction of phosphoryl flow is always from ATP to glucose. For example, the first step in the metabolism of glucose in the glycolytic pathway is the phosphorylation of glucose by ATP. We will now consider why ATP and the other molecules in Table 7.1 have such high free energies of hydrolysis.

7.1.2 Adenosine triphosphate (ATP)

As seen in Table 7.1, the standard free energy of hydrolysis of the phosphate ester bond in ATP at pH 7 is -35 kJ. To understand why so much energy is released during the hydrolysis reaction, we must consider the structure of the ATP molecule (Fig. 7.2). Notice that at pH 7, all of the phosphate groups are ionized. This produces electrostatic repulsion between the negatively charged phosphates. Reactions during which phosphate is removed from ATP will be favored because the electrostatic repulsion is decreased as a result of the hydrolysis.

Transfer of a phosphoryl group to other acceptors besides water
Any group that is electronegative (e.g., the hydroxyl groups in sugars) can attack the

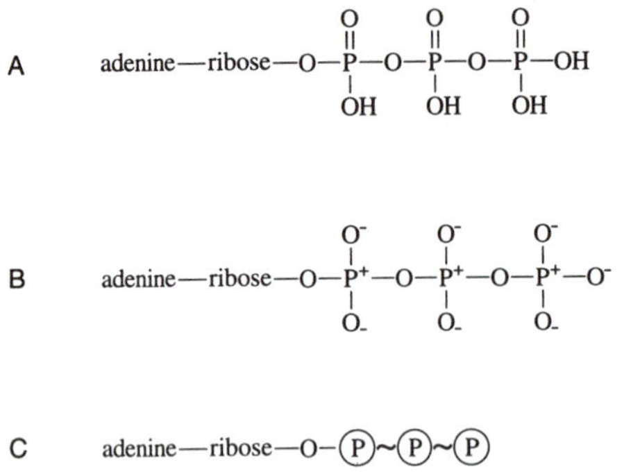

A   adenine—ribose—O—P—O—P—O—P—OH
with three double-bonded O atoms above and OH, OH, OH below

B   adenine—ribose—O—P$^+$—O—P$^+$—O—P$^+$—O$^-$
with O$^-$ above and O$_-$ below each phosphorus

C   adenine—ribose—O—(P)~(P)~(P)

Fig. 7.2 Structure of ATP. ATP has three phosphate groups. (A) Unionized. (B) All of the phosphate groups are shown as ionized (pH 7), giving them a net negative charge. Note that the phosphate–oxygen double bond is drawn as a semipolar bond, which takes into account the electronegativity of the oxygen and the low propensity of phosphorus to form double bonds. The structure predicts strong electrostatic repulsion between the phosphate groups because they bear a net negative charge. The electrostatic repulsion favors transfer of phosphate to a nucleophile (e.g., water). (C) ATP drawn with "squiggles" to show the bonds with high free energy of hydrolysis. The phosphates are drawn as phosphoryl groups as illustrated in Fig. 7.1.

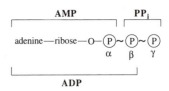

adenine—ribose—O—(P)~(P)~(P)
with labels α, β, γ beneath; AMP and PP$_i$ bracketed above; ADP bracketed below

Fig. 7.3 ATP drawn with "squiggles". A high free energy of hydrolysis (i.e., high group transfer potential) is denoted by a squiggle. In ATP, two of the phosphate ester bonds have a high free energy of hydrolysis due to electrostatic repulsion. The phosphates are labelled α, β, and γ, starting with the one nearest to the ribose. If the α phosphate is attacked, then AMP is transferred and pyrophosphate (PP$_i$) is displaced. If the phosphate is attacked then ADP is usually transferred and inorganic phosphate (P$_i$) is displaced. However, some enzymes catalyze the transfer of the pyrophosphoryl group and displace the AMP. If the γ phosphate is attacked, then the phosphoryl group is transferred and ADP is displaced. All four reactions take place and the specificity depends upon the enzyme that catalyzes the reaction.

electropositive phosphorus (shown in Fig. 7.2) resulting in phosphoryl group transfer, provided the appropriate enzyme is present to catalyze the reaction. In this way, ATP can phosphorylate many different compounds. Enzymes that catalyze phosphoryl group transfer reactions are called *kinases*. In summary, then, the high free energy of hydrolysis of ATP is a good predictor for the tendency of ATP to donate a phosphoryl group to nucleophiles such as the hydroxyl groups in sugars. Much of the biochemistry of ATP is directly related to this tendency. We will return to this point later, but first we must further discuss the "squiggle" and the "high-energy bond."

The "squiggle"

As mentioned, one denotes a high negative free energy of hydrolysis, and thus a greater group transfer potential, with the symbol of a "squiggle" (∼). ATP is usually drawn with two squiggles because there are two phosphate ester bonds with a high free

energy of hydrolysis (Fig. 7.3). Note: the squiggle does not refer to the energy in the phosphate bond but rather to the free energy of hydrolysis. To make the distinction clear, consider the definition of bond energy. Bond energy is the energy required to break a bond. It is not the energy released when a bond is broken. In fact, the P–O bond energy is about +413 kJ (100 kcal). Compare this to the −35 kJ of hydrolysis energy.

ATP as a donor of AMP, ADP, pyrophosphoryl, or phosphoryl groups

ATP can donate other parts of the molecule besides the phosphoryl group and this becomes important for the synthesis of many polymers (e.g., proteins, polysaccharides, and nucleic acids) as well as other biochemical reactions. This is because, when these groups are transferred to an acceptor molecule, the acceptor molecule becomes energized for biosynthetic reactions. Let us look again at the structure of ATP and examine some of the group transfer reactions that it can undergo (Fig. 7.3). The phosphates are labelled α, β, and γ, when counting from the ribose moiety, with α being the phosphate closest to the ribose. If the α phosphate is attacked, then

AMP is the group transferred and PP_i is the leaving group. For example, this occurs during protein synthesis, which is discussed in Section 7.2.2. If the β phosphate is attacked, then ADP is usually the group transferred and P_i is the leaving group. ADP–sugar derivatives are precursors to some polysaccharides. Some enzymes catalyze the transfer of the pyrophosphoryl group when the β phosphate is attacked. For example, enzymes that synthesize phosphoenolpyruvate from pyruvate make an enzyme–pyrophosphate derivative by transferring the pyrophosphoryl group from ATP to the enzyme (Section 8.12.2). Another example is the synthesis of phosphoribosylpyrophosphate (PRPP) during which the pyrophosphoryl group is transferred from ATP to the C1 of ribose-5-phosphate (Section 9.2.2). If the γ phosphate is attacked, then the phosphoryl group is transferred, forming phosphorylated derivatives and ADP as the leaving group. This is a very common reaction in metabolism. As stated, all of the above reactions can occur. The reaction that takes place depends upon which enzyme is the catalyst, because it is the enzyme that determines the specificity of the attack.

7.1.3 Phosphoenolpyruvic acid

Another high-energy phosphoryl donor is phosphoenolpyruvate (PEP). In fact, PEP is a more energetic phosphoryl donor than ATP and will donate the phosphoryl group to ADP to make ATP with release of 62–35 or 27 kJ/mol (under standard conditions, pH 7) (Table 7.1). As discussed in Section 7.3.4, this is an important source of ATP. PEP also donates the phosphoryl group to sugars during sugar transport in the phosphotransferase (PTS) system (Chapter 15).

Why does PEP have a high free energy of hydrolysis?
The reason why PEP has a high group transfer potential is different from that for ATP. Consider the reaction written below and illustrated in Fig. 7.4.

$$PEP + H_2O \longrightarrow pyruvic\ acid + P_i$$

$$\Delta G'_o \approx -61\ kJ/mol$$

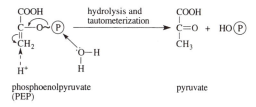

Fig. 7.4 Conversion of phosphoenolpyruvate to pyruvate. This is an enol–keto tautomerization. The removal of the phosphoryl group allows the electrons to shift into the keto form. Energy is released because the keto form has less free energy and is more stable than the enol form. A comparison of the difference in the bond energies of pyruvic acid and enolpyruvic acid shows that pyruvic acid has 76 kJ more bond energy than enolpyruvic acid. The keto form of pyruvic acid is more stable than the enol form by 42-50 kJ/mole (10–12 kcal/mole).

Enolpyruvic acid		Pyruvic acid	
C–O	293 kJ	C–C	247 kJ
O–H	460 kJ	C–H	364 kJ
C=C	418 kJ	C=O	636 kJ
	1,171 kJ		1,247 kJ

The hydrolysis removes the phosphate and allows the enol form of pyruvic acid to tautomerize into the keto form. Energy is released because the keto form is more stable than the enol form. One can account for the energy released during hydrolysis by the difference in bond energies between the keto and enol forms of pyruvic acid. A summation of bond energies reveals that the keto form has 76 kJ/mol (18 kcal/mol) more bond energy than the enol form. Recall that bond energy is the energy required to break a bond and therefore it is equal to the energy released when the bond is formed. Hence the formation of a molecule with higher bond energy will result in the release of energy. Thus, the hydrolysis of the phosphate ester bond results in the release of energy because it promotes the enol-keto tautomerization.

7.1.4 Acyl derivatives of phosphate and coenzyme A

Acyl derivatives of phosphate and coenzyme A also have high energy. An acyl group is a derivative of a carboxylic acid and has the

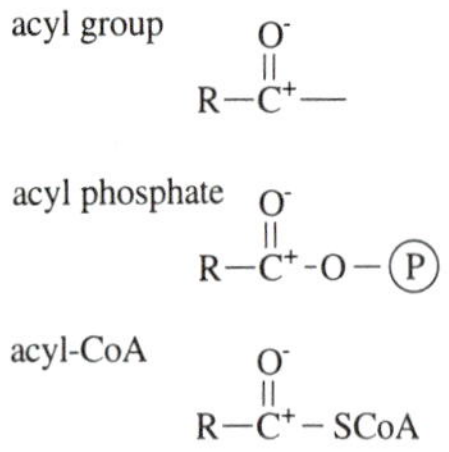

acyl group

$$R-\overset{O^-}{\underset{}{C^+}}-$$

acyl phosphate

$$R-\overset{O^-}{\underset{}{C^+}}-O-\text{\textcircled{P}}$$

acyl-CoA

$$R-\overset{O^-}{\underset{}{C^+}}-SCoA$$

Fig. 7.5 Structures of the acyl group, acyl-CoA and acyl-phosphates. Notice that the carbonyl group is polarized and the carbon is subject to nucleophilic attack during acyl group transfer reactions. The phosphate ester and thioester bonds in the acyl phosphates and acyl-CoAs have a high free energy of hydrolysis.

structure shown in Fig. 7.5, where R may be an alkyl or aryl group. The acyl derivatives of coenzyme A (CoA) and of phosphate are also shown in Fig. 7.5.

Why do acyl derivatives have a high free energy of hydrolysis?

The acyl derivatives of phosphate esters and thioesters have high group transfer potential because they do not resonate well. Consider the resonance forms of a normal ester and a phosphate ester illustrated in Fig. 7.6. In a normal ester, when two electrons shift from the oxygen to form a double bond during resonance, oxygen acquires a positive charge. In acyl-CoA and acyl-phosphate derivatives the oxygen attached to R' is replaced by a phosphorus or sulfur atom (Fig. 7.6). Resonance is hindered in phosphate esters because

$$R-\overset{O}{\underset{}{C}}-O-R' \rightleftharpoons R-\overset{O^-}{\underset{}{C}}=O^+-R'$$

$$R-\overset{O}{\underset{}{C}}-O-\overset{}{\underset{O_{\cdot}}{P^+}}-O-R' \xrightarrow{\text{unlikely}} R-\overset{O^-}{\underset{}{C}}=O^+-\overset{}{\underset{O_{\cdot}}{P^+}}-O-R'$$

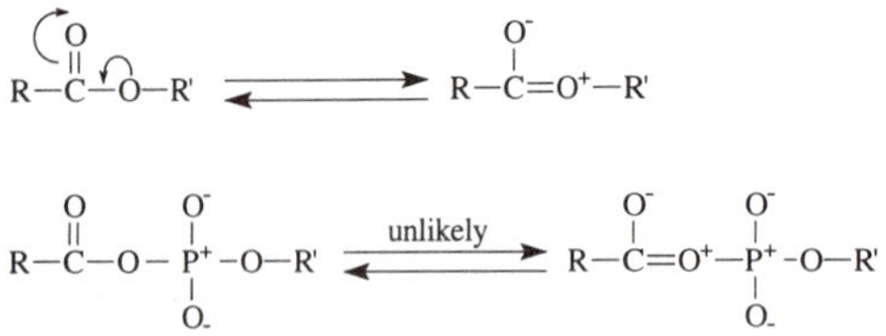

Fig. 7.6 Resonance of an oxygen ester. Note that electrons shift from the oxygen in the C–OR' to form a double bond. This is unlikely in phosphate esters because the phosphorus atom bears a positive charge and prevents electrons from shifting in from the oxygen atom to form the double bond. Resonance is made less likely in thioesters because the sulfur atom does not form double bonds.

the positive charge on the phosphorus atom prevents electrons from shifting in from the oxygen to form a double bond, which would leave two adjacent positive centers. Thioesters also do not resonate as well as normal esters. The reason for this is that sulfur forms double bonds poorly. The poor resonance of the phosphate and thioesters is in sharp contrast to the high resonance of the free carboxylate group that forms when the phosphoryl group or CoA is transferred to an acceptor molecule. Thus, the hydrolysis of the acyl derivatives of phosphate and coenzyme A is energetically favored because it leads to products stabilized by resonance with respect to the reactants.

The importance of acyl derivatives in group transfer reactions

Acyl derivatives are very versatile. Depending on the specificity of the enzyme catalyzing the reaction, they donate *acyl groups, CoA groups,* or *phosphoryl groups*. For example: fatty acids and proteins are synthesized via acyl group transfer reactions; CoA transfer takes place during fermentative reactions in bacteria; and acyl phosphates donate their phosphoryl groups to ADP to form ATP. Enzymes that catalyze the transfer of acyl groups are called *transacylases*. Those that transfer CoA or phosphoryl groups are called *CoA transferases* and *kinases*, respectively.

7.2 The central role of group transfer reactions in biosynthesis

Group transfer reactions are central to all of metabolism because biological molecules such as proteins, lipids, carbohydrates, and nucleic acids, are synthesized as a result of group transfer reactions. The high-energy donors are ATP or other nucleotide derivatives, acyl-CoA, and acyl-phosphates.

7.2.1 How ATP can be used to form amide linkages, thioester bonds, and ester bonds

As examples, we will consider the cases where ATP is used to form amide linkages, thioes-

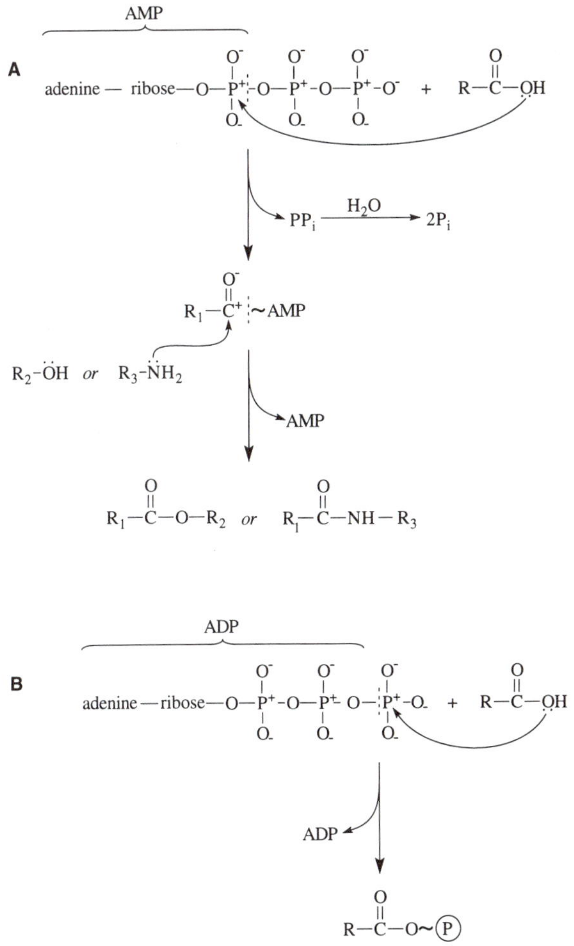

Fig. 7.7 ATP provides the energy to make ester and amide linkages. (A) A carboxyl attacks the α phosphate of ATP, displacing pyrophosphate, and forming the AMP derivative. The hydrolysis of pyrophosphate catalyzed by pyrophosphatase drives the reaction to completion. The AMP is then displaced by an attack on the carbonyl carbon by a hydroxyl or amino group, resulting in the transfer of the acyl group to form the ester or the substituted amide. (B) This is similar to (A) except that the attack is on the γ phosphate of ATP, displacing ADP, and forming the acyl phosphate.

ters, and esters. In all of these reactions, a carboxyl group accepts either AMP or phosphate from ATP in a group transfer reaction to form a *high-energy intermediate*. The high-energy intermediate is an acyl derivative with a high group transfer potential. The acyl-AMP or acyl-phosphate can donate the acyl group in a subsequent reaction forming an amide, ester, or thioester bond, depending upon whether the attacking nucleophile is N:, O:, or S:, respectively. Consider the reactions in Fig. 7.7. Notice

the sequence of reactions. First there is a displacement on either the α or γ phosphate of ATP to form the acyl derivative. Then there is a displacement on the carbonyl carbon to transfer the acyl moiety to the nucleophile to form the ester or amide bond. Notice that a pyrophosphatase is associated with reactions in which pyrophosphate is displaced. Because of the high free energy of hydrolysis of the pyrophosphate bond, the group transfer reaction is driven to completion.

7.2.2 How ATP is used to form peptide bonds during protein synthesis

As an example of the principles described in Sections 7.2 and 7.2.1, we will consider the formation of peptide bonds during protein synthesis. The acyl donor is made by derivatizing the α-carboxyl group on the amino acid with AMP using ATP as the AMP donor (Fig. 7.8). The acyl-AMP is the high-energy intermediate. In the next reaction, the acyl group is transferred to transfer RNA (tRNA) so that the carboxyl group becomes derivatized with tRNA. The acyl-tRNA also has a high group transfer potential. Thus, energy has flowed from ATP to aminoacyl-AMP to aminoacyl-tRNA. The aminoacyl-tRNA is attacked by the nucleophilic nitrogen of an amino group from another amino acid and the acyl portion is transferred to the amino group forming the

peptide bond. This last reaction takes place on the ribosome. The reaction goes to completion because of the large difference in group transfer potential between the amino acyl-tRNA and the peptide that is formed. These reactions exemplify the principle that the energy to make covalent bonds (in this case a peptide bond) derives from a series of group transfer reactions starting with ATP, in which ATP provides the energy to make high-energy intermediates that serve as group donors. In this way, all of the large complex molecules (i.e., proteins, nucleic acids, polysaccharides, lipids, and so on) are synthesized.

7.3 ATP synthesis by substrate-level phosphorylation

We have seen how ATP can drive the synthesis of biological molecules via a coupled

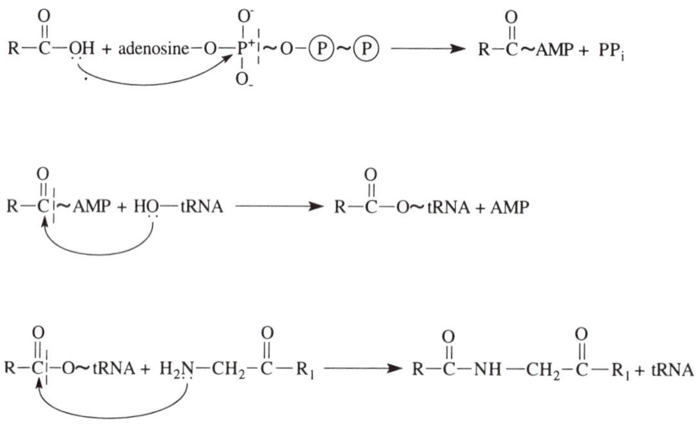

Fig. 7.8 Formation of a peptide bond. Peptide bonds are formed as a result of a series of group transfer reactions. The first reaction is the transfer of AMP from ATP to the carboxyl group of the amino acid. In this reaction the α phosphorus of ATP is attacked by the OH in the carboxyl group. Recall that the P=O bonds in ATP are semipolar and the phosphorus is an electropositive center. Most of the group transfer potential of ATP is trapped in the product and the reaction is freely reversible. However, the reaction is driven to completion by the hydrolysis of the pyrophosphate (not shown). The second reaction is the displacement of the AMP by tRNA. This is not done for energetic reasons, but rather because the tRNA is an adaptor molecule that aids in placing the amino acid in the correct position with respect to the mRNA on the ribosome. The synthesis of the aminoacyl-tRNA is reversible, indicating that the group transfer potential of the aminoacyl-tRNA is similar to that of the aminoacyl-AMP. The third reaction, which takes place on the ribosome, is the displacement of the tRNA by the amino group of a second amino acid, resulting in the synthesis of a peptide bond. This reaction proceeds with the release of a relatively large amount of free energy and is irreversible. (On the ribosome, it is the amino group of the incoming aminoacyl-tRNA at the "a" site that attacks the carbonyl of the resident aminoacyl-tRNA at the "p" site. The polypeptide is thus transferred from the "p" site to the "a" site.)

series of group transfer reactions. But how is ATP itself made? The answer depends upon whether the ATP is synthesized in the membranes or in the cytosol. In the membranes, the phosphorylation of ADP is coupled to oxidation–reduction reactions via the generation of an electrochemical gradient of protons which is then used to phosphorylate ADP via the membrane ATP synthase. That process is called oxidative phosphorylation or electron transport phosphorylation, and is discussed in Sections 3.6.2, 3.7.1 and 4.5. ATP in the soluble part of the cell is made by phosphorylating ADP by a process called *substrate-level phosphorylation*. We can define a substrate-level phosphorylation as the phosphorylation of ADP in the soluble part of the cell using a high-energy phosphoryl donor. Substrate level phosphorylations are catalyzed by enzymes called *kinases*. During a substrate-level phosphorylation, an oxygen in the β phosphate of ADP acts as a nucleophile and bonds to the phosphate phosphorous in the high-energy donor. The phosphoryl group is transferred to ADP making ATP. Consider the phosphoryl group transfer from an acyl phosphate to ADP (Fig. 7.9). Phosphoryl donors for ATP synthesis during substrate level phosphorylations include 1.3-bisphosphoglycerate (BPGA), phosphoenolpyruvate (PEP), acetyl-phosphate, and succinyl-CoA plus inorganic phosphate. These high-energy phosphoryl donors are listed in table 7.1. The four major substrate-level phosphorylations are listed in table 7.2. The metabolic pathways in which they occur are:

1. The substrate-level phosphorylations using BPGA and PEP take place during glycolysis;

2. The succinyl-CoA reaction is part of the citric acid cycle;

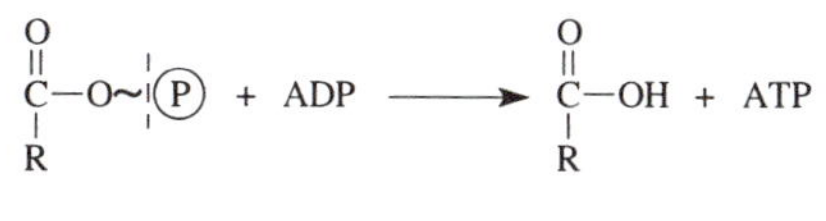

Fig. 7.9 A substrate level phosphorylation. This is an example of an acyl phosphate donating a phosphoryl group to ADP. The carboxylic acid is displaced. An example is the phosphorylation of ADP by 1,3-bisphosphoglycerate.

Table 7.2. Four substrate level phosphorylations

1,3-BPGA + ADP	$\longrightarrow$ 3-PGA + ATP
PEP + ADP	$\longrightarrow$ Pyruvic acid + ATP
Acetyl-P + ADP	$\longrightarrow$ Acetic acid + ATP
Succinyl-CoA + P$_i$ + ADP	$\longrightarrow$ Succinic acid + ATP + CoASH

3. Acetyl-phosphate is formed from acetyl-CoA, itself formed from pyruvate. This is an important source of ATP in anaerobically growing bacteria.

The synthesis of ATP via substrate-level phosphorylation first requires the synthesis of one of the high-energy phosphoryl donors listed in Table 7.2. All of the reactions that synthesize a high-energy molecule are oxidations. The single exception is the synthesis of phosphoenopyruvate, which results from a dehydration. The synthesis of the phosphoryl donors and the substrate level phosphorylations of ADP are described next. The following reactions are extremely important because they produce ATP, hence their emphasis in this chapter. They will be referred to again in subsequent chapters in the context of the pathways in which they participate.

7.3.1 1,3-Bisphosphoglycerate

During the degradation of sugars in the metabolic pathway (glycolysis), the 6-carbon sugar glucose is cleaved into two 3-carbon fragments called phosphoglyceraldehyde (PGALD) (Chapter 8). The phosphoglyceraldehyde is then oxidized to 1,3-bisphosphoglycerate, using inorganic phosphate as the source of phosphate and NAD$^+$ as the electron acceptor (Fig. 7.10). The reaction is catalyzed by *phosphoglyceraldehyde dehydrogenase*. The 1,3-bisphosphoglycerate then donates a phosphoryl group to ADP, in a substrate level phosphorylation, to form ATP and 3-phosphoglycerate in a reaction catalyzed by *phosphoglycerate kinase* (Fig. 7.11).

Phosphoglyceraldehyde is an important intermediate in all of the sugar catabolic pathways, not simply glycolysis, because it is oxidized to a high-energy phosphoryl donor

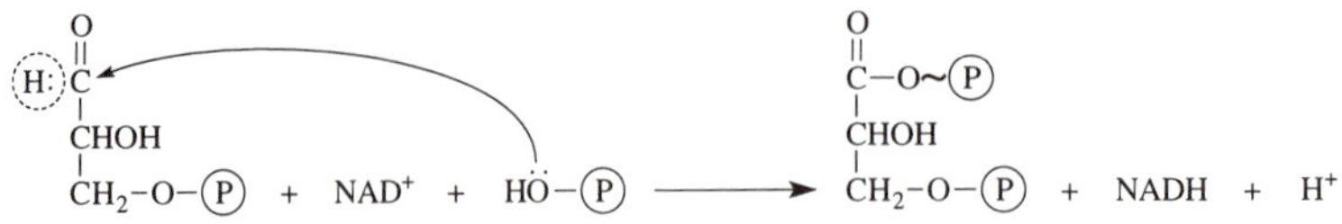

O
||
H:C
|
CHOH
|
CH$_2$-O-(P) + NAD$^+$ + HO-(P) $\longrightarrow$

3-PGALD   **P$_i$**

O
||
C-O~(P)
|
CHOH
|
CH$_2$-O-(P) + NADH + H$^+$

BPGA

Fig. 7.10 Oxidation of phosphoglyceraldehyde (PGALD). The incorporation of inorganic phosphate (P$_i$) into a high-energy phosphoryl donor occurs during the oxidation of phosphoglyceraldehyde (PGALD). The product is the acyl phosphate, 1,3-bisphosphoglycerate (BPGA). Energy that would normally be released as heat is trapped in 1,3-bisphosphoglycerate because inorganic phosphate rather than water is the nucleophile. As a consequence, an acyl phosphate rather than a free carboxylic acid is formed.

using inorganic phosphate. When coupled to the phosphoglycerate kinase reaction, it is the only way to synthesize net ATP from inorganic phosphate and ADP during the degradation of sugars to pyruvate.

7.3.2 Acetyl-phosphate

Acetyl-phosphate can be made from pyruvate via acetyl-CoA. The sequence is:

Pyruvate $\longrightarrow$ acetyl-CoA $\longrightarrow$ acetyl-phosphate.

These reactions are extremely important for fermenting bacteria because they produce acetyl-phosphate, which is a phosphoryl donor for ATP synthesis (Chapter 13). The acetate that is produced from the acetyl-phosphate is excreted into the medium. The conversion of pyruvate to acetyl-CoA will be discussed first.

Formation of acetyl-CoA from pyruvate
Acetyl-CoA is usually made by decarboxylating pyruvate, which is a key intermediate in

the breakdown of sugars. There are three well-characterized enzyme systems in the bacteria that decarboxylate pyruvate to acetyl-CoA. One is found in aerobic bacteria (and mitochondria) and is called *pyruvate dehydrogenase*. It usually is not present in anaerobically growing bacteria. The other two are found only in bacteria growing anaerobically. These are *pyruvate–formate lyase* and *pyruvate–ferredoxin oxidoreductase* (Fig. 7.12). Acetyl-CoA formed aerobically using pyruvate dehydrogenase is not a source of acetyl-phosphate but usually enters the citric acid cycle (Chapter 8).

An important difference between pyruvate dehydrogenase and the other two enzymes is that the pyruvate dehydrogenase reaction produces NADH. This can be disadvantageous

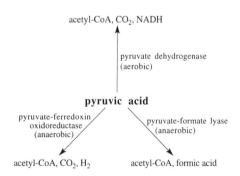

acetyl-CoA, CO$_2$, NADH

pyruvate dehydrogenase
(aerobic)

pyruvic acid

pyruvate-ferredoxin
oxidoreductase
(anaerobic)

pyruvate-formate lyase
(anaerobic)

acetyl-CoA, CO$_2$, H$_2$ acetyl-CoA, formic acid

O
||
C-O~(P)
|
H-C-OH + ADP $\longrightarrow$
|
CH$_2$O-(P)

1,3-BPGA

O
||
C-OH
|
H-C-OH + ATP
|
CH$_2$O-(P)

3-PGA

Fig. 7.11 The phosphoglycerate kinase reaction. ADP carries out a nucleophilic attack on the phosphoryl group of 1,3 bisphosphoglycerate (BPGA) displacing the free carboxylic acid (3-phosphoglycerate, 3-PGA).

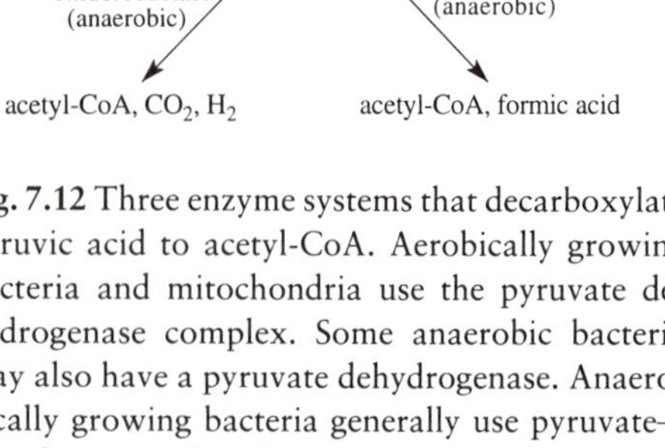

Fig. 7.12 Three enzyme systems that decarboxylate pyruvic acid to acetyl-CoA. Aerobically growing bacteria and mitochondria use the pyruvate dehydrogenase complex. Some anaerobic bacteria may also have a pyruvate dehydrogenase. Anaerobically growing bacteria generally use pyruvate–ferredoxin oxidoreductase or pyruvate–formate lyase instead of pyruvate dehydrogenase.

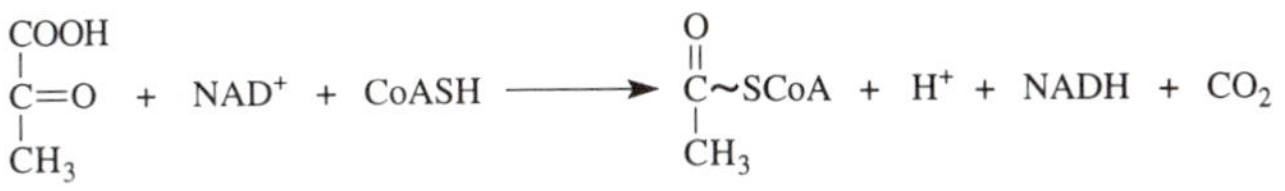

$$\begin{array}{c} \text{COOH} \\ | \\ \text{C=O} \\ | \\ \text{CH}_3 \end{array} + \text{NAD}^+ + \text{CoASH} \longrightarrow \begin{array}{c} \text{O} \\ || \\ \text{C} \sim \text{SCoA} \\ | \\ \text{CH}_3 \end{array} + \text{H}^+ + \text{NADH} + \text{CO}_2$$

Fig. 7.13 The pyruvate dehydrogenase reaction. In this reaction, the two electrons that bond the carboxyl group to the rest of the molecule are transferred by the enzyme to NAD^+. At the same time, coenzyme A attaches to the carbonyl group to form the acylated coenzyme A derivative. If the oxidation were to take place using the :OH from water instead of the :SH from CoASH to supply the fourth bond to the carbonyl carbon, then the product would be acetic acid and a great deal of energy would be lost as heat. However, the thioester of the carboxyl group cannot resonate as well as the free carboxyl group and, for this reason, the energy that would have normally been released during the oxidation is 'trapped' in the acetyl-CoA, a molecule with a high group transfer potential.

to anaerobically growing bacteria because there is often no externally provided electron acceptor to re-oxidize the NADH.

1. The pyruvate dehydrogenase reaction

Bacteria that are respiring aerobically use pyruvate dehydrogenase[2] to decarboxylate pyruvic acid to acetyl-CoA. This is an enzyme reaction that is also found in mitochondria. The reaction is shown in Fig. 7.13. A more detailed description of the pyruvate dehydrogenase reaction can be found in Section 8.6.

2. Pyruvate–ferredoxin oxidoreductase

Most anaerobically growing bacteria use different enzymes to decarboxylate pyruvate to acetyl-CoA. One of these enzymes is *pyruvate–ferredoxin oxidoreductase* which is found in the clostridia, sulfate-reducing bacteria, and some other anaerobes. The enzyme catalyzes a reaction similar to pyruvate dehydrogenase except that the electron acceptor is not NAD^+. Instead, it is an iron–sulfur protein called ferredoxin (Fig. 7.14). An important feature of the pyruvate–

ferredoxin oxidoreductase in fermenting bacteria is that the enzyme is coupled to a second enzyme called *hydrogenase*. The hydrogenase catalyzes the transfer of electrons from reduced ferredoxin to H^+ to form hydrogen gas. The importance of the hydrogenase reaction is that it re-oxidizes the reduced ferredoxin. Re-oxidation of reduced ferredoxin and NADH in fermenting bacteria relies on the use of protons or endogenously produced organic compounds as electron sinks. (See the discussion of fermentations in Chapter 13.) The hydrogenase reaction accounts for the hydrogen gas produced during fermentations. The ferredoxin-linked decarboxylation of pyruvate to acetyl-CoA is reversible and is used for autotrophic CO_2 fixation in certain anaerobic bacteria (Sections 12.1.2 and 12.1.7).

3. Pyruvate–formate lyase

Pyruvate–formate lyase is an enzyme found in certain fermenting bacteria (e.g., the enteric bacteria and certain lactic acid bacteria). In the reaction, the electrons stay with the

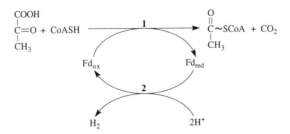

Fig. 7.14 The pyruvate–ferredoxin oxidoreductase reaction and the hydrogenase. The electrons travel from the ferredoxin (Fd) to protons via the enzyme hydrogenase. Enzymes: 1, pyruvate–ferredoxin oxidoreductase; 2, hydrogenase.

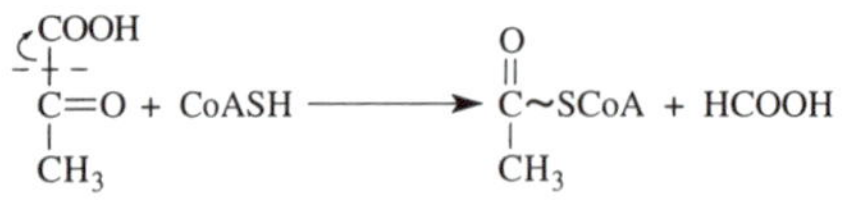

$$\begin{array}{c} \underset{|}{\overset{\text{COOH}}{C}} \\ \overset{+}{\underset{|}{C}}{=}O + \text{CoASH} \longrightarrow \overset{O}{\overset{\|}{C}}{\sim}\text{SCoA} + \text{HCOOH} \\ \underset{|}{\overset{}{CH_3}} \qquad\qquad \underset{|}{\overset{}{CH_3}} \end{array}$$

Fig. 7.15 The pyruvate–formate lyase reaction. Part of the molecule becomes oxidized and part becomes reduced. The part that becomes reduced is the carboxyl group that leaves as formic acid.

carboxyl that is removed, and therefore formate is formed instead of carbon dioxide. One of the advantages to using pyruvate–formate lyase is that reduced ferredoxin or NADH are not produced and the electrons are disposed of as part of the formate. The reaction is illustrated in Fig. 7.15.

Formation of acetyl phosphate from acetyl-CoA

Acetyl-phosphate is made from acetyl-CoA by a displacement of the CoASH by inorganic phosphate in a reaction catalyzed by *phosphotransacetylase*. (Bacteria growing aerobically usually oxidize the acetyl-CoA in the citric acid cycle.)

Phosphotransacetylase

$$\text{Acetyl-CoA} + P_i \longrightarrow \text{acetyl-P} + \text{CoASH}$$

(Acetyl-P can also be made directly from pyruvate and inorganic phosphate using *pyruvate oxidase*. This is a flavoprotein enzyme found in certain *Lactobacillus* species. See Section 13.9.)

Formation of ATP from acetyl-phosphate

The acetyl-phosphate then donates the phosphoryl group to ADP in a substrate level phosphorylation catalyzed by *acetate kinase*.

Fermenting bacteria that oxidize pyruvate to acetate use both the phosphotransacetylase and the acetate kinase, and derive an ATP from the process. These reactions are widespread among fermenting bacteria.

Acetate kinase

$$\text{Acetyl-P} + \text{ADP} \longrightarrow \text{acetic acid} + \text{ATP}$$

7.3.3 Succinyl-CoA

Succinyl-CoA is made by the oxidative decarboxylation of α-ketoglutarate, a reaction that occurs in the citric acid cycle (Fig. 7.16). It is strictly analogous in its mechanism and cofactor requirements to the oxidative decarboxylation of pyruvate by pyruvate dehydrogenase. The enzyme that carries out the oxidation of α-ketoglutarate is called alpha-ketoglutarate dehydrogenase. Succinyl-CoA then drives the synthesis of ATP from inorganic phosphate and ADP in a reaction catalyzed by the citric acid cycle enzyme succinate thiokinase *succinate thiokinase*. The reaction is as follows,

$$\text{Succinyl-CoA} + \text{ADP} + P_i$$
$$\longrightarrow \text{succinate} + \text{CoASH} + \text{ATP}$$

Succinyl-phosphate is not a free intermediate. Perhaps the CoASH is transferred from succinyl-CoA to the enzyme where it is displaced by phosphate. The phosphorylated enzyme would then be the phosphoryl donor for ADP in ATP synthesis. It should be pointed out that, whereas bacteria and plants produce ATP from succinyl-CoA, animals produce GTP instead. Succinyl-CoA is important not merely for ATP synthesis but also as a precursor for heme synthesis.

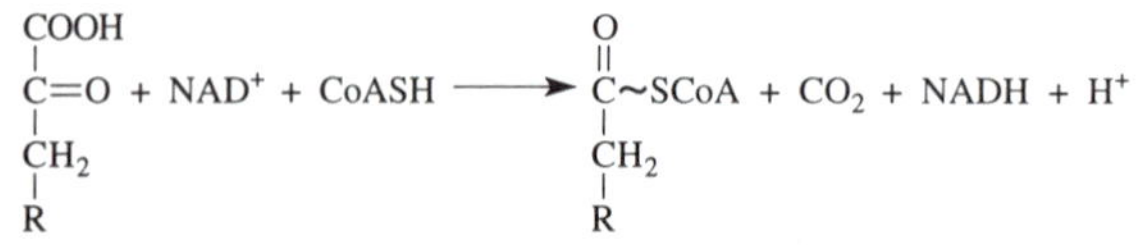

$$\begin{array}{c} \underset{|}{\overset{\text{COOH}}{}} \\ \underset{|}{\overset{|}{C}}{=}O + \text{NAD}^+ + \text{CoASH} \longrightarrow \overset{O}{\overset{\|}{C}}{\sim}\text{SCoA} + CO_2 + \text{NADH} + \text{H}^+ \\ \underset{|}{\overset{}{CH_2}} \qquad\qquad\qquad\qquad\qquad \underset{|}{\overset{}{CH_2}} \\ \underset{}{\overset{}{R}} \qquad\qquad\qquad\qquad\qquad\qquad\quad \underset{}{\overset{}{R}} \\ \qquad\text{α-ketoglutarate} \qquad\qquad\qquad\qquad \text{succinyl-CoA} \end{array}$$

Fig. 7.16 The α-ketoglutarate dehydrogenase reaction. Notice that the substrate molecule resembles pyruvic acid. The difference is that in pyruvic acid, the R group is H, whereas in α-ketoglutaric acid, it is CH_2–COOH.

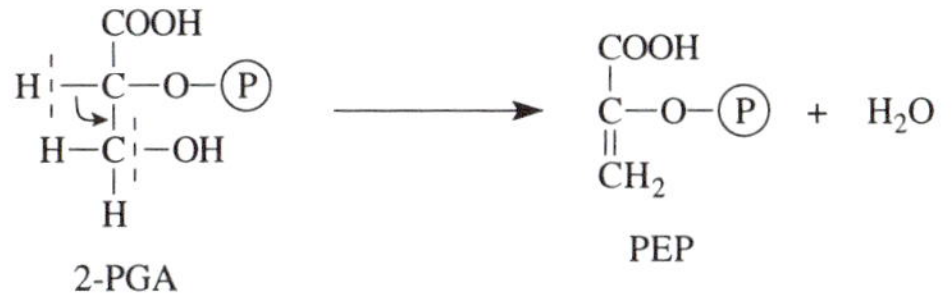

COOH
|
H—C—O—(P)
|
H—C—OH
|
H

2-PGA

$\longrightarrow$

COOH
|
C—O—(P) + H₂O
‖
CH₂

PEP

Fig. 7.17 The enolase reaction. 2-Phosphoglycerate is dehydrated to phosphoenolpyruvate. Isotope exchange studies suggest that the first step is the removal of a proton from C2 to form a carbanion intermediate which loses the hydroxyl and becomes phosphoenolpyruvate. 2-PGA, 2-phosphoglycerate; PEP, phosphoenolpyruvate.

7.3.4 Phosphoenolpyruvate

When cells are growing on sugars using the glycolytic pathway, they make phosphoenolpyruvate from 2-phosphoglycerate (2-PGA), an intermediate in the breakdown of the sugars. The reaction is a dehydration and is catalyzed by the enzyme *enolase* (Fig. 7.17). Phosphoenolpyruvate then phosphorylates ADP in a reaction catalyzed by *pyruvate kinase*:

PEP + ADP $\longrightarrow$ pyruvate + ATP

The enolase and pyruvate kinase reactions are important in energy metabolism because they serve to regenerate the ATP that is used to phosphorylate the sugars during the initial stages of sugar catabolism. However, they cannot account for the synthesis of net ATP from ADP and inorganic phosphate, since the phosphate in the PEP originated from ATP (or PEP), rather than from inorganic phosphate. Phosphoenolpyruvate is also necessary for the synthesis of muramic acid and certain amino acids (Chapters 9 and 10).

7.4 Summary

Biochemical reactions in the cytosol are driven by high-energy molecules. There are several high-energy molecules (e.g., ATP, BPGA, PEP, acetyl-P, acetyl-CoA, and succinyl-CoA). They are called high-energy molecules because they have a bond with a

high free energy of hydrolysis. The reasons for this depend upon the structure of the whole molecule, not on any particular bond. High free energies of hydrolysis can be due to electrostatic repulsion between adjacent phosphate groups and/or diminished resonance. The term "high-energy bond" is a misnomer because the free energy of hydrolysis is not bond energy. The term "group transfer potential" refers to the negative of the free energy of hydrolysis and is a useful concept when comparing the tendency of chemical groups to be transferred to attacking nucleophiles. Thus, ATP has a high phosphoryl group transfer potential (i.e., around 35 kJ under standard conditions, pH 7), whereas glucose-6-phosphate has a low phosphoryl group transfer potential (i.e., around 14 kJ). Hence, ATP will transfer the phosphoryl group to glucose with the release of 35 − 14 = 21 kJ of energy.

Group transfer reactions can occur with conservation of energy to form high-energy intermediates, which themselves can be group donors in coupled reactions. This explains how ATP can provide the energy to drive a series of coupled chemical reactions that result in the synthesis of nucleic acids, proteins, polysaccharides, lipids, and so on.

The formation of a high-energy molecule usually involves an oxidation of an aldehyde (phosphoglyceraldehyde) or the oxidative decarboxylation of a β-keto carboxylic acid (pyruvate or succinate). In substrate-level phosphorylation, the energy of the redox reaction is trapped in an acyl-phosphate or an acyl-CoA derivative. This is in contrast to respiratory phosphorylation where the energy from the redox reaction is trapped in a Δp which drives ATP synthesis.

Phosphoenolpyruvate is not formed as a result of an oxidation–reduction reaction but from a dehydration. However, the phosphate in phosphoenolpyruvate was already present in 2-phosphoglycerate, having been previously donated by ATP or phosphenolpyruvate during the sugar phosphorylations. Therefore, synthesis of ATP from phosphenolpyruvate does not represent the formation of ATP from ADP and inorganic phosphate, but rather the regeneration of ATP that was used previously to phosphorylate the sugars.

Study Questions

1. What is the definition of group transfer potential? Why do ATP, acyl phosphates, and PEP have a high phosphoryl group transfer potential?

2. Write a series of hypothetical reactions in which PEP drives the synthesis of A–B from A, B, and ADP.

3. What are two features that distinguish substrate level phosphorylations from electron transport phosphorylation?

4. What features do the synthesis of BPGA, acetyl-CoA, and succinyl-CoA have in common with each other but not with PEP?

5. What do the synthesis of acetyl-CoA and succinyl-CoA have in common?

6. Write a series of reactions that result in the synthesis of a substituted amide or an ester from a carboxylic acid. Use ATP as the source of energy. How is protein synthesis a modification of this reaction?

REFERENCES AND NOTES

1. For a discussion of high-energy molecules, see, Ingraham, L. L. 1962. *Biochemical Mechanisms* John Wiley and Sons, Inc. New York.

2. Dehydrogenases are enzymes that catalyze oxidation–reduction reactions in which hydrogens as well as electrons are transferred. They are named after one of the substrates (e.g., pyruvate dehydrogenase).

3. With rare exceptions, sugars must be phosphorylated in order to be metabolized.

8

Central Metabolic Pathways

The central metabolic pathways are those pathways that provide the precursor metabolites to all of the other pathways. They are the pathways for the metabolism of carbohydrates and carboxylic acids, such as C_4 dicarboxylic acids and acetic acid. The major carbohydrate pathways are the *Embden–Meyerhof–Parnas pathway* (also called the EMP pathway or glycolysis), the *pentose phosphate pathway* (PPP), and the *Entner–Doudoroff pathway* (ED). The Entner–Doudoroff pathway has been found only among the prokaryotes. The three pathways differ in many ways, but two generalizations can be made:

1. All three pathways convert glucose to phosphoglyceraldehyde, albeit by different routes.

2. The phosphoglyceraldehyde is converted to pyruvate via reactions that are the same in all three pathways.

From an energetic point of view, the reactions that convert phosphoglyceraldehyde to pyruvate are extremely important because they generate ATP from inorganic phosphate and ADP. This is because there is an oxidation in which inorganic phosphate is incorporated into an acyl-phosphate (i.e., the oxidation of phosphoglyceraldehyde to 1,3-bisphosphoglycerate). The 1,3-bisphosphoglycerate then donates the phosphoryl group to ADP in a substrate level phosphorylation.

The fate of the pyruvate that is formed during the catabolism of carbohydrates depends on whether the cells are respiring. If the organisms are respiring, then the pyruvate that is formed by the carbohydrate catabolic pathways is oxidized to acetyl-CoA, which is subsequently oxidized to carbon dioxide in the *citric acid cycle*. The latter generally operates only during aerobic respiration. If fermentation rather than respiration is taking place, then the pyruvate is converted to fermentation end products such as alcohols, organic acids, and solvents, rather than oxidized in the citric acid cycle. Fermentations are discussed in Chapter 13.

An overview of the carbohydrate catabolic pathways and their relationship to one another and to the citric acid cycle is shown in Fig. 8.1. Several points can be made about this figure. Notice that there are three substrate level phosphorylations, two during carbohydrate catabolism and one in the citric acid cycle. Furthermore, there are six oxidation reactions, one in glycolysis, one in the pyruvate dehydrogenase reaction, and four in the citric acid cycle. These oxidations produce NADH (primarily) and $FADH_2$. The NADH and $FADH_2$ must be reoxidized in order to regenerate the NAD^+ and FAD that are required for the oxidations. The route of re-oxidation and the energy yield depend upon whether the organism is respiring or fermenting. During respiration, the NADH and $FADH_2$ are re-oxidized via electron

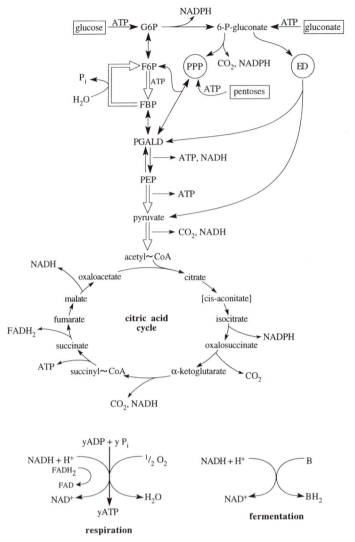

Fig. 8.1 Relationships between the major carbohydrate pathways and the citric acid cycle. The pathway from glucose-6-phosphate to pyruvate is the Embden–Meyerhof–Parnas pathway (glycolysis). The pentose phosphate pathway (PPP) and the Entner-Doudoroff pathway (ED) branch from 6-phosphogluconate. Both of these pathways intersect with the glycolytic pathway at phosphoglyceraldehyde. All the carbohydrate pathways produce pyruvate which is oxidized to acetyl-CoA. In aerobically-growing organisms, the acetyl-CoA is oxidized to CO_2 in the citric acid cycle. The electrons from NAD(P)H and $FADH_2$ are transferred to the electron transport chain in respiring organisms with the formation of ATP. In fermenting cells, the NADH is re-oxidized by an organic acceptor (B) that is generated during catabolism. The citric acid cycle does not operate as an oxidative pathway during fermentative growth. G6P, glucose-6-phosphate; F6P, fructose-6-phosphate; FDP, fructose-1,6-bisphosphate; PGALD, 3-phosphoglyceraldehyde; PEP, phosphoenolpyruvate.

transport with the formation of a Δp. (The Δp is used for ATP synthesis via respiratory phosphorylation as explained in Chapter 4.) In fermenting cells, most of the NADH is re-oxidized in the cytosol by an organic acceptor, but ATP is not made.[1] The different pathways for the re-oxidation of NADH in fermenting bacteria are discussed in Chapter 13. The student will notice that the citric acid cycle generates a great deal of NADH and $FADH_2$. The re-oxidation of the NADH and $FADH_2$ requires adequate amounts of electron acceptor, such as is provided to respiring organisms. In fact, the oxidative citric acid cycle as illustrated in Fig. 8.1 is coupled to respiration, and during fermentative growth it becomes modified into a reductive pathway (Section 8.9).

8.1 Glycolysis

It is best to think of glycolysis as occurring in two stages:

Stage 1. This stage catalyzes the splitting of the glucose molecule (C_6) into two phosphoglyceraldehyde (C_3) molecules. It consists of four consecutive reactions. Two ATPs are used per glucose metabolized, and these donate the phosphoryl groups that become the phosphates in phosphoglyceraldehyde (The phosphoglyceraldehyde eventually becomes phosphoenolpyruvate in stage two, and the phosphate that originated from ATP is returned to ATP in the pyruvate kinase step, thus regenerating the ATP.)

Stage 2. This stage catalyzes the oxidation of phosphoglyceraldehyde to pyruvate. It consists of five consecutive reactions. Stage 2 generates four ATPs per glucose metabolized, hence the net yield of ATP is two. Stage 2 reactions are not unique to glycolysis and also occur when pyruvate is formed from phosphoglyceraldehyde in the pentose phosphate pathway and the Entner–Doudoroff pathway, accounting for ATP synthesis in these pathways.

Stage 1: glucose + 2 ATP $\longrightarrow$ 2 PGALD + 2 ADP

Stage 2: 2 PGALD + 2 P_i + 4 ADP + 2 NAD$^+$

$\longrightarrow$ 2 pyruvate + 4 ATP + 2 NADH + 2H$^+$

Sum: glucose + 2 ADP + 2 P_i + 2 NAD$^+$

$\longrightarrow$ 2 pyruvate + 2 ATP + 2 NADH + 2H$^+$

The reactions are summarized in Fig. 8.2. The pathway begins with the phosphoryla-tion of glucose to form glucose-6-phosphate (reaction 1). The phosphoryl donor is ATP in a reaction catalyzed by hexokinase. The ATP is regenerated from phosphoenolpyruvate in stage 2. Some bacteria phosphorylate glucose during transport into the cell via the phosphotransferase (PTS) system, in which case the phosphoryl donor is phosphoenolpyruvate. (See Section 15.3.4 for a discussion of the phosphotransferase system.) The glucose-6-phosphate (G6P) isomerizes to fructose-6-phosphate (F6P) in a reaction catalyzed by the enzyme isomerase (reaction 2). The isomerization is an electron shift where two electrons from the C2 carbon reduce the C1 aldehyde of the glucose-6-phosphate molecule to an alcohol (Section 8.1.3). The fructose-6-phosphate is phosphorylated at the expense of ATP to fructose-1,6-bisphosphate (FBP) by the enzyme fructose-6-phosphate kinase (reaction 3). The ATP used to phosphorylate fructose-6-phosphate is also regenerated from phosphoenolpyruvate in stage 2. The fructose-1,6-bisphosphate is split into phosphoglyceraldehyde (PGALD) and dihydrooxyacetone phosphate (DHAP) by fructose-1,6-bisphosphate aldolase (reaction 4). The splitting of fructose-1,6-bisphosphate is facilitated by the electron attracting keto group at C2, thus rationalizing the isomerization of glucose-6-phosphate to fructose-6-phosphate (Section 8.1.3). The dihydroxyacetone phosphate is isomerized to phosphoglyceraldehyde (reaction 5), in a reaction similar to the earlier isomerase reaction (reaction 2). Thus, stage 1 produces two moles of phosphoglyceraldehyde per mole of glucose.

In stage two, both moles of phosphoglyceraldehyde are oxidized to 1,3-bisphosphoglycerate (also called disphosphoglycerate, DPGA) (reaction 6). The bisphosphoglycerate serves as the phosphoryl donor for a substrate level phosphorylation catalyzed by the enzyme phosphoglycerate kinase (reaction 7). At this point, two ATPs are made, one from each of the two bisphosphoglycerates. The product of the phosphoglycerate kinase reaction is 3-phosphoglycerate (3-PGA). The two moles of 3-phosphoglycerate are converted to two moles of 2-phosphoglycerate (2-PGA) (reaction 8), which are dehydrated to two moles of phosphoenolpyruvate (PEP)

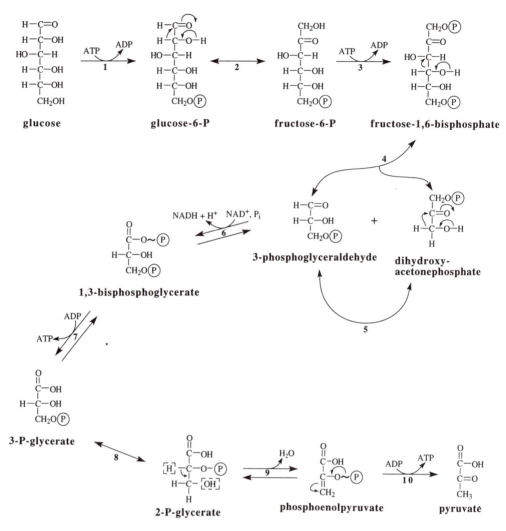

Fig. 8.2 Glycolysis. Enzymes: (1) hexokinase; (2) isomerase; (3) phosphofructokinase; (4) fructose-1,6-bisphosphate aldolase; (5) triosephosphate isomerase; (6) triosephosphate dehydrogenase; (7) phosphoglycerote kinase; (8) mutase; (9) enolase; (10) pyruvate kinase.

(reaction 9). The phosphoenolpyruvate serves as the phosphoryl donor in a second site substrate level phosphorylation to form two more moles of ATP and two moles of pyruvate (reaction 10). Notice that the phosphate in the phosphoglycerate originated from ATP during the phosphorylations in stage 1 (reactions 1 and 3). Thus, the net synthesis of ATP from ADP and inorganic phosphate in glycolysis is coupled to the oxidation of phosphoglyceraldehyde to 3-phosphoglycerate (reactions 6 and 7). For a more detailed discussion of substrate level phosphorylations, see Section 7.3.

8.1.1 Glycolysis as an anabolic pathway

The glycolytic pathway serves not only to oxidize carbohydrate to pyruvate and to phosphorylate ADP, but also provides precursor metabolites for many other pathways. Figure 8.3 summarizes the glycolytic reactions and points out only a few of the branch points to other pathways. For example, glucose-6-phosphate is a precursor to polysaccharides, pentose phosphates, and aromatic amino acids, fructose-6-phosphate is a precursor to amino sugars (e.g., muramic acid and glucos-amine found in the cell wall), dihydroxy-

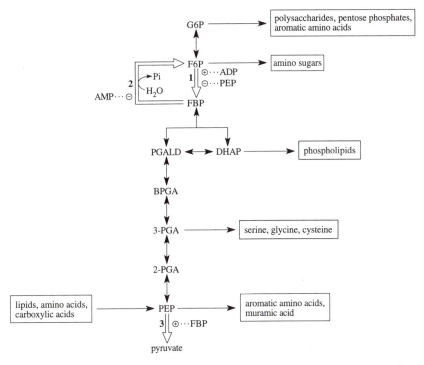

Fig. 8.3 Glycolysis as an anabolic pathway and its regulation in *E. coli*. The rationale for the pattern of regulation is that when the ADP and AMP levels are high, the ATP levels are low and therefore glycolysis is stimulated. Steady-state levels of intermediates are maintained by positive and negative feedback inhibition.

acetone phosphate is a precursor to phospholipids, and to the amino acids glycine, serine, and cysteine, and phosphoenolpyruvate is a precursor to aromatic amino acids and to the lactyl portion of muramic acid. When the organisms are not growing on carbohydrate, they must synthesize these glycolytic intermediates from other carbon sources. Figure 8.3 shows that some of the carbon from amino acids, carboxylic acids (organic acids), and lipids is converted to phosphoenolpyruvate from which the glycolytic intermediates can be synthesized. Also, pyruvate can serve as a carbon source and therefore must be converted to glycolytic intermediates. However, it can be seen that the glycolytic pathway can be reversed from phosphoenolpyruvate only to fructose-1,6-bisphosphate (FBP), and not at all from pyruvate. This is because the pyruvate kinase and phosphofructokinase reactions are physiologically irreversible due to the high free energy in the phosphoryl donors with respect to the phosphorylated products. *Therefore, to reverse glycolysis the kinase reactions are bypassed.* The conver-

sion of fructose-1,6-bisphosphate to fructose-6-phosphate requires fructose-1,6-bisphosphate phosphatase (Fig. 8.4). There are also alternative ways to convert pyruvate directly to phosphoenolpyruvate without using pyruvate kinase. These are discussed in Section 8.12.2.

8.1.2 Regulation of glycolysis

Figure 8.3 also illustrates the regulation of glycolysis in *E. coli*. Two key enzymes in regulating the *directionality* of carbon flow are phosphofructokinase (reaction 1) and fructose-1,6-bisphosphate phosphatase (reaction 2), which catalyze physiologically irreversible steps. The kinase catalyzes the phosphorylation of fructose-6-phosphate to fructose-1,6-bisphosphate, whereas the phosphatase catalyzes the dephosphorylation of fructose-1,6-bisphosphate to fructose-6-phosphate. Models for the regulation of glycolysis are based primarily on *in vitro* studies of the allosteric properties of the enzymes. Important effector molecules are AMP and ADP. When both of these are high, ATP is low,

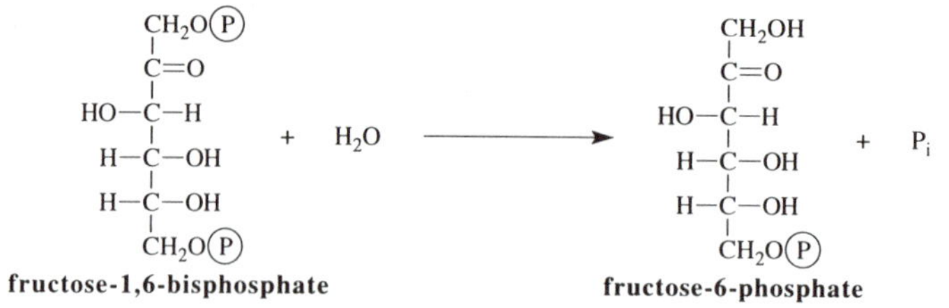

$$
\begin{array}{ccc}
\begin{array}{c}
\text{CH}_2\text{O}\,\text{P} \\
| \\
\text{C=O} \\
| \\
\text{HO−C−H} \\
| \\
\text{H−C−OH} \\
| \\
\text{H−C−OH} \\
| \\
\text{CH}_2\text{O}\,\text{P} \\
\textbf{fructose-1,6-bisphosphate}
\end{array}
& \quad + \text{ H}_2\text{O} \quad \longrightarrow \quad &
\begin{array}{c}
\text{CH}_2\text{OH} \\
| \\
\text{C=O} \\
| \\
\text{HO−C−H} \\
| \\
\text{H−C−OH} \\
| \\
\text{H−C−OH} \\
| \\
\text{CH}_2\text{O}\,\text{P} \\
\textbf{fructose-6-phosphate}
\end{array}
\quad + \text{ P}_i
\end{array}
$$

Fig. 8.4 The fructose-1,6-bisphosphatase reaction.

since they are both derived from ATP. That is to say,

$$ATP \longrightarrow ADP + P_i$$

$$ATP \longrightarrow AMP + PP_1$$

Thus, high ADP and AMP concentrations are a signal that the ATP levels are low. (Allosteric activation and inhibition is discussed in Chapter 6.) Since glycolysis produces ATP, it makes sense to stimulate glycolysis when the ATP levels are low. *E. coli* accomplishes this by allosterically activating the phosphofructokinase with ADP, which, as mentioned, is at a higher concentration when the ATP levels are low. At the same time that glycolysis is stimulated by ADP, the reversal of glycolysis is slowed by AMP which is also at a higher concentration when the ATP levels are low. The reason for this is that AMP inhibits the fructose-1,6-bisphosphate phosphatase reaction. The student may notice that the sum of the phosphofructokinase and fructose-1,6-bis-phosphatase reaction is the hydrolysis of ATP, i.e., ATPase activity. The stimulation of the phosphofructokinase by ADP and the inhibition of the phosphatase by AMP prevents the unnecessary hydrolysis of ATP. Glycolysis is not only regulated by AMP and ADP in *E. coli*, but also by phosphoenolpyruvate and fructose-6-phosphate. As indicated in Fig. 8.3, the phosphofructokinase is feedback inhibited by phosphoenolpyruvate. This can be considered an example of end-product inhibition. The pyruvate kinase, another physiologically irreversible reaction, is positively regulated by fructose-1,6-bisphosphate, which is an example of a precursor metabolite activating a later step in the pathway. (Feedback inhibition and precursor activation are discussed in Sections 6.1.1 and 6.1.2.)

8.1.3 The chemical bases for the isomerization and aldol cleavage reactions in glycolysis

It is important to understand the chemistry of metabolic reactions as well as to learn the pathways and their physiological role. To this end, the isomerization and aldol cleavage reactions will be explained because they are common reactions that we will see later in other pathways. Consider the isomerization of glucose-6-phosphate to fructose-6-phosphate. The rationale for this isomerization is that it creates an electron attracting keto group at C2 of the sugar, and the electron attracting keto group is necessary to break the bond between C2 and C3 in the aldolase reaction. These reactions are shown in Fig. 8.5. The isomerization can be viewed as the oxidation of C2 by C1, because two electrons shift from C2 to C1. This happens in two steps. A hydrogen dissociates from C2 and two electrons shift in to form the *cis*-enediol. Then, the hydrogen in the C2 hydroxyl dissociates and two electrons shift in, forcing the two electrons in the double bond to go to the C1. The result is fructose-6-phosphate. The fructose-6-phosphate becomes phosphorylated to fructose-1,6-bisphosphate. The fructose-1,6-bisphosphate is split by the aldolase when the keto group on C2 pulls electrons away from the C–C bond between C3 and C4, as two electrons shift in from the hydroxyl on C4 (Fig. 8.5). The products of the split are phosphoglyceraldehyde and dihydroxyacetone phosphate. A second isomerization

glucose-6-P → cis-enediol → fructose-6-P → (ATP) → fructose-bisphosphate

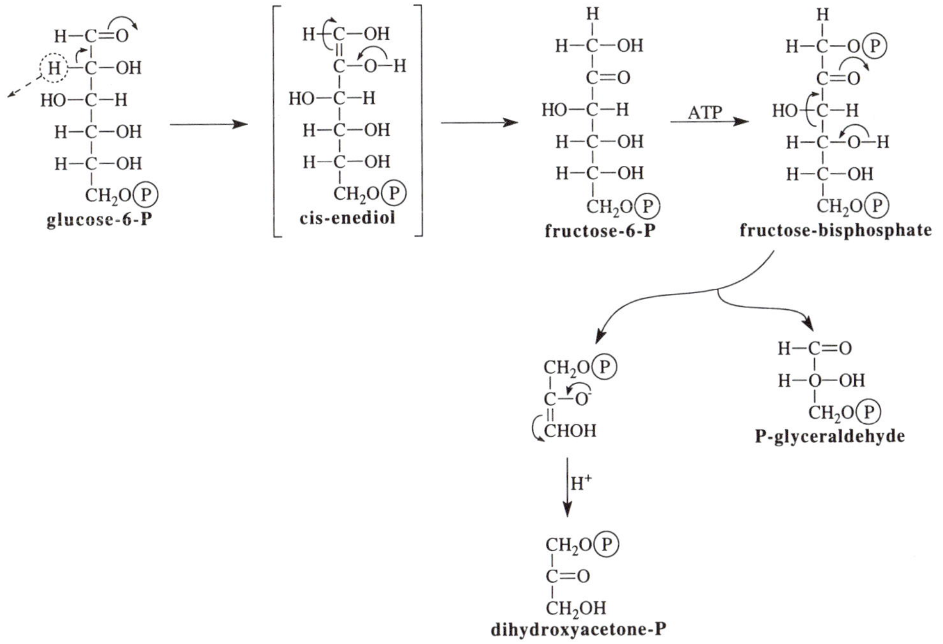

fructose-bisphosphate → dihydroxyacetone-P (CH$_2$O(P), C=O, CH$_2$OH) and P-glyceraldehyde (H—C=O, H—O—OH, CH$_2$O(P))

CH$_2$O(P) / C—O / CHOH →(H$^+$)→ dihydroxyacetone-P

Fig. 8.5 Making two phosphoglyceraldehydes from glucose-6-phosphate. Glucose-6-phosphate itself cannot be split because there is no electron-attracting group to withdraw the electrons from the C–C bond between C3 and C4. An electron-withdrawing keto group is created on C2 when glucose-6-phosphate is isomerized to fructose-6-phosphate.

converts the dihydroxyacetone phosphate to phosphoglyceraldehyde via the same mechanism as the isomerization between glucose-6-phosphate and fructose-6-phosphate. In this way, two phosphoglyceraldehydes can be formed from glucose-6-phosphate.

8.1.4 Why are the glycolytic intermediates phosphorylated?

In glycolysis, the phosphorylation of ADP by inorganic phosphate is due to two reactions that take place in stage two (i.e., the oxidation of the C1 aldehyde of 3-phosphoglyceraldehyde to the acyl phosphate) and the subsequent transfer of the phosphoryl group to ADP. The first reaction is catalyzed by triose-phosphate dehydrogenase and the second reaction is catalyzed by phosphoglycerate kinase. Given that these are the steps in which net ATP is made from ADP and inorganic phosphate, one can ask why the other intermediates are phosphorylated. Phosphorylation of the intermediates requires the use of two ATPs in stage 1, which are

simply regenerated in the pyruvate kinase step. There is probably more than one reason why all of the intermediates are phosphorylated. One reason may be because the kinase reactions in stage one are irreversible and therefore drive the reactions rapidly in the direction of pyruvate. Another reason has to do with the physiological role of the pathway. Glycolysis is not simply a pathway for the oxidation of glucose and the provision of ATP. Very importantly, the glycolytic pathway also provides phosphorylated precursors to many other pathways. In subsequent chapters, we will study these interconnections with other pathways. However, although phosphorylated sugar pathways are the rule, there are exceptions. There is a partly nonphosphorylated pathway and a completely nonphosphorylated pathway for the Entner–Doudoroff pathway. They are described in Section 8.5.4.

8.2 The fate of NADH

If the NADH were not re-oxidized to NAD$^+$, then all pathways (including glycolysis) that require NAD^+ would stop. Clearly, glycolysis must be coupled to pathways that reoxidize NADH back to NAD^+. Bacteria have three ways to reoxidize NADH: respiration, fermentation, and the hydrogenase reaction.

Respiration (aerobic or anaerobic)

$$NADH + H^+ + B + yADP + yP_i$$
$$\longrightarrow NAD^+ + BH_2 + yATP$$

where y is the number of coupling sites and B is the terminal electron acceptor.

Fermentation (anaerobic)

$$NADH + H^+ + B\ (organic) \longrightarrow NAD^+ + BH_2$$

Hydrogenase (anaerobic)

$$NADH + H^+ \longrightarrow H_2 + NAD^+$$

Aerobic and anaerobic respiration are discussed in Chapter 4. In the absence of respiration, NADH can be re-oxidized in the cytosol via fermentation discussed in Chapter 13. A third way to re-oxidize NADH is via the enzyme hydrogenase in the cytosol. Hydrogenases that use NADH as the electron donor are found in fermenting bacteria. However, the oxidation of NADH with the production of hydrogen gas generally proceeds only when the hydrogen gas concentration is kept low (e.g., during growth with hydrogen gas utilizers). This is because the equilibrium favors the reduction of NAD^+. Interspecies hydrogen transfer is discussed in Section 13.4.1.

8.3 Why write NAD^+ instead of NAD, and NADH instead of $NADH_2$?

Oxidized nicotinamide adenine dinucleotide is written as NAD^+, and the reduced form is written NADH, not $NADH_2$. To understand why, we must examine the structures (Fig. 8.6). Notice that the molecule can accept two electrons but only one hydrogen. That

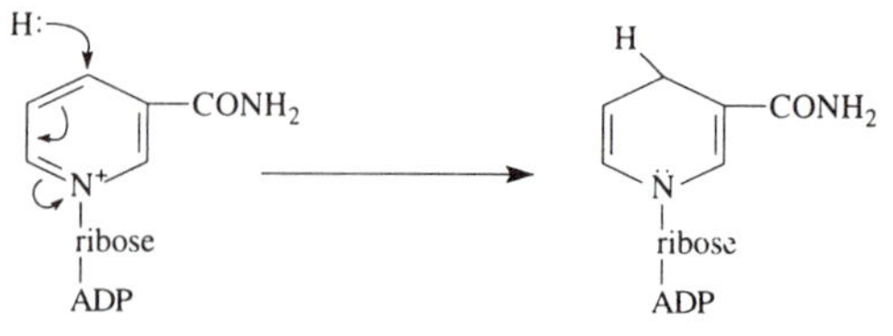

Fig. 8.6 The structures of NAD^+, NADH, and nicotinamide. NAD^+ is a derivative of nicotinamide, to which ADP-ribose is attached to the nitrogen of nicotinamide. In the oxidized form the nitrogen has four bonds and carries a positive charge. NAD^+ accepts two electrons but only one hydrogen (hydride ion) to become NADH. The second hydrogen removed from the electron donor (the reductant) is released into the medium as a proton.

is why it is written as $NADH + H^+$. The oxidized molecule is written NAD^+ because the nitrogen carries a formal positive charge.

8.4 The pentose phosphate pathway

Another important pathway for carbohydrate metabolism is the pentose phosphate pathway. The pentose phosphate pathway is important first because it produces the pentose phosphates which are the precursors to the ribose and deoxyribose in the nucleic acids, and second because it provides erythrose phosphate, which is the precursor to the aromatic amino acids, phenylalanine, tyrosine, and tryptophan. Also, the NADPH produced in the pentose phosphate pathway is a major source of electrons for biosynthesis in most of the pathways in which reductions occur.[2] The pathway is important to learn for yet one more reason. Several of the reactions of the pentose phosphate pathway are the same as the reactions in the Calvin cycle, which is used by many autotrophic organisms to incorporate CO_2 into organic carbon (Chapter 12).

The overall reaction of the pentose phosphate pathway is:

(1) $G6P + 6\,NADP^+$

$\longrightarrow 3\,CO_2 + PGALD + 6\,NADPH + 6\,H^+$

8.4.1 The reactions of the pentose phosphate pathway

The pentose phosphate pathway is complex and can be best learned by dividing the reactions into three stages. Stage one consists of oxidation–decarboxylation reactions. The CO_2 and NADH are produced in stage one. Stage two consists of isomerization reactions that make the precursors for stage three. Stage three reactions are sugar rearrangements. The phosphoglyceraldehyde is produced in stage three.

Stage 1: Oxidation–decarboxylation reactions

The oxidation–decarboxylation reactions are shown in Fig. 8.7. These reactions oxidize the aldehydic C1 in glucose-6-phosphate to a carboxyl and remove it as carbon dioxide. Glucose actually exists as a ring structure which forms because the aldehyde group at C1 reacts with the C5 hydroxyl group forming a hemiacetal. The C1 is therefore not a typical

aldehyde in that it does not react in the Schiff test and does not form a bisulfite addition product. Nevertheless, it is easily oxidized. The glucose-6-phosphate is oxidized by $NADP^+$ to 6-P-gluconolactone by glucose-6-phosphate dehydrogenase (reaction 1). The lactone is then hydrolyzed to 6-P-gluconate by gluconolactonase (reaction 2). Recall that during the oxidation of phosphoglyceraldehyde, inorganic phosphate is added and 1,3-bisphosphoglycerate is formed. The energy of oxidation is trapped in the acyl-phosphate rather than being lost as heat. A subsequent substrate level phosphorylation recovers the energy of oxidation in the form of ATP. In the oxidation discussed here, water contributes the second oxygen in the carboxyl group and a phosphorylated derivative is not formed. This means that the energy from the oxidation is lost as heat, and the reaction is physiologically irreversible, as is often the case for the first reaction in a metabolic pathway. The product of the oxidation, 6-P-gluconate, is then oxidized on

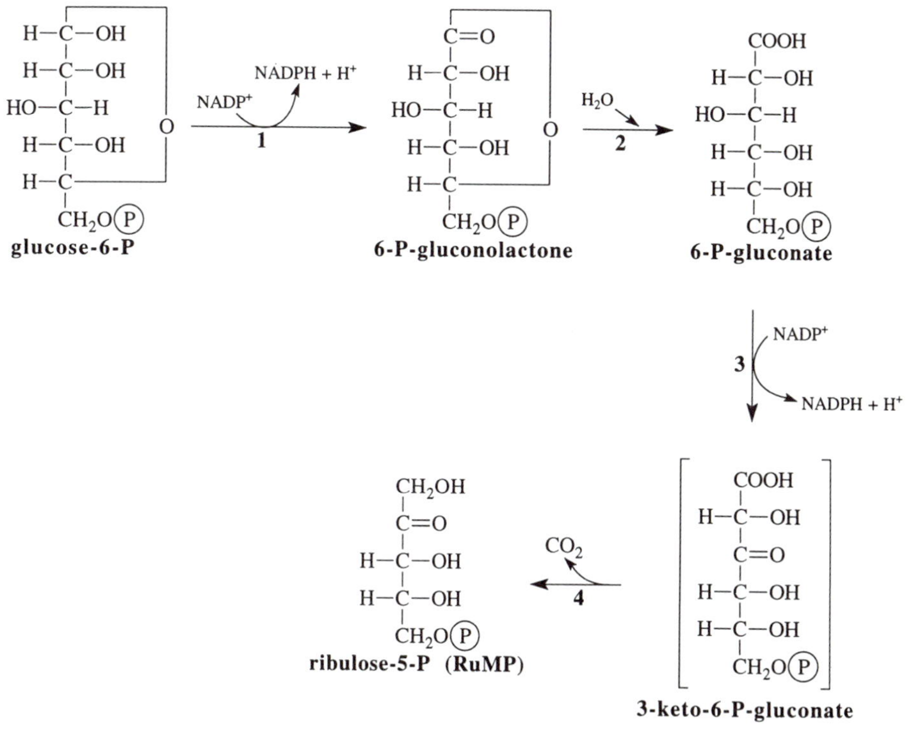

Fig. 8.7 The oxidation–decarboxylation reactions. Enzymes: (1) glucose-6-phosphate (G6P) dehydrogenase; (2) phosphogluconate dehydrogenase. 6-P-gluconolactone, which is the immediate product of G6P oxidation, is not shown. The lactone is hydrolyzed by gluconolactonase to 6-P-gluconate.

163

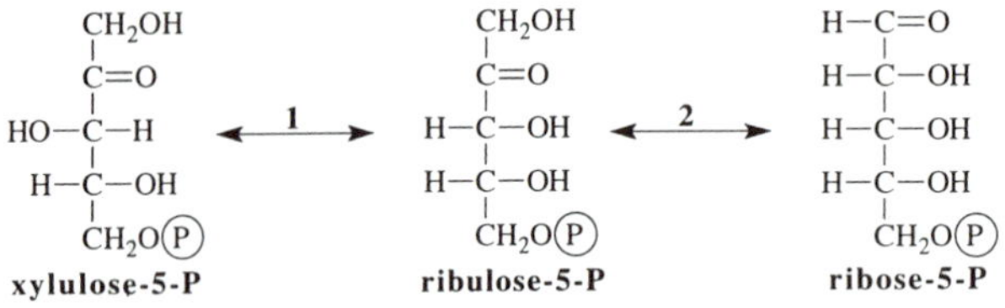

Fig. 8.8 The isomerization reactions. Enzymes: (1) ribulose-5-phosphate epimerase; (2) ribose-5-phosphate isomerase.

the C3 to generate a keto group β to the carboxyl (reaction 3). A β-decarboxylation then occurs generating ribulose-5-phosphate (reaction 4). The mechanism of β-decarboxylations is described in Section 8.10.2. Therefore, the products of stage 1 are carbon dioxide, 2 NADPH, and the five carbon sugar ribulose-5-phosphate. The rest of the pathway continues with ribulose-5-phosphate.

Stage 2: The isomerization reactions

During the second stage, some of the ribulose-5-phosphate is isomerized to ribose-5-phosphate and to xylulose-5-phosphate. Isomers are molecules having the same chemical formula but different structural formulae. That is to say, their parts have been switched around. For example, the chemical formula for ribulose-5-phosphate is $C_5H_{11}O_8P$. Ribose-5-phosphate and xylulose-5-phosphate have the same chemical formula. However, their structures are different (Fig. 8.8).

One of the isomerases in an epimerase. The epimerase catalyzes a movement of the hydroxyl group from one side of the C3 in ribulose-5-phosphate to the other. The product is xylulose-5-phosphate (the epimer[3] of ribulose-5-phosphate). The other isomerase converts ribulose-5-phosphate to ribose-5-phosphate.

Stage 3: The sugar rearrangement reactions

Stage three of the pentose-phosphate pathway involves sugar rearrangement reactions. There are two basic types of reactions. One kind transfers a *two*-carbon fragment from a ketose to an aldose. The enzyme that catalyzes the transfer of the two-carbon fragment is called a *transketolase* (TK). A second kind of reaction transfers a *three*-carbon fragment from a ketose to an aldose. The enzyme that catalyzes the transfer of a three-carbon fragment is called a *transaldo-*

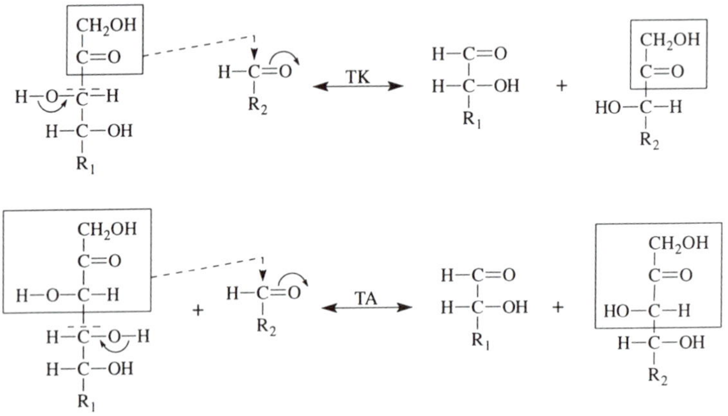

Fig. 8.9 The transketolase and transaldolase reactions. Enzymes: TK, transketolase; TA, transaldolase. The donor is always a ketose with the keto group on C2, and the hydroxyl on C3 on the "left." In the transketolase reaction, a C_2 unit is transferred with its bonding electrons to the carbonyl group on an aldehyde acceptor. The transaldolase transfers a three-carbon fragment. In the transketolase reaction, the newly-formed alcohol group is on the "left," which means that the products of both the transketolase and transaldolase reactions can act as donors in a subsequent transfer.

lase (TA). The rule is that the donor is always a ketose (with the OH group of the third carbon "on the left," as in xyulose-5-phosphate) and the acceptor is always an aldose. This rule is important to learn because we shall see other transketolase and transaldolase reactions later. Knowing the requirements will make it easier to remember the reactions. The transketolase and transaldolase reactions are summarized in Fig. 8.9.

Summarizing the pentose phosphate pathway

Figure 8.10 summarizes the pentose phosphate pathway. Reactions 1–3 comprise the oxidative decarboxylation reactions of stage one. Three moles of glucose-6-P must be oxidized in order to produce three moles of CO_2 and one mole of phosphoglyceraldehyde. Therefore, stage one produces three moles of ribulose-5-P. Reactions 4 and 5 are the

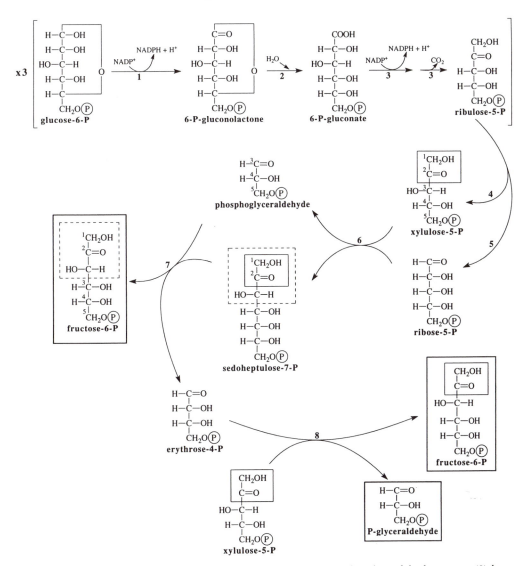

Fig. 8.10 The pentose phosphate pathway. Enzymes: (1) glucose-6-phosphate dehydrogenase; (2) lactonase; (3) 6-phosphogluconate dehydrogenase; (4) ribulose-5-phosphate epimerase; (5) ribose-5-phosphate isomerase; (6, 8) transketolase; (7) transaldolase. The two-carbon moiety transferred by the transketolase is shown in the boxed area. The three-carbon fragment transferred by the transaldolase is shown in the dashed box. The product of the glucose-6-phosphate dehydrogenase reaction is the lactone, which is unstable and hydrolyzes spontaneously to the free acid. However, there is a specific lactonase that catalyzes the reaction.

isomerization reactions of stage 2 in which the three moles of ribulose-5-phosphate are converted to one mole of ribose-5-phosphate and two moles of xylulose-5-P. Reactions 6, 7, and 8 comprise stage 3. Reaction 6 is a transketolase reaction in which a xylulose-5-P (C_5) transfers a two-carbon moiety to ribose-5-P (C_5) with the formation of sedoheptulose-7-P (C_7) and phosphoglyceraldehyde (C_3). The two-carbon moiety is highlighted as a boxed area. Reaction 7 is a transaldolase reaction in which the sedoheptulose-7-P transfers a three-carbon moiety to the phosphoglyceraldehyde to form erythrose-4-P (C_4) and fructose-6-P (C_6). The three carbon moiety is highlighted as a dashed box. Reaction 8 is a transketolase reaction in which xylulose-5-P transfers a two-carbon moiety to the erythrose-4-P forming phosphoglyceraldehyde and fructose-6-P. You will notice that the sequence of reactions is

Transketolase ⟶ transaldolase

⟶ transketolase.

The result is that three moles of glucose-6-P are converted to two moles of fructose-6-P and one mole of phosphoglyceraldehyde. The two moles of fructose-6-P become glucose-6-phosphate by isomerization and the net result is the conversion of one mole of glucose-6-P to one mole of phosphoglyceraldehyde, three moles of carbon dioxide, and six moles of NADPH. This is shown in the carbon balance below.

Later we will study the Calvin cycle, which is a pathway by which many organisms can grow on CO_2 as the sole source of carbon (Chapter 12). In the Calvin cycle, CO_2 is first reduced to phosphoglyceraldehyde. The phosphoglyceraldehyde is then converted to pentose phosphates. The Calvin cycle resembles the pentose phosphate pathway in all respects except there is no transaldolase. Instead, the erythrose-4-phosphate formed in the transketolase reaction is converted to sedoheptulose-7-phosphate by an alternate route which runs irreversibly in the direction of sedoheptulose-7-phosphate. The consequence of this fact is that, whereas the pentose phosphate pathway converts pentose phosphates reversibly to phosphoglyceraldehyde, the Calvin cycle operates only in the direction of pentose phosphates.

The pentose phosphate pathway serves important biosynthetic functions. Notice that stage 1 (the oxidative decarboxylation reactions) and stage 2 (the isomerization reactions) generate the pentose phosphates required for nucleic acid synthesis. Stage 1 also produces NADPH which is used in several biosynthetic pathways. Stage 3 generates the erythrose-4-phosphate necessary for aromatic amino acid biosynthesis.

Some bacteria rely completely on the pentose phosphate pathway for sugar catabolism

Thiobacillus novellus and *Brucella abortus* lack both stage one of the Embden–Meyerhof–Parnas pathway and the enzymes of the

---

Carbon balance for the pentose phosphate pathway

Oxidative decarboxylation   3 glucose-6-P ⟶ 3 ribulose-5-P + 3CO_2
       $3C_6$ $3C_5$ $3C_1$

Isomerizations   3 ribulose-5-P ⟶ 2 xylulose-5-P + ribose-5-P
       $3C_5$ $2C_5$ C_5

Transketolase   xylulose-5-P + ribose-5-P ⟶ sedoheptulose-7-P + phosphoglyceraldehyde
       C_5 C_5 C_7 C_3

Transaldolase   sedoheptulose-7-P + phosphoglyceraldehyde ⟶ fructose-6-P + erythrose-4-P
       C_7 C_3 C_6 C_4

Transketolase   xylulose-5-P + erythrose-4-P ⟶ fructose-6-P + phosphoglyceraldehyde
       C_5 C_4 C_6 C_3

Sum: glucose-6-P ⟶ phosphoglyceraldehyde + 3CO_2
       C_6 C_3 $3C_1$

Entner–Doudoroff pathway. These organisms use only an oxidative pentose phosphate pathway to grow on glucose. They oxidize the glucose to phosphoglyceraldehyde via the pentose phosphate pathway. The phosphoglyceraldehyde is then oxidized to pyruvate via reactions that are the same as stage two of the EMP pathway, and then the pyruvate is oxidized to CO via the citric acid cycle.

Relationship of the pentose phosphate pathway to glycolysis

The pentose phosphate pathway and glycolysis interconnect at phosphoglyceraldehyde and fructose-6-phosphate (Fig. 8.1). Thus organisms growing on pentoses can make hexose phosphates. Furthermore, because stages 2 and 3 of the pentose phosphate pathway are reversible, it is possible to synthesize pentose phosphates from phosphoglyceraldehyde and avoid the oxidative decarboxylation reactions of stage 1. This would uncouple pentose phosphate synthesis from NADPH production and confer a possibly advantageous metabolic flexibility on the cells.

8.5 The Entner–Doudoroff pathway

Many prokaryotes have another pathway for the degradation of carbohydrates called the Entner–Doudoroff or ED pathway. The other pathways that we have been studying are common to all cells, whether they be prokaryotes or eukaryotes. But the Entner–Doudoroff pathway has been found only in prokaryotes. The pathway is widespread, particularly among the aerobic gram-negative bacteria. It is usually not found among anaerobic bacteria, perhaps because of the low ATP yields discussed below. Most bacteria degrade sugars via the Embden–Meyerhof–Parnas pathway but when grown on certain compounds (e.g., gluconic acid), they use the Entner–Doudoroff pathway. However, some strictly aerobic bacteria cannot carry out stage one of the Embden–Meyerhof–Parnas pathway and rely entirely on the Entner–Doudoroff pathway for sugar degradation (Table 8.1). The overall reaction

Table 8.1 Distribution of the Embden–Meyerhoff–Parnas (EMP) and Entner–Doudoroff (ED) pathways in certain bacteria.

Bacterium	EMP	ED
Arthrobacter species	+	−
Azotobacter chroococcum	+	−
Alcaligenes eutrophus	−	+
Bacillus species	+	−
Escherichia coli and other enteric bacteria*	+	−
Pseudomonas species	−	+
Rhizobium species	−	+
Thiobacillus species	−	+
Xanthomonas species	−	+

*Organisms such as *E. coli* synthesize the enzymes of the ED pathway when growing on gluconate.
Source: Gottschalk, G. 1986. *Bacterial Metabolism.* Springer-Verlag, New York, Berlin.

for the Entner–Doudoroff pathway is:

$$\text{Glucose} + NADP^+ + NAD^+ + ADP + P_i$$
$$\longrightarrow 2 \text{ pyruvic acid} + NADPH + 2H^+$$
$$+ NADH + ATP$$

It can be seen that the pathway catalyzes the same overall reaction as the Embden–Meyerhoff–Parnas pathway (i.e., the oxidation of one mole of glucose to two moles of pyruvic acid) except that only one ATP is made, and one NADPH and one NADH are made instead of two NADHs. The reason why only one ATP is made is that only one phosphoglyceraldehyde is made from glucose (Fig. 8.11).

8.5.1 The Entner–Doudoroff reactions

The first oxidation is the oxidation of the aldehyde group in glucose-6-phosphate to the carboxyl in 6-P-gluconate (Fig. 8.11, reaction 2). These are the same enzymatic reactions that oxidize glucose-6-phosphate in the pentose phosphate pathway, and proceed through the gluconolactone. This oxidation is catalyzed by the same enzyme that oxidizes glucose-6-phosphate in the pentose phosphate pathway; that is, glucose-6-phosphate dehydrogenase. The pathway diverges from the

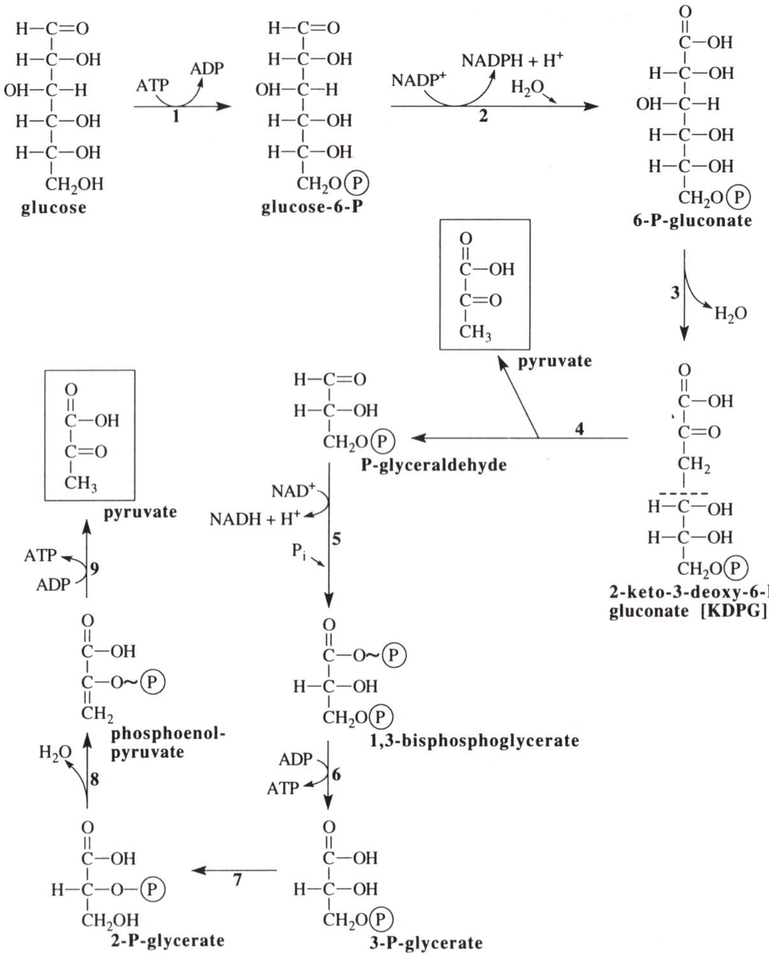

Fig. 8.11 The Entner–Doudoroff pathway. Because there is only one PGALD formed, there is only one ATP made. The enzymes unique to this pathway are the 6-phosphogluconate dehydratase (reaction 3) and the KDPG aldolase (reaction 4). The other enzymes are present in the pentose phosphate pathway and the glycolytic pathway. Enzymes: (1) hexokinase; (2) glucose-6-phosphate dehydrogenase; (3) 6-phosphogluconate dehydratase; (4) KDPG aldolase; (5) triose phosphate dehydrogenase; (6) PGA kinase; (7) mutase; (8) enolase; (9) pyruvate kinase.

pentose phosphate pathway at this point because some of the 6-P-gluconate is dehydrated to 2-keto-3-deoxy-6-P-gluconate (KDPG), rather than being oxidized to ribulose-5-phosphate (reaction 3). The KDPG is split by KDPG aldolase to pyruvate and phosphoglyceraldehyde (reaction 4). The phosphoglyceraldehyde is oxidized to pyruvate in a sequence of reactions identical to those in stage two of the EMP pathway (reactions 5–9).

Reaction 3, which is the dehydration of 6-phosphogluconate to KDPG, takes place via an enol intermediate that tautomerizes to KDPG. This is illustrated in Fig. 8.12. In this way, it is similar to the dehydration of 2-phosphoglycerate to phosphoenolpyruvate by enolase described in Section 7.3.4. However, in phosphoenolpyruvate, the enol derivative is stabilized by the phosphate group and the tautomerization does not take place until the phosphoryl group is transferred.

8.5.2 Physiological role for the Entner–Doudoroff pathway

Since the Entner–Doudoroff pathway produces only one ATP, one can ask why the pathway

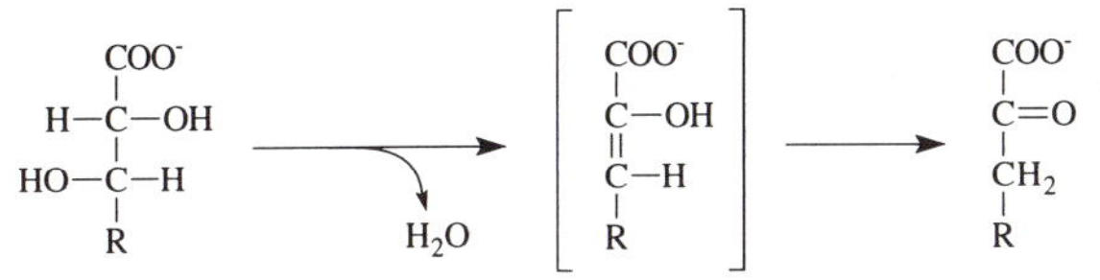

Fig. 8.12 Dehydration of a carboxylic acid with hydroxyl groups in the α and β positions. The dehydration of a carboxylic acid with hydroxyl groups in both the α and β positions leads to the formation of an enol which tautomerizes to the keto compound. That is because the hydroxyl on the C3 leaves with its bonding electrons and the electrons bonded to the hydrogen on the C2 shift in to form the double bond. This happens when 6-phosphogluconate is dehydrated to 2-keto-3-deoxy-6-phosphogluconate in the Entner–Doudoroff pathway, and when 2-phosphoglycerate is dehydrated to phosphoenolpyruvate during glycolysis. The phosphoenolpyruvate tautomerizes to pyruvate when the phosphate is removed during the kinase reaction.

is so common in the bacteria. Whereas hexoses are readily degraded by the Embden–Meyerhoff–Parnas pathway, aldonic acids (aldoses oxidized at the aldehydic carbon) such as gluconate, are not, but can be degraded via the Entner–Doudoroff pathway. (Aldonic acids occur in nature and can be an important nutrient.) An example of this occurs when *E. coli* is transferred from a medium containing glucose as the carbon source to one in which gluconate is the source of carbon. Growth on gluconate results in the induction of three enzymes: a gluconokinase that makes 6-P-gluconate from the gluconate (at the expense of ATP), 6-P-gluconate dehydratase, and KDPG aldolase. Thus, *E. coli* uses the Entner–Doudoroff pathway to grow on gluconate and the Embden–Meyerhoff–Parnas pathway to grow on glucose.

Some bacteria do not have a complete Embden–Meyerhoff–Parnas pathway and rely on the Entner–Doudoroff pathway for hexose degradation

Several prokaryotes (e.g., pseudomonads) do not make phosphofructokinase or fructose bisphosphate aldolase. Hence, they cannot carry out glucose oxidation using the Embden– Meyerhoff–Parnas pathway. Instead, they use the Entner–Doudoroff (ED) pathway. See table 8.1 for the distribution of the EMP and ED pathways. Some of the pseudomonads even oxidize glucose to gluconate before degrading it via the ED pathway, instead of making glucose-6-phosphate. The oxidation of glucose to gluconate may confer

a competitive advantage, since it removes glucose, which is more readily utilizable by other microorganisms.

8.5.3 A partly nonphosphorylated Entner–Doudoroff pathway

A modified ED pathway has been found in the archaeon *Halobacterium saccharovorum*, and in several bacteria, including members of the genera *Clostridium*, *Alcaligenes*, *Achromobacter*, and in *Rhodopseudomonas sphaeroides*. The pathway is characterized as having nonphosphorylated intermediates prior to 2-keto-3-deoxygluconate. The first reaction is the oxidation of glucose to gluconate via an NAD^+-dependent dehydrogenase. The gluconate is then dehydrated to 2-keto-3-deoxygluconate by a gluconate dehydratase. The 2-keto-3-deoxygluconate is phosphorylated by a special kinase to form KDPG, which is metabolized to pyruvate using the ordinary ED reactions. Those bacteria that have a modified ED pathway can use it for the catabolism of gluconate or glucose if the glucose dehydrogenase is present.

8.5.4 A completely nonphosphorylated Entner–Doudoroff pathway

A completely nonphosphorylated Entner–Doudoroff pathway has been reported for three thermoacidophilic archaea: *Sulfolobus*, *Thermoplasma*, and *Pyrococcus*.[4,5] These organisms do not possess key enzymes of the

EMP pathway. For example, they do not have phosphofructokinase activity. The pathway proposed for the anaerobic *Pyrococcus furiosus* is shown in Fig. 8.13. Maltose is split

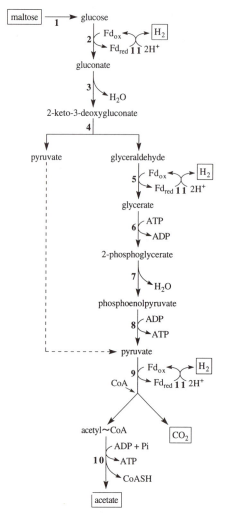

Fig. 8.13 Proposed pathway for maltose fermentation to acetate, CO_2, and H_2 in *Pyrococcus furiosus*. Enzymes: (1) α-glucosidase; (2) glucose:-ferredoxin oxidoreductase; (3) gluconate dehydratase (not demonstrated); (4) KDG aldolase; (5) glyceraldehyde:ferredoxin oxidoreductase; (6) glycerate kinase; (7) enolase; (8) pyruvate kinase; (9) pyruvate:ferredoxin oxidoreductase; (10) ADP-dependent acetyl-CoA synthetase; (11) hydrogenase. CoA, coenzyme A; Fd_{ox}, oxidized ferredoxin; Fd_{red}, reduced ferredoxin. (From Schafer, T. and P. Schonheit, 1992. Maltose fermentation to acetate, CO_2, and H_2 in the anaerobic hyperthermophilic archaeon *Pyrococcus furiosus*: Evidence for the operation of a novel sugar fermentation pathway. *Arch. Microbiol.* **158**:188–202.)

into glucose. The glucose is oxidized to gluconate which is dehydrated to 2-keto-3-deoxygluconate. The 2-keto-3-deoxygluconate is split by an aldolase into pyruvate and glyceraldehyde. The glyceraldehyde is oxidized to glycerate which is phosphorylated to phosphoglycerate at the expense of an ATP. The 2-phosphoglycerate is dehydrated to phosphoenolpyruvate, which donates the phosphoryl group to ADP to regenerate the ATP used to phosphorylate the glycerate. There is no net ATP made in the formation of pyruvate. (This is most unusual. The sugar catabolic pathways that were described earlier in this chapter synthesize one or two net ATPs per pyruvate made, depending upon whether the pathway yields one or two phosphoglyceraldehydes.) Both pyruvate molecules are oxidized to acetyl-CoA and CO_2 by pyruvate:ferredoxin oxidoreductase. (See Section 7.3.2 for a description of this reaction.) Finally, an ADP-dependent acetyl-CoA synthetase catalyzes a reaction in which ADP is phosphorylated and acetate is formed. Thus, there is one ATP made per acetate. This, of course, is a substrate level phosporylation. This last step, which is the energy-yielding step in the pathway, has been found only in some archaea and not in any bacteria thus far. Recall that bacteria usually use phosphotransacetylase and acetate kinase to make ATP from acetyl-CoA (Section 7.3.2). Another unusual feature of the pathway is that NAD^+ is not used as an electron acceptor. The electron acceptor for the three oxidations is suggested to be ferredoxin, which transfers the electrons to protons generating hydrogen gas. Disposing of electrons in hydrogen gas allows fermenting cells to produce more acetate, with the formation of more ATP (from acetyl-P). It is an advantage for an anaerobe to dispose of its electrons via ferredoxin-linked dehydrogenases rather than NADH-linked hydrogenases because ferredoxin is at a sufficiently low potential that the equilibrium favors production of hydrogen gas. This is not the case for NADH-linked dehydrogenases. The electron acceptor for the oxidations in the aerobic *Sulfolobus* and *Thermoplasma* (which have a similar pathway) is $NAD(P)^+$, which is re-oxidized via respiration.[6]

The nonphosphorylated ED pathway does not produce net ATP during the catabolism of sugars to pyruvate. One consequence of this is that the pathway proceeds with a much larger decrease in free energy when compared to the EMP pathway which conserves energy in the form of two ATPs per glucose. Because of this, the nonphosphorylated ED pathway is irreversible and cannot be used for gluconeogensis from pyruvate. The following enzymes of the nonphosphorylated ED pathway are irreversible: glucose dehydrogenase, gluconate dehydratase, glyceraldehyde dehydrogenase, and glycerate kinase. However, archaea such as *Pyrococcus furiosus* can grow on pyruvate and therefore synthesize carbohydrates from pyruvate.[7] They are able to do this because all archaea thus far studied have the enzymes of the EMP pathway and use the pathway for gluconeogenesis from pyruvate. [The EMP pathway has two irreversible reactions: the pyruvate kinase reaction and the phosphofructokinase reaction. To reverse glycolysis from pyruvate, these reactions are not used (Sections 8.1.1 and 8.12.2)].

The study of archaeal metabolism has recently received wider attention, compared to the long history of research with bacteria, and there have been many rewarding findings which suggest the presence of several metabolic features distinct from those of the bacteria, as the example of the nonphosphorylated ED pathway illustrates. Other examples include the novel ether-linked lipids described in Chapter 1 and their biosynthesis, discussed in Chapter 9. Additional features of archaeal metabolism that appear to be unique to these microorganisms include the synthesis of methane, and the presence of novel coenzymes for acetate and methane metabolism. These aspects are discussed in Chapter 12.

8.6 The oxidation of pyruvate to acetyl-CoA: The pyruvate dehydrogenase reaction

Pyruvate is the common product of sugar catabolism in all of the major carbohydrate catabolic pathways (Fig. 8.1). We now examine the metabolic fate of pyruvate. What happens to the pyruvate depends upon whether the organism is respiring aerobically or fermenting. Fermentative metabolism of pyruvate is considered in Chapter 13. The aerobic oxidation of pyruvate will be considered here. Pyruvate is first oxidized to acetyl-CoA and CO_2. The oxidation of pyruvate to acetyl-CoA and CO_2 during aerobic growth is carried out by the enzyme complex *pyruvate dehydrogenase*, which is widespread in both prokaryotes and in eukaryotes. The overall reaction is:

$$CH_3COCOOH + NAD^+ + CoASH$$

$$\longrightarrow CH_3COSCoA + CO_2 + NADH + H^+$$

The pyruvate dehydrogenase complex is a very large enzyme complex (in *E. coli* about 1.7 times the size of the ribosome) located in the mitochondria of eukaryotic cells and in the cytosol of prokaryotes. The pyruvate dehydrogenase from *E. coli* consists of 24 molecules of enzyme E1 (pyruvate dehydrogenase), 24 molecules of enzyme E2 (dihydrolipoate transacetylase), and 12 molecules of enzyme E3 (dihydrolipoate dehydrogenase). Several very important cofactors are involved. The cofactors are thiamine pyrophosphate (TPP) derived from the vitamin thiamine, flavin adenine dinucleotide (FAD) derived from the vitamin riboflavin, lipoic acid (RS_2), nicotinamide adenine dinucleotide (NAD^+) derived from the vitamin nicotinamide, and coenzyme-A, derived from the vitamin pantothenic acid.[8] The large size of the complex is presumably designed to process the heavy stream of pyruvate that is generated during the catabolism of sugars and other compounds. As described below, the pyruvate dehydrogenase complex catalyzes a short metabolic pathway rather than simply a single reaction. The individual reactions carried out by the pyruvate dehydrogenase complex are as follows (Fig. 8.14).

Step 1. Pyruvate is decarboxylated to form "active acetaldehyde" bound to TPP (Fig. 8.14). The reaction is catalyzed by pyruvate dehydrogenase (E1). (The mechanism of this reaction is described in Section 13.10 and Fig. 13.9.)

Step 2. The "active acetaldehyde" is oxidized to the level of carboxyl by the disulfide in

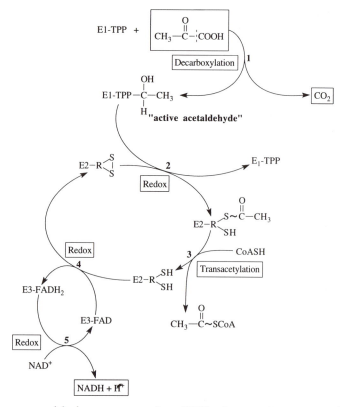

Fig. 8.14 The pyruvate dehydrogenase complex (PDH). Step 1. Pyruvate is decarboxylated to "active acetaldehyde." The decarboxylation requires thiamine pyrophosphate (TPP). Step 2: The "active acetaldehyde" is oxidized to an acylthioester with a high acyl group transfer potential. The oxidant is reduced lipoic acid ($R-S_2$). Step 3: The lipoylacylthioester transfers the acetyl group to coenzyme A (CoASH) to form acetyl-CoA. Step 4: The reduced lipoic acid is reoxidized by FAD, which in turn is reoxidized by NAD^+ (Step 5). The products of the reaction are acetyl CoA, CO_2, and NADH. Enzymes: E1, pyruvate dehydrogenase; E2, dihydrolipoate transacetyase; E3, dihydrolipoate dehydrogenase.

lipoic acid. The disulfide of the lipoic acid is reduced to a sulfhydryl. During the reaction, TPP is displaced and the acetyl group is transferred to the lipoic acid. The reaction is also catalyzed by pyruvate dehydrogenase.

Step 3. A transacetylation occurs where lipoic acid is displaced by CoASH forming acetyl-CoA and reduced lipoic acid. The reaction is catalyzed by dihydrolipoate transacetylase, E_2.

Step 4. The lipoic acid is oxidized by dihydrolipoate dehydrogenase, E_3-FAD.

Step 5. The E_3-FADH$_2$ transfers the electrons to NAD^+.

All of the intermediates remain bound to the complex and are passed from one active site to another. Presumably, this has the advantage inherent in all multienzyme complexes (i.e., there is no dilution of intermediates in the cytosol, and side-reactions are minimized). The student should refer to Section 1.2.6 for a discussion of multienzyme complexes in the cytoplasm.

8.6.1 Physiological control

The pyruvate dehydrogenase reaction, which is physiologically irreversible, is under metabolic control by several allosteric effectors (Fig. 8.15). The *E. coli* pyruvate dehydrogenase is feedback inhibited by the products it

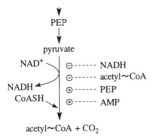

Fig. 8.15 Regulation of pyruvate dehydrogenase in *E. coli*. The activity of the enzyme *in vitro* is modified by several effector molecules. NADH and acetyl-CoA are negative effectors, and PEP and AMP are positive effectors.

forms, acetyl-CoA and NADH. This can be rationalized as ensuring that the enzyme produces only as much acetyl-CoA and NADH as can be used immediately. It is also stimulated by phosphoenolpyruvate (the precursor to pyruvate), presumably signaling the dehydrogenase that more pyruvate is on the way. It is also stimulated by AMP, which signals low ATP. The stimulation by AMP probably reflects the fact that the oxidation of the product acetyl-CoA in the citric acid cycle is a major source of ATP (via respiratory phosphorylation).

8.7 The citric acid cycle

The acetyl-CoA that is formed by pyruvate dehydrogenase is oxidized to CO_2 in the citric acid cycle (Fig. 8.1). The overall reaction is:

$$\text{Acetyl-CoA} + \text{ADP} + P_i + \text{FAD} + 2H_2O$$

$$+ \text{NADP}^+ + 2\text{NAD}^+$$

$$\longrightarrow 2CO_2 + \text{ATP} + \text{FADH}_2 + \text{NADPH}$$

$$+ 2\text{NADH} + 3H^+ + \text{COASH}$$

Notice that there are four oxidations per acetyl-CoA producing two NADH, one NADPH, and one $FADH_2$, and one substrate level phosphorylation producing ATP. The cycle usually operates in conjunction with respiration that re-oxidizes the NAD(P)H and $FADH_2$. Other names for this pathway are the tricarboxylic acid (TCA) cycle and the Krebs cycle. The latter name honors Sir Hans

Krebs, who did much of the pioneering work and proposed the cycle in 1937.

8.7.1 The individual reactions of the citric acid cycle

The pathway is outlined in Fig. 8.16. Reaction 1 is the addition of the acetyl group from acetyl-CoA to oxaloacetate to form citrate. In this reaction, the methyl group of acetyl-CoA acts as a nucleophile and bonds to the carbon in the keto group of oxaloacetate (OAA). The reaction is driven to completion by the hydrolysis of the thioester bond of acetyl-CoA, which has a high free energy of hydrolysis. Reaction 1 is catalyzed by citrate synthase. This enzyme operates irreversibly in the direction of citrate. The oxaloacetate acts catalytically in the cycle, and if it is not regenerated or replenished the pathway stops. In reaction 2, catalyzed by aconitase, the citrate is dehydrated to *cis*-aconitate which remains bound to the enzyme. Reaction 3 (also catalyzed by aconitase) is the rehydration of *cis*-aconitate to form isocitrate, an isomer of citrate. In reaction 4 (isocitrate dehydrogenase), the isocitrate is oxidized to oxalosuccinate. This oxidation creates a keto group β to the carboxyl group. The creation of the keto group is necessary for the decarboxylation which takes place in the next reaction. (β-Keto decarboxylations are explained in Section 8.10.2.) Reaction 5 (isocitrate dehydrogenase) is the decarboxylation of oxalosuccinate to α-ketoglutarate. Reaction 6 (α-ketoglutarate dehydrogenase) is the oxidative decarboxylation of α-ketoglutarate to succinyl-CoA. This is an α-decarboxylation in contrast to a β-decarboxylation. It is a complex reaction and requires the same cofactors as does the decarboxylation of pyruvate to acetyl-CoA. Reaction 7 (succinate thiokinase) is a substrate level phosphorylation resulting in the formation of ATP from ADP and inorganic phosphate. This reaction was described in Section 7.3.3. Reaction 8 (succinate dehydrogenase) is the oxidation of succinate to fumarate, catalyzed by a flavin enzyme. Succinate dehydrogenase is the only citric acid cycle enzyme which is membrane-bound. In bacteria, it is part of the

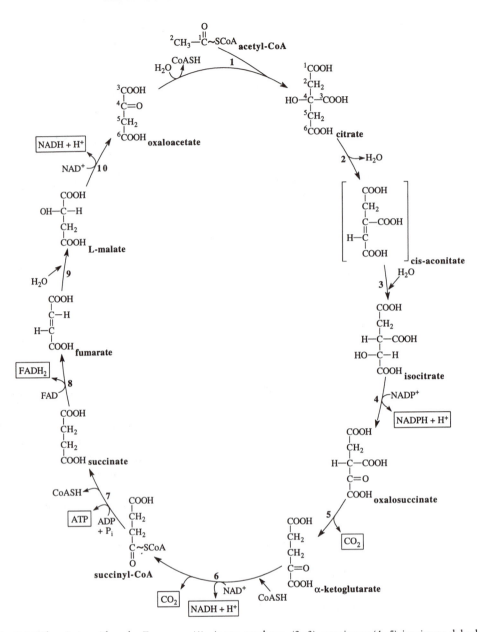

Fig. 8.16 The citric acid cycle. Enzymes: (1) citrate synthase; (2, 3) aconitase; (4, 5) isocitrate dehydrogenase; (6) α-ketoglutarate dehydrogenase; (7) succinate thiokinase; (8) succinate dehydrogenase; (9) fumarase; (10) malate dehydrogenase. *Cis*-aconitate is drawn in parentheses because it is an enzyme-bound intermediate.

cell membrane and transfers electrons directly to quinone in the respiratory chain. (See the description of the electron transport chain in Section 4.3.) Reaction 9 (fumarase) is the hydration of fumarate to malate. Finally, the oxaloacetate is regenerated by oxidizing the malate to oxaloacetate in reaction 10 (malate dehydrogenase). The citric acid cycle proceeds in the direction of acetyl-CoA oxidation because of two irreversible steps: the citrate synthase and the α-ketoglutarate dehydrogenase reactions.

Summing up the citric acid cycle

When one examines the reactions in the citric acid cycle, it can be seen that there is no net synthesis. In other words, all of the carbon that enters the cycle exits as CO_2. This is made clear by writing a carbon balance. In the carbon balance written below, C_2 represents the two-carbon molecule acetyl-CoA, C_6 represents citrate or isocitrate, C_5 represents α-ketoglutarate, and C_4 represents either succinate fumarate, malate, or oxaloacetate. Of course, C_1 represents carbon dioxide.

Carbon balance for citric acid cycle

$$C_2 + C_4 \longrightarrow C_6$$
$$C_6 \longrightarrow C_5 + C_1$$
$$C_5 \longrightarrow C_4 + C_1$$

sum: $C_2 \longrightarrow C_1$

8.7.2 Regulation of the citric acid cycle

The citric acid cycle is feedback inhibited by several intermediates that can be viewed as end products of the pathway. In gram-negative bacteria, the citrate synthase is allosterically inhibited by NADH, and in facultative anaerobes such as *E. coli*, also by α-ketoglutarate. The inhibition of the citrate synthase by NADH may be a way to prevent oversynthesis of NADH. The inhibition by α-ketoglutarate can also be viewed as an example of end-product inhibition, in this case to prevent overproduction of the amino acid glutamate, which is derived from α-ketoglutarate.[9] The citrate synthase from gram-positive bacteria and eukaryotes is not sensitive to NADH and α-ketoglutarate, but is inhibited by ATP, another end product of the citric acid pathway. Recall the discussion in Chapter 6 emphasizing that the pattern of regulation of a particular pathway need not be the same in different bacteria.

8.7.3 The citric acid cycle as an anabolic pathway

The citric acid cycle reactions provide precursors to 10 of the 20 amino acids found in proteins. It is therefore a multifunctional pathway and is not used simply for the oxidation of acetyl-CoA. Succinyl-CoA is

necessary for the synthesis of the amino acids L-lysine and L-methionine. Succinyl-CoA is also a precursor to tetrapyrroles, which are the prosthetic groups in several proteins, including cytochromes and chlorophylls. Oxaloacetate is a precursor to the amino acid aspartate, which itself is the precursor to five other amino acids. In some bacteria, fumarate is also a precursor to aspartate. α-Ketoglutarate is the precursor to the amino acid glutamate, which itself is the precursor to three other amino acids. The biosynthesis of amino acids is described in Chapter 9. However, since the citric acid cycle requires a constant level of oxaloacetate in order to function, net synthesis of these molecules requires replenishment of the oxaloacetate. This is discussed in Section 8.8.

8.7.4 Distribution of the citric acid cycle

The citric acid cycle is present in most heterotrophic bacteria growing aerobically. However, not all aerobic bacteria have a complete citric acid cycle. For example, organisms that grow on C_1 compounds (methane, methanol, and so on, described in Chapter 12) lack α-ketoglutarate dehydrogenase and carry out a reductive pathway as described in Section 8.9. An oxidative citric acid cycle is not necessary for these organisms since acetyl-CoA is not an intermediate in the oxidation of the C_1 compounds. Although the oxidative citric acid cycle is a pathway associated with aerobic bacteria, it is present in certain anaerobes. These are the group II sulfate reducers, which are anaerobes using sulfate as an electron acceptor. They have a modified oxidative citric acid cycle discussed in Section 11.2.2.

8.8 Carboxylations that replenish oxaloacetate: The pyruvate and phosphoenolpyruvate carboxylases

Because the citric acid cycle intermediates are constantly being removed to provide precursors for biosynthesis, they must be replaced (Section 8.7.3). Failure to do this

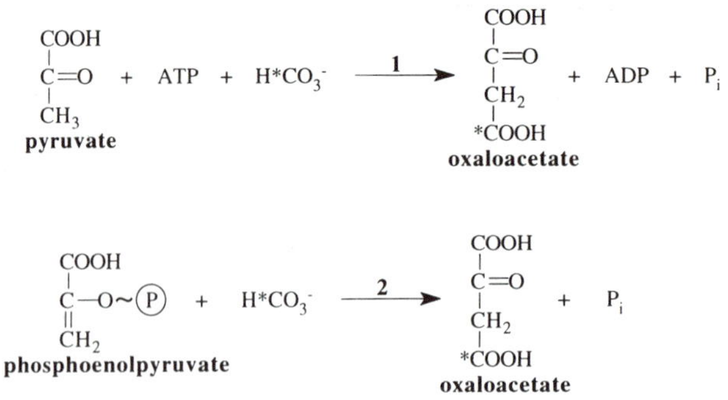

$$
\begin{array}{ll}
\text{COOH} & \text{COOH} \\
| & | \\
\text{C}=\text{O} \;+\; \text{ATP} \;+\; \text{H*CO}_3^- \quad \xrightarrow{\;1\;} \quad & \text{C}=\text{O} \;+\; \text{ADP} \;+\; \text{P}_i \\
| & | \\
\text{CH}_3 & \text{CH}_2 \\
\textbf{pyruvate} & | \\
& \text{*COOH} \\
& \textbf{oxaloacetate}
\end{array}
$$

$$
\begin{array}{ll}
\text{COOH} & \text{COOH} \\
| & | \\
\text{C}-\text{O}\sim\textcircled{P} \;+\; \text{H*CO}_3^- \quad \xrightarrow{\;2\;} \quad & \text{C}=\text{O} \;+\; \text{P}_i \\
\| & | \\
\text{CH}_2 & \text{CH}_2 \\
\textbf{phosphoenolpyruvate} & | \\
& \text{*COOH} \\
& \textbf{oxaloacetate}
\end{array}
$$

Fig. 8.17 Carboxylation reactions that replenish the supply of oxaloacetate. Enzymes: (1) pyruvate carboxylase; (2) PEP carboxylase. Bacteria may have one or the other.

would decrease the level of oxaloacetate that is necessary for the citrate synthase reaction, and thus for the continuation of the cycle. If the organism is growing on amino acids or organic acids (e.g., malate), then replenishment of oxaloacetate is not a problem, since these molecules are easily converted to oxaloacetate (Fig. 8.27). If the carbon source is a sugar (e.g., glucose), then the carboxylation of pyruvate or phosphoenolpyruvate replenishes the oxaloacetate (Fig. 8.17). Two enzymes that carry out the carboxylation of phosphoenolpyruvate and pyruvate are *PEP carboxylase* and *pyruvate carboxylase*, which are widespread among the bacteria. A bacterium will have one or the other. PEP

carboxylase is not found in animal tissues or in fungi.

8.8.1 Regulation of PEP carboxylase

In *E. coli*, PEP carboxylase is an allosteric enzyme that is positively regulated by acetyl-CoA and negatively regulated by aspartate (Fig. 8.18). Presumably, if the oxaloacetate levels drop, then acetyl-CoA will accumulate and result in the activation of the PEP carboxylase. This should produce more oxaloacetate. Aspartate can slow down its own synthesis by negatively regulating PEP carboxylase.

8.9 Modification of the citric acid cycle into a reductive (incomplete) cycle during fermentative growth

In the presence of air, the citric acid cycle operates as an oxidative pathway coupled to aerobic respiration in respiratory organisms. Since fermenting organisms are not carrying out aerobic respiration, it seems best not to have an oxidative pathway that produces so much NADH and FADH$_2$. On the other hand, the reactions that make oxaloacetate, succinyl-CoA, and α-ketoglutarate are necessary because these molecules are required for the biosynthesis of amino acids and tetrapyrroles. (See Section 8.7.3 and Fig. 8.27.) The solution to the problem is to convert the citric acid cycle from an oxidative into a reductive pathway. The reductive pathway

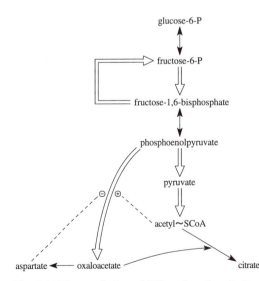

Fig. 8.18 The regulation of PEP carboxylase in *E. coli*. The carboxylase is positively regulated by acetyl-CoA and negatively regulated by aspartate.

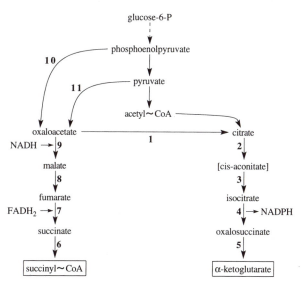

Fig. 8.19 The reductive citric acid pathway in fermenting bacteria. There are two "arms." One route oxidizes citrate to α-ketoglutarate. A second route reduces oxaloacetate to succinyl-CoA. The enzyme α-ketoglutarate dehydrogenase is missing. The enzyme fumarate reductase replaces succinate dehydrogenase. Enzymes: (1) citrate synthase; (2, 3) aconitase; (4, 5) isocitrate dehydrogenase; (6) succinate thiokinase; (7) fumarate reductase; (8) fumarase; (9) malate dehydrogenase; (10) PEP carboxylase; (11) pyruvate carboxylase.

is also referred to as an incomplete citric acid cycle. Fermenting bacteria have little or no activity for the enzyme α-ketoglutarate dehydrogenase. Thus, the pathway is blocked between α-ketoglutarate and succinyl-CoA, and cannot operate in the oxidative direction (Fig. 8.19). Succinyl-CoA is made by reversing the reactions between oxaloacetate and succinyl-CoA, using the enzyme fumarate reductase instead of succinate dehydrogenase. The latter enzyme is replaced by fumarate reductase under anaerobic conditions. These reactions consume 4H. If one includes the reactions from citrate to α-ketoglutarate which produce 2H, the net result is that the reductive pathway consumes 2H. The reductive citric acid pathway is found not only in fermenting bacteria, but also in some other bacteria (including the enteric bacteria) that are carrying out anaerobic respiration using nitrate as the electron acceptor.[10] The reason for this is that oxygen induces the synthesis of α-ketoglutarate dehydrogenase in certain facultative anaerobes, and under anaerobic conditions, the enzyme levels are very low.

One of the consequences of an incomplete citric acid cycle is that acetate is excreted as a by-product of sugar metabolism during anaerobic growth. That is because some of the acetyl-CoA is converted to acetate, concomitant with the formation of an ATP. These reactions, which are an important source of ATP, are discussed in Chapter 13. It should be pointed out that some strict anaerobes (e.g., the green photosynthetic sulfur bacteria), have a reductive citric acid pathway that is "complete" in that it reduces oxaloacetate to citrate. The pathway, called the reductive tricarboxylic acid pathway, is a CO_2 fixation pathway used for autotrophic growth and differs in some key enzymological reactions from the pathways discussed here. The reductive tricarboxylic acid pathway is described in Section 12.1.9.

8.10 Chemistry of some of the reactions in the citric acid cycle

This section presents a rational basis for the chemistry of some of the key reactions in the citric acid cycle. Similar reactions are seen in other pathways.

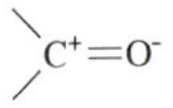

$$\diagdown C^+ = O^-$$
$$\diagup$$

Fig. 8.20 The carbonyl group is polarized. Electrons are attracted by the oxygen, leaving a partial positive charge on the carbon.

8.10.1 Acetyl-CoA condensation reactions

Acetyl-CoA is a precursor for the biosynthesis of many different molecules besides citrate. These include lipids (Chapter 9), and various fermentation end products (Chapter 13). The reason why acetyl-CoA is so versatile is that it undergoes condensations at both the methyl and carboxyl end of the molecule. Condensations at both ends of acetyl-CoA can be understood in terms of the chemistry of the polarized carbonyl group. The electrons in the carbonyl group are not shared equally by the carbon and oxygen; rather, the oxygen is much more electronegative than the carbon and pulls the electrons in the double bond closer to itself. That is to say, the C=O group is polarized making the carbon slightly positive (Fig. 8.20).

Because the oxygen in the carbonyl group is electron attracting, there is a tendency to pull electrons away from the C–H bond in the carbon adjacent to the carbonyl. This results in the formation of an enolate ion which acts as a nucleophile (Fig. 8.21). The enolate anion seeks electrophilic centers (e.g., the carbon atoms in carbonyl groups). Thus, acetyl-CoA can be a nucleophile at its methyl end and attack other carbonyl groups, even other acetyl-CoA molecules. In the formation of citric acid, the methyl group of acetyl-CoA attacks a carbonyl group in oxaloacetate to form citric acid. At the same time, the thioester linkage to coenzyme A is hydrolyzed, driving the reaction to completion. Later, we will examine other pathways in which the methyl carbon of acetyl-CoA attacks a carbonyl.

Another result of the polarization of the carbonyl group is that the carbon in the carbonyl is electron deficient and subject to attack by nucleophiles that seek a positive center (i.e., it is electrophilic). Thus, acetyl-CoA undergoes condensations at the carboxyl end in reactions in which the CoASH is displaced during a nucleophilic displacement and the acetyl portion is transferred to the nucleophile (Fig. 8.22). For example, this occurs during fatty acid synthesis and during butanol fermentations (Chapters 9 and 13), as well as in several other pathways. Because of the reactivity of acetyl-CoA at both the methyl and carboxyl end, the molecule is widely used in building larger molecules.

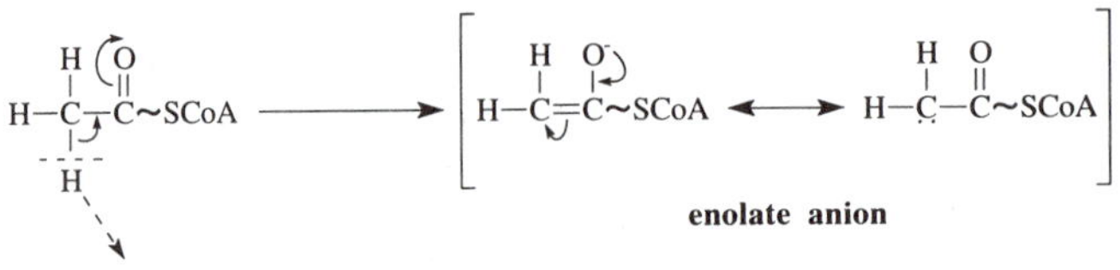

Fig. 8.21 The methylene carbon of acetyl-CoA can act as a nucleophile. Because of the electron-attracting ability of the carbonyl group, a hydrogen dissociates from the methylene carbon forming an enolate anion which resonates. Electrons can shift to the methylene carbon which then seeks a positive center (e.g., a carbonyl group). Because of this, acetyl-CoA will form covalent bonds to the carbon of carbonyl groups in condensation reactions.

Fig. 8.22 Acetyl group transfer. Because the carbonyl group in acetyl-CoA is polarized, it is subject to nucleophilic attack. The result is that the acetyl group is transferred to the nucleophile and CoASH is displaced.

8.10.2 Decarboxylation reactions

Oxalosuccinate is a β-ketocarboxylic acid (i.e., the carboxyl group is beta to a keto group). The decarboxylation of β-ketocarboxylic acids occurs throughout metabolism. Another β-decarboxylation occurs in the pentose phosphate pathway. When 6-phosphogluconate is oxidized by NADP$^+$, a 3-keto intermediate is formed (Fig. 8.7). The decarboxylation of β-ketocarboxylic acids is relatively straightforward. A single enzyme is required, and the cofactor requirements are met by Mn^{2+} or Mg^{2+}. A β-decarboxylation is shown in Fig. 8.23. The β-keto group attracts electrons facilitating the breakage of the bond holding the carboxyl group to the molecule.

Decarboxylation of α-ketoglutarate

α-Ketoglutarate is an α-keto carboxylic acid and its decarboxylation is more complex than that of a β-keto carboxylic acid. The mechanism is the same as for the decarboxylation of pyruvate, another α-keto carboxylic acid, and is described in more detail in Section 13.10. Pyruvate and α-ketoglutarate dehydrogenases are similar in that they can be separated into three components (i.e., the TPP-containing decarboxylase, the FAD-containing dihydrolipoyl dehydrogenase and the dihydrolipoyl transacetylase). Other α-ketocarboxylic acid dehydrogenases exist (e.g., for the catabolism of α-ketoacids derived from the degradation of the branched chain amino acids, leucine, isoleucine, and valine).

8.11 The glyoxylate cycle

We now come to a second pathway central to the metabolism of acetyl-CoA, the glyoxylate cycle, also called the glyoxylate bypass (Fig. 8.24). The glyoxylate cycle is required by aerobic bacteria to grow on fatty acids and acetate. (Plants and protozoa also have the glyoxylate cycle. However, it is absent in animals.) The glyoxylate cycle resembles the citric acid cycle except that it bypasses the two decarboxylations in the citric acid cycle. For this reason, the acetyl-CoA is not oxidized to CO$_2$. Examine the summary of the glyoxylate cycle shown in Fig. 8.24. The glyoxylate cycle shares with the citric acid cycle the reactions that synthesize isocitrate from acetyl-CoA. The two pathways diverge at isocitrate. In the glyoxylate cycle the isocitrate is cleaved to succinate and glyoxylate by the enzyme *isocitrate lyase* (reaction 1). The glyoxylate condenses with acetyl-CoA to form malate, in a reaction catalyzed by *malate synthase* (reaction 2). The malate synthase reaction is of the same type as the citrate synthase (i.e., the methylene carbon of acetyl-CoA attacks the carbonyl group in glyoxylate). The reaction is driven to completion by the hydrolysis of the coenzyme A thioester bond just as during citrate synthesis. The malate replenishes the oxaloacetate, leaving one succinate and NADH as the products. The cells incorporate the succinate into cell material by first oxidizing it to oxaloacetate from which phosphoenolpyruvate is synthesized (Section 8.12).

Carbon balance for the glyoxylate cycle

The carbon balance is written below. The net result of the glyoxylate cycle is the condensation of two molecules of acetyl-CoA to form succinate.

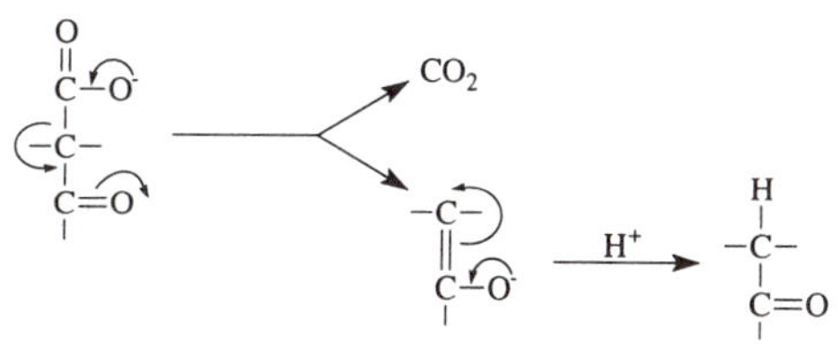

Fig. 8.23 The decarboxylation of a β-keto carboxylic acid. The keto group attracts electrons causing an electron shift and the breakage of the C–C bond holding the carboxyl group to the molecule. The decarboxylation of oxalosuccinate is physiologically reversible. Notice the resemblance to the aldol cleavage shown in Fig. 8.5.

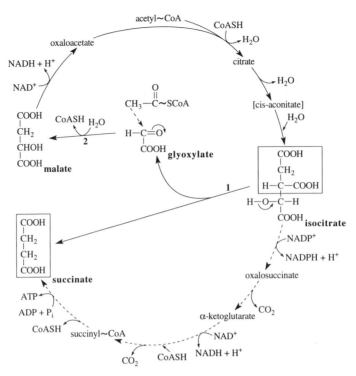

Fig. 8.24 The glyoxylate cycle. Enzymes: (1) isocitrate lyase; (2) malate synthase. The dotted arrows represent the reactions of the citric acid cycle that are bypassed.

Carbon balance for the glyoxylate pathway

$$C_2 + C_4 \longrightarrow C_6$$

$$C_6 \longrightarrow C_4 + C_2$$

$$C_2 + C_2 \longrightarrow C_4$$

$$C_2 + C_2 \longrightarrow C_4$$

8.11.1 Regulation of the glyoxylate cycle

As Fig. 8.24 illustrates, isocitrate is at a branch point for both the citric acid cycle and the glyoxylate cycle. That is to say, it is a substrate for both the isocitrate lyase and the isocitrate dehydrogenase. What regulates the fate of the isocitrate? In *E. coli* the isocitrate dehydrogenase activity is partially inactivated by phosphorylation when cells are grown on acetate.[11] The regulation of enzyme activity by covalent modification is discussed in Chapter 6, and other examples are listed in Table 6.1. Acetate also induces the enzymes of the glyoxylate cycle. Therefore, in the presence of acetate, isocitrate lyase activity increases while the isocitrate dehydrogenase is partially inactivated. However, the K_m for the isocitrate lyase is high relative to the concentrations of isocitrate. This means that the isocitrate lyase requires a high intracellular concentration of isocitrate. It is presumed that the partial inactivation of isocitrate dehydrogenase increases the concentration of isocitrate resulting in an increase in flux through the glyoxylate cycle.

8.12 Formation of phosphoenolpyruvate

Organic acids such as lactate, pyruvate, acetate, succinate, malate, amino acids, and so on, can be used for growth because pathways exist to convert them to phosphoenolpyruvate, which is a precursor to the glycolytic intermediates. The phosphoenolpyruvate is generally made in two ways. It can be made via the decarboxylation of oxaloacetate or via the phosphorylation of pyruvate. These two reactions are described next.

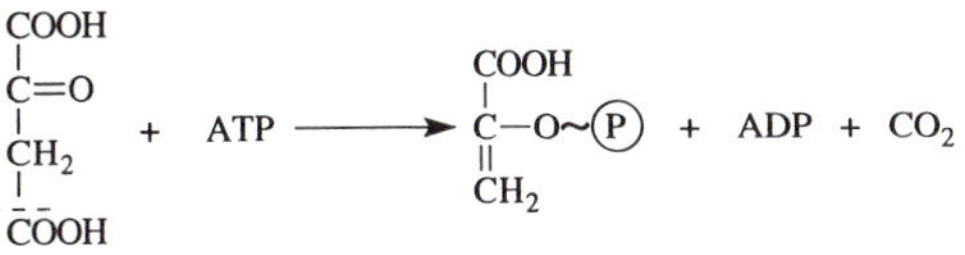

$$\begin{array}{l} \text{COOH} \\ | \\ \text{C=O} \\ | \\ \text{CH}_2 \\ | \\ \text{COOH} \end{array} + \text{ATP} \longrightarrow \begin{array}{l} \text{COOH} \\ | \\ \text{C}-\text{O}{\sim}\text{(P)} \\ \| \\ \text{CH}_2 \end{array} + \text{ADP} + \text{CO}_2$$

Fig. 8.25 The PEP carboxykinase reaction. The PEP carboxykinase generally operates in the direction of PEP synthesis, although during fermentation in anaerobes it can work in the direction of oxaloacetate, which is subsequently reduced to the fermentation end product, succinate.

8.12.1 Formation of phospho-enopyruvate from oxaloacetate

This reaction is an ATP-dependent decarboxylation catalyzed by PEP carboxykinase shown in Fig. 8.25. The enzyme is widespread and accounts for phosphoenolpyruvate synthesis in both eukaryotes and prokaryotes. Although PEP carboxykinase is an important enzyme for the synthesis of phosphoenolpyruvate from oxaloacetate, it catalyzes the synthesis of oxaloacetate from phosphoenolpyruvate in some anaerobic bacteria. For example, anaerobic bacteria that ferment glucose to succinate may use this enzyme to carboxylate phosphoenolpyruvate to oxaloacetate and then in other reactions reduce the oxaloacetate to succinate.[12]

8.12.2 Formation of phosphoenolpyruvate from pyruvate

Prokaryotes are able to synthesize phosphoenolpyruvate by phosphorylating pyruvate instead of converting the pyruvate to oxaloacetate and then decarboxylating the oxaloacetate. This is necessary for many bacteria that cannot synthesize oxaloacetate from pyruvate. Prokaryotes that fall into the latter category do not have a glyoxylate pathway and also lack pyruvate carboxylase. For example, E. coli does not have a glyoxylate cycle unless growing on acetate, and it has PEP carboxylase instead of pyruvate carboxylase. Furthermore, there are strict anaerobes (e.g., methanogens and green sulfur photosynthetic bacteria) that grow autotrophically converting CO_2 to acetyl-CoA, which is carboxylated to pyruvate (Sections 12.1.4 and 12.1.9). They do not have a glyoxylate cycle and must phosphorylate the pyruvate to form phosphoenolpyruvate. The phosphorylation of pyruvate to phosphoenolpyruvate is described next.

PEP synthetase and pyruvate-phosphate dikinase reactions

Prokaryotes that convert pyruvate directly to phosphoenolpyruvate use one of two enzymes. They are PEP synthetase and pyruvate-phosphate dikinase. These enzymes are found in prokaryotes and plants, but not in animals. The reactions are illustrated in Fig. 8.26.

PEP synthetase

$$\text{Pyruvate} + \text{ATP} \longrightarrow \text{PEP} + \text{AMP} + \text{P}_i$$

Pyruvate-phosphate dikinase

$$\text{Pyruvate} + \text{ATP} + \text{P}_i \longrightarrow \text{PEP} + \text{AMP} + \text{PP}_i$$

Pyrophosphatase reaction

$$\text{PP}_i + \text{H}_2\text{O} \longrightarrow 2\text{P}_i$$

In the PEP synthetase reaction, a pyrophosphoryl group is transferred from ATP to the enzyme. One phosphate is removed by hydrolysis, leaving a phosphorylated enzyme. The phosphorylated enzyme then donates the phosphoryl group to pyruvate to form PEP. The reaction catalyzed by the pyruvate-phosphate dikinase is similar, except that instead of the phosphate being hydrolytically removed from the pyrophosphorylated

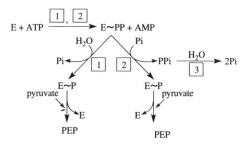

Fig. 8.26 The PEP synthetase and pyruvate-phosphate dikinase reactions. Enzymes: (1) PEP synthetase; (2) pyruvate-phosphate dikinase; (3) pyrophosphatase.

enzyme, it is transferred to inorganic phosphate to form pyrophosphate. The pyrophosphate is hydrolyzed by a pyrophosphatase pulling the reaction to completion. The net result from both reactions is the same (i.e., the sum of the pyruvate-phosphate dikinase and pyrophosphatase reactions is the same as the PEP synthetase reaction). In either case, the synthesis of phosphoenolpyruvate from pyruvate requires the hydrolysis of two phosphodiester bonds with a high free energy of hydrolysis.

in the glyoxylate cycle. Reaction 1 is isocitrate lyase. Reaction 2 is malate synthase. Fatty acids and acetate are converted to acetyl-CoA that enters the citric acid cycle and the glyoxylate cycle. Dicarboxylic acids and amino acids are eventually degraded to citric acid cycle intermediates. Sugars can be catabolized via the Embden–Meyerhof–Parnas pathway, the pentose phosphate pathway, or the Entner–Doudoroff pathway. All of the sugar pathways intersect at phosphoglyceraldehyde.

8.13 Summary of the relationships between the pathways

Figure 8.27 illustrates the relationship between the different pathways discussed in this chapter. Reactions 1 and 2 are key enzymes

8.14 Summary

The nearly ubiquitous pathway for glucose degradation is the Embden–Meyerhof–Parnas pathway, also called the glycolytic pathway. The pathway oxidizes one mole of glucose

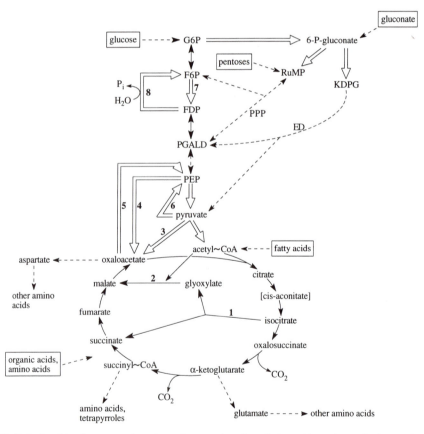

Fig. 8.27 Relationship between the glyoxylate cycle, the citric acid cycle and the major carbohydrate pathways. Enzymes: (1) isocitrate lyase; (2) malate synthase; (3) pyruvate carboxylase; (4) PEP carboxylase; (5) PEP carboxykinase; (6) PEP synthetase or pyruvate phosphodikinase; (7) phosphofructokinase; (8) fructose-1,6-bisphosphatase.

to two moles of pyruvate, and produces two moles of NADH and two moles of ATP. There is only one oxidation, and that is the oxidation of phosphoglyceraldehyde to phosphoglycerate. An intermediate, 1,3-bisphosphoglycerate, is formed, which donates a phosphoryl group to ADP in a substrate level phosphorylation. Since two phosphoglyceraldehydes are formed from one glucose, two ATPs are made. The pathway can be considered as two stages. Stage one generates two phosphoglyceraldehydes. Stage two is the oxidation of phosphoglyceraldehyde to pyruvate. Glycolysis not only provides ATP, NADH, and pyruvate, but its intermediates are used for biosynthesis in other pathways. This will become more evident in subsequent chapters when we examine other metabolic pathways. The pathway cannot be reversed from pyruvate to glucose-6-phosphate because of two irreversible reactions, the pyruvate kinase reaction (phosphoenolpyruvate to pyruvate) and the phosphofructokinase reaction (fructose-6-phosphate to fructose-1,6-bisphosphate). Both these reactions proceed so far in the direction of products that they are not physiological reversible. However, the pathway can be reversed from phosphoenolpyruvate provided fructose-1,6-bisphosphate phosphatase is present to convert fructose-1,6-bisphosphate to fructose-6-phosphate.

A second sugar-catabolizing pathway is the Entner–Doudoroff (ED) pathway. The ED pathway is widespread among prokaryotes where it can be used for growth on gluconate. An intermediate of this pathway is phosphoglyceraldehyde which is oxidized to pyruvate using the same reactions that occur in stage 2 of glycolysis. Because only one phosphoglyceraldehyde is produced from glucose-6-phosphate in the ED pathway, only one ATP is made. Several bacteria lack key enzymes in the first stage of the EMP pathway (i.e. phosphofructokinase and fructose-1,6-bisphosphate aldolase) and rely on the ED pathway for growth on glucose. Interestingly, modifications of the ED pathway exist in a few bacteria and archaea in which some of the intermediates are not phosphorylated. There is even a completely nonphosphorylated ED pathway in some archaea.

Among the many pathways that intersect with the second stage of glycolysis is the pentose phosphate pathway. The pentose phosphate pathway oxidizes glucose-6-phosphate to pentose phosphates. The pentose phosphates can be used for the synthesis of nucleic acids, or be converted via the transaldolase and transketolase reactions to phosphoglyceraldehyde. The pathway also produces NADPH for biosynthetic reductions, and erythrose-4-phosphate for aromatic amino acid biosynthesis. The pentose phosphate pathway is also used when pentoses are the source of carbon and energy.

In aerobic prokaryotes pyruvate is oxidatively decarboxylated via pyruvate dehydrogenase to acetyl-CoA, CO_2, and NADH. The acetyl-CoA enters the citric acid cycle where it is oxidized completely to CO_2. The electrons generated during the oxidations are transferred to the electron transport chain, where a Δp is created. The citric acid cycle is also an anabolic pathway. Several biosynthetic pathways draw citric acid cycle intermediates out of the pools. For example, oxaloacetate and α-ketoglutarate are used for amino acid biosynthesis, and succinyl-CoA is used for tetrapyrrole, lysine, and methionine synthesis. Oxaloacetate is also used to synthesize phosphoenolpyruvate for glucogenesis. Since oxaloacetate is used catalytically in the cycle, it must be replenished during growth. When organisms are growing on carbohydrates, the oxaloacetate is replenished by carboxylating pyruvate or phosphoenolpyruvate. Growth on proteins or amino acids poses no problem since they are degraded to citric acid cycle intermediates.

In order to use pyruvate as a source of cell carbon, it must be converted to phosphoenolpyruvate. The reason for this is that phosphoenopyruvate is a precursor to several metabolites, including the intermediates of glycolysis. One route is the ATP-dependent carboxylation of pyruvate to oxaloacetate (pyruvate carboxylase) combined with the ATP-dependent decarboxylation to phosphoenolpyruvate using the enzyme PEP carboxykinase. Two ATPs are therefore used to convert one pyruvate to phosphoenolpyruvate. A second route is the direct phosphorylation of pyruvate to phosphoenolpyruvate using either PEP synthetase or pyruvate-phosphate dikinase combined with a pyrophosphatase. The products are phosphoenolpyruvate, AMP, and P_i. Thus, regardless of the pathway used, two ATPs are required to synthesize phosphoenolpyruvate from pyruvate.

Bacteria growing aerobically on fatty acids or acetate use the glyoxylate cycle for net incorporation of C_2 units into cell material. This pathway is a modification of the citric acid cycle, in which the two decarboxylation reactions are bypassed. The decarboxylation reactions are bypassed using two enzymes unique to the glyoxylate cycle (i.e., isocitrate lyase and malate synthetase). The net result is that acetyl-CoA is converted to succinate rather than to carbon dioxide.

The citric acid cycle is usually present only during aerobic growth. However, some sulfate reducers, which are anaerobic bacteria that use sulfate as an electron acceptor, have an oxidative citric acid cycle. Under anaerobic conditions, prokaryotes have a modified pathway called the reductive citric acid pathway. In the reductive citric acid pathway there are two major changes. The α-ketoglutarate dehydrogenase activity is low or missing, and fumarate reductase replaces succinate dehydrogenase. The result is that oxaloacetate is reduced to succinyl-CoA, and citrate is oxidized to α-ketoglutarate. Thus, the intermediates necessary for biosynthesis of amino acids and tetrapyrroles are made, but the number of reductions exceed the number of oxidations.

Study Questions

1. Suppose you isolated a mutant that did not grow on glucose as the sole source of carbon. However, the mutant did grow on the glucose when it was supplemented with succinate, fumarate, or malate. An examination of broken cell extracts revealed that all of the citric acid cycle enzymes were present. What might be the metabolic defect in the mutant?

2. There are two physiologically irreversible reactions in the EMP pathway starting with glucose and ending with pyruvate. Which ones are they? How are they regulated in E. coli?

3. Do you think that the glyoxylate cycle and the citric acid cycle can operate at the same time in an organism growing on acetic acid? How is this regulated?

4. Fermenting bacteria have little or no α-ketoglutarate dehydrogenase activity, and therefore have a block in the citric acid cycle. Under these conditions, the pathway operates reductively instead of oxidatively. Draw a reductive citric acid pathway showing how these organisms make oxaloacetate, succinyl-CoA, and α-ketoglutarate. Why is it important to be able to synthesize these compounds under all growth conditions?

5. Which nonglycolytic enzyme is necessary to reverse the pathway from PEP to G6P?

6. Some bacteria (e.g., *Brucella abortus*), lack key enzymes in the EMP and the ED pathway, but they can still grow on glucose. *Brucella abortus* is an animal pathogen that causes spontaneous abortion in cattle and can also infect humans, causing fever, headache, and joint pains. It has a citric acid cycle. Write a series of reactions showing how the pentose phosphate pathway can result in glucose oxidation completely to CO_2 without the involvement of stage one of the EMP pathway, or the ED pathway.

7. The ED pathway is only 50% as efficient as the EMP pathway in making ATP from glucose. Why is that?

8. E. coli mutants lacking G6P dehydrogenase can be grown on glucose as the sole source of carbon. Since this is the enzyme that catalyzes the entrance of G6P into the pentose phosphate pathway, how can these mutants make NADPH or pentose phosphates?

9. Show how the pentose phosphate pathway might oxidize glucose completely to CO_2 without using the citric acid cycle.

10. The combination of phosphofructokinase and fructose-1,6-bisphosphatase has ATPase activity. Write these reactions

and sum them to satisfy yourself that this is so. How does *E. coli* ensure that these reactions do not use up all of the ATP in the cell?

11. When *E. coli* is grown on pyruvate as its sole source of carbon, it does not synthesize the glyoxylate cycle enzymes. How is the pyruvate converted to PEP? [Hint: For this question and the ones that follow, start by examining Fig. 8.27.]

12. Suppose you are growing a bacterial culture on ribose as its only source of carbon. Write a series of reactions that will convert ribose to glucose-6-phosphate. Why is it important to be able to synthesize glucose-6-phosphate?

13. Suppose you are growing a culture on gluconate. How would the cells convert the gluconate to phosphoglyceraldehyde if they lacked 6-phosphogluconate dehydrogenase? How might they make ribose-5-phosphate and erythrose-4-phosphate? Why is it necessary to have a supply of the latter two compounds?

14. Suppose you are growing a bacterial culture on succinate as the source of carbon. Write the reactions by which succinate is converted to PEP. Why is it important to be able to synthesize PEP?

15. How might cells growing on lactate make PEP without using the glyoxylate pathway, PEP synthetase, or pyruvate-phosphate dikinase? (This occurs in the liver during gluconeogenesis.)

REFERENCES AND NOTES

1. During some fermentations a portion of the NADH is reoxidized by fumarate via membrane bound electron carriers and a Δp is created. This is called fumarate respiration. An example of fumarate respiration occurs during mixed acid fermentation discussed in Section 13.10 and propionate fermentation discussed in Section 13.7.

2. However, some bacteria (e.g., *E. coli*), also contain an enzyme called transhydrogenase, which catalyzes the reduction of $NADP^+$ by NADH. Therefore, for those bacteria the pentose phosphate pathway is not essential for NADPH synthesis.

3. Sugars that differ in the configuration at a single asymmetric carbon are called *epimers*. For example, xylulose-5-phosphate and ribulose-5-phosphate are epimers of each other at C3.

4. Danson, M. J. 1988. Archaebacteria: The comparative enzymology of their central metabolic pathways, p. 166–231. In: *Advances in Microbial Physiology*, Vol. 29. A. H. Rose and D. W. Tempest (eds.). Academic Press, New York.

5. Schaffer, T., and P. Schonheit. 1992. Maltose fermentation to acetate, CO_2, and H_2 in the anaerobic hyperthermophilic archaeon *Pyrococcus furiosus*: Evidence for the operation of a novel sugar fermentation pathway. Arch. Microbiol. 158:188-202.

6. Schafer, G., S. Anemuller, R. Moll, W. Meyer, and M. Lubben. 1990. Electron transport and energy conservation in the archaebacterium *Sulfolobus acidocaldarius*. FEMS Microbiol. Rev. 75:335–348.

7. Schafer, T., and P. Schonheit. 1993. Gluconeogenesis from pyruvate in the hyperthermophilic archaeon *Pyrococcus furiosus*: Involvement of reactions of the Embden–Meyerhof pathway. Arch. Microbiol. 159:354–363.

8. Vitamins are growth factors that cannot be made by animals. During the early years of vitamin research, vitamins were divided into two groups: the fat-soluble vitamins were placed into one group and the water-soluble vitamins (B complex and vitamin C) into another. Vitamins are generally assayed according to specific diseases that they cure in animals fed on a vitamin deficient diet. We now know that the water-soluble vitamin most responsible for stimulating growth in rats is riboflavin (B_2). Vitamin B_6 (pyridoxine) prevents rat facial dermatitis. Pantothenic acid cures chick dermatitis. Nicotinic acid (a precursor to nicotinamide) cures human pellagra. (The symptoms of pellagra are weakness, dermatitis, diarrhea, mental disorder, and death.) Vitamin C (or ascorbic acid) prevents scurvy. Folic acid and vitamin B_{12} (cobalamin) prevent anemia. Vitamin D or calciferol (fat soluble) prevents rickets. Vitamin E or tocopherol (fat soluble) is required for rats for full-term pregnancy and prevents sterility in male rats. Vitamin K (fat soluble) is necessary for normal blood clotting. A deficiency in vitamin A (fat soluble) leads to dry skin, conjunctivitis of the eyes, night blindness, and retardation of growth. Male rats fed a diet deficient in vitamin A do not form sperm and become blind. Vitamin A (retinol) is an important component of the light-sensitive pigment in the rod cells of the eye. In addition, animals cannot make polyunsaturated fatty acids ("essential fatty acids') because they lack the enzyme to desaturate monounsaturated fatty acids. A diet deficient in polyunsaturated fatty acids leads to poor growth, skin lesions, impaired fertility,

and kidney damage. Lipoic acid is synthesized by animals and is therefore not a vitamin. However, certain bacteria and other microorganisms require lipoic acid for growth. It was soon recognized that lipoic acid was required for the oxidation of pyruvic acid. The chemical identification of lipoic acid followed the purification of 30 mg from 10 tons (!) of water-soluble residue of liver. It was done by Lester Reed at the University of Texas in 1949, with collaboration from scientists at Ely Lilly and Co. Lipoic acid is a growth factor for some bacteria. It is an eight-carbon saturated fatty acid in which carbons 6 and 8 are joined by a disulfide bond to form a ring when the molecule is oxidized. The reduced molecule has two sulfhydryl groups. The coenzyme functions in the oxidative decarboxylation of alpha-keto acids.

9. Danson, M. J., S. Harford, and P. D. J. Weitzman. 1979. Studies on a mutant form of *Escherichia coli* citrate synthase desensitized to allosteric effectors. *Eur. J. Biochem.* **101**:515–521.

10. Stewart, V. 1988. Nitrate respiration in relation to facultative metabolism in Enterobacteria. *Microbiol. Rev.* **52**:190–232.

11. LaPorte, D. C. 1993. The isocitrate dehydrogenase phosphorylation cycle: Regulation and enzymology. *J. Cell. Biochem.* **51**:14–18.

12. Podkovyrov, S. M. and J. G. Zeikus. 1993. Purification and characterization of phosphoenolpyruvate carboxykinase, a catabolic CO_2-fixing enzyme, from *Anaerobiospirillum succiniciproducens*. *J. Gen. Microbiol.* **139**:223–228.

9

Metabolism of Lipids, Nucleotides, Amino Acids, and Hydrocarbons

The central pathways, i.e., glycolysis, the pentose phosphate pathway, the Entner–Doudoroff pathway, and the citric acid cycle, provide precursors to all of the other metabolic pathways. This chapter focuses on some of those metabolic pathways that feed off the central pathways, and are necessary for lipid, protein, and nucleic acid metabolism.

9.1 Lipids

Lipids are a structurally heterogeneous group of substances that share the common property of being highly soluble in nonpolar solvents (e.g., methanol, chloroform, and so on) and relatively insoluble in water. Essentially all of the lipids in prokaryotes are in the membranes. The major membrane lipids in bacteria and eukaryotes are phospholipids consisting of fatty acids esterified to glycerol phosphate derivatives, and are called phosphoglycerides. Archaea can also have phosphoglycerides but their structure and mode of synthesis is different. The structures of archaeal lipids are summarized in Section 1.2.5 and their synthesis in Section 9.1.3. The metabolism of fatty acids will be discussed first.

9.1.1 Fatty acids

Types found in bacteria

Fatty acids are chains of methylene carbons with a carboxyl group at one end. They can be branched (methyl), saturated, unsaturated, or hydroxylated. Some examples are shown in table 9.1. Fatty acids differ in the number of carbon atoms, the number of double bonds they contain, where the double bonds are placed in the molecule, and whether the molecule is branched as is tuberculostearic, or whether it is cyclopropane (e.g., lactobacillic). However, some generalizations can be made about the fatty acids. They are usually 16 or 18 carbons long and are either saturated or have one double bond. Generally, gram-positive bacteria are richer in branched chain fatty acids than are gram-negative bacteria.

The role of fatty acids

Fatty acids do not occur free in bacteria but are covalently attached to other molecules. Most of the fatty acids are esterified to glycerolphosphate derivatives to make *phosphoglycerides* which are an important structural component of membranes. Fatty acids are also esterified to carbohydrate. For example, the *lipid A* portion of lipopolysaccharide consists of fatty acids esterified to glucosamine (Section 10.2.1). Fatty acids can also be esterified to protein. For example, gram-negative bacteria have a *lipoprotein* that is covalently attached to the peptidoglycan and protrudes into the outer membrane. The lipoprotein appararently is important for the stability of the outer membrane (Section

Table 9.1 Some fatty acids

Number of carbon atoms	Fatty acid	
16	$CH_3(CH_2)_{14}COOH$	Palmitic
18	$CH_3(CH_2)_{16}COOH$	Stearic
18	$CH_3(CH_2)_7CH=CH(CH_2)_7COOH$	Oleic
18	$CH_3(CH_2)_4CH=CHCH_2CH=CH(CH_2)_7COOH$	Linoleic
9	$CH_3(CH_2)_5CH-CH(CH_2)_9COOH$ $\backslash\,/$ CH_2	Lactobacillic
19	$CH_3(CH_2)_7CH(CH_2)_8COOH$ $\vert$ CH_3	Tuberculostearic

1.2.3). Because fatty acids are nonpolar, they anchor the molecules to which they are attached to the membrane. Furthermore, whether a fatty acid is unsaturated, saturated, or branched has important significance for the physical properties of the membrane. Unsaturated fatty acids and branched chain fatty acids make the membrane more fluid, which is a necessary condition for membrane function.

β-Oxidation of fatty acids

Many bacteria (e.g., pseudomonads, various bacilli, and *E. coli*) can grow on long chain fatty acids. The fatty acids are oxidized to acetyl-CoA via a pathway called *β-oxidation* (Fig. 9.1). For this to take place the fatty acid is first converted to the acyl-CoA derivative in a reaction catalyzed by *acyl-CoA synthetase*. This takes place in two steps. First, pyrophosphate is displaced from ATP to make the AMP derivative of the carboxyl group (acyl adenylate) which remains tightly bound to the enzyme (Fig. 9.1, reaction 1). Then the AMP is displaced by CoASH to make the fatty acyl-CoA. The activation of a carboxyl group by formation of an acyl adenylate is common in metabolism and is discussed in Section 7.2.1. For example, amino acids are activated for protein synthesis in this way. As with other reactions in which pyrophosphate is displaced from ATP, the reaction is driven to completion by hydrolysis of the pyrophosphate. A keto group is then generated β to the carboxyl. The sequence of reactions leading to the keto group are: (1) an oxidation to form a double bond catalyzed by acyl-CoA dehydrogenase (reaction 2); (2)

hydration of the double bond to form the hydroxyl catalyzed by 3-hydroxyacyl-CoA hydrolyase (reaction 3); and (3) oxidation of the hydroxyl to form the keto group,

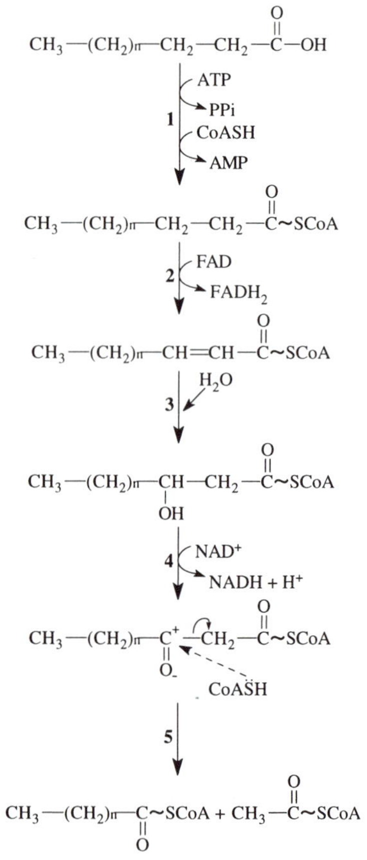

Fig. 9.1 Beta oxidation of fatty acids. Enzyme: (1) acyl-CoA synthetase; (2) fatty acyl-CoA dehydrogenase; (3) 3-hydroxyacyl-CoA hydrolyase; (4) L-3-hydroxyacyl-CoA dehydrogenase; (5) β-ketothiolase.

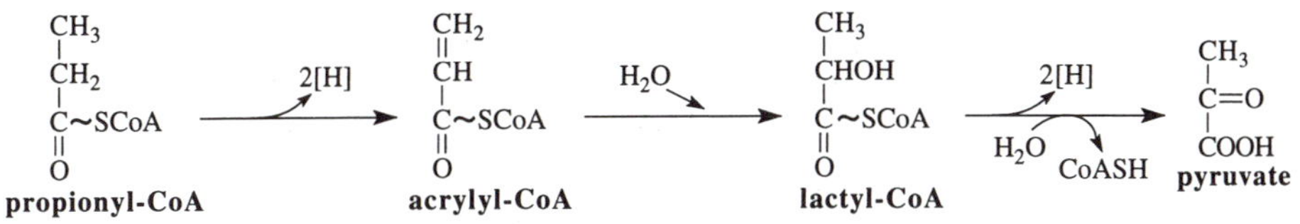

Fig. 9.2 Oxidation of propionyl-CoA to pyruvate by *E. coli*.

catalyzed by L-3-hydroxyacyl-CoA dehydrogenase (reaction 4). The chemistry is the same as the conversion of succinate to oxaloacetate in the citric acid cycle (Fig. 8.16). The carbonyl of the β-keto acyl-CoA is then attacked by CoASH, displacing an acetyl-CoA catalyzed by β-ketothiolase (reaction 5). Usually the bacterium is growing aerobically and the acetyl-CoA is oxidized to carbon dioxide in the citric acid cycle. The acetyl-CoA can also be a source of cell carbon when it is assimilated via the glyoxylate cycle (Section 8.11). The displacement of the acetyl-CoA results in the generation of a fatty acid acyl-CoA that is recycled through the β-oxidation pathway.

If the fatty acid has an even number of carbon atoms, then the entire chain is degraded to acetyl-CoA. However, if the fatty acid is an odd-chain fatty acid, then the last fragment is propionyl-CoA rather than acetyl-CoA. *E. coli* oxidizes the propionyl-CoA to acrylyl-CoA, which is hydrated to lactyl-CoA. The lactyl-CoA is oxidized to pyruvate, which can be oxidized via the citric acid cycle. Those reactions are shown Fig. 9.2.

The synthesis of fatty acids

An examination of metabolic pathways reveals that catabolic pathways are usually different from synthetic pathways. In other words, biosynthesis is not simply due to reversing the reactions of catabolism. This is true because there is usually at least one irreversible step in the catabolic pathway. For example, glycolysis has two irreversible reactions, the phosphofructokinase and pyruvate kinase reactions, and these must be bypassed in order to reverse glycolysis. Often, an entirely different set of reactions is used for biosynthesis. This is seen in fatty acid metabolism where the biosynthetic pathway consists of reactions completely different

from the β-oxidation pathway.[1] *The biosynthetic pathway differs from the β-oxidation pathway in the following ways:*

1. The reductant is NADPH, rather than NADH.

2. Biosynthesis requires CO_2 and proceeds via a carboxylated derivative of acetyl-CoA, malonyl-CoA.

3. The acyl carrier is not CoA, as it is in the degradative pathway, but a protein called the acyl carrier protein (ACP). The acyl carrier protein is a small protein (MW = 10,000 in *E. coli*) having a residue similar to CoASH attached to one end.[2]

In eukaryotic cells, fatty acid biosynthesis takes place in the cytosol and the degradation takes place in the matrix of the mitochondria. Since prokaryotes do not have similar organelles, both synthesis and degradation take place in the same compartment (i.e., the cytosol). The fatty acid biosynthetic enzymes exist as multienzyme complexes in eukaryotes, including yeast. However, in *E. coli*, they are present in the cytosol as separate enzymes.

The sequence of reactions begins with the carboxylation of acetyl-CoA to make malonyl-CoA (Fig. 9.3A, reaction 1). The carboxylation requires ATP, a vitamin (biotin) and is catalyzed by acetyl-CoA carboxylase. (Acetyl-CoA carboxylase consists of four subunit proteins. These are biotin carboxylase, biotin carboxyl carrier protein, and two proteins that carry out the trans-carboxylation of the carboxy group from biotin to acetyl-CoA.) Then the CoA is displaced by the acyl carrier protein, ACP, to form malonyl-ACP in a reaction catalyzed by malonyl-CoA:ACP transacetylase (reaction 2). A similar reaction, catalyzed by acetyl-CoA:ACP transacetylase, displaces CoA from

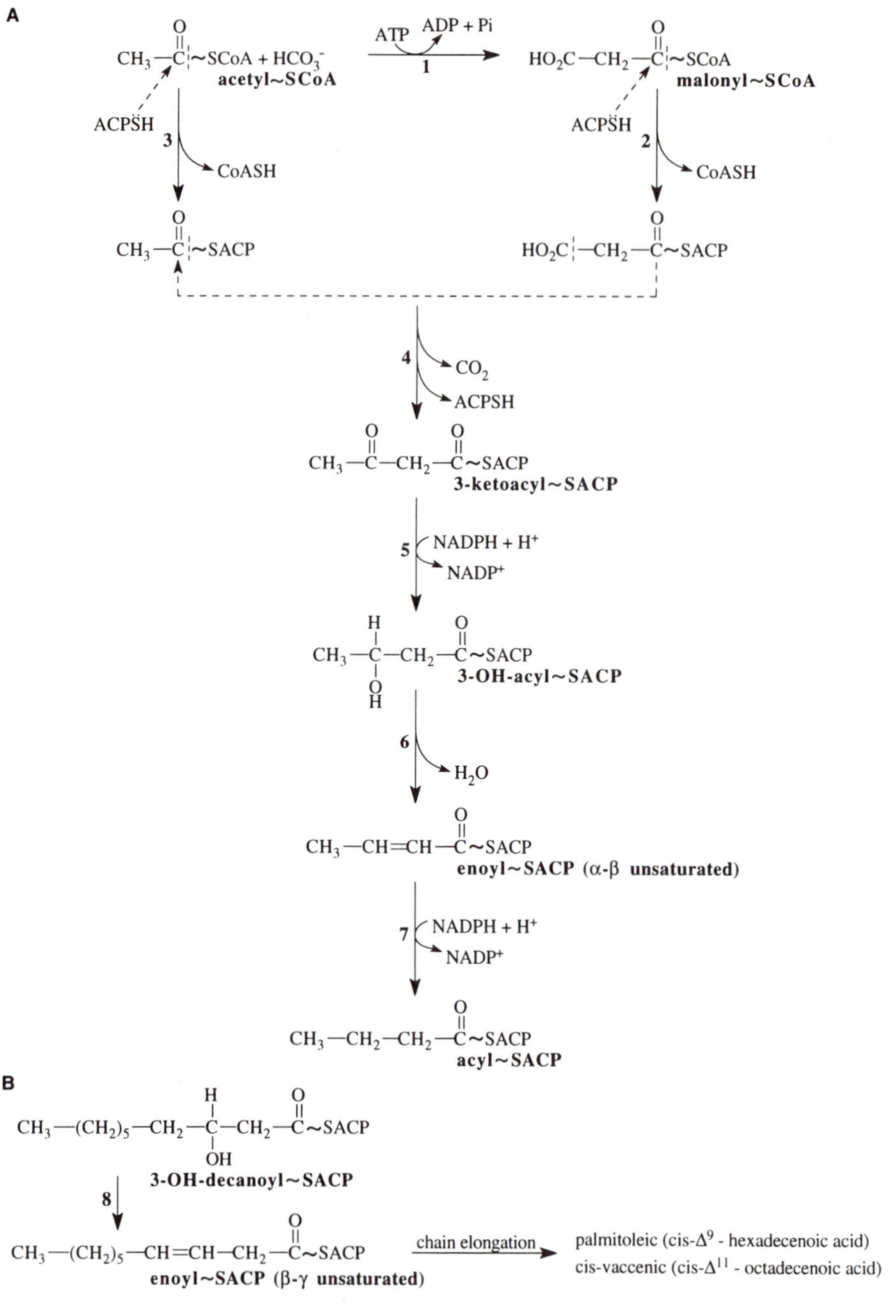

Fig. 9.3 Biosynthesis of fatty acids. Enyzmes: (1) acetyl-CoA carboxylase; (2) malonyl transacetyl-ase; (3) acetyl transacetylase; (4) 3-ketoacyl-ACP synthase; (5) 3-ketoacyl-ACP reductase; (6) β-hydroxyacyl-ACP dehydrase; (7) enoyl-ACP reductase; (8) 3-hydroxydecenoyl-ACP dehydrase. (A) synthesis of saturated fatty acids; (B) synthesis of unsaturated fatty acids.

another molecule of acetyl-CoA to form the acetyl-ACP derivative (reaction 3). [Whether there is a separate acetyl-CoA:ACP trans-acetylase is not certain. Reaction 3 can be catalyzed by β-ketoacyl-ACP synthase III (acetoacetyl-ACP synthase) which has trans-acetylase activity.] The methylene carbon in malonyl-ACP acts as a nucleophile and displaces the ACP from acetyl-ACP to form the β-ketoacyl-ACP derivative in a reaction catalyzed by β-ketoacyl-ACP synthase (reaction 4). At the same time, the newly added carboxyl is removed, driving the reaction to completion. (In *E. coli*, either acetyl-CoA or acetyl-ACP can be used for the condensation with malonyl-ACP because there is more than one condensing enzyme.)[3] The β-keto acyl-ACP is then reduced to the hydroxy derivative by an NADPH-dependent β-ketoacyl-ACP reductase (reaction 5), and dehydrated to the α-β derivative by the β-hydroxylacyl-ACP dehydrase to form the unsaturated acyl-ACP derivative (reaction 6). (The double bond is between C2, the carbon α to the carboxyl carbon, and C3, the carbon β to the carboxyl carbon.) The unsaturated acyl-ACP derivative is then reduced by enoyl-ACP reductase to the saturated acyl-ACP (reaction 7). The acyl-ACP chain is elongated by a series of identical reactions initiated by the attack of malonyl-ACP on the carboxyl end of the growing acyl-ACP chain displacing the ACP. Fatty acid synthesis must be regulated because long-chain acyl-ACPs do not accumulate. However, the mechanism of regulation has not been elucidated.

When the acyl-ACP chain is completed, the acyl portion is immediately transferred to membrane phospholipids by the glycerol phosphate acyltransferase reactions described in Section 9.1.2. Of course, not all of the fatty acids are incorporated into phospholipids. For example, some of the acyl-ACPs are used for lipid A biosynthesis (Section 10.2.2). These include β-hydroxymyristoyl-ACP, lauroyl-ACP, and myristoyl-ACP.

Synthesis of unsaturated fatty acids

If the fatty acid is unsaturated, it is almost always monounsaturated (one double bond). Polyunsaturated fatty acids are more typical of eukaryotic organisms.[4] Unsaturated fatty acids are formed in two different ways depending upon the organism, the *anaerobic* pathway and the *aerobic* pathway. The anaerobic pathway is restricted to prokaryotes. It is widespread in bacteria, being found in *Clostridium*, *Lactobacillus*, *Escherichia*, *Pseudomonas*, the cyanobacteria, and the photosynthetic bacteria. The aerobic pathway is found in *Bacillus*, *Mycobacterium*, *Corynebacterium*, *Micrococcus*, and in eukaryotes.

In the anaerobic pathway, a special dehy-

drase desaturates the C_{10} hydroxyacyl-ACP intermediate to the trans-α,β-decenoyl-ACP which is isomerized while bound to the enzyme to cis-β,γ-ACP (Fig. 9.3B). In the β-γ derivative the double bond is between C3 and C4, whereas in the α-β derivative the double bond is between C2 and C3. *Very importantly, the cis-β,γ does not serve as a substrate for the enoyl reductase.* The consequence of this fact is that the cis-β,γ derivative is not reduced. Instead, it is elongated leading to a fatty acid with a double bond. Two common monounsaturated fatty acids in bacteria synthesized in this way are palmitoleic acid which has 16 carbons (cis-Δ^9-hexadecenoic acid) and cis-vaccenic acid which has 18 carbons (cis-Δ^{11}-octadecenoic acid).[5] In palmitoleic acid the double bond is between C9 and C10 in the final product. This is because six carbons are added to the C1 of the C10 β-γ derivative. Since vaccenic acid is two carbons longer, the double bond is between C11 and C12.

The aerobic pathway makes use of special desaturases (oxidases) that introduce a double bond into the completed fatty acyl-CoA or fatty acyl-ACP derivative. The enzyme system requires molecular oxygen and NADPH. Bacteria that use the aerobic pathway make oleic acid from its C_{18} saturated progenitor, stearic acid, rather than cis-vaccenic. Oleic acid has a double bond between C9 and C10 and is called cis-Δ^9-octadecenoic acid.

9.1.2 Phospholipid synthesis in bacteria

Phospholipids are lipids with covalently attached phosphate groups. They are an important constituent of cell membranes. Because the phosphate groups are ionized, phospholipids always have negatively charged groups. There may also be positively charged groups (e.g., the protonated amino group in phosphatidyl enthanolamine). There are several different types of phospholipids. However, the ones that concern us here are the major phospholipids in bacteria. These are phospholipids that contain fatty acids and phosphate esterified to glycerol, and usually some other molecule (e.g., an amino acid,

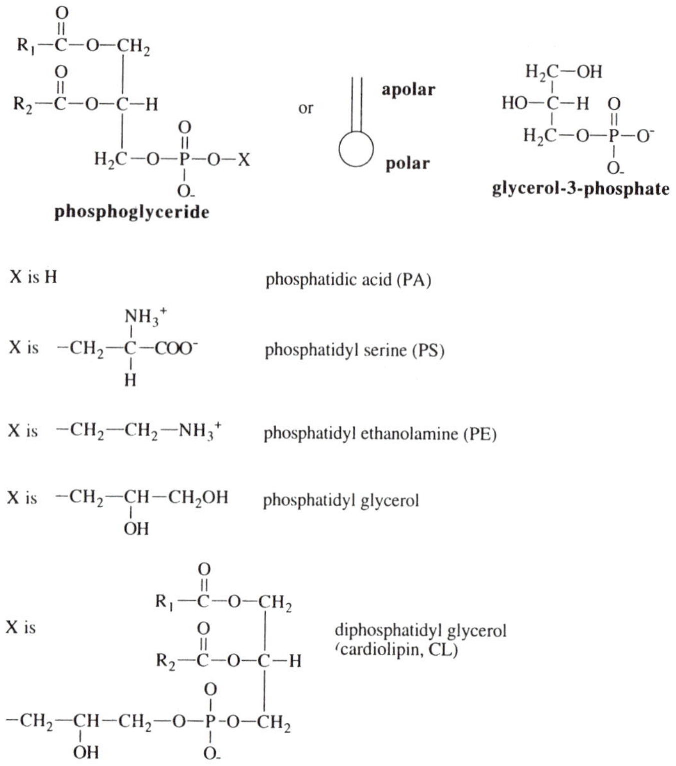

Fig. 9.4 Some common phosphoglycerides in bacteria. R represents a fatty acyl moiety esterified to the glycerol phosphate. Note that phospholipids have an apolar and a polar end. The configuration of glycerol-3-phosphate and the glycerol-3-phosphate moiety of phosphoglycerides is called *sn*-glycerol-3-phosphate and belongs to the L-stereochemical series. Horizontal bonds are projected to the front and vertical bonds behind the plane of the page. Thus, the H and OH are in front of the plane of the page and C1 and C3 are behind.

an amine, or a sugar covalently bound to the phosphate). They are called *phosphoglycerides*. The structure of a typical phosphoglyceride is shown later (Fig. 9.4). The fatty acids in positions C1 and C2 within the same molecule need not be the same.

Phospholipids and the structure of the cell membrane

A major structural feature of phospholipids is that they are amphibolic. That is to say, one part of the molecule is hydrophobic (apolar) and another part is hydrophilic (polar). The hydrophobic area (tail) is the part where the fatty acids are located and the hydrophilic area (head) is the end containing the phosphate and its attached group. The amphibolic nature of the phospholipids explains their orientation in membranes. Phospholipids are oriented in membranes as a bilayer with the charged head groups pointing out into the aqueous phase and the hydrophobic fatty acid tails interacting with each other in a hydrophobic interior. The cell membrane is completed by proteins (Fig. 1.15). Some properties of phospholipid membranes that should be remembered are: (1) they are permeable only to water, gases, and small hydrophobic molecules; (2) they have a low ionic conductance and are capable of doing work when ions (especially protons and sodium ions) are transported through them via special carriers along an electrochemical gradient; (3) the lipid portion of membranes must remain fluid in order for the membrane to function; and (4) the fluidity of the membrane is due to the presence of unsaturated fatty acids or branched chain (methyl) fatty acids in the phospholipids.

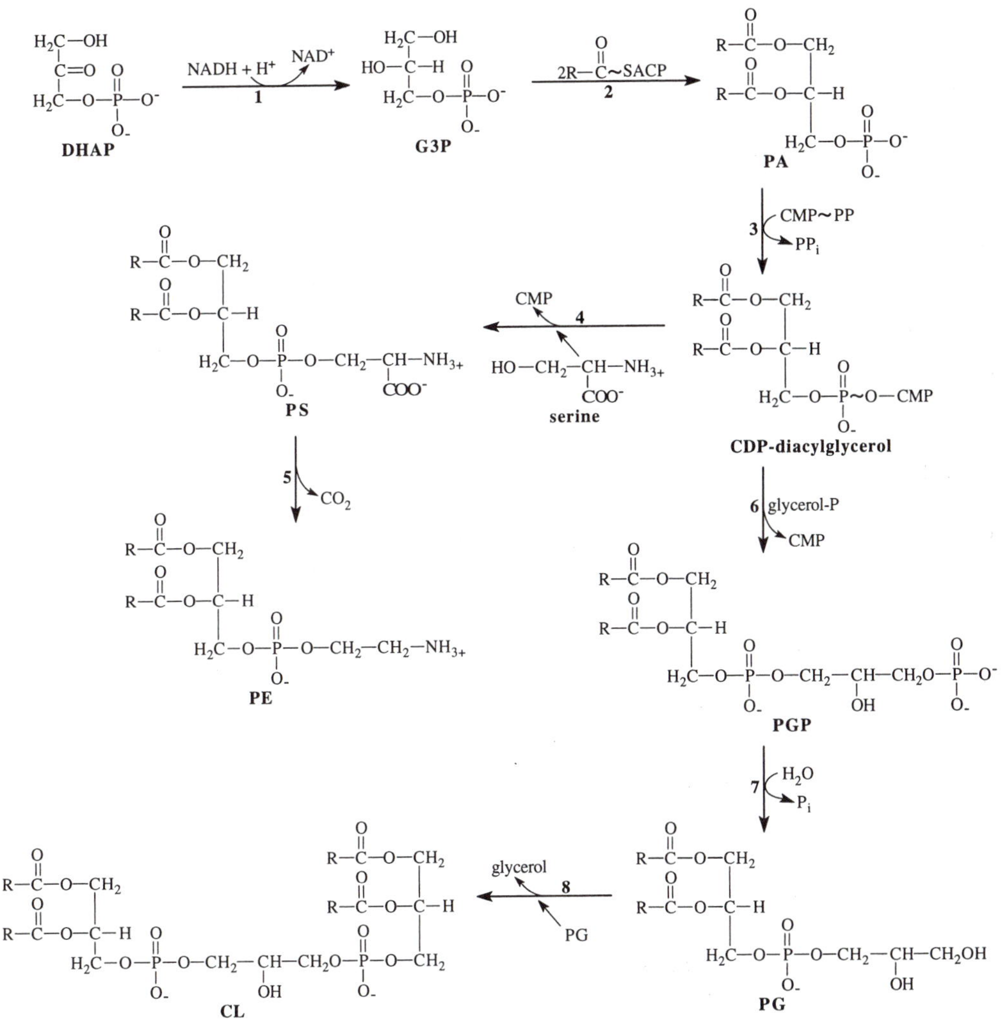

Fig. 9.5 Phosphoglyceride synthesis. The biosynthetic pathway for phosphoglycerides branches off glycolysis at DHAP. The G3P dehydrogenase is a soluble enzyme but the G3P acyl transacylase and all subsequent reactions take place in the cell membrane. The fatty acyl-ACP derivatives are made in the cytosol, diffuse to the membrane, and transfer the acyl portion to the phospholipid on the inner surface of the membrane. DHAP, dihydroxyacetone phosphate; G3P, glycerol phosphate; PA, phosphatidic acid; PS, phosphatidylserine; PE, phosphatidylethanolamine; PGP, phosphatidylglycerolphosphate; PG, phosphatidylglycerol; CL, cardiolipin (diphosphatidylglycerol). Enzymes: (1) G3P dehydrogenase; (2) G3P acyltransferase and 1-acyl-G3P acyltransferase; (3) CDP-diglyceride synthase; (4) phosphatidyl-serine synthase; (5) phosphatidylserine decarboxylase; (6) PGP synthase; (7) PGP phosphatase; (8) CL synthase.

The synthesis of the phosphoglycerides

Phosphoglyceride synthesis starts with dihydroxyacetone phosphate (DHAP), an intermediate in glycolysis (Fig. 9.5). Step 1 is the reduction of dihydroxyacetone phosphate to glycerol phosphate by the enzyme *glycerol phosphate dehydrogenase*. Step 2 is the transfer to glycerol phosphate of fatty acids from fatty acyl-S-ACP (newly synthesized as described above). The reactions are catalyzed by membrane-bound enzymes called *G3P acyl transferases*. (Some bacteria are also capable of using acyl-SCoA derivatives as the acyl donor.) The first phospholipid made is phosphatidic acid, which is glycerol phosphate esterified to two fatty acids. The other phospholipids are made from phosphatidic acid. Step 3 is a reaction in which the phosphate on the phosphatidic acid reacts with cytidine triphosphate (CTP) and displaces PP$_i$ to form CDP-diacylglycerol. The reaction is catalyzed by *CDP-diglyceride synthase*. The formation of CDP-diacylglycerol is driven to completion by the hydrolysis of the PP$_i$ catalyzed by a pyrophosphatase. Step 4 is the displacement of CMP by serine catalyzed by *phosphatidylserine synthase* (PS synthase). Step 5 is the decarboxylation of phosphatidylserine to yield phosphatidylethanolamine, which is the major phospholipid in several bacteria. Bacteria make other phospholipids from CDP-diglyceride. The displacement of CDP by α-glycerolphosphate yields phosphatidylglycerol phosphate (step 6). Phosphatidylglycerol phosphate is dephosphorylated to yield phosphatidylglycerol (step 7). The latter displaces glycerol from a second molecule of phosphatidylglycerol to form diphosphatidylglycerol (cardiolipin) (step 8). Note that modification of the head groups in all of these cases occurs via a nucleophilic attack by a hydroxyl on the electropositive phosphorous in the phosphate group.

9.1.3 Synthesis of archaeal lipids

The student should review the structure of archaeal lipids and membranes described in Section 1.2.5. The lipids of archaea differ in the following two ways from those of bacteria:

1. Instead of fatty acids, archaeal lipids have long chain alcohols called isopranyl alcohols.

2. The linkage to glycerol is via an ether bond rather than an ester bond.

The metabolic pathway for the synthesis of the glycerol ethers in the archaea is not well understood at all. Figure 9.6 illustrates a recent proposal for *Halobacterium halobium*. The glycerol backbone is synthesized from either glycerol-3-phosphate or dihydroxyacetone phosphate. The alcohol is believed to be derived from geranylgeranyl-PP, which is synthesized from acetyl-CoA via mevalonic acid in a well-known pathway that is widespread among the bacteria and higher

organisms. (The alcohol group is esterified to the pyrophosphate.) The C1 hydroxyl on glycerol-3-phosphate or the dihydroxyacetone phosphate nucleophilically attacks the geranylgeranyl-PP displacing pyrophosphate and forming the ether linkage. If dihydroxyacetone phosphate is used then the monoalkenyl-DHAP is formed. If glycerol-3-P is used then the monoalkenylglycerol-1-P is formed. The monoalkenylglycerol-1-phosphate, which is formed either directly or in two steps, reacts with another geranylgeranyl-PP to form the dialkenyl-glycerol-1-phosphate. It is proposed that in *Methanobacterium* and *Sulfolobus* glycerol-1-phosphate, rather than dihydroxacetone phosphate or glycerol-3-phosphate, attacks the geranylgeranyl-PP. The rest of the pathway is unknown but involves a reduction of the double bonds in the hydrocarbons, the attachment of the polar head groups, and the formation of the tetraether lipids. It is not known whether the condensation between the alkyl groups to form the tetraethers occurs before or after substitution with polar head groups.

9.2 Nucleotides

9.2.1 Nomenclature and structures

A nucleotide is a molecule containing a pyrimidine or purine, a sugar (ribose or deoxyribose), and phosphate. If there were no phosphate, it would be called a nucleoside. Thus, nucleotides are nucleoside phosphates. If one phosphate is present, then the molecule is called a nucleoside monophosphate. If two phosphates are present, then it is called a nucleoside diphosphate, and so on. The three major pyrimidines are cytosine, thymine, and uracil. The corresponding nucleosides are called cytidine, thymidine, and uridine, respectively. The cytidine nucleotides are called cytidine monophosphate (CMP), cytidine diphosphate (CDP), and cytidine triphosphate (CTP). A similar naming system is used for the uridine nucleotides (UMP, UDP, UTP) and the thymidine nucleotides (TMP,

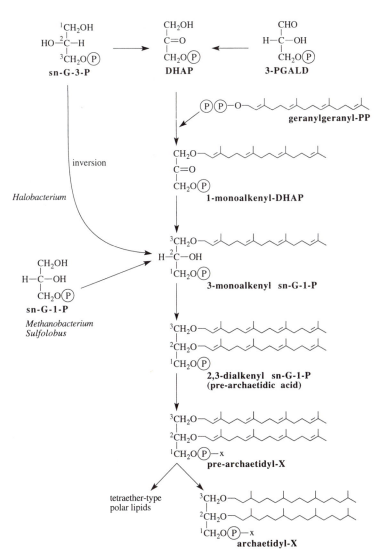

Fig. 9.6 Postulated pathway for the synthesis of glycerol ether lipids in archaea. Two pathways are depicted, one for *Halobacterium* and one for *Methanobacterium* and *Sulfolobus*. It is proposed that a nucleophilic displacement takes place on geranylgeranyl-PP displacing the pyrophosphate and forming the ether linkage. The molecule that condenses with the geranylgerany-PP might be glycerol-3-phosphate, glycerol-1-phosphate, or dihydroxyacetone phosphate. The rest of the pathway is unknown. (From Koga, Y., M. Nishihara, H. Morii, and M. Akagawa-Matsushita. 1993. Ether polar lipids of methanogenic bacteria: Structures, comparative aspects, and biosynthesis. *Microbiol. Rev.* 57:164–182.)

TDP, TTP). If the sugar is deoxyribose rather than ribose, then the nucleotide is written dCTP and so on. The two major purines are adenine and guanine. The nucleosides are adenosine and guanosine. The nucleotides are adenosine mono-, di-, and triphosphate and guanosine mono-, di-, and triphosphate. There are even separate names for the nucleoside monophosphates. For example, adenosine monophosphate is also called adenylic acid. Guanosine monophosphate is guanylic acid. We also have cytidylic acid, uridylic acid, and thymidylic acid. The structures of the purines, pyrimidines, and sugars are shown in Fig. 9.7. Notice that deoxyribose differs from ribose in having a hydrogen substituted for the hydroxyl group on C2 of the sugar.

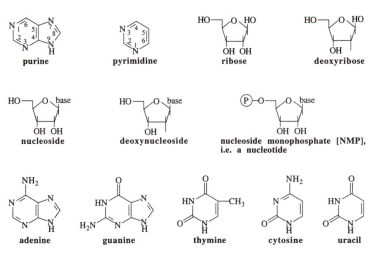

Fig. 9.7 The structures of the nucleotides and their subunits. The bases are attached to the C1 of the sugars and the phosphates to the C5. The nucleoside di- and triphosphates are not shown.

9.2.2 The synthesis of the pyrimidine nucleotides

The pyrimidine part of the nucleotide is made from aspartic acid, ammonia, and carbon dioxide (Fig. 9.8). Step 1 (Fig. 9.9) is the biotin-dependent synthesis of carbamoyl phosphate from HCO_3^-, glutamine, and ATP. The enzyme that catalyzes the reaction is called *carbamoyl phosphate synthetase*.[6] It requires two ATPs. One ATP is used to form a carboxylated biotin intermediate, and the second ATP is used to phosphorylate the carboxyl group. Notice that carbamoyl phosphate is an acyl phosphate and therefore has high group transfer potential. In step 2, the carbamoyl group is transferred to aspartate as the phosphate is displaced by the nitrogen atom of aspartate. The product is *N*-carbamoylaspartate. The enzyme that catalyzes step 2 is *aspartate transcarbamylase* (ATCase), which is feedback inhibited by the end product, CTP. Step 3 is the cyclization of *N*-carbamoylaspartate with loss of water to form the first pyrimidine, dihydroorotate. Step 4 is the oxidation of dihydroorotate to orotate. In step 5 the orotate displaces PP$_i$ from phosphoribosyl pyrophosphate (PPRP) to form the first nucleotide, orotidine-5′-phosphate, also called orotidylate. The reaction is driven to completion by the hydrolysis of PP$_i$ catalyzed by a pyrophosphatase. Note that the PRPP itself is synthesized from

ribose-5-phosphate and ATP via a phosphoryl group transfer reaction from ATP to the C1 carbon of ribose-5-phosphate. Orotodine-5′-phosphate is decarboxlated in step 6 to uridine monophosphate (UMP). Phosphorylation of UMP using ATP as the donor yields UDP and UTP (steps 7 and 8). CTP is synthesized from UTP by an ATP-dependent amination using NH_4^+ (step 9).[7]

The CTP can be dephosphorylated to CDP, which is the precursor to the deoxyribonucleotides dCTP and dTTP. The sequence of reactions is:

$$CDP \longrightarrow dCDP \longrightarrow dCTP \longrightarrow$$

$$dUTP \longrightarrow dUMP \longrightarrow dTMP \longrightarrow$$

$$dTDP \longrightarrow dTTP$$

Also, UDP can be reduced to dUDP which can be converted to dUTP. Thus, there are two routes to dUTP, one from CDP and the other (a minor route) from UDP.

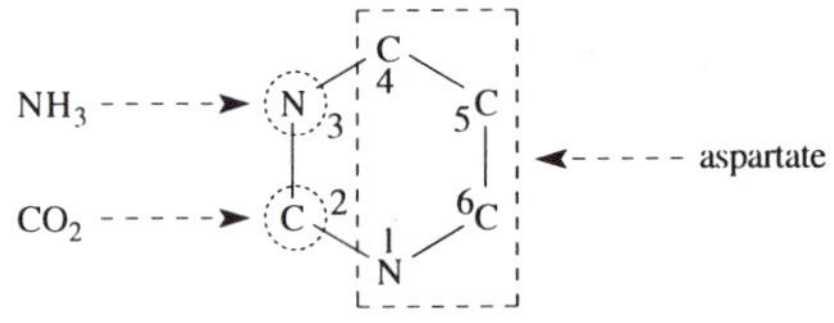

Fig. 9.8 Origins of atoms in the pyrimidine ring. C2 is derived from CO_2 and N3 comes from ammonia, via carbamoyl phosphate. The rest of the atoms are derived from aspartate.

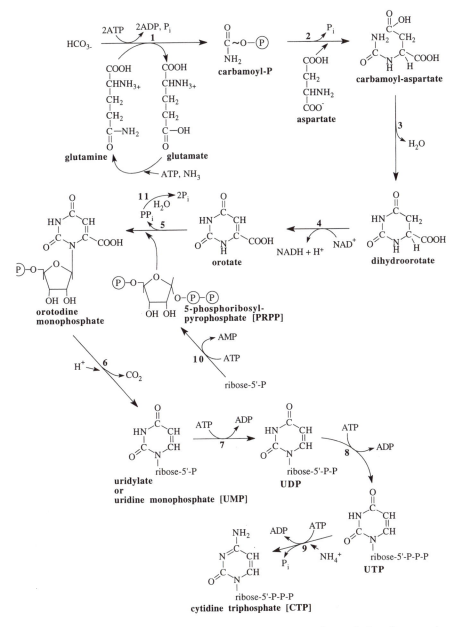

Fig. 9.9 The biosynthesis of pyrimidine nucleotides. Enzymes: (1) carbamoyl phosphate synthetase; (2) aspartate transcarbamoylase; (3) dihydroorotase; (4) dihydroorotate dehydrogenase; (5) orotate phosphoribosyl transferase; (6) orotidine-5-phosphate decarboxylase; (7) nucleoside monophosphate kinase; (8) nucleoside diphosphate kinase; (9) CTP synthetase; (10) PRPP synthetase; (11) pyrophosphatase.

9.2.3 The synthesis of purine nucleotides

A purine is drawn in Fig. 9.10 showing the origin of the atoms. The synthesis of the purine ring requires as precursors, glutamine and aspartic acid (to donate amino groups), glycine, carbon dioxide, and a C_1 unit at the oxidation state of formic acid. The latter three donate all of the carbon atoms in the purine. The C_1 unit at the oxidation level of formic acid can come from formic acid itself, or from serine.

Enzymatic reactions

The biosynthetic pathway for purines is shown in Fig. 9.11. The purine molecule is

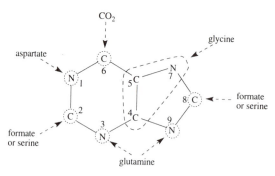

Fig. 9.10 Metabolic origins of atoms in purines.

synthesized in stages while attached to ribose phosphate, which in turns comes from PRPP. Step 1 is the attachment of an amino group to PRPP, donated by glutamine. The product is 5-phosphoribosylamine. This is a reaction in which the amide nitrogen of glutamine displaces the pyrophosphate in a nucleophilic displacement. The pyrophosphate that is released has a high free energy of hydrolysis and the amination is driven to completion by the hydrolysis of the pyrophosphate by pyrophosphatase. In step 2, glycine is added to the amino group of phosphoribosylamine to form 5-phosphoribosyl-glycineamide in an ATP-dependent step. In this reaction the carboxyl group of glycine is added to the amino group of phosphoribosylamine to make the amide bond. The ATP first phosphorylates the carboxyl group forming an acyl phosphate. Then the nitrogen in the amino group displaces the phosphate forming the amide bond. The student should review these kinds of reactions in Section 7.2.1. Step 3 is the formylation of the amino group on the glycine moiety. This requires a molecule to donate the formyl group (i.e., a molecule with high formyl group transfer potential). The donor of the formyl group is formyl-tetrahydrofolic acid (formyl-THF). The product is 5-phosphoribosyl-N-formylglycineamide. In step 4 the amide group is changed into an amidine, the nitrogen being donated by glutamine in an ATP-dependent reaction. The product is 5-phosphoribosyl-N-formylglycineamidine. In step 5 the 5-phosphoribosyl-N-formylglycineamidine cyclizes to 5-phosphoribosyl-5-aminoimidazole in a reaction that requires ATP. Probably the carbonyl

is phosphorylated and the phosphate is displaced by the nitrogen to form the C–N bond. The cyclized product tautomerizes to form the amino group on aminoimidazole ribonucleotide. Reaction 6 is the carboxylation of 5-phosphoribosyl-5-aminoimidazole to form 5-phosphoribosyl-5-aminoimidazole-4-carboxylic acid. In step 7 aspartic acid combines with 5-phosphoribosyl-5-amino-imidazole-4-carboxylic acid to form 5-phosphoribosyl-4-(N-succinocarboxyamide)-5-amionoimidazole in an ATP-dependent step. In step 8 fumarate is removed, leaving the nitrogen, to form 5-phosphoribosyl-4-carboxamide-5-aminoimidazole, which is formylated in step 9 to form 5-phosphoribosyl-4-carboxamide-5-formaminoimidazole. The latter is cyclized in step 10 to inosinic acid (IMP). The purine itself is called hypoxanthine. Inosinic acid is the precursor to all of the purine nucleotides.

Biosynthesis of AMP and GMP from IMP

IMP is converted to AMP by substitution of the carbonyl oxygen at C6 with an amino group (Fig. 9.12). The amino donor is aspartate and the reaction requires GTP. AMP can then be phosphorylated to ADP using ATP as the phosphoryl donor. The ADP can be reduced to dADP or phosphorylated to ATP via substrate level phosphorylation or respiratory phosphorylation. GMP is synthesized by an oxidation of IMP at C2 followed by an amination at that carbon. The oxidation produces a carbonyl group which becomes aminated at the expense of ATP.

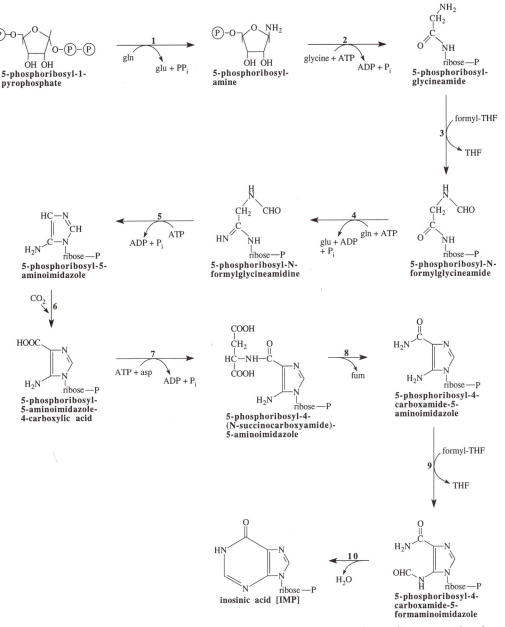

Fig. 9.11 Biosynthesis of purine nucleotides. Enzymes: (1) PRPP amidotransferase; (2) phosphoribosylglycineamide synthetase; (3) phosphoribosylglycineamide formyltransferase; (4) phosphoribosylformylglycineamidine synthetase; (5) phosphoribosyl-aminoimidazole synthetase; (6) phosphoribosylaminoimidazole carboxylase; (7) phosphoribosylaminoimidazole succinocarboxamide synthetase; (8) adenylosuccinate lyase; (9) phosphoribosylaminoimidazolecarboxamide formyltransferase; (10) IMP cyclohydrolase. gln, glutamine; glu, glutamate; asp, aspartate; fum, fumarate; THF, tetrahydrofolate.

9.2.4 The role of tetrahydrofolic acid

Derivatives of tetrahydrofolic acid (THF; synthesized from the B vitamin folic acid) are coenzymes that carry single carbon groups and are very important in metabolism. For example, the C_1 units used in purine biosynthesis are carried by derivatives of tetrahydrofolic acid. For an explanation of how THF is used, examine Fig. 9.13. The C_1 units can be donated to THF from serine which donates its β-carbon at the oxidation level of

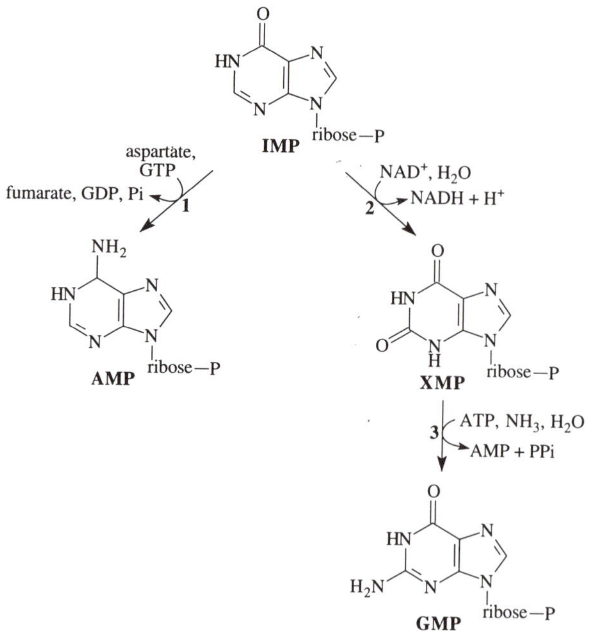

Fig. 9.12 Synthesis of AMP and GMP from IMP. Enzymes: (1) adenylosuccinate synthetase and adenylosuccinate lyase; (2) IMP dehydrogenase; (3) GMP synthetase. IMP, inosinic acid; AMP, adenylic acid; XMP, xanthylic acid; GMP, guanylic acid.

formaldehyde to THF to form methylene-THF and glycine. (This is also the pathway for glycine biosynthesis as described in Section 9.3.1.) Methylene-THF is then oxidized to *formyl-THF* which donates the formyl group in purine biosynthesis. Bacteria can also use formate for purine biosynthesis. The formate is attached to THF to make formyl-THF in an ATP-dependent reaction. THF is also an important methyl carrier. Instead of methylene-THF being oxidized to formyl-THF, it is reduced to *methyl-THF*. Methyl-THF donates the methyl group in various biosynthetic reactions, including the biosynthesis of methionine. *Sulfanilamide* and its derivatives, called sulfonamides, (the "sulfa drugs"), inhibit the formation of folic acid and therefore inhibit purine biosynthesis in bacteria. Sulfonamides resemble *p*-aminobenzoic acid, an intermediate in folic acid biosynthesis and the sulfonamides inhibit the enzyme that utilizes *p*-aminobenzoic acid for folic acid synthesis. The sulfonamides are toxic to bacteria and not to animals because animals obtain their folic acid from their diet, whereas most bacteria synthesize folic acid.

9.2.5 Synthesis of deoxyribonucleotides

The deoxyribonucleotides are synthesized from the ribonucleoside diphosphates by a reductive dehydration catalyzed by *ribonucleoside diphosphate reductase* (Fig. 9.14). The electron donor is a sulfhydryl protein called thioredoxin that obtains its electrons from NADPH.

9.3 Amino acids

There are 20 different amino acids in proteins. Inspection of Table 9.2 reveals that six of the 20 amino acids are synthesized from oxaloacetate and four from α-ketoglutarate, two intermediates of the citric acid cycle. In addition, succinyl-CoA donates a succinyl group in the formation of intermediates in the biosynthesis of lysine, methionine, and diaminopimelic acid. (The succinate is removed at a later step and does not appear in the product.) This again emphasizes that not only is the citric acid cycle important for energy generation, but also for biosynthesis.

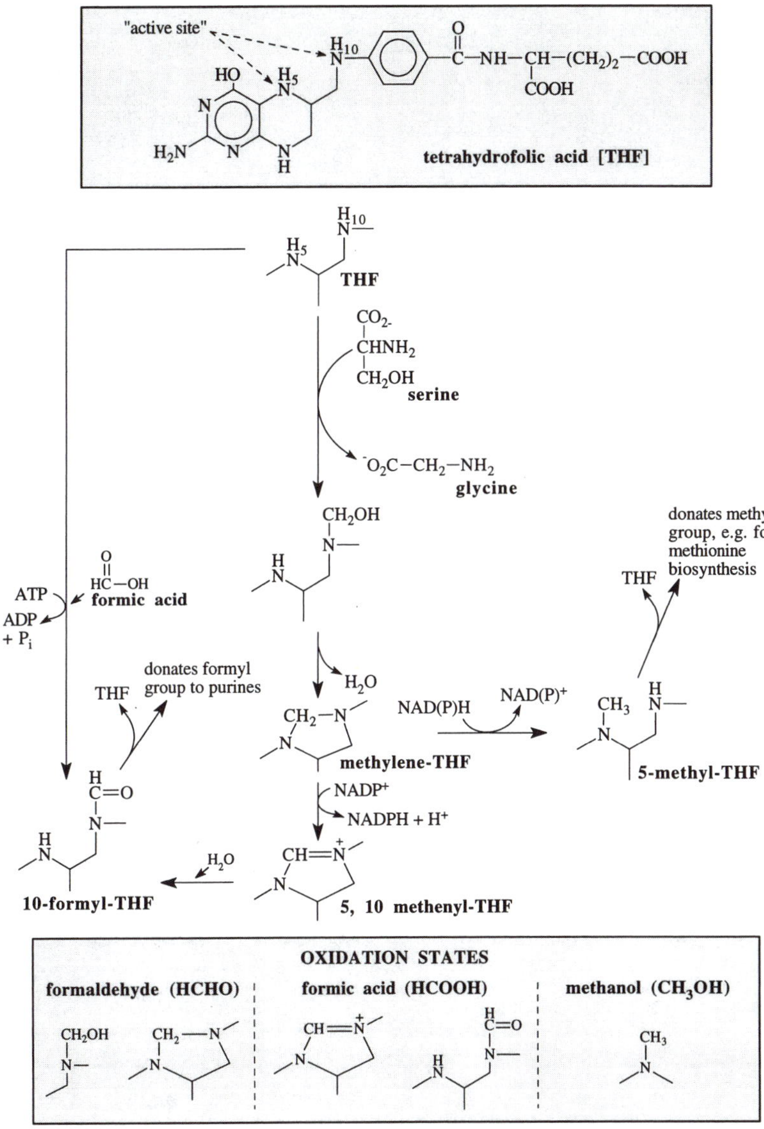

Fig. 9.13 Tetrahydrofolic acid (THF) as a C_1 carrier. THF is the reduced form of the vitamin folic acid. THF and its derivatives are important coenzymes that function in many pathways to carry one-carbon units. The one-carbon units can be at the oxidation level of formic acid (HCOOH), formaldehyde (HCHO), or methyl (CH_3^-). It is essential for purine biosynthesis because it transfers groups at the oxidation level of formic acid in two separate steps. The most common precursor of the C_1 unit is the amino acid serine, which donates the C_1 unit as formaldehyde to THF. The product can then be oxidized to the level of formate to form 10-formyl-THF. Some bacteria can use formic acid itself as the source of the C_1 unit. Methylene-THF can also be reduced to methyl-THF which serves as a methyl donor.

The table also points out that three glycolytic intermediates (pyruvate, 3-PGA, and PEP) are precursors to seven more amino acids. The pentose phosphate pathway provides erythrose-4-P, which is a precursor to aromatic amino acids. (It is actually used with

PEP in the synthesis of the aromatic amino acids.) Finally, 5-phosphoribosyl-1-pyrophosphate (PRPP), which donates the sugar for nucleotide synthesis, is also the precursor to the amino acid histidine. Therefore, the three central pathways (i.e., glycolysis, the pentose

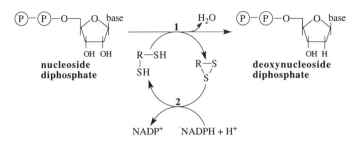

Fig. 9.14 Formation of deoxynucleotides by ribonucleoside diphosphate reductase. The deoxy-nucleotides are made from the nucleoside diphosphates. The OH on C2 of the ribose leaves with its bonding electrons to form water. The OH is replaced by a hydride ion (H:⁻) donated by reduced thioredoxin, a dithioprotein. Thioredoxin is re-reduced by thioredoxin reductase, a flavo-protein. Thioredoxin reductase, in turn, accepts electrons from NADPH. Enzymes: (1) ribonucleo-tide diphosphate reductase; (2) thioredoxin reductase. R-$(SH)_2$, thioredoxin.

phosphate pathway, and the citric acid cycle) provide the precursors to all of the amino acids. In this discussion, we will examine the biosynthesis and catabolism of a select group of amino acids in order to illuminate some principles of amino-acid metabolism.

9.3.1 Synthesis

Glutamate and glutamine synthesis
The ability to synthesize glutamate and glutamine is of extreme importance because this is the only route for incorporation of inorganic nitrogen into cell material. All

Table 9.2 Precursors for amino-acid biosynthesis

Precursor	Amino acid
Pyruvic acid	L-alanine, L-valine, L-leucine
Oxaloacetic acid	L-aspartate, L-asparagine, L-methionine, L-lysine, L-threonine, L-isoleucine
α-Ketoglutaric acid	L-glutamate, L-glutamine, L-arginine, L-proline
3-PGA	L-serine, glycine, L-cysteine
PEP and erythrose-4-P	L-phenylalanine, L-tyrosine, L-tryptophan
PRPP and ATP	L-histidine

inorganic nitrogen must first be converted to ammonia which is then incorporated as an amino group into glutamate and glutamine. (The synthesis of ammonia from nitrate and from nitrogen gas is described in Chapter 11.) The amino group is then donated from these amino acids to all the other nitrogen-containing compounds in the cell. Glutamate is the amino donor for most of the amino acids, and glutamine is the amino donor for the synthesis of purines, pyrimidines, amino sugars, histidine, tryptophan, asparagine, NAD^+, and p-aminobenzoate.

Glutamate is synthesized by two alternate routes. One requires the enzyme *glutamate dehydrogenase*, and the second requires two enzymes: *glutamine synthetase* (GS) and *glutamate synthase* (glutamine oxoglutarate aminotransferase or *GOGAT* reaction).

Glutamate dehydrogenase catalyzes the reductive amination of α-ketoglutarate (Fig. 9.15). However, it should be pointed out that because of its high K_m for ammonia, the glutamate dehydrogenase reaction can be used for glutamate synthesis only when ammonia concentrations are high (>1 mM); otherwise ammonia is incorporated into glutamate via glutamine. This is the situation in many natural environments. Under conditions of low ammonia concentration, bacteria use two enzymes in combination for ammonia incorporation. One is L-glutamine synthetase (Fig. 9.16, reaction 1). This enzyme incorporates ammonia into glutamate to form glutamine. The glutamine synthetase reaction uses ATP which drives the reaction to

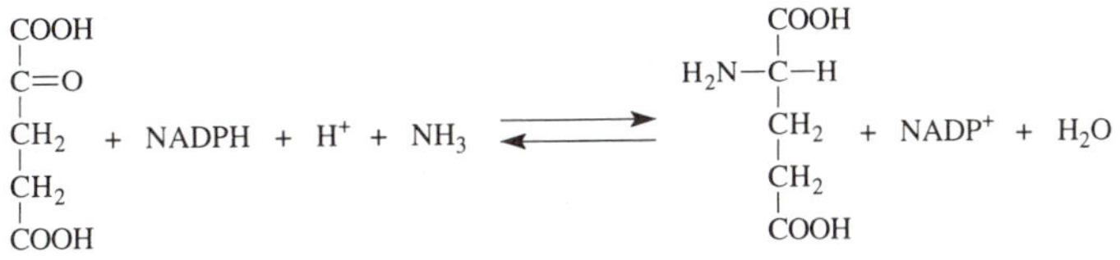

$$
\begin{array}{c}
\text{COOH} \\
| \\
\text{C}{=}\text{O} \\
| \\
\text{CH}_2 \quad + \text{ NADPH } + \text{ H}^+ + \text{ NH}_3 \\
| \\
\text{CH}_2 \\
| \\
\text{COOH}
\end{array}
\rightleftharpoons
\begin{array}{c}
\text{COOH} \\
| \\
\text{H}_2\text{N}-\text{C}-\text{H} \\
| \\
\text{CH}_2 \quad + \text{ NADP}^+ + \text{ H}_2\text{O} \\
| \\
\text{CH}_2 \\
| \\
\text{COOH}
\end{array}
$$

Fig. 9.15 The glutamate dehydrogenase reaction.

completion. The other enzyme is glutamate synthase (reaction 2). This enzyme transfers the newly incorporated ammonia from glutamine to α-ketoglutarate to form glutamate. Notice in Fig. 9.16 that glutamine is used catalytically. This is also seen in the reactions written below.

The glutamine synthetase and glutamate synthase reactions:

The use of ATP drives the reaction to completion.

As mentioned, glutamine is an amino donor for many compounds, including pyrimidines and purines. Therefore, glutamine synthetase is an indispensable enzyme unless the medium is supplemented with glutamine. Glutamate dehydrogenase, however, is not necessary for growth, since the organism can synthesize glutamate from glutamine and

$$\text{Glutamate} + \text{ATP} + \text{NH}_3 \longrightarrow \text{glutamine} + \text{ADP} + \text{P}_i$$

$$\text{Glutamine} + \alpha\text{-ketoglutarate} + \text{NADPH} + \text{H}^+ \longrightarrow 2 \text{ glutamate} + \text{NADP}^+$$

$$\alpha\text{-Ketoglutarate} + \text{ATP} + \text{NH}_3 + \text{NADPH} + \text{H}^+ \longrightarrow \text{glutamate} + \text{ADP} + \text{P}_i + \text{NADP}^+$$

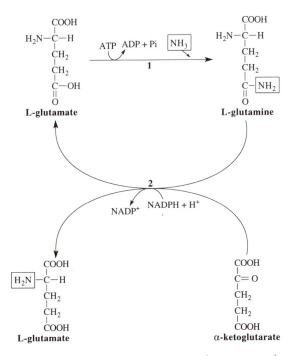

Fig. 9.16 The GS and GOGAT reactions. Enzymes: (1) L-glutamine synthetase; (2) glutamine: α-oxoglutarate aminotransferase, also called the GOGAT enzyme, or glutamate synthase.

α-ketoglutarate using glutamate synthase. The regulation of glutamine synthetase activity is complex and discussed in Section 17.4.2.

Transamination reactions from glutamate are required for the synthesis of the other amino acids

The synthesis of the other amino acids requires that glutamate donates an amino group to an α-keto carboxylic acid. The enzyme that catalyzes the amino group transfer is called a *transaminase*. A generalized transamination reaction is shown in Fig. 9.17. Note that after the glutamate has donated its amino group it becomes α-ketoglutarate again, which then becomes aminated once more to form another molecule of glutamate. Thus, glutamate can be thought of as a conduit through which ammonia passes into other amino acids. Transaminases use a coenzyme called pyridoxal phosphate (derived from vitamin B$_6$) to carry the amino group from glutamate to the α-keto carboxylic acids.

Synthesis of aspartate and alanine

Aspartate is synthesized via a transamination from glutamate to oxaloacetate, and alanine

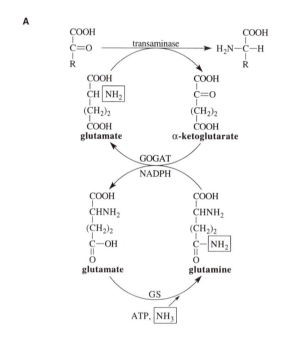

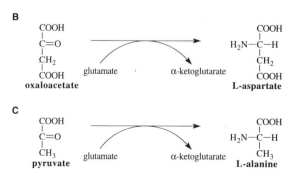

Fig. 9.17 The transamination reaction. (A) Glutamate donates an amino group to an α-keto carboxylic acid to form an α-amino acid. The glutamate becomes α-ketoglutarate, which is aminated either via the glutamate dehydrogenase or the GS and GOGAT reactions. (B) Formation of L-aspartate from oxaloacetate. (C) Formation of L-alanine from pyruvate.

is made via a transamination from glutamate to pyruvate (Fig. 9.17).

Synthesis of serine, glycine, and cysteine
Serine is synthesized from 3-phosphoglycerate (Fig. 9.18). The phosphoglycerate is first oxidized to the α-keto acid, phosphohydroxypyruvate. A transamination, using glutamate as the donor, converts the phosphohydroxypyruvate to phosphoserine. Then the phosphate is hydrolytically removed to form L-serine. The hydrolysis of the phosphate ester bond drives the reaction to completion. Serine is a precursor to both glycine and cysteine. Glycine is formed by transfer of the β-carbon of serine to tetrahydrofolic acid (Fig. 9.13). The route to cysteine is initiated when serine accepts an acetyl group from acetyl-CoA to become O-acetylated. Sulfide then displaces the acetyl group from O-acetylserine forming cysteine (see Fig. 11.2).

9.3.2 Catabolism of amino acids

A scheme showing the overall pattern of carbon flow during amino-acid catabolism is shown in Fig. 9.19. The first reaction is always the removal of the amino group to generate the α-keto acid which eventually enters the citric acid cycle. All 20 amino acids are degraded to seven intermediates that enter the citric acid cycle. These seven intermediates are: pyruvate, acetyl-CoA, acetoacetyl-CoA, α-ketoglutarate, succinyl-CoA, fumarate, and oxaloacetate. Acetoacetyl-CoA itself is a precursor to acetyl-CoA.

Removal of the amino group
There are several ways that amino groups can be removed from amino acids during their catabolism (Fig. 9.20). Usually amino acids are oxidatively deaminated to their corresponding keto acid. The oxidation may be catalyzed by a nonspecific *flavoprotein oxidase* that oxidizes any one of a number of amino acids and feeds the electrons directly into the electron transport chain (Fig. 9.20A). These oxidases can be D-amino oxidases as well as L-amino oxidases. The D-amino acid oxidases are useful because several biological molecules (e.g., peptidoglycan and certain antibiotics) have D-amino acids. There are

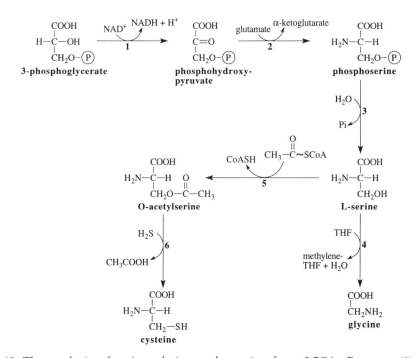

Fig. 9.18 The synthesis of serine, glycine, and cysteine from 3-PGA. Enzymes: (1), phosphoglycerate dehydrogenase; (2) phosphoserine aminotransferase; (3) phosphoserine phosphatase; (4) serine hydroxymethyltransferase; (5) serine transacetylase; (6) O-acetylserine sulfhydrylase.

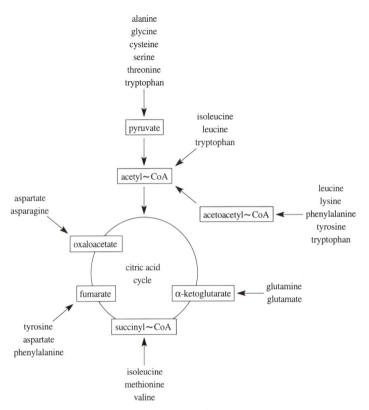

Fig. 9.19 Fates of the carbon skeletons of amino acids. Amino acid carbon can be used to synthesize all of the cell components or be oxidized to CO_2. In the absence of the glyoxylate cycle, carbon entering at acetyl-CoA cannot be used for net glucogenesis except in some strict anaerobic bacteria that can carboxylate acetyl-CoA to pyruvate (Chapter 12).

also *NAD(P)$^+$-linked dehydrogenases* which are more specific than the flavoprotein oxidases (Fig. 9.20B). The dehydrogenases catalyze the reversible reductive amination of the keto acid. A transaminase coupled to a dehydrogenase can lead to the deamination of amino acids (Fig. 9.20C). In the example shown in Fig. 9.20C, a nonspecific transaminase transfers an amino group to pyruvate-forming L-alanine. The amino group is released as ammonia and pyruvate is regenerated by L-alanine dehydrogenase. α-Ketoglutarate and its dehydrogenase serve the same function in amino-acid catabolism. Certain amino acids are deaminated by specific deaminases (Fig. 9.20D). In these cases, a redox reaction is not involved. These amino acids include serine, threonine, aspartate, and histidine. Some deaminations (i.e., the deamination of serine and threonine) proceed by the elimination of water.

9.4 Aliphatic hydrocarbons

Many bacteria can grow on long-chain hydrocarbons, for example, alkanes (C_{10}–C_{18}). A few bacteria (mycobacteria, flavobacteria, *Nocardia*) grow on short-chain hydrocarbons (C_2–C_8). The hydrocarbons exist as droplets of oil outside of the cell, and a major problem in their utilization is to transfer these water-insoluble molecules from the oil layer across the cell wall to the cell membrane, where they can be metabolized. Some bacteria have cell walls containing glycolipids in which the hydrocarbons can dissolve and be transported to the membrane. *Acinetobacter* strains secrete particles resembling the outer membrane in which the hydrocarbons dissolve. The particles with the dissolved hydrocarbon fuse with the outer membrane and the hydrocarbons are then transferred to the cell membrane. Once in the cell membrane, the degradation of the alkanes requires

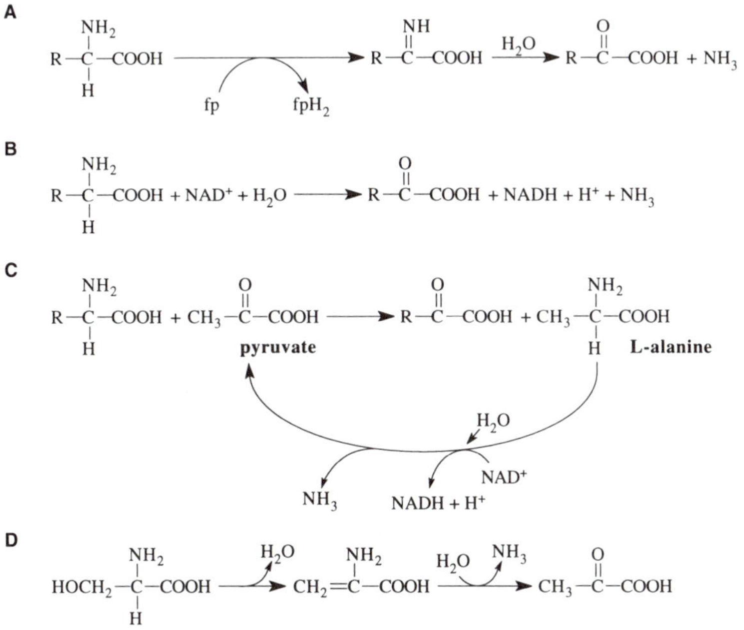

A

$$R-\underset{\underset{H}{|}}{\overset{\overset{NH_2}{|}}{C}}-COOH \xrightarrow{\quad fp \quad fpH_2 \quad} R-\overset{\overset{NH}{||}}{C}-COOH \xrightarrow{\ H_2O\ } R-\overset{\overset{O}{||}}{C}-COOH + NH_3$$

B

$$R-\underset{\underset{H}{|}}{\overset{\overset{NH_2}{|}}{C}}-COOH + NAD^+ + H_2O \longrightarrow R-\overset{\overset{O}{||}}{C}-COOH + NADH + H^+ + NH_3$$

C

$$R-\underset{\underset{H}{|}}{\overset{\overset{NH_2}{|}}{C}}-COOH + CH_3-\overset{\overset{O}{||}}{C}-COOH \longrightarrow R-\overset{\overset{O}{||}}{C}-COOH + CH_3-\underset{\underset{H}{|}}{\overset{\overset{NH_2}{|}}{C}}-COOH$$

pyruvate   L-alanine

H_2O

NH_3 NAD^+

$NADH + H^+$

D

$$HOCH_2-\underset{\underset{H}{|}}{\overset{\overset{NH_2}{|}}{C}}-COOH \xrightarrow{\ H_2O\ } CH_2{=}\overset{\overset{NH_2}{|}}{C}-COOH \xrightarrow[\ NH_3\ ]{\ H_2O\ } CH_3-\overset{\overset{O}{||}}{C}-COOH$$

Fig. 9.20 The removal of amino groups from amino acids during catabolism. (A) Amino acid oxidases are flavoproteins that are specific for L or D amino acids, but generally not for the particular amino acid. Electrons are transferred from the flavoprotein to the electron transport chain. (B) An amino acid dehydrogenase. These are NAD(P) linked and more specific for the amino acid than are the flavoprotein oxidases. (C) Transamination catalyzed by a non-specific transaminase that uses pyruvate as the amino acid acceptor. The amino group is removed from the alanine by alanine dehydrogenase. α-Ketoglutarate can also accept amino groups and be deaminated by α-ketoglutarate dehydrogenase. (D) The deamination of serine by dehydratase.

their hydroxylation using molecular oxygen and an enzyme called *monooxygenase*.

9.4.1 Degradative pathways

Hydroxylation at one end

After the hydrocarbon dissolves in the membrane, it is hydroxylated at one end using molecular oxygen in a reaction catalyzed by a monooxygenase. Membrane-bound enzymes then oxidize the long-chain alcohol to the carboxylic acid (Fig. 9.21A). The carboxylic acid is derivatized with CoASH in an ATP-dependent reaction and oxidized via β-oxidation in the cytoplasm. There must exist an alternative mechanism that does not require oxygen because aliphatic hydrocarbons can be oxidized to CO_2 by certain sulfate reducing bacteria in the absence of oxygen.[8] However, the mechanism is not known. In the sequence shown in Fig. 9.21A, a short electron transport chain carries

electrons from NADH via flavoprotein and an FeS protein to a cytochrome called P-450. The P-450 is in a complex with the hydrocarbon substrate and is called a co-substrate. The reduced P-450 binds O_2 and activates it, presumably converting it to bound O_2^- (superoxide) or O_2^{2-} (peroxide) which attacks the C1 carbon on the hydrocarbon. One of the oxygen atoms replaces a hydrogen on the hydrocarbon and becomes a hydroxyl group, while the other oxygen atom is reduced to H_2O. Some bacteria use an iron protein called rubredoxin instead of P-450 to activate the oxygen. Electrons flow from NADH through a rubredoxin:NADH oxidoreductase to rubredoxin.

Hydroxylation on the penultimate carbon

An alternative degradative pathway is found in *Nocardia* species (Fig. 9.21B). The hydroxylation takes place on the second carbon forming a secondary alcohol which is then

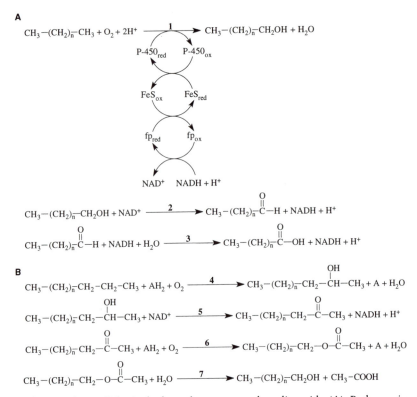

Fig. 9.21 Oxidation of an aliphatic hydrocarbon to a carboxylic acid. (A) Pathway in yeast and *Corynebacterium* species. The hydrocarbon is hydroxylated by oxygen which is activated by cytochrome P-450. The second oxygen atom is reduced to water. Other bacteria, (e.g., *Pseudomonas* species) use rubredoxin instead of P-450. In the latter case the electron transport is from NADH to rubredoxin:NADH oxidoreductase, to rubredoxin. The primary alcohol is oxidized to an aldehyde, which is then oxidized to a carboxylic acid. The carboxylic acid is oxidized via the β-oxidation pathway. (B) pathway in *Nocardia* species. The C2 carbon is hydroxylated forming a secondary alcohol which is then oxidized to a ketone. A second monooxygenase reaction converts the ketone into an acetylester. The acetylester is hydrolyzed to acetate and a long-chain alcohol which is oxidized as in (A). Enzymes: (1,4,6), monooxygenase; (2,5) alcohol dehydrogenase; (3) aldehyde dehydrogenase; (7) acetylesterase.

oxidized to form a ketone. Then, in an unusual reaction, a second monooxygenase reaction creates an acetylester which is subsequently hydrolyzed to acetate and the long-chain alcohol. The alcohol is oxidized to the carboxylic acid which is degraded via the β-oxidation pathway.

9.5 Summary

Many aerobic bacteria can grow on long-chain fatty acids. The major pathway for degradation is the β-oxidative pathway. First, the fatty acid is converted to the acyl-CoA derivative in an ATP-dependent reaction catalyzed by acyl-CoA synthetase. The first reaction is a displacement of PP_i from ATP to form the acyl-AMP which remains bound to the enzyme. Then the AMP is displaced by CoASH to form the acyl-CoA. The reaction is driven to completion by hydrolysis of the PP_i catalyzed by pyrophosphatase. Thus, the activation of fatty acids requires the equivalent of two ATP molecules. A keto group is then introduced β to the carboxyl via an oxidation, hydration, and oxidation to form the β-ketoacyl-CoA derivative. Finally, the β-ketoacyl-CoA derivative is cleaved with CoASH to yield acetyl-CoA. The process is repeated as two-carbon fragments are sequentially removed as acetyl-CoA from the fatty acid. If the fatty acid is an odd-chain fatty acid, then the last fragment is propionyl-CoA, which can be oxidized to pyruvate. Because fatty acids are degraded to acetyl-CoA, growth on fatty acids uses the same metabolic pathways as growth on acetyl-CoA.

Fatty acid synthesis occurs via an entirely different set of reactions. Acetyl-CoA is carboxylated to malonyl-CoA. A transacetylase then converts malonyl-CoA to malonyl-ACP. Acetyl-CoA is also transacylated to acetyl-ACP. Malonyl-ACP then condenses with acetyl-ACP (or acetyl-CoA, depending upon which synthase is used) to form a β-ketoacyl-ACP. The condensation with malonyl-ACP is accompanied by a decarboxylation reaction that drives the condensation. The β-ketoacyl-ACP is reduced via NADPH, dehydrated, and reduced again to form the saturated acyl-ACP. The acyl-ACP is elongated by condensation with malonyl-ACP.

There are two systems found among the bacteria for synthesizing unsaturated fatty acids: an anaerobic and an aerobic pathway. In the anaerobic pathway the C_{10} hydroxyacyl-ACP is desaturated and elongated rather than being desaturated and reduced. In the aerobic pathway, which is found in eukaryotes and certain aerobic bacteria, unsaturated fatty acids are synthesized in a pathway requiring oxygen and NADPH. The double bond is introduced into the completed fatty acyl-ACP or CoA derivative by special desaturases (oxidases) that require O_2 and NAD(P)H as substrates. During the introduction of the double bond, four electrons are transferred to O_2 to form two H_2O. Two electrons are contributed by NAD(P)H and two electrons are derived from the fatty acid when the double bond is formed. Bacteria that use this pathway make oleic acid rather than cis-vaccenic acid, which is made in the anaerobic pathway. They are both C_{18} carboxylic acids, but in oleic acid the double bond is between C9 and C10, whereas in cis-vaccenic acid, it is between C11 and C12.

The phosphoglycerides are synthesized from the fatty acyl-ACP derivatives and glycerol phosphate, the latter being the reduced product of dihydroxyacetone phosphate. The phosphoglycerides differ with respect to the group that is substituted on the phosphate. The major phospholipid in cell membranes from several bacteria is phosphatidylethanolamine. Substitution on the phosphate begins with the displacement of PP_i from CTP by the phosphate on phosphatidic acid, the product being CDP-diglyceride. The PP_i is hydrolyzed by a pyrophosphatase, which makes the pathway irreversible. CMP is then displaced by serine to form phosphatidylserine. A decarboxylation leads to phosphatidylethanolamine. Phosphatidylglycerolphosphate is formed by displacing CMP from CDP-diacylglyceride with glycerol phosphate. A subsequent dephosphorylation produces phosphatidylglycerol. Diphosphatidylglycerol is made from two phosphatidylglycerol molecules. The transfer of the fatty acyl groups to the glycerol and all subsequent steps in phospholipid synthesis take place in the cell membrane. Archaeal phospholipids are synthesized from either dihydroxyacetone phosphate or glycerol phosphate. The precursor to the alcohol portion is geranylgeranylpyrophosphate which forms an ether linkage to the glycerol backbone.

The ribose and deoxyribose moieties of the nucleotides are derived from 5-phosphoribosyl-1-pyrophosphate (PPRP), which is synthesized from ribose-5-phosphate and ATP. In this reaction, the OH on the C-1 of ribose-5-phosphate displaces AMP from ATP to form the pyrophosphate derivative. It is therefore a pyrophosphoryl group transfer rather than the usual phosphoryl or AMP group transfer. The pyrophosphate itself is displaced from PPRP by orotic acid during pyrimidine synthesis or by an amino group from glutamine during purine biosynthesis. A pyrophosphatase hydrolyzes the pyrophosphate to inorganic phosphate, thus driving nucleotide synthesis. Whenever pyrophosphate is hydrolyzed to inorganic phosphate, the equivalent of two ATPs are necessary to restore both of the phosphates to ATP. Pyrimidine and purine biosynthesis differ in that the pyrimidines are made separately and then attached to the ribose phosphate, whereas the purine ring is built piece by piece while attached to the ribose phosphate. The deoxyribonucleotides are formed by reduction of the ribonucleotide diphosphates.

All of the phosphates in the nucleotides are donated by ATP. The α phosphate is derived from ribose-5-phosphate which in turn is synthesized from glucose-6-phosphate. In some bacteria (e.g., *E. coli*), the phosphate in glucose-6-phosphate may come from PEP during transport into the cell via the phosphotransferase system, whereas in other bacteria it is transferred to glucose from ATP via hexokinase or glucokinase. The β and γ phosphates are derived from kinase reactions in which ATP is the donor. The inorganic phosphate is incorporated into ATP via substrate level phosphorylation or electron transport phosphorylation. Since the various nucleotide triphosphates provide the energy for the synthesis of nucleic acids, protein, lipids, and polysaccharides, as well as other reactions, it is clear that ATP fuels all of the biochemical reactions in the cytosol.

Cells must synthesize both glutamate and glutamine in order to incorporate ammonia into cell material. The glutamate donates amino groups to the amino acids and the glutamine donates amino groups to purines, pyrimidines, amino sugars, and some amino acids. Glutamine is synthesized by the ATP-dependent amination of glutamate. This is catalyzed by glutamine synthetase. The glutamine then donates the amino group to α-ketoglutarate to form glutamate. The latter reaction is catalyzed by glutamate synthase, also known as the GOGAT enzyme. Glutamate can also be synthesized by the reductive amination of α-ketoglutarate by glutamate dehydrogenase, a reaction which requires high concentrations of ammonia.

The degradation of amino acids occurs by pathways different from the biosynthetic ones. The first step is the removal of the amino group either by an amino acid oxidase, an amino acid dehydrogenase, or a deaminase. The carbon skeleton eventually enters the citric acid cycle.

Hydrocarbon catabolism begins with a hydroxylation to form the alcohol, which is oxidized to the carboxylic acid. The carboxylic acid is degraded via the β-oxidative pathway. Some bacteria use an alternative pathway to initiate hydrocarbon degradation,

in which the hydrocarbon is oxidized to a ketone which is eventually hydrolyzed to acetate and a long-chain alcohol.

Study Questions

1. Fatty acid synthesis is not simply the reverse of oxidation. What features distinguish the two pathways from each other?

2. What is it about the structure of phospholipids that causes them to form bilayers spontaneously?

3. Glycerol can be incorporated into phospholipids. Write a pathway showing the synthesis of phosphatidic acid from glycerol.

4. The C3 of glycerol becomes the C2 and C8 of purines. Write a sequence of reactions showing how this can occur.

5. Show how three carbons of succinic acid can become C4, C5, and C6 of pyrimidines. What happens to the fourth carbon from succinate?

6. What drives the condensation reaction in fatty acid synthesis?

7. Write a reaction sequence by which bacteria incorporate the nitrogen from ammonia into glutamate and glutamine when ammonia concentrations are low. How might this occur when ammonia concentrations are high? What is the fate of the nitrogen incorporated into glutamate, and that incorporated into glutamine?

8. ATP drives the synthesis of phospholipids. Write the reactions showing the incorporation of glycerol into phosphatidyl serine. Focus on those steps that require a high-energy donor. How is ATP involved? (You must account for the synthesis of CTP.)

9. Bacteria that utilize aliphatic hydrocarbons as a carbon and energy source are usually aerobes. What is the explanation for the requirement for oxygen?

REFERENCES AND NOTES

1. Magnunson, K., S. Jackowski, C. O. Rock, and J. E. Cronan, Jr. 1993. Regulation of fatty acid biosynthesis in *Escherichia coli*. *Microbiol. Rev.* 57:522–542.

2. CoASH = P-AMP-pantothenic acid-β-mercapto-ethylamine-SH. Acyl carrier protein = protein- pantothenic acid-β-mercaptoethylamine-SH. Coenzyme A has a phosphorylated derivative of AMP (AMP-3′-phosphate) attached via a pyrophosphate linkage to the vitamin pantothenic acid (a B_2 vitamin) which is covalently bound to β-mercaptoethylamine via an amide linkage. The β-mercaptoethylamine provides the SH group at the end of the molecule. In the acyl carrier protein, the AMP is missing and the pantothenic acid is bound directly to the protein. Therefore, the functional end of the acyl carrier protein is identical to CoASH but the end that binds to the enzymes is different.

3. There are actually three synthases and three possible routes to acetoacetyl-ACP in *E. coli*. In one pathway, β-ketoacyl-ACP synthase III catalyzes the condensation of acetyl-CoA with malonyl-ACP. A defect in synthase III leads to overproduction of 18-carbon fatty acids, whereas overproduction of synthase III leads to a decrease in the average chain lengths of the fatty acids synthesized. The decrease in the fatty acid chain lengths in strains overproducing synthase III has been rationalized by assuming that synthase III is primarily active in the initial condensation of acetyl-CoA and malonyl-ACP, and that the increased levels of synthase III stimulate the initial condensation reaction and divert malonyl-ACP from the terminal elongation reactions. In a second initiation pathway, acetyl-ACP is a substrate for β-ketoacetyl-ACP synthase I or II. In a third pathway, which is not thought to be physiologically significant under most growth conditions, malonyl-ACP is decarboxylated by synthase I and the resultant acetyl-ACP condenses with malonyl-ACP. The different synthases appear to be involved in determining the types of fatty acids that are made. In particular, synthases I and II which catalyze condensations in both saturated and unsaturated fatty acid synthesis, appear to have specific roles in the synthesis of unsaturated fatty acids. For example, mutants of *E. coli* that lack synthase I do not make any unsaturated fatty acids, suggesting that only synthase I is capable of catalyzing the elongation of *cis*-3-decenoyl-ACP. Mutations in synthase II lead to an inability to synthesize *cis*-vaccenate, in agreement with the finding that synthase II can elongate palmitoleoyl ACP but synthase I cannot.

4. Animals cannot introduce more than one double bond during the synthesis of fatty acids and therefore require polyunsaturated fatty acids in their diet.

5. Naturally occurring fatty acids are mostly *cis* with respect to the configuration of the double bond, i.e.

$$\begin{array}{c} \text{---C=C---} \\ |\quad| \\ \text{H}\quad\text{H} \end{array}$$

This cis configuration produces a bend of about 30° in the chain, whereas the *trans* configuration is a straight chain as is the saturated.

Unsaturated (*cis*)   Saturated

6. In addition to the carbamoyl synthetase described here, vertebrates have a carbamoyl synthetase that combines NH_4^+, CO_2, 2 ATP, and H_2O to form carbamoyl phosphate that is used to convert NH_4^+ to urea in the urea cycle.

7. In mammals the amino group is donated from the amide group of glutamine. In *E. coli* the enzyme can use either NH_4^+ or glutamine. In both cases the requirement of ATP can be explained by the formation of a phosphate ester with the carbonyl oxygen. The amino group then displaces the phosphate.

8. Aeckersberg, F., F. Bak, and F. Widdel. 1991. Anaerobic oxidation of saturated hydrocarbons to CO_2 by a new type of sulfate-reducing bacterium. *Arch. Microbiol.* 156:5–14.

10

Cell Wall Biosynthesis

Studying cell wall synthesis in bacteria is instructive in showing how logistic problems of extracellular biosynthesis are solved. For example, the subunits of the cell wall polymers are synthesized as water-soluble precursors in the cytosol, but polymerization takes place on the outer membrane surface. How do the subunits traverse the lipid barrier in the cell membrane to the sites of polymer assembly? A second problem concerns the final stages of peptidoglycan synthesis. During peptidoglycan synthesis the newly polymerized glycan chains become cross-linked by peptide bonds on the outside cell surface. What is the source of energy for making the peptide cross-links at a site where there is no ATP?

10.1 Peptidoglycan

10.1.1 Structure

Peptidoglycan is a heteropolymer of glycan chains cross-linked by amino acids (Section 1.2.3). The peptidoglycan is a huge molecule, since it surrounds the entire cell and appears to be covalently bonded throughout. A schematic drawing of how this might look in gram-negative bacteria is shown in Fig. 10.1 (see also Fig. 1.8). Peptidoglycan confers strength to the cell wall, and if one were to enzymatically destroy the integrity of the peptidoglycan (with lysozyme) or prevent its

synthesis (with antibiotics), then the cell is likely to swell through the weak areas and lyse as a result of the internal turgor pressures.

The chemical composition of peptidoglycan

Peptidoglycan is made of glycan strands of alternating residues of N-acetylmuramic acid and N-acetylglucosamine linked by β-1,4 glycosidic bonds between the C1 of N-acetylmuramic acid and the C4 of N-acetylglucosamine (Fig. 10.2). N-Acetylmuramic acid is a modified form of N-acetylglucosamine in which a lactyl group has been attached to the C3 carbon. Attached to each N-acetylmuramic acid is a tetrapeptide. The tetrapeptide is L-alanyl-γ-D-glutamyl-L-R$_3$-D-alanine. The amino acid in position 3 varies with the species of bacterium. Gram-negative bacteria generally have *meso*-diaminopimelic acid. (However, some spirochetes contain ornithine instead of *meso*-diaminopimelic acid.) In contrast, there is much more variability in the amino acids in position 3 in gram-positive bacteria (Fig. 10.2).

Cross-linking

The tetrapeptide chains are cross-linked to each other by peptide bonds (Fig. 10.3A,B). There is a great deal of variability in the composition of the cross-links between the different groups of bacteria. In fact, the amino acid composition and location of the

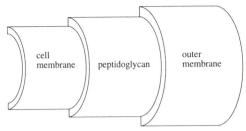

Fig. 10.1 The topological relationship of the peptidoglycan to the cell membrane and rest of the cell wall. In gram-negative bacteria such as *E. coli*, the peptidoglycan is a thin layer sandwiched between the inner and outer membranes. In gram-positive bacteria there is no outer membrane and the peptidoglycan is a thick layer usually covalently bonded to other molecules (e.g., teichoic acids).

cross-links have been used for taxonomic purposes. In most instances the peptide bridge is from the carboxyl in the terminal D-alanine in one tetrapeptide to an amino group in the amino acid in the L-R$_3$ position in another tetrapeptide. In some bacteria the cross-linking is direct, as for example, between

D-alanine and diaminopimelic acid in gram-negative bacteria and many *Bacillus* species. However, in most gram-positive bacteria there is a bridge of one or more amino acids. Some examples are a bridge of five glycine residues in *Staphylococcus aureus*, three L-alanines and one L-threonine in *Mirococcus roseus*, three glycines and two L-serines in *Staphylococcus epidermidis*, and so on. Sometimes the bridge is from the terminal D-alanine to α-carboxyl of D-glutamic acid of another tetrapeptide. Since this is a connection between two carboxyl groups, a bridge of amino acids containing a diamino acid is necessary.

10.1.2 Synthesis

Peptidoglycan is made in several stages: (1) the precursors to the peptidoglycan are UDP derivatives of the amino sugars that are made in the cytosol; (2) the amino sugars are then transferred to a lipid carrier in the membrane which carries the amino sugars across the

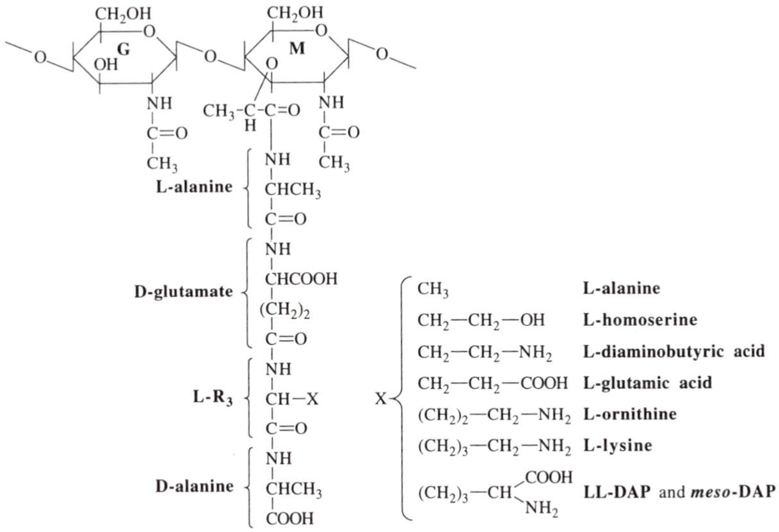

Fig. 10.2 The disaccharide–peptide subunit in peptidoglycan. The glycan backbone consists of alternating residues of N-acetylglucosamine (G) and N-acetylmuramic acid (M) linked β-1,4. The carboxyl group of the lactyl moiety in N-acetylmuramic acid is substituted with a tetrapeptide. The amino acids in the tetrapeptide are usually L-alanine, D-glutamic, L-R$_3$ (residue 3), which is an amino acid that varies with the species, and D-alanine. The peptide linkages are all α except for that between D-glutamic and the amino acid in postion 3, which is γ linked. X can be any one of a large number of side chains, some examples of which are shown. The α carboxyl of glutamic acid can be free, an amide, or substituted (e.g., by glycine).

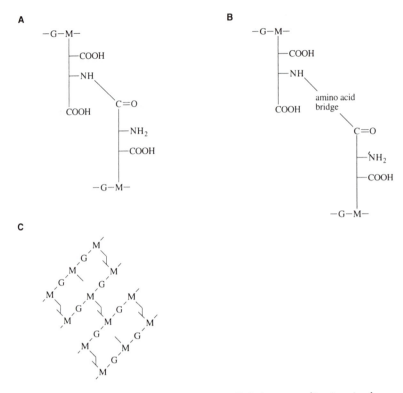

Fig. 10.3 Peptidoglycan cross-linking. (A) Direct cross-link between diaminopimelate and D-alanine as it occurs in gram-negative bacteria. (B) An amino acid bridge between two tetrapeptides as in many gram-positive bacteria. Sometimes the bridge is between the C-terminal D-alanine and the α-carboxyl of D-glutamic acid. When this occurs, there must be a diamino acid in the bridge. (C) Schematic drawing of cross-linked peptidoglycan. Tetrapeptides are indicated by straight lines. The lengths of the glycan chain in some bacteria has been measured to be about 20–100 sugar residues, and apparently vary in length within the same bacterium. The degree of cross-linking depends upon the bacterium. In *E. coli* about 50% of the peptide chains are cross-linked. This is considered a peptidoglycan with relatively few cross-links. In some bacteria, about 90% of the chains are cross-linked. G, N-acetylglucosamine; M, N-acetylmuramic acid.

membrane; (3) the peptidoglycan is polymerized on the outer surface of the membrane; and (4) a transpeptidation reaction cross-links the peptidoglycan.

Synthesis of the UDP derivatives:
UDP-N-acetylglucosamine and
UDP-N-acetylmuramyl-pentapeptide

The two amino sugars that are precursors to the peptidoglycan are N-acetylglucosamine and N-acetylmuramamic acid. Both amino sugars are made from fructose-6-phosphate (Fig. 10.4). In step 1, glutamine donates an amino group to fructose-6-phosphate, converting it to glucosamine-6-phosphate. Then in step 2, a transacylase transfers an

acetyl group from acetyl-CoA to the amino group on glucosamine-6-phosphate to make N-acetylglucosamine-6-phosphate, which is isomerized in step 3 to N-acetylglucosamine-1-phosphate. The latter attacks UTP displacing pyrophosphate in step 4 to form UDP-N-acetylglucosamine. The reaction is driven to completion by a pyrophosphatase. Some of the UDP-N-acetylglucosamine (UDP-GlcNAc) is used as the precursor to the N-acetylglucosamine in peptidoglycan and some is converted to UDP-N-acetylmuramic acid (UDP-MurNAc). The UDP-GlcNAc is converted to UDP-MurNAc by the addition of a lactyl group to the sugar in step 5. In this reaction, the C3 OH of the sugar displaces the

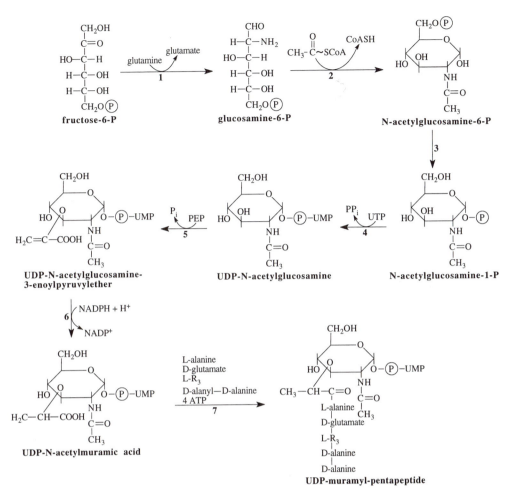

Fig. 10.4 Synthesis of *N*-acetylglucosamine and *N*-acetylmuramyl-pentapeptide. Enzymes: (1) glutamine:fructose-6-phosphate aminotransferase; (2) glucosamine phosphate transacetylase; (3) *N*-acetylglucosamine phosphomutase; (4) UDP-*N*-acetylglucosamine pyrophosphorylase; (5) enoylpyruvate transferase; (6) UDP-*N*-acetylenolpyruvoylglucosamine reductase. Step 7: The UDP-*N*-acetylmuramic acid is converted to the pentapeptide derivative by the sequential additions of L-alanine, D-glutamate, L-R₃, and D-alanyl-D-alanine by separate enzymes.

phosphate from the α carbon of phosphoenol-pyruvate forming the enol pyruvate ether derivative of the UDP-N-acetylmuramic acid. Then, in step 6, the enol derivative is reduced to the lactyl moiety by NADPH. The UDP-MurNAc is converted into UDP-MurNAc-pentapeptide by the sequential addition of five amino acids, L-alanine, D-glutamate, L-R₃ (residue 3), and the dipeptide D-alanyl-D-alanine in step 7. Each reaction is catalyzed by a separate enzyme and requires ATP to activate the carboxyl group of the amino acid. The activated carboxyl is probably an acyl-

phosphate. The products are ADP and inorganic phosphate. (The fifth D-alanine is removed during the cross-linking reaction described later.) The D-alanine-D-alanine is made by separate enzymes. The first of these is a racemase which converts L-alanine to D-alanine. Then an ATP-dependent D-alanyl-D-alanyl synthetase makes D-alanyl-D-alanine from two D-alanines. The racemase and synthetase are inhibited by the antibiotic D-cycloserine. The MurNAc-pentapeptide is transferred to the lipid carrier in the membrane as described next.

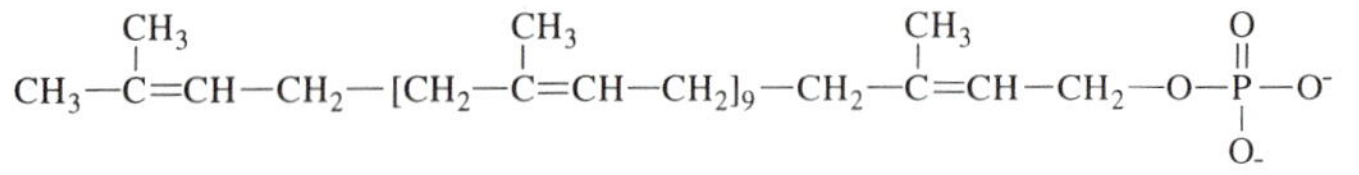

$$CH_3-\overset{\overset{\displaystyle CH_3}{|}}{C}=CH-CH_2-[CH_2-\overset{\overset{\displaystyle CH_3}{|}}{C}=CH-CH_2]_9-CH_2-\overset{\overset{\displaystyle CH_3}{|}}{C}=CH-CH_2-O-\overset{\overset{\displaystyle O}{||}}{\underset{\underset{\displaystyle O_-}{|}}{P}}-O^-$$

Fig. 10.5 The structure of undecaprenyl phosphate. Undecaprenyl phosphate is a C_{55} isoprenoid phosphate that carries precursors to peptidoglycan, lipopolysaccharide, and teichoic acids through the cell membrane.

Reactions in the membrane

The lipid carrier is called *undecaprenyl phosphate* or *bactoprenol*. Undecaprenyl phosphate is a C_{55} isoprenoid phosphate whose structure is shown in Fig. 10.5. Undecaprenyl phosphate not only serves as a carrier for peptidoglycan precursors but also serves as a carrier for the precursors of other cell wall polymers (e.g., lipopolysaccharide and teichoic acids). Interestingly, eukaryotes also use an isoprenoid phosphate (dolichol phosphate) to carry oligosaccharide subunits across the endoplasmic reticulum (ER) membrane to be attached to glycoproteins in the ER lumen. Dolichol phosphate is larger than undecaprenyl phosphate but has the same structure. The nucleotide sugars diffuse to the membrane where undecaprenyl phosphate

(lipid-P) attacks the UDP-MurNAc-pentapeptide displacing UMP (Fig. 10.6, step 1). The product is lipid-PP-MurNAc-pentapeptide. The GlcNAc is then transferred from UDP-GlcNAc to the MurNAc on the lipid carrier (step 2). This occurs when the C4 hydroxyl in the MurNAc attacks the C1 carbon in UDP-GlcNAc displacing the UDP. The product is the disaccharide precursor to the peptidoglycan, lipid-PP-MurNAc (pentapeptide)-GlcNAc. The lipid-disaccharide moves to the other side of the membrane, presumably by diffusion (step 3). On the outside surface of the membrane, the lipid–disaccharide is transferred to the growing end of the acceptor glycan chain (step 4). Step 4 is a transglycosylation where the C4 hydroxyl of the incoming GlcNAc attacks the C1 of

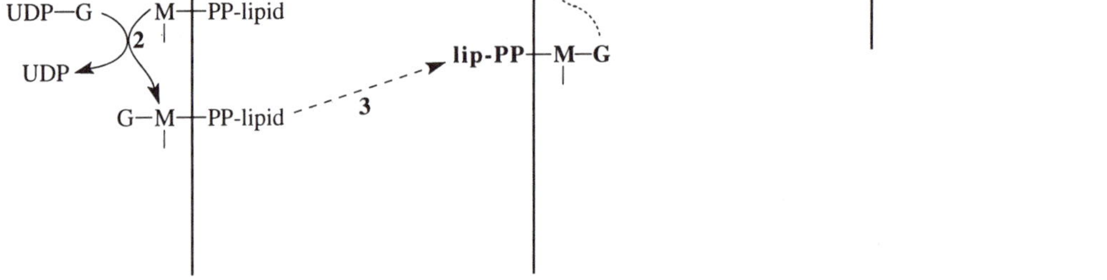

Fig. 10.6 Extension of the glycan chain during peptidoglycan synthesis. Step 1: The MurNac-pentapeptide (M) is transferred to the phospholipid carrier (undecaprenyl phosphate) on the cytoplasmic side of the cell membrane. Step 2: The GlcNAc (G) is transferred to the MurNAc-pentapeptide to form the disaccharide-PP-lipid precursor. Step 3: The disaccharide-PP-lipid precursor moves to the external face of the membrane. Step 4: A transglycosylase transfers the incoming disaccharide to the growing glycan, displacing the lipid-PP (lip-PP) from the growing chain. Thus, the growing chain remains anchored to the membrane by the lipid carrier at the site of the transglycosylase. Step 5: The lipid-PP released from the growing chain is hydrolyzed to lipid-P by a membrane-bound pyrophosphatase. Note that the glycan chain grows at the reducing end (i.e., displacements occur on the C1 of muramic acid).

216

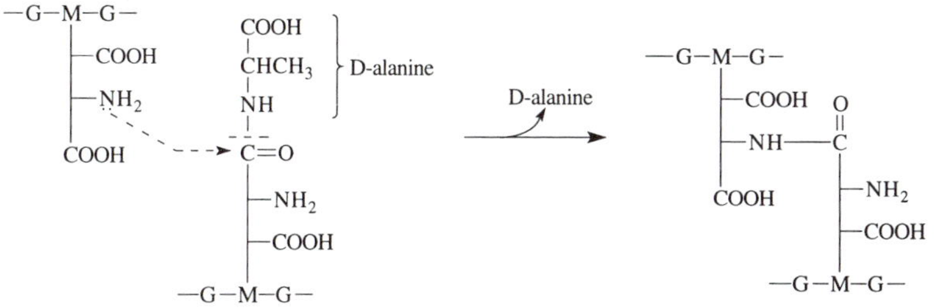

$$-G-M-G-$$

（Figure 10.7 chemical structures showing transpeptidation reaction with D-alanine, COOH, CHCH₃, NH, C=O, NH₂, COOH groups）

Fig. 10.7 The transpeptidation reaction. This is a nucleophilic displacement where the nucleophilic nitrogen attacks the carbonyl displacing the terminal D-alanine The reaction is inhibited by penicillin. Terminal D-alanine residues in chains not participating in cross-linking are removed by a D-alanine carboxypeptidase.

the MurNAc in the glycan, displacing the lipid-PP from the growing glycan chain. This reaction is catalyzed by a membrane-bound enzyme called transglycosylase. Notice that the growing glycan chain remains anchored via the lipid carrier to the membrane at the site of the transglycosylase. The lipid-PP released from the growing glycan chain is hydrolyzed by a membrane-bound pyrophosphatase to lipid-P and P_i (step 5). This reaction is very important because it helps to drive the transglycosylation reaction to completion, since the hydrolysis of the phosphodiester results in the release of substantial energy. This hydrolysis also regenerates the lipid-P which is necessary for continued growth of the peptidoglycan as well as other cell wall polymers (e.g., lipolysaccharide and teichoic acids). The hydrolysis is inhibited by the antibiotic bacitracin.

Making the peptide cross-link
The problem of providing the energy to make the peptide cross-link outside of the cell membrane is solved by using a reaction called transpeptidation. In the transpeptidation reaction, an :NH₂ group from the diamino acid in the 3 position (e.g. DAP) attacks the

carbonyl carbon in the peptide bond holding the two D-alanine residues together in the pentapeptide, and displaces the terminal D-alanine. The result is a new peptide bond (Fig. 10.7). The transpeptidation reaction is inhibited by penicillin.

10.2 Lipopolysaccharide

10.2.1 Structure

Lipopolysaccharide (LPS) is a complex polymer of polysaccharide and lipid in the outer membrane of gram-negative bacteria. The outer membrane is described in more detail in Section 1.2.3. The best characterized lipopolysaccharides are those of *E. coli* and *Salmonella typhimurium* (Fig. 10.8). The lipopolysaccharides of other bacteria are very similar.

As seen in Fig. 10.8, lipopolysaccharide is composed of three parts: (1) a hydrophobic region called lipid A, which is composed of a backbone of two glucosamine residues linked β 1,6 and esterified via the hydroxyl groups to fatty acids; (2) a core polysaccharide region, whose composition is similar in all

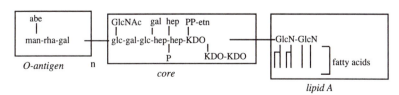

Fig. 10.8 The LPS of *Salmonella typhimurium*. abe, abequose; man, mannose; rha, rhamnose; gal, galactose; GlcNAc, N-acetylglucosamine; glc, glucose; hep, heptose (L-glycero-D-mannoheptose); KDO, 3-deoxy-D-mannooctulosonic acid; etn, ethanolamine; P, phosphate; GlcN, glucosamine.

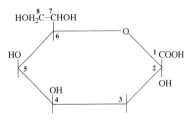

Fig. 10.9 The structure of KDO. The linkage between KDO residues has been reported to be a glycosidic linkage between a C2 hydroxyl in one KDO with either the C4 or C5 hydroxyl in a second KDO. The linkage to lipid A has been suggested to be from the C2 hydroxyl of KDO to the C6 hydroxyl of lipid A.

the Enterobacteriaceae and which is connected to lipid A via 3-deoxy-D-manno-octulosonate (KDO);[1] and (3) a distal polysaccharide region connected to the core. The distal polysaccharide region is sometimes called the O-antigen or the repeat oligosaccharide. It is made up of repeating units

of four to six sugars that vary considerably in composition between different strains of bacteria. The repeating oligosaccharide may have as many as 30 units. The structure of KDO is shown in Fig. 10.9. The arrangement of the LPS in the outer envelope is shown in Fig. 10.10 which shows that the LPS is anchored in the outer envelope by the hydrophobic lipid A region while the repeating oligosaccharide protrudes into the medium.

The fatty acids in lipid A

Four identical fatty acids are attached directly to the glucosamine in lipid A from *E. coli* and *Salmonella typhimurium*. These are a C_{14} hydroxy fatty acid, β-hydroxymyristic acid (3-hydroxytetradecanoic acid), linked via ester bonds to the 3' hydroxyls of the glucosamine, and via amide bonds to the nitrogen of the glucosamine. These fatty acids appear to be found uniquely in lipid A and

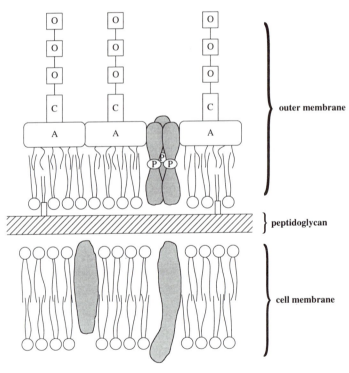

Fig. 10.10 The arrangement of the LPS in the outer membrane of some gram-negative bacteria. A, lipid A; C, core; O, oligosaccharide; P, porin. The outer envelope in the enteric gram-negative bacteria is asymmetric with the lipopolysaccharide confined to the outer leaflet of the lipid bilayer. However, some bacteria (e.g., penicillin-sensitive strains of *Neisseria* and *Treponema*) also have phospholipid in the outer leaflet. Phospholipids in the outer membrane make the bacterium more sensitive to hydrophobic antibiotics. (See Section 1.2.3 for a more complete discussion of the outer membrane.)

their presence in a bacterium implies the presence of a lipopolysaccharide-containing lipid A. Esterified to the hydroxyls of two of the β-hydroxymyristic acids are long-chain saturated fatty acids. In *E. coli*, these are lauric acid (C_{12}) and myristic acid (C_{14}). The fatty acids attached to the glucosamine anchor the lipopolysaccharide into the outer membrane.

10.2.2 Synthesis of the lipopolysaccharide

Lipid A

The lipid A portion of the lipopolysaccharide is synthesized from UDP-GlcNAc, which as described in Fig. 10.4, is made from fructose-6-phosphate.[2] A model for lipid A synthesis in *E. coli* proposed by Raetz is shown in Fig. 10.11.[3] It is believed that lipid A is made in the cytoplasmic membrane by peripheral membrane proteins. β-OH myristic acid is transferred from an ACP derivative to C3 of UDP-GlcNAc to form the monoacyl derivative. (Recall that fatty acids are synthesized as ACP-derivatives, Chapter 9.) Then, the acetate from the nitrogen on C2 is removed and a second molecule of β-OH myristic is transferred to the nitrogen to form the 2,3-diacyl derivative. Some of the latter loses UMP and is converted to 2,3-diacylglucosamine-1-P which condenses with UDP-2,3-diacylglucosamine to form the disaccharide linked in β 1,6 linkage. A phosphate is added to form the 1,4 diphosphate derivative and this is then modified by the addition of KDO and the esterification of fatty acids (lauryl and myristoyl) to the OH of the β-hydroxymyristic moieties.

Core

The core in the *Enterobacteriaceae* contains an inner region consisting of KDO, heptose, ethanolamine, and phosphate, and an outer region that consists of hexoses. The biosynthesis of the inner region is not fully understood. The outer core region grows as hexose units are donated one at a time from nucleoside diphosphate derivatives to the non-reducing end of the growing glycan chain attached to the KDO. The addition of each

sugar is catalyzed by a specific glycosyl transferase, which is membrane bound. The core–lipid A portion of the LPS is translocated across the cell membrane to the periplasmic surface where the LPS is completed by attachment of the O-antigen (Fig. 10.13).[4]

O-Antigen

The O-antigen region is synthesized by a mechanism that is different from that of core synthesis (Fig. 10.12). Whereas the core is synthesized via the addition of sugars one at a time to the growing end of the glycan chain, the O-antigen is synthesized as a separate polymer on a lipid carrier and then transferred as a unit to the core. The lipid carrier is undecaprenyl phosphate (i.e., the same molecule that carries the peptidoglycan precursors across the membrane). First, the repeat unit of the O-antigen is synthesized on the lipid carrier. This is done by a series of consecutive reactions in which a sugar moiety is transferred from a nucleoside diphosphate carrier to the nonreducing end of the growing repeat unit (Fig. 10.12, steps 1–4). Each of these reactions is catalyzed by a different enzyme. Then the repeat unit is transferred to the growing oligosaccharide chain (step 5). The lipid-PP of the acceptor oligosaccharide chain is displaced and enters the lipid pyrophosphate pool where it is hydrolyzed to lipid-P by a bacitracin-sensitive enzyme. Finally, the completed oligosaccharide chain is transferred to the lipid A–core (step 6), displacing the lipid pyrophosphate.

How the lipopolysaccharide might be assembled

Figure 10.13 depicts a model for how the LPS might be assembled. The O-antigen tetrasaccharide subunit is probably synthesized on the lipid carrier on the cytoplasmic side of the membrane and then moves to the periplasmic side, where it is added to the growing O-antigen anchored to the membrane by its lipid carrier. The lipid carrier on the growing oligosaccharide is displaced. The core–lipid A region may also be assembled on the cytoplasmic surface and translocated to the periplasmic surface. The final completion of the lipopolysaccharide would take place

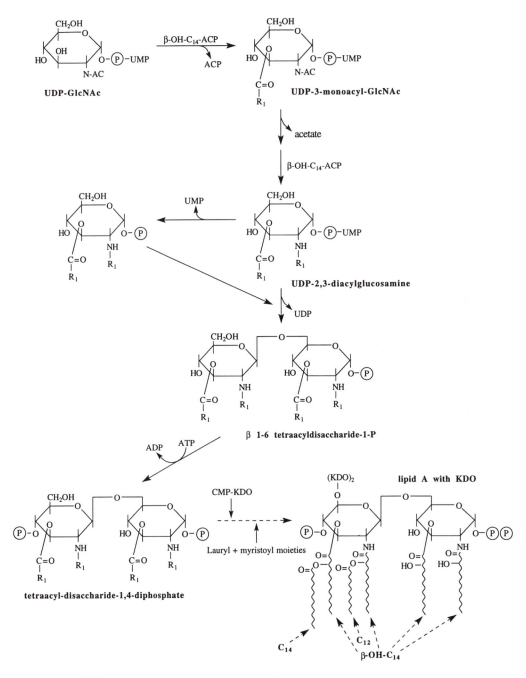

Fig. 10.11 Model for synthesis of lipid A in *E. coli.* Biosynthesis is thought to occur in the cytoplasmic membrane catalyzed by peripheral membrane proteins that have access to the soluble precursors. The final modifications (e.g., the attachments of the KDO residues and the lauryl and myristic acids) are not well characterized. (Adapted from Raetz, C. R. H. 1987. Structure and biosynthesis of lipid A in *Escherichia coli,* Vol. 1, pp. 498–503. In: Escherichia *coli* and Salmonella typhimurium: *Cellular and Molecular Biology.* F. C. Neidhardt, J. L. Ingraham, K. B. Low, B. Magasanik, M. Schaechter and H. E. Umbarger (eds.). American Society for Microbiology, Washington, DC.)

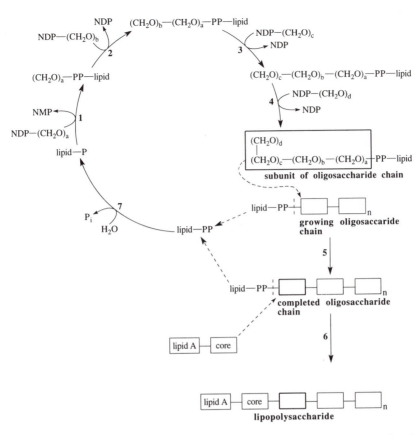

Fig. 10.12 Synthesis of O-antigen and attachment to core in *S. typhimurium*. The O-antigen is polymerized on the same phospholipid carrier that is used for peptidoglycan synthesis. Each sugar is added from a nucleotide diphosphate derivative using a specific sugar transferase (reactions 1–4). The nucleotide that is used depends upon the sugar and the specific glycosyl transferase. For example, in *S. typhimurium* the nucleotide precursors are: UDP-galactose, TDP-rhamnose, GDP-mannose, and CDP-abequose, in that order. The first sugar that is added (e.g., galactose in *S. typhimurium*) is transferred as a phosphorylated derivative to the lipid carrier. Hence the NMP (i.e., UMP in the case of *E. coli* and *S. typhimurium*) is released. When the O-antigen tetrasaccharide subunit is finished, it is transferred to the growing O-antigen chain by an enzyme called O-antigen polymerase (reaction 5). The completed O-antigen is transferred from its lipid carrier to core–lipid A by an enzyme called O-antigen:lipopolysaccharide ligase (reaction 6). The lipid-PP that is displaced is hydrolyzed by a bacitracin-sensitive phosphatase (reaction 7). All of the reactions take place in the cell membrane. There exist immunoelectron microscopy data to suggest that the ligase reaction and perhaps the polymerase reaction take place on the periplasmic surface of the cell membrane.[4] Presumably, the lipid carrier ferries the subunits across the cell membrane. It is not known how the LPS crosses the periplasm to enter the outer envelope.

on the periplasmic surface, where the O-antigen is transferred to the core–lipid A, displacing the lipid-PP. The lipid-P is regenerated via a phosphatase and enters a common pool of lipid-P also used in the biosynthesis of peptidoglycan. It is not known how the lipopolysaccharide moves through the periplasm into the outer envelope.

10.3 Summary

There are two major problems which must be solved in order to synthesize the cell wall. One is how to move the precursors through the cell membrane and the second is how to make peptide bonds outside of the cell membrane far from the cellular ATP pools.

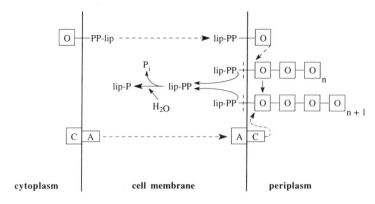

Fig. 10.13 A model for LPS assembly. The O-antigen subunit is synthesized on the cytoplasmic surface and then moves on the lipid carrier to the periplasmic surface of the membrane. A polymerase then transfers the O-antigen subunit to the growing O-antigen chain. The core–lipid A is also synthesized on the cytoplasmic surface and is translocated to the periplasmic surface. The O-antigen is transferred to the core to complete the LPS. The entire LPS is translocated to the outer membrane. O, O-antigen; C, core; lip, lipid carrier; A, lipid A.

Bacteria employ a lipid carrier, undecaprenyl phosphate, to carry the cell wall precursors through the membrane. Interestingly, eukaryotes employ a similar compound, dolichol phosphate, to move oligosaccharide precursors through the ER membrane to synthesize glycoproteins. The peptidoglycan peptide cross-link forms as a result of a transpeptidation reaction. During the transpeptidation reaction, an amino group from the diamino acid (e.g., DAP) displaces a terminal D-alanine. This is an exchange of a peptide bond for one that was made in the cytosol at the expense of ATP. ATP also provides the energy to make the glycosidic linkages in the cell wall polymers because ATP drives the synthesis of the sugar-PP-lipid intermediates. The formation of the glycosidic linkage is a straightforward displacement of the lipid pyrophosphate from the C1 carbon of the N-acetylmuramic acid by the OH on the incoming C4 carbon of N-acetylglucosamine. The reaction is driven to completion by the subsequent hydrolysis of the lipid pyrophosphate to the lipid phosphate and inorganic phosphate.

Lipopolysaccharide synthesis can be thought of as occurring in four stages: (1) the synthesis of the lipid A portion; (2) synthesis of the core region by adding one sugar at a time to lipid A from nucleoside diphosphate precursors; (3) synthesis of the complete repeat

oligosaccharide; and (4) attachment of the repeat oligosaccharide to the core. All of these events are associated with the membrane. It appears that the lipid A and core portions are synthesized on the cytoplasmic side of the membrane and then translocated in an unknown manner to the periplasmic surface. The oligosaccharide (O-antigen) subunits are assembled on undecaprenyl pyrophosphate and transferred to a growing oligosaccharide chain. When the oligosaccharide is complete, it is attached to the core. How the lipopolysaccharide enters the outer envelope is not known.

Study Questions

1. O-Antigen synthesis requires a lipid carrier, but core synthesis does not. Offer a plausible explanation for the difference.

2. During the synthesis of the pentapeptide in the peptidoglycan precursor, an ATP is expended to make the peptide bond as each amino acid is added to the growing pentapeptide. The products are ADP and P_i. Write a plausible mechanism by which ATP is used to provide the energy to make a peptide bond between two amino acids. (Note: m-RNA and ribosomes are not involved. The addition of each amino acid

is catalyzed by a separate enzyme specific for that amino acid.)

3. Peptidoglycan and lipid A share a common pathway early in their syntheses. Outline the early stages of both pathways up to the branch point.

4. Carriers of subunit moieties play important roles in biosynthesis. Usually, the carriers are involved in more than one pathway. What carrier is common to both peptidoglycan and lipopolysaccharide synthesis? What two carriers are common to phospholipid and lipid A biosynthesis?

5. Important enzymes in cell wall peptidoglycan synthesis are membrane-bound transglycosidases that transfer carbohydrate subunits to the growing polymer. What ensures that the growing end of the polymer remains at the site of the transglycosidase?

6. Show how ATP drives the synthesis of UDP-GlcNAc from glucose. Focus on reactions in which phosphoryl and nucleotide groups are transferred. You must show how ATP drives the synthesis of UTP. There are two phosphate groups in MurNAc(pentapeptide)-PP-lipid. Show how one of them is derived from ATP. What eventually happens to this phosphate?

REFERENCES AND NOTES

1. This used to be called 2-keto-3-deoxyoctonate.

2. Rick, P. D. 1987. Lipopolysaccharide biosynthesis. In: Escherichia coli *and* Salmonella typhimurium: *Cellular and Molecular Biology.* Vol. 1, p. 648–662. F. C. Neidhardt, J. L. Ingraham, K. B. Low, M. Magasanik, M. Schaechter and H. E. Umbarger (eds.). American Society for Microbiology, Washington, DC.

3. Raetz, C. R. H. 1987. Structure and biosynthesis of lipid A . in *Escherichia coli.* In: Escherichia coli *and* Salmonella typhimurium: *Cellular and Molecular Biology.* F. C. Neidhardt, J. L. Ingraham, K. B. Low, M. Magasanik, M. Schaechter and H. E. Umbarger (eds.). American Society for Microbiology, Washington, DC.

4. Mulford, C. A. and M. J. Osborn. 1983. An intermediate step in translocation of lipopolysaccharide to the outer membrane of *Salmonella typhimurium. Proc. Natl. Acad. Sci. USA* 80:1159–1163.

11

Inorganic Metabolism

Inorganic molecules such as derivatives of sulfur, nitrogen, and iron are used by prokaryotes in a variety of metabolic ways related to energy metabolism and biosynthesis.

1. There are *assimilatory pathways* in which inorganic nitrogen and sulfur are incorporated into sulfur and/or nitrogen-containing organic material (e.g., amino acids and nucleotides). Most prokaryotes can do this. In addition, many prokaryotes can also utilize nitrogen gas as a source of nitrogen, a process called *nitrogen fixation*.

2. There are *dissimilatory pathways* in which inorganic compounds are used instead of oxygen as electron acceptors, a process called *anaerobic respiration*. The reduced products are excreted into the environment. During anaerobic respiration, a Δp is created in the same way as during aerobic respiration (Section 4.6). For example, many facultative anaerobes can use nitrate as an electron acceptor, reducing it to ammonia or nitrogen gas. Several obligate anaerobes use sulfate as an electron acceptor, reducing it to hydrogen sulfide. There also exist bacteria that can use Fe^{3+} or Mn^{4+} as an electron acceptor during anaerobic growth.[1] The latter organisms are responsible for most of the reduction of iron and manganese that takes place in sedimentary organic matter (e.g., in lake sediments). However, only a few of the iron and manganese reducers have been isolated and little is known about their physiology.

3. There are *oxidative pathways* in which inorganic compounds such as H_2, NH_3, NO_2^-, S^o, H_2S, and Fe^{2+}, rather than organic compounds, are oxidized as a source of electrons and energy. Organisms that derive their energy and electrons for biosynthesis in this way are called *chemolithotrophs*.[2]

11.1 Assimilation of nitrate and sulfate

Many bacteria can grow in media in which the only sources of nitrogen and sulfur are inorganic nitrate salts and sulfate salts. The nitrate is reduced to ammonia and the ammonia is incorporated into the amino acids glutamine and glutamate using the GS/GO-GAT system (Section 9.3.1). Glutamate and glutamine are the sources of amino groups for the other nitrogen-containing organic compounds. The sulfate is reduced to H_2S which is immediately incorporated into the amino acid cysteine using the O-acetylserine pathway described in Section 9.3.1. Cysteine, in turn, is the source of sulfur for other organic molecules (e.g., methionine, coenzyme A (CoASH) and acyl carrier protein (ACPSH)

Nitrate assimilation
Since the oxidation state for the nitrogen in nitrate (NO_3^-) is $+5$ and for the nitrogen in ammonia (NH_3) is -3, eight electrons must

be transferred to nitrate in order to reduce it to ammonia. In bacteria, this takes place in the soluble portion of the cell. The enzymes involved in this reduction are called *soluble nitrate reductase* and *soluble nitrite reductase*. The electron transport pathway is shown in Fig. 11.1. The ammonia that is formed is incorporated into glutamine via *glutamine synthase* (GS). The glutamine then serves as the amino donor for purine, pyrimidine, and amino sugar biosynthesis (Sections 9.22, 9.23, and 10.1.2), as well as for the synthesis of glutamate (the GOGAT enzyme). Glutamate is the amino donor for amino acid biosynthesis via the transamination reactions described in Section 9.3.1.

Sulfate assimilation

Many bacteria can use sulfate (SO_4^{2-}) as their principal source of sulfur. The sulfate is first reduced to sulfide (H_2S and HS^-) and then incorporated into cysteine.[3] Since the oxidation level of sulfur in SO_4^{2-} is $+6$ and in S^{2-} is -2, a total of eight electrons are required to reduce sulfate to sulfide. The first step in the reduction is the formation of adenosine-5'-phosphosulfate (APS) catalyzed by the enzyme ATP sulfurylase (Fig. 11.2, reaction 1). Here sulfate acts as a nucleophile and displaces pyrophosphate (PP_i). (See Section 7.1.1 for a discussion of nucleophilic displacements.) The pyrophosphate is subsequently hydrolyzed to inorganic phosphate, thus driving the synthesis of APS to completion.

There is a sound thermodynamic reason for making the AMP derivative of sulfate prior to its reduction. Attaching the sulfate to AMP raises the reduction potential making APS a better electron acceptor than free sulfate. (The E_o' for the reduction potential of sulfate to sulfite is very low, $E_o' = -520\,\text{mV}$, such that its reduction even by H_2, $E_o' = -420\,\text{mV}$, is endergonic.) The second reaction is the phosphorylation of APS to form adenosine-3'-phosphate-5'-phosphosulfate (PAPS), catalyzed by APS kinase (reaction 2). The reaction is an attack by the 3' hydroxyl of the ribose of APS on the terminal phosphate of ATP displacing ADP. The PAPS is then reduced to sulfite with the release of AMP-3'-phosphate (reaction 3). The reductant is a sulfhydryl protein called thioredoxin which in turn accepts electrons from NADPH.[4] The sulfite is then reduced by NADPH to hydrogen sulfide (H_2S) (reaction 4). Hydrogen sulfide is very toxic and does not accumulate. The sulfide enzymatically displaces acetate from O-acetylserine to form cysteine (Section 9.3.1). The AMP-3'-phosphate is hydrolyzed to AMP and P_i, thus helping to drive the overall reaction to completion. Note that three ATPs are used to reduce sulfate to sulfide, two to make the PAPS derivative, and a third to phosphorylate the AMP released from AMP-3'-phosphate to ADP (reaction 3). (The ADP is then converted to ATP via respiratory phosphorylation or substrate level phosphorylation.)

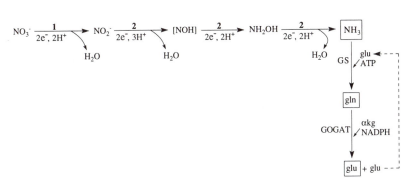

Fig. 11.1 Assimilatory nitrate reduction. This pathway is present in all bacteria that reduce nitrate to ammonia, which is then incorporated into cell material. The enzymes are found in the cytosol and are not coupled to ATP formation. Nitrate is reduced via two-electron steps to nitrite, nitroxyl, hydoxylamine, and ammonia. The ammonia is incorporated into organic carbon via glutamine synthetase (GS) and the GOGAT enzyme, or via glutamate dehydrogenase (Section 9.3.1). Enzymes: (1) nitrate reductase; (2) nitrite reductase. glu, glutamate; α-kg, α-ketoglutarate.

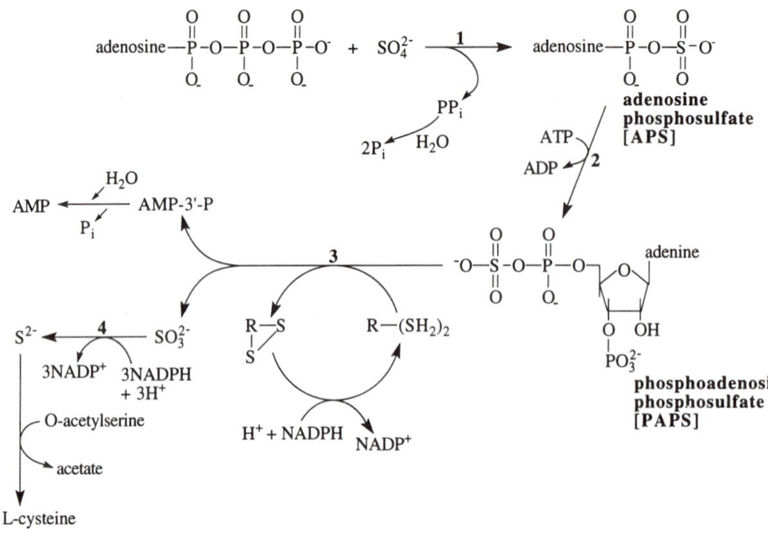

Fig. 11.2 Assimilatory sulfate reduction. Enzymes: (1) ATP sulfurylase; (2) APS phosphokinase; (3) PAPS reductase; (4) sulfite reductase; (5) O-acetylserine sulfhydrylase. R-$(SH_2)_2$ is reduced thioredoxin.

11.2 Dissimilation of nitrate and sulfate

In the *dissimilatory* pathways the nitrate and sulfate are used as electron acceptors during anaerobic respiration. The reduced products are excreted rather than being incorporated into cell material. Whereas many facultative anaerobes are capable of dissimilatory nitrate reduction and employ this pathway when oxygen is not available, the use of sulfate as an electron acceptor is restricted to obligately anaerobic bacteria called *sulfate reducers*.

11.2.1 Dissimilatory nitrate reduction

Dissimilatory nitrate reduction (nitrate respiration) is usually facultative and occurs as a substitute for aerobic respiration when the oxygen levels become very low. It takes place in membranes and a Δp is usually made. The products of nitrate respiration can be nitrite, ammonia, or nitrogen gas.

Denitrification

When the nitrate (or nitrite) is reduced to nitrogen gas (or nitric oxide gas, NO, or

nitrous oxide gas, N_2O), the process is called *denitrification*, which can be an ecologically important drain of nitrogen from the soil. Denitrification occurs in the soil when conditions become anaerobic (e.g., in water-logged soil). The electron transport pathway for denitrification by *Paracoccus denitrificans* is summarized in Section 4.7.2.

11.2.2 Dissimilatory sulfate reduction

A general description of the sulfate reducers

The sulfate reducers are heterotrophic strict anaerobes that grow in anaerobic muds, mostly in anaerobic parts of fresh water, and in sea water. They carry out anaerobic respiration during which sulfate is reduced to H_2S. Some can be grown autotrophically with H_2 as the source of electrons and SO_4^{2-} as the electron acceptor. Formerly the sulfate reducers were believed to use as carbon sources only a limited variety of compounds, (e.g., formate, lactate, pyruvate, malate, fumarate, ethanol, and a few other simple compounds). Recently, however, it has been realized that, depending upon the species, many other carbon sources can be used, including straight chain alkanes and a variety of aromatic compounds. Sulfate reducers comprise a very diverse group of organisms that include both gram-positive and gram-negative bacteria, as well as archaea. An example of the latter is *Archaeoglobus*, a hyperthermophile isolated from sediments near hydrothermal vents. Gram-positive spore-forming sulfate reducers belong to the genus *Desulfotomaculum*, which is very diverse. The most prominent of the gram-negative sulfate reducers belong to the genus *Desulfovibrio*, which is also phylogenetically diverse.

Traditionally, the sulfate reducers are divided into two physiological groups, I and II. Those in group I cannot oxidize acetyl-CoA to CO_2 and therefore excrete acetate when growing on certain carbon sources (e.g., lactate or ethanol). The group I genera include *Desulfovibrio* and most *Desulfotomaculum* species. *Group II organisms · can oxidize acetyl-CoA to CO_2*. Group II sulfate reducers are found in several genera, including *Desulfotomaculum* and *Desulfobacter*.

There exist two pathways for oxidizing acetyl-CoA anaerobically to CO_2: a modified citric acid cycle and the acetyl-CoA pathway. *Desulfobacter* has a modified citric acid cycle that resembles that found in aerobes except that: (1) instead of a citrate synthase there is an ATP-citrate lyase; and (2) the NAD^+-linked α-ketoglutarate dehydrogenase is replaced by a ferredoxin-dependent enzyme. The pathway is called the reductive tricarboxylic acid pathway, although it can be used in the oxidative direction (Section 12.1.9). Other sulfate reducers (e.g., *Desulfobacterium autotrophicum*, *Desulfotomaculum acetooxidans*, and the archaeon, *Archaeoglobus fulgidus*), oxidize acetyl-CoA to CO_2 using the acetyl-CoA pathway described in Section 12.1.3. Autotrophic CO_2 fixation by facultatively autotrophic sulfate reducers using these pathways in the reductive direction is described in Sections 12.1.3 and 12.1.4. It should also be pointed out that many sulfate reducers are known to be able to ferment pyruvate to acetate, or to acetate and propionate in the absence of sulfate. This is described in Section 13.8.

The path of electrons to sulfate in *Desulfovibrio*

Desulfovibrio carries out an anaerobic respiration during which electrons flow in the cell membrane to sulfate as the terminal electron acceptor, reducing it to H_2S. Electron flow is coupled to the generation of a Δp which is used for ATP synthesis via respiratory phosphorylation. As stated above, these electrons may come from the oxidation of organic compounds (e.g., lactate). Since dissimilatory sulfate reduction takes place in membranes, involves cytochromes, and generates a Δp, it is very different from the assimilatory pathway which is a soluble pathway and does not generate a Δp or ATP. A pathway of electron transport has been proposed for the genus *Desulfovibrio* (Fig. 11.3). In the cytoplasm, lactate is oxidized to pyruvate yielding two electrons (Fig. 11.3, reaction 1). The oxidation of lactate to pyruvate is catalyzed by a membrane-bound lactate dehydrogenase, which is probably a flavoprotein. The pyruvate is then oxidized to acetyl-CoA and CO_2 by pyruvate-ferredoxin oxidoreductase, an enzyme found in other anaerobes (reaction 2). (See Section 7.3.2 for a description of the pyruvate–ferredoxin oxidoreductase reaction.) The acetyl-CoA is used to generate ATP via a substrate-level phosphorylation, using two enzymes common in bacteria, *phosphotransacetylase* and *acetate kinase* (reactions 3 and 4). (See Section 7.3.2 for a description of the phosphotransacetylase and acetate kinase reactions.) Since each lactate that is oxidized to acetyl-CoA yields four electrons, two lactates must be oxidized to provide the eight electrons to reduce one sulfate to sulfide. The model proposes that the electrons are transferred from lactate dehydrogenase and pyruvate–ferredoxin oxidoreductase to a cytoplasmic hydrogenase and then to H^+ producing H_2 (reaction 5) and the H_2 diffuses out of the cell into the periplasm. In the periplasm, the H_2 is oxidized by a periplasmic hydrogenase and the electrons are transferred to cytochrome (cyt c_3) (reaction 6). From cyt c_3 the electrons travel through a series of membrane-bound electron carriers to APS reductase and sulfite reductase in the cytosol (reactions 7, 9 and 10). A pyrophosphatase

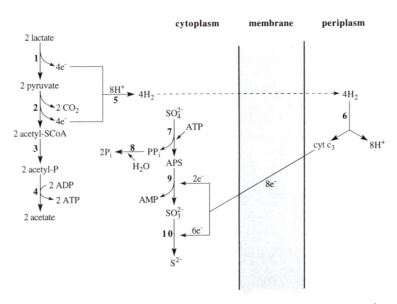

Fig. 11.3 Pathway for dissimilatory sulfate reduction in *Desulfovibrio*. Enzymes: (1) lactate dehydrogenase; (2) pyruvate–ferredoxin oxidoreductase; (3) phosphotransacetylase; (4) acetate kinase; (5) cytoplasmic hydrogenase; (6) periplasmic hydrogenase; (7) ATP sulfurylase; (8) pyrophosphatase; (9) APS reductase; (10) sulfite reductase.

pulls the sulfurylation of ATP to completion (reaction 8). Note that according to the scheme proposed in Fig. 11.3, the inward flow of electrons across the membrane leaves the protons from the hydrogen on the outside, thus generating a Δp. An examination of the scheme reveals that the Δp is necessary for growth. The two ATPs made via substrate level phosphorylation from the two moles of acetyl-CoA are used up in reducing the SO_4^{2-}. This follows because, after reduction to sulfite, AMP is produced and it requires the energy equivalent of two ATPs to make one ATP from one AMP. Thus, without using the Δp, there would be no ATP left over for growth. Some strains of *Desulfovibrio* can grow on CO_2 and acetate as the sole sources of carbon, and the Δp produced by the redox reaction between H_2 and SO_4^{2-} is the sole source of energy. There are also facultatively autotrophic strains of sulfate reducers that grow on CO_2, H_2, and SO_4^{2-} (e.g., *Desulfobacterium autotrophicum*). All of these strains derive their ATP from the Δp created during sulfate reduction. It should be pointed out that some reservations have been expressed as to whether free H_2 is actually an electron carrier during lactate oxidation by sulfate in *Desulfovibrio*.[5]

11.3 Nitrogen fixation

From an ecological point of view one of the most important metabolic processes carried out by prokaryotes is nitrogen fixation (i.e., the reduction of N_2 to NH_3). As far as is known, eukaryotes have not evolved this capability. Since fixed nitrogen is usually limiting for plant growth, the ability of prokaryotes to fix nitrogen is necessary to maintain the food chain. The ammonia that is produced via nitrogen fixation is incorporated into cell material using glutamine synthetase and glutamate synthase (Section 9.3.1). It used to be thought that nitrogen fixation was restricted to a few bacteria such as *Azotobacter*, *Rhizobium*, *Clostridium*, and the cyanobacteria. It is now realized that nitrogen fixation is a capability widespread among many different families of bacteria and also occurs in archaea. The enzyme responsible for nitrogen fixation, nitrogenase, is very similar in the different bacteria and the nitrogen fixation genes have homologous regions. This has led to the suggestion that the nitrogenase gene may have been transferred laterally between different groups of bacteria. Organisms that fix nitrogen encompass a wide range of physiological types and include aerobes, anaerobes, facultative anaerobes, autotrophs, heterotrophs, and phototrophs. Nitrogen fixation takes place when N_2 is the only, or major, source of nitrogen because the genes for nitrogen fixation are repressed by exogenously supplied sources of fixed nitrogen (e.g., ammonia). The signal transduction pathway responsible for repression is discussed in Chapter 17. It is a remarkable fact that biological nitrogen reduction takes place at all. The nitrogen molecule is so stable that very high pressures and temperatures in the presence of inorganic catalysts are necessary to make it reactive in non-biological systems. Industrially, nitrogen gas is reduced to ammonia using the Haber process which requires 200 atm and 800°C. Yet prokaryotes carry out the reduction at atmospheric pressures and ordinary temperatures.

11.3.1 The nitrogen-fixing systems

The biological nitrogen fixing systems are listed below.[6-8]

1. *Rhizobium and Bradyrhizobium in symbiotic relationships with leguminous plants* (soybeans, clover, alfalfa, string beans, peas, i.e., plants that bear seeds in pods). The bacteria infect the roots of the plants and stimulate the production of *root nodules*, within which the bacteria fix nitrogen. The plant responds by feeding the bacteria organic nutrients made during photosynthesis.

2. *Nonleguminous plants in symbiotic relationships with nitrogen fixing bacteria.* For example, the water fern *Azolla* makes small pores in its fronds within which a nitrogen-fixing cyanobacterium, *Anaebaena azollae*, lives. The *Azolla–Anabaena* symbiotic system is used to enrich rice paddies with fixed nitrogen. Another example is the alder tree, which has nitrogen-fixing nodules containing *Frankia*, a bacterium resembling the streptomycetes. A third example is *Azospirillum lipoferum*, which is a N_2-fixing rizosphere bacterium that is found around the roots of tropical grasses.

3. *Many free-living soil and aquatic prokaryotes.* As indicated in the introduction, many different prokaryotes fix nitrogen.

They include *Azotobacter, Clostridium*, certain species of *Desulfovibrio*, the photosynthetic bacteria, and various cyanobacteria. Nitrogen fixation is not confined to (eu)bacteria. Recently, some methanogens have been reported to be nitrogen fixers.[9]

Nitrogen fixation is sensitive to oxygen. The enzyme that fixes nitrogen, *nitrogenase*, is inhibited by oxygen. Thus for many prokaryotes, nitrogen fixation takes place only under anaerobic or microaerophilic conditions. However, some prokaryotes can fix nitrogen while growing in air. As described later, they have evolved systems to protect the nitrogenase from oxygen.

11.3.2 The nitrogen fixation pathway

Nitrogenase
As mentioned, the enzyme that reduces nitrogen gas is called *nitrogenase*[10]. The major nitrogenase in nitrogen-fixing organisms is a molybdenum-containing enzyme that consists of two multimeric proteins. One of these is usually called the molybdenum–iron protein (MoFe protein). It is also known as dinitrogenase, or component I. The second protein is called the iron protein (Fe protein), and is also known as dinitrogenase reductase, or component II. Both of the proteins contain FeS centers.

The MoFe protein is a tetramer $(\alpha_2\beta_2)$ of four polypeptides. When it is extracted with certain solvents, a cofactor, called the iron–molybdenum cluster (FeMoco), is removed. The cofactor contains approximately one-half of the iron and labile sulfide of the protein. Thus, the MoFe protein contains FeMoco plus additional FeS centers.

The Fe protein is a dimer (γ_2) of two identical polypeptides. The dimer contains a single Fe_4S_4 cluster which is responsible for reducing FeMoco during nitrogen fixation.

Twenty-one genes have been identified to be necessary for the expression and regulation of the nitrogenase enzyme system in *Klebsiella pneumoniae*. They are called the *nif* genes. The presence of sufficient NH_4^+ represses the synthesis of nitrogenase. Other nitrogen sources (nitrates, amino acids, urea) also

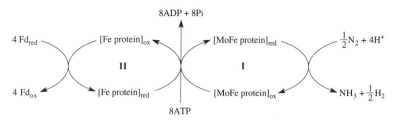

Fig. 11.4 The nitrogenase reaction. The enzyme system consists of two components. Component I is called the molybdenum–iron protein (MoFe protein) or dinitrogenase. Component II is called the iron protein (Fe protein) or dinitrogenase reductase. Both of the proteins contain FeS centers. A low potential reductant, either ferrodoxin or flavodoxin, reduces component II which transfers the electrons to component I. Component I reduces N_2. There is always some H_2 produced. ATP is required despite the fact that the overall reduction of N_2 by ferredoxin or flavodoxin is an exothermic reaction.

suppress the synthesis of nitrogenase, probably by producing ammonia. The *K. pneumoniae* system has been used as a model for the regulation of expression of the nitrogen-fixing genes (Chapter 17).

The nitrogenase reaction

The nitrogenase reaction is a series of reductions during which 0.5 mole of N_2 and 1 mole of H^+ are reduced to 1 mole of NH_3 and 0.5 mole of H_2. (Fig. 11.4).

Nitrogenase reaction:

$$4e^- + 0.5N_2 + 4H^+ + 8ATP \longrightarrow$$

$$NH_3 + 0.5H_2 + 8ADP + 8P_i$$

Since the oxidation state of N_2 is 0 and the oxidation state of the nitrogen in NH_3 is -3, this requires three electrons per nitrogen atom. A fourth electron is transferred to a proton to reduce it to hydrogen gas. The electrons are transferred one at a time in an ATP-dependent reaction from the Fe_4S_4 cluster in the Fe protein to the MoFe cluster in the MoFe protein and from there to N_2. The details of the electron transport pathway through the proteins and the role of ATP are not well understood.[11] However, it is clear that the hydrolysis of at least two ATP molecules is required per electron transferred between the two proteins. Therefore, about 16 moles of ATP are necessary to convert one mole of nitrogen gas to two moles of ammonia. This is a great deal of energy (about 800 kJ). Recall that only two moles of ATP

are generated during the fermentation of one mole of glucose to lactic acid, and 38 moles of ATP are produced during the complete oxidation of one mole of glucose to carbon dioxide and water. Therefore, an organism growing on nitrogen gas must consume a large fraction of the ATP that it produces in order to reduce the nitrogen to ammonia. Not surprisingly, bacteria do not fix nitrogen gas if an alternative source of nitrogen is present. As stated previously, this is because the nitrogen fixation genes are repressed when nitrogen sources other than N_2 are available.

During nitrogen reduction, hydrogen gas is always produced. In other words, some of the electrons go to protons as well as to nitrogen gas. The production of hydrogen gas appears to be wasteful of electrons and ATP. Indeed, some bacteria (e.g., *Azotobacter*) are very good at scavenging the hydrogen gas with a hydrogenase and re-utilizing it so that very little is produced during nitrogen fixation.

Other nitrogenases

Although the nitrogenase described above is certainly the major one, other nitrogenases have recently been discovered. For example, *Azotobacter vinelandii* can synthesize three nitrogenases. The molybdenum-containing nitrogenase (nitrogenase I) is made when the organism is grown in media containing molybdenum. Nitrogenase II is a vanadium-containing nitrogenase that is synthesized when the cells are grown in media lacking molybdenum but containing vanadium. Instead of

FeMoco, the nitrogenase contains a vanadium cofactor called FeVaco. When both molybdenum and vanadium are lacking in the media, *Azotobacter* makes a third nitrogenase, called nitrogenase III, which requires only Fe.

The source of electrons for nitrogen reduction

The nitrogenases are reduced by ferredoxins (FeS proteins) or flavodoxins (flavoproteins) in most known systems. These electron carriers have midpoint potentials sufficiently low to reduce nitrogenase (-400 to -500 mV). The source of electrons for the ferredoxins and flavodoxins varies with the metabolism of the organism. For example, during heterotrophic anaerobic growth the oxidation of pyruvate to acetyl-CoA and carbon dioxide generates reduced ferredoxin (pyruvate:ferredoxin oxidoreductase, Section 7.3.2) or flavodoxin (pyruvate: flavodoxin oxidoreductase) that donates electrons to nitrogenase. (*Clostridium pasteurianum* uses the ferredoxin enzyme, whereas *Klebsiella pneumoniae* uses the flavodoxin enzyme.) The path of electrons to nitrogenase in aerobic and phototrophic bacteria and in cyanobacteria is not as well understood, but is thought to involve ferredoxin or flavodoxin as the immediate electron donor. The ferredoxin or flavodoxin might be reduced by NAD(P)H (or some other electron carrier) generated during metabolism, for example, during the oxidation of carbohydrate. However, a source of energy (i.e., the protonmotive force) would be necessary to drive the reduction of ferredoxin and flavodoxin by NAD(P)H (reversed electron transport, Section 4.5) because the midpoint potential of the $NAD(P)^+/NAD(P)H$ couple is -320 mV as compared to -400 to -500 mV for the ferredoxins and flavodoxins. Alternatively, the cyanobacteria and green sulfur bacteria could use light energy to reduce the electron donor for nitrogenase during photosynthetic noncyclic electron flow. For example, it has been suggested that photosystem I and some electron donor other than water might reduce nitrogenase in the heterocyst, and similarly that a light-generated reductant might drive the reduction

of nitrogenase in the green sulfur bacteria. Recall that these two photosystems have reaction centers that produce a reductant at a sufficiently low potential to reduce ferredoxin (Sections 5.3 and 5.4).

Protecting the nitrogenase from oxygen

All nitrogenases are rapidly inactivated by oxygen *in vitro*. Some nitrogen-fixing microorganisms are strict anaerobes (e.g., the clostridia or nitrogen-fixing sulfate reducers). Others are facultative anaerobes (i.e., they can grow aerobically or anaerobically). These fix nitrogen only when they are living anaerobically (e.g., the purple photosynthetic bacteria and *Klebsiella* spp.). Some microorganisms are microaerophilic (i.e., can grow only in low levels of oxygen). Some of these are nitrogen fixers and can grow on nitrogen gas under microaerophilic conditions. But what about strict aerobes or the cyanobacteria which produce oxygen in the light? Various strategies have evolved to protect the nitrogenase in those microorganisms that fix nitrogen in air.[12,13]

The *Azotobacter* species have a very active respiratory system. It is suggested that they utilize oxygen rapidly enough to lower the intracellular concentrations in the vicinity of the nitrogenase. These organisms are also able to protect their nitrogenase from inactivation by associating it with protective proteins. For example, the nitrogenase of *Azotobacter* can be isolated as an air-tolerant complex with a redox protein called the Shethna, FeS II, or protective protein.

The rhizobia in root nodules exist in plant vesicles in the inner cortex of the nodule as modified cells called bacteroids. Using oxygen-sensitive microelectrodes, it can be shown that the oxygen concentrations in the inner cortex are much lower than in the surrounding tissue. The oxygen levels in the vicinity of the bacteroids are controlled by a boundary of densely packed plant cells between the inner and outer cortex. The control of oxygen access to the inner cortex is achieved by regulating the intercellular spaces, which are either air-filled or contain variable amounts of water, within the boundary layer. However, the bacteroids are dependent upon oxygen for respiration, and within the nodule a plant protein called leghemoglobin binds oxygen and delivers it to the bacteroids.

Some unicellular cyanobacteria temporarily separate photosynthesis from nitrogen fixation. For example, cultures of *Gloethece* or *Synechococcus* grown under light–dark cycles fix nitrogen only in the dark (i.e., when they are not producing oxygen).

It is not understood how the unicellular cyanobacteria protect their nitrogenase from atmospheric oxygen. However, filamentous cyanobacteria protect their nitrogenase by fixing nitrogen in special cells called heterocysts which are discussed next.

Heterocysts

Filamentous cyanobacteria protect their nitrogenase by differentiating special nitrogen fixing cells called heterocysts[14] (Fig. 11.5). The heterocyst differs from the vegetative cells in:

1. Possessing the nitrogenase enzymes;

2. Having only photosystem 1 (PS1);

3. Being surrounded by a thick cell wall consisting of glycolipid and polysaccharide;

4. Not dividing.

The ATP made during cyclic photophosphorylation by photosystem I is used to fix the nitrogen. But where do the heterocysts get the reducing power? What happens is that the heterocysts fix N_2 and feed the rest of the filament reduced nitrogen (Fig. 11.5). In turn, the heterocyst is fed carbohydrate made from CO_2 by the vegetative cells. The heterocyst oxidizes the carbohydate, reducing ferredoxin which, in turn, reduces the nitrogenase. Therefore, this is a complex situation where two different cell types in the filament are feeding each other.

Since the heterocysts do not have photosystem 2, they do not produce oxygen. However, there is still the problem of protecting the nitrogenase from atmospheric oxygen. It has been suggested that the crystalline glycolipid and polysaccharide cell wall may present a diffusion barrier to oxygen.

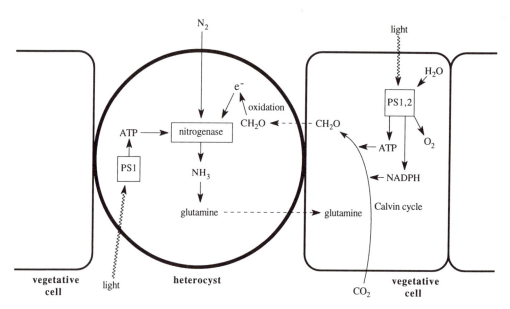

Fig. 11.5 Heterocyst interactions with vegetative cells in *Anabaena*. The heterocyst reduces dinitrogen to ammonia which is then incorporated into glutamine via glutamine synthetase. The glutamine then enters the vegetative cells where it serves as a source of fixed nitrogen for growth. The vegetative cells fix carbon dioxide into carbohydrate using the Calvin cycle. Some of the carbohydrate enters the heterocyst where it serves as a source of carbon and NADPH. The NADPH reduces the nitrogenase via ferredoxin:NADP oxidoreductase and ferredoxin. ATP is made via cyclic photophosphorylation in the heterocyst using PS1. Since PS2 is lacking in the heterocyst, oxygen is not produced there.

Table 11.1 Chemoautotrophs

Bacterial group	Typical species	Electron donor	Electron acceptor	Carbon source	Product
Hydrogen-oxidizing	*Alcaligenes eutrophus*	H_2	O_2	CO_2	H_2O
Carbon-monoxide oxidizing (carboxydobacteria)	*Pseudomonas carboxydovorans*	CO	O_2	CO_2	CO_2
Ammonium-oxidizing	*Nitrosomonas europaea*	NH_4^+	O_2	CO_2	NO_2^-
Nitrite-oxidizing	*Nitrobacter winogradskyi*	NO_2^-	O_2	CO_2	NO_3^-
Sulfur-oxidizing	*Thiobacillus thiooxidans*	$S, S_2O_3^{2-}$	O_2	CO_2	SO_4^{2-}
Iron-oxidizing	*Thiobacillus ferrooxidans*	Fe^{2+}	O_2	CO_2	Fe^{3+}
Methanogenic	*Methanobacterium thermoautotrophicum*	H_2	CO_2	CO_2	CH_4
Acetogenic	*Acetobacterium woodii*	H_2	CO_2	CO_2	CH_3COOH

Source: Schlegel, H. G., and H. W. Jannasch. 1992. Prokaryotes and their habitats, pp. 75–125. In: *The Prokaryotes*, Vol. I. A. Balows, H. G. Trüper, M. Dworkin, W. Harder and K.-H. Schleifer (eds.). Springer-Verlag, Berlin.

11.4 Lithotrophy

While most organisms derive energy from oxidizing organic nutrients (chemoorgano-trophs) or from the absorption of light (phototrophs), there exist many prokaryotes that derive energy from the oxidation of inorganic compounds such as H_2, CO, NH_3, NO_2^-, H_2S, S^o, $S_2O_3^{2-}$, or Fe^{2+}. This type of metabolism is called *lithotrophy* and the organisms are called *lithotrophs*.[15]

11.4.1 The lithotrophs

The lithotrophs are physiologically diverse and exist among several different groups of bacteria and archaea. Many of the lithotrophs are aerobes (i.e., they carry out electron transport with oxygen as the terminal electron acceptor). However, some are facultative anaerobes using nitrate or nitrite as the electron acceptor when oxygen is unavailable, and a few are obligate anaerobes that use sulfate or CO_2 as the electron acceptor. Most of the lithotrophs are autotrophs (i.e., their sole or major source of carbon is CO_2). They are called *chemoautotrophs* or *chemolitho-autotrophs*. The lithotrophs vary with regard to the autotrophic CO_2 fixation pathway that they use (Chapter 12). Other lithotrophs are facultatively heterotrophic. The facultative heterotrophs include all of the bacterial hydrogen-oxidizers, some sulfur-oxidizing thio-bacilli, and some thermophilic iron-oxidizing bacteria. Some representative species are listed in Table 11.1.

Table 11.2 is a list of midpoint potentials of the inorganic substrates at pH 7. Theoretically, all of these organisms, with the possible exception of the hydrogen oxidizers and the CO oxidizers, must carry out reversed electron transport because the electron donor is more electropositive than the $NAD^+/$-NADH couple ($E_o' = -0.32$ V). Reversed electron flow is driven by the Δp. Because of the relatively small ΔE_h between the inorganic electron donor and oxgyen, and the need to reverse electron transport, the energy yields, and therefore the cell yields, are relatively

Table 11.2 Some mid-point potentials of inorganic reductants used by chemoautotrophs

Bacteria	Couple	$E_o'^*$ (V)
Carboxydobacteria	CO_2/CO	−0.54
Hydrogen bacteria	H^+/H_2	−-0.41
Sulfur bacteria	SO_4^{2-}/SO_3^{2-}	−0.28
	S/H_2S	−0.25
	SO_3^{2-}/S	+0.05
Nitrite oxidizers	NO_3^-/NO_2^-	+0.42
Ammonium oxidizers	NO_2^-/NH_4^+	+0.44
Iron bacteria	Fe^{3+}/Fe^{2+}	+0.78

*For comparison, the standard potential at pH 7 for O_2/H_2O is +0.815 V.

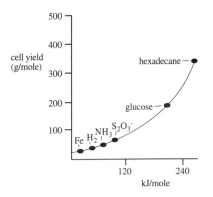

Fig. 11.6 Cell yields versus available energy in inorganic and organic electron sources. (Adapted from Brock, T. D. and M. T. Madigan. 1991. *Biology of Microorganisms*. Reprinted by permission of Prentice-Hall, Englewood Cliffs, NJ.)

small compared to growth on organic substrates (Fig. 11.6).

Aerobic hydrogen-oxidizing bacteria and carboxydobacteria

The hydrogen-oxidizing bacteria are usually facultative and can live either autotrophically or heterotrophically. However, some always require organic carbon for growth and are called chemolithoheterotrophs. The hydrogen oxidizers can be found in aerobic environments where H_2 is available. The hydrogen gas itself is produced as a by-product of nitrogen fixation (e.g., in the rhizosphere of nitrogen-fixing plants and in cyanobacterial blooms). Hydrogen gas is also produced in anaerobic environments via fermentations, where some of it escapes into the aerobic atmosphere. (However, most of the H_2 produced anaerobically is utilized by the sulfate reducers and methanogens.) Among the hydrogen-oxidizing bacteria are some that can grow on carbon monoxide (CO) as the sole source of energy and carbon using oxygen, or in some cases nitrate (denitrifiers) as the electron acceptor. They are called *carboxydobacteria*. The hydrogen-oxidizing bacteria and the carboxydobacteria are represented by several genera, including representatives from *Pseudomonas*, *Arthrobacter*, *Bacillus*, and *Rhizobium*. Anaerobic hydrogen oxidizers include some sulfate-reducing bacteria and some archaea, including the methanogens when growing auto-

trophically on CO_2, and certain sulfur-dependent archaea that use elemental sulfur as the electron acceptor, reducing it to hydrogen sulfide.

Ammonia-oxidizing bacteria

Bacteria that oxidize ammonia as a source of energy are called nitrifiers.[16] There are five genera of nitrifiers. They are *Nitrosomonas*, *Nitrosococcus*, *Nitrosospira*, *Nitrosolobus*, and *Nitrosovibrio*. All of the nitrifiers are aerobic obligate chemolithoautotrophs that assimilate CO_2 via the Calvin cycle. Ammonia that is produced in the anaerobic niches by deamination of amino acids, urea, or uric acid, or via dissimilatory nitrate reduction diffuses into the aerobic environment where it is oxidized by the nitrifiers. The nitrifiying bacteria are often found at the aerobic/anaerobic interfaces where they capture the ammonia as it diffuses from the anaerobic environments, as well as in the more highly aerobic parts of the soil and water.

Nitrosomonas oxidizes ammonia to nitrite. Along with *Nitrobacter* which oxidizes nitrite to nitrate, it is responsible for a major portion of the conversion of ammonia to nitrate, a process called *nitrification*. The first oxidation is the oxidation of ammonia to hydroxylamine catalyzed by *ammonia monooxygenase*, (AMO). The reaction is:

$$2H^+ + NH_3 + 2e^- + O_2$$
$$\longrightarrow NH_2OH + H_2O$$

In this reaction, one oxygen atom is incorporated into hydroxylamine and the other is reduced to water. The oxidation of hydroxylamine to nitrite is catalyzed by the enzyme *hydroxylamine oxidoreductase* (HAO).

$$NH_2OH + H_2O \longrightarrow HONO + 4e^- + 4H^+$$

Of the four electrons removed from hydroxylamine, two are used in the ammonia monooxygenase reaction and the other two are transferred to oxygen via cytochrome oxidase:

$$2H^+ + 0.5O_2 + 2e^- \longrightarrow H_2O$$

A proposed electron transport scheme showing the topological arrangement of the electron carriers is shown in Fig.11.7. The electron transport scheme shown in Fig. 11.7 is speculation based upon the known location of the enzymes and their redox potentials. The scheme proposes that ammonia oxidation to hydoxylamine takes place in the cytoplasm, although it is not known whether this occurs in the cytoplasm or the periplasm. (Ammonia monooxygenase is in the cell membrane but it is unknown whether the substrate binding site is exposed to the cytoplasm or the periplasm.) Assuming cytoplasmic oxidation, the hydroxylamine diffuses across the cell membrane to the periplasm where it is oxidized by the hydroxylamine oxidoreductase, a periplasmic enzyme. The electrons travel from hydroxylamine oxidoreductase to a periplasmic cytochrome c. Here the electron transport pathway branches.

Two electrons travel to ubiquinone and then to the ammonia monooxygenase enzyme. The other electrons travel to oxygen via cytochrome aa_3, which is presumed to act as a proton pump. The Δp that is created as a result of the oxidation of ammonia and hydroxylamine and the reduction of oxygen is the sole source of energy for these bacteria.

Nitrite-oxidizing bacteria

The nitrite oxidizers are *Nitrobacter*, *Nitrococcus*, *Nitrospina*, and *Nitrospira*. They are aerobic obligate chemolithoautotrophs, with the exception of *Nitrobacter*, which is a facultative autotroph (i.e., it can also be grown heterotrophically). The details of the electron transport scheme have not been fully elucidated. However a proposed model is shown in Fig. 11.8. Electrons travel from nitrite to oxygen via a periplasmic cytochrome c. There is a thermodynamic problem

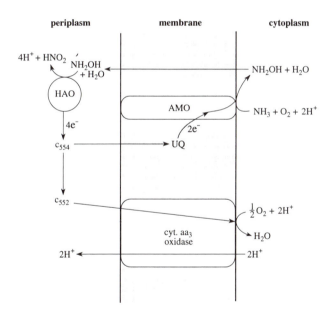

Fig.11.7 Model for the electron transport scheme in *Nitrosomonas*. Ammonia is oxidized to hydroxylamine by the enzyme ammonia monooxygenase (AMO). Although the oxidation of ammonia is depicted as occurring in the cytoplasm, it has not been ruled out that the oxidation takes place in the periplasm. The reaction requires two electrons to reduce one of the oxygen atoms to water. The hydroxylamine diffuses across the membrane to the periplasm where it is oxidized to nitrite by a complex cytochrome called hydroxylamine oxidoreductase (HAO). The electrons are passed to a periplasmic cytochrome c and from there to ubiquinone in the membrane. Electrons travel from ubiquinone in two branches. One branch passes two electrons to the ammonia monooxygenase and the second branch leads to oxygen via cytochromes c and aa_3. All four electrons end up in water. (Adapted from Hooper, A. B. 1989. Biochemistry of the nitrifying lithoautotrophic bacteria, pp. 239–265. In: *Autotrophic Bacteria*. H. G. Schlegel and B. Bowien, (eds.). Springer-Verlag, Berlin.)

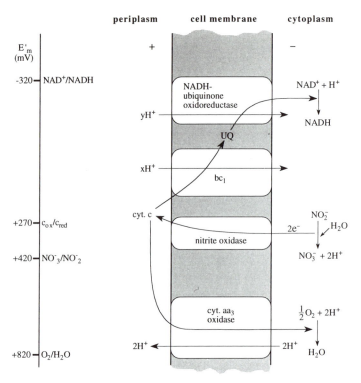

Fig. 11.8 A model for electron transport in *Nitrobacter*. The electrons travel from nitrite via membrane-bound nitrite oxidase to cytochrome c which is at a more negative potential. It is proposed that nitrite oxidation takes place on the cytoplasmic side of the membrane and that the membrane potential drives the electrons transmembrane to cytochrome c. From cytochrome c the electrons diverge. Most travel to oxygen at a more positive potential and a Δp is created. The coupling site is the cytochrome aa_3 oxidase which is a proton pump. Other electrons travel to NAD^+ which is at a more negative potential. The Δp drives the electrons in reverse flow to NAD^+. This is accomplished by coupling electron transport with the return of protons down the proton potential to the cytoplasmic side. The scheme presumes the presence of a bc_1 complex as well as a reversible NADH dehydrogenase complex. (Adapted from Nicholls, D. G. and S. J. Ferguson. 1992. *Bioenergetics 2*. Academic Press, London.)

here. Cytochrome c exists at a midpoint potential ($E_{m,7}$) of +270 mV, whereas the midpoint potential of the nitrate/nitrite couple is more electropositive at +420 mV. Because electrons do not spontaneously flow towards the lower redox potential, energy must be provided to drive the electrons over the 150 mV difference in order for nitrite to reduce cytochrome c. Nichols and Ferguson suggested that the membrane potential drives the electrons from nitrite to cytochrome c.[17] It was proposed that nitrite oxidation takes place on the cytoplasmic surface of the membrane, and that the electrons flow across the membrane to cytochrome c located on the periplasmic side, which is typically 170 mV more positive than the cytoplasmic side (Fig. 11.8). In this way, the membrane

potential lowers the potential difference between the nitrate/nitrite couple and the c_{ox}/c_{red} couple by approximately 170 mV. The student will recognize this model as *reversed electron transport* driven by the membrane potential (which is consumed in the process). The role of the membrane potential in driving electron transport from nitrite to cytochrome c is consistent with the observation that experimental procedures that lower the membrane potential (e.g., incubation with proton ionophores) decrease electron transfer from nitrite to oxygen in inverted membrane vesicles from *Nitrobacter*. The electrons then flow from cytochrome c back across the membrane to oxygen through cytochrome aa_3 oxidase driven by a favorable midpoint potential difference of about $+440\,mV$. A Δp is created by the outward pumping of protons by the cytochrome aa$_3$ oxidase. Electron flow is also reversed from cytochrome c to NAD$^+$ through two coupling sites: a ubiquinone–cytochrome c oxidoreductase (probably a bc$_1$ complex), and a reversible NADH–ubiquinone oxidoreductase. The model proposes that reversed electron flow to NAD$^+$ is coupled to the *influx* of protons (i.e., it is driven by the Δp).

Sulfur-oxidizing prokaryotes

The sulfur-oxidizing prokaryotes[18,19] include the photosynthetic sulfur oxidizers (Chapter 5) and the nonphotosynthetic sulfur bacteria. It is the latter which concern us here. They include the colorless sulfur bacteria such as *Beggiatoa* and *Thiothrix*, bacteria belonging to the genus *Thiobacillus*, and an archaeon belonging to the genus *Sulfolobus*. The thiobacilli are the most prominent of the sulfur oxidizers, and their sulfur oxidation pathways are the most well studied.

Sulfur compounds commonly used as sources of energy and electrons include hydrogen sulfide (H_2S), elemental sulfur (S^o), and thiosulfate ($S_2O_3^{2-}$), all of which can be oxidized to sulfate. Although most sulfur bacteria are aerobes, a few can be grown anaerobically using nitrate as the electron acceptor. The sulfur bacteria can be found in nature growing near sources of sulfur [e.g., sulfur deposits, sulfide ores, hot sulfur springs, sulfur mines, and coal mines that are sites of iron pyrite (FeS_2) deposits]. Sulfide-oxidizers can sometimes be found in large accumulations in thin layers between the aerobic and anaerobic environments. The reason they are sometimes confined to the aerobic/anaerobic interface is that H_2S, which is produced anaerobically by sulfate reducers, is rapidly oxidized by oxygen. Thus, the sulfide-oxidizing bacteria grow where oxygen levels are relatively low in order to compete with the rapid chemical oxidation of sulfide. Some of the sulfur bacteria, particularly those growing around sulfur and coal mines which produce sulfuric acid, are acidophiles. An example is *Thiobacillus thiooxidans*, which can grow at a pH of 1.

The sulfur prokaryotes are usually auto-trophic, but can also grow heterotrophically. For example, *Sulfolobus* is an aerobic faculta-tive autotrophic thermophilic archaeon that oxidizes S^o or H_2S to sulfuric acid. It can be isolated from hot acid sulfur springs. Some sulfur bacteria (e.g., *Beggiatoa*) can be grown only mixotrophically (i.e., using H_2S as the energy source and organic carbon as the source of carbon).

There has been much confusion in the literature regarding the intermediate stages of sulfur oxidation. Probably, more than one pathway exists. Two pathways that are receiving much research attention are de-scribed below.

In *Thiobacillus versutus*, thiosulfate oxida-tion takes place in the periplasm on a multienzyme complex (Fig. 11.9). No free intermediates are released and both atoms of thiosulfate are oxidized to sulfate. Protons are released in the periplasm during the oxidations and the electrons are transferred *electrogenically* from the periplasmic side of the membrane to the cytoplasmic side through cytochrome c_{552} and then cytochrome aa$_3$ oxidase to oxygen. (See Section 3.2.2 for a discussion of electrogenic movement of elec-trons.) A proton current is maintained by the release of protons in the periplasm during the oxidations and their consumption in the cytoplasm during the reduction of oxygen. The expected H^+/O ratio (H^+ produced in the periplasm or translocated from the cytoplasm to the periplasm per oxygen atom reduced) can be obtained from Fig. 11.9 (and confirmed experimentally). The ratio is 2.5 for thiosulfate oxidation. ATP is synthesized by a proton-translocating ATP synthase driven by the Δp. If one assumes that the H^+/ATP ratio for the ATP synthase is three (a consensus value), then the maximum P/O ratio would be 2.5/3 or 0.83 for thiosulfate oxidation. (See Section 4.5.2 for a discussion of the relationship between the size of the proton current and the upper limit of ATP that can be made.)

In *Thiobacillus tepidarius*, the pathway of oxidation of thiosulfate to sulfate (called the polythionate pathway) begins in the peri-plasm, but the bulk of the oxidations take place in the cytoplasm. Two molecules of thiosulfate are oxidized to tetrathionate in

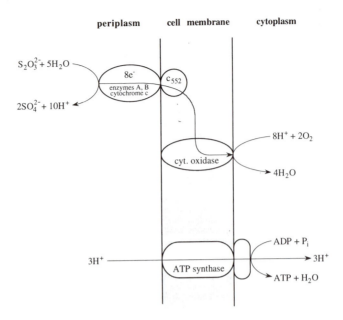

Fig. 11.9 Model for thiosulfate oxidation in *Thiobacillus versutus*. Thiosulfate is oxidized to sulfate in the periplasm by a multienzyme complex consisting of enzyme A, enzyme B, and cytochromes c. The electrons are electrogenically transferred to oxygen via a membrane-bound cytochrome c_{552} and a cytochrome oxidase (cytochrome aa_3). A Δp is created by the release of protons in the periplasm via the oxidations, the consumption of protons in the cytoplasm during oxygen reduction, and electrogenic flow of electrons across the membrane to oxygen. A proton-translocating ATP synthase makes ATP. (Adapted from Kelly, D. P. 1989. Physiology and biochemistry of unicellular sulfur bacteria, pp. 193–217. In: *Autotrophic Bacteria*. H. G. Schlegel and B. Bowien (eds.). Springer-Verlag, Berlin.)

the periplasm (Fig. 11.10). A periplasmic cytochrome c accepts the two electrons generated from the thiosulfate oxidation and transfers these electrons to a membrane-bound cytochrome c. The tetrathionate is believed to be transported into the cell where it is oxidized to four sulfates via the intermediate sulfite (generating 14 electrons). The 14 electrons generated during the tetrathionate oxidations in the cytoplasm are transfered to a proton-translocating ubiquinone–cytochrome b complex in the cell membrane, which reduces membrane-bound cytochrome c. Cytochrome c transfers all of the electrons to cytochrome oxidase (probably cytochrome o), which reduces oxygen. Because the oxidations of tetrathionate and sulfite are cytoplasmic rather than periplasmic, the proton current is due to proton translocation by the ubiquinone–cytochrome b complex (coupled to the cytoplasmic oxidations), rather than to periplasmic oxidations, as is the case for *T. versutus*.

Some thiobacilli can incorporate a substrate level phosphorylation when oxidizing sulfur. The substrate level phosphorylation occurs at the level of sulfite. The sulfite reacts with adenosine monophosphate (AMP) to form adenosine phosphosulfate (APS), in a reaction catalyzed by APS reductase. The APS is oxidized to sulfate producing ADP or ATP, using either ADP sulfurylase or ATP sulfurylase.

APS reductase

$$SO_3^{2-} + AMP \longrightarrow APS + 2e^-$$

ADP sulfurylase

$$APS + P_i \longrightarrow ADP + SO_4^{2-}$$

ATP sulfurylase

$$APS + PP_i \longrightarrow ATP + SO_4^{2-}$$

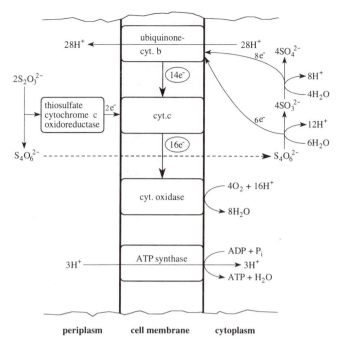

Fig. 11.10 A model for thiosulfate oxidation in *Thiobacillus tepidarius*. Thiosulfate is oxidized to tetrathionate in the periplasm by a thiosulfate–cytochrome c oxidoreductase. It is hypothesized that the tetrathionate is transported into the cell in symport with protons where it is oxidized to sulfate. The electrons travel to oxygen via a proton-translocating ubiquinone–cytochrome b system (probably a bc_1 complex) and cytochrome oxidase, which appears to be a cytochrome o. (Adapted from Kelly, D. P. 1989. Physiology and biochemistry of unicellular sulfur bacteria, pp. 193–217. In: *Autotrophic Bacteria*. H. G. Schlegel and B. Bowien (eds). Springer-Verlag, Berlin and Smith, D. W. and W. R. Strohl. 1991. Sulfur-oxidizing bacteria, pp. 121–146. In: *Variations in Autotrophic Life*. J. M. Shively and L. L. Barton (eds.). Academic Press, New York.)

Recall that ATP sulfurylase and APS reductase are used by dissimilatory sulfate reducers to reduce sulfate to sulfite in reactions that consume ATP (Section 11.2.2). By running these reactions in the oxidative direction, sulfur oxidizers can use the energy in pyrophosphate to make ATP.

Figure 11.11 summarizes the inorganic sulfur oxidation pathways. Elemental sulfur exists as an octet ring of insoluble sulfur (S_8^0). It is first activated by reduced glutathione (GSH) to form a linear polysulfide (G-S-S_8-H). Sulfide (S^{2-}) also reacts with GSH and is oxidized to linear polysulfide. The sulfur atoms are removed from the polysulfide one at a time during the oxidation to sulfite (SO_3^{2-}).

Another elemental sulfur oxidation pathway has been reported for some thiobacilli, in which S^0 is oxygenated by a sulfur oxygenase,

Sulfur oxygenase

$$S^0 + O_2 + H_2O \longrightarrow H_2SO_3$$

However, the oxygenase reaction cannot account for sulfur oxidation under anaerobic conditions when the oxidant provided is nitrate (*T. denitrificans*) or Fe^{3+} (*T. ferrooxidans*). Furthermore, the oxygenase reaction cannot conserve energy for the cell, since the electrons do not enter the respiratory chain. (Even though the periplasm might be acidified with H_2SO_3 and thus generate a ΔpH, this by itself cannot generate net ATP in a growing cell because protons entering via the ATP synthase must not accumulate in the cytoplasm The extrusion of protons from the cells or the utilization of protons to form water requires electron transport.)

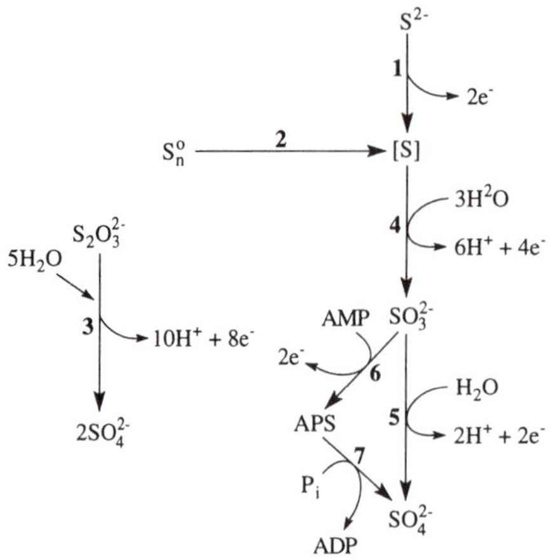

Fig. 11.11 A summary of sulfur oxidation pathways (1) oxidation of sulfide to linear polysulfide [S]; (2) conversion of elemental sulfur to linear polysulfide; (3) thiosulfate multienzyme complex; (4) sulfur oxidase; (5) sulfite oxidase; (6) APS reductase; (7) ADP sulfurylase. (Adapted from Gottschalk, G. 1985. *Bacterial Metabolism.* Springer-Verlag, Berlin.)

Iron-oxidizing bacteria

A few bacteria derive energy from the aerobic oxidation of ferrous ion to ferric ion.[20] Most of these are also acidophilic sulfur oxidizers (i.e., they oxidize sulfide to sulfuric acid). The acidophilic iron oxidizers can be found growing at the sites of geological deposits of iron sulfide minerals [e.g., pyrite (FeS_2) and chalcopyrite ($CuFeS_2$)] where water and oxygen are also present. The iron-sulfide minerals are uncovered during mining operations, and the presence of acid mine water at these sites is due to the growth of the iron-sulfide oxidizers. Since Fe^{2+} is rapidly oxidized chemically by oxygen to Fe^{3+} at neutral pH but only slowly at acid pH, the acidic environment is conducive to growth of the iron oxidizers. An example is *Thiobacillus ferrooxidans*, which is an autotroph able to use either ferrous ion or inorganic sulfur compounds as a source of energy and electrons. *Thiobacillus ferrooxidans* can be grown on ferrous ion (e.g., $FeSO_4$) at optimal external pH values between 1.8 and 2.4. It maintains an internal pH of around 6.5 and therefore a ΔpH of about 4.5. The membrane potential ($\Delta\Psi$) at low pH is reversed and energy in the Δp is due entirely to the pH

component (Section 3.3). (As discussed in Section 14.1.3, the inversion of the $\Delta\Psi$ is necessary for maintenance of the pH and may be due to electrogenic influx of K^+.)

1. Growth on ferrous sulfate

Figure 11.12 illustrates a model for the electron transport pathway in *T. ferrooxidans* growing on the aerobic oxidation of ferrous sulfate. The pathway proposes that extracellular Fe^{2+} is oxidized to Fe^{3+} by a Fe^{3+} complex in the outer membrane. The electrons travel to a periplasmic cytochrome c. There is also a copper protein called rusticyanin in the periplasm, which is thought to be part of the electron transport scheme. From cytochrome c the electrons flow across the membrane via cytochrome oxidase (cytochrome a_1) to oxygen, generating a $\Delta\Psi$, inside negative. (The inversion of the $\Delta\Psi$ may be due to electrogenic influx of K^+ as discussed in Section 14.1.3.) The oxidation of ferrous ion drives the consumption of protons in the cytoplasm during oxygen reduction. The ΔpH is maintained because the rate of respiration and proton consumption matches the rate at which protons enter the cell through the ATPase or through leakage. From an energetic point of view, one would not expect any proton pumping by the cytochrome oxidase. The reason for this is that the difference in mid-point potentials between the Fe^{3+}/Fe^{2+} couple (pH 2) and the O_2/H_2O (presumed to be at pH 6.5)[21] is very small, perhaps only 0.08 V or less.[22] Under these circumstances, a two electron transfer would generate at the most only 0.16 eV. A simple calculation reveals that this would not be enough energy to pump protons against the pH gradient. At an external pH of 2 and an internal pH of 6.5, the ΔpH is 4.5. This is equivalent to 0.06(4.5) or 0.27 V at 30°C (eq. 3.10). Thus, each proton would have to be energized by approximately 0.27 eV volts in order to be pumped out of the cell, even in the presence of a small $\Delta\Psi$, inside positive, which may be on the order of +0.01 to +0.02 V.

ATP is synthesized by a membrane H^+-translocating ATP synthase driven by a Δp of approximately −250 mV. Electrons from Fe^{2+} must also move towards a lower redox potential in order to generate NAD(P)H.

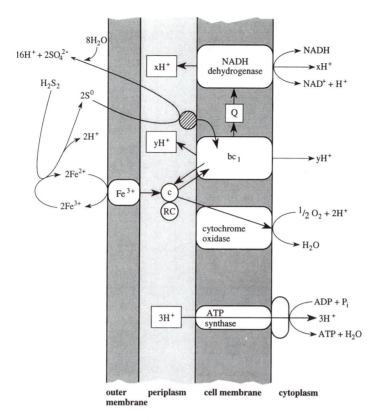

Fig. 11.12 Electron transfer in *Thiobacillus ferrooxidans*. Ferrous ion is oxidized to Fe^{3+} by a Fe^{3+} complex in the outer membrane, and the electrons flow through periplasmic cytochrome c and a copper protein, rusticyanin (RC) to a membrane cytochrome oxidase of the a_1 type. The Δp is maintained by the inward flow of electrons (contributing towards a negative $\Delta\Psi$) and the consumption of cytoplasmic protons during oxygen reduction, maintaining a ΔpH. The consumption of two protons per two electrons in the cytoplasm and the uptake of three protons through the ATP synthase, indicates that a maximum of 2/3 of an ATP can be made per oxidation of two ferrous ions. When the bacteria are oxidizing iron pyrite (FeS_2), the Fe^{3+} that is produced outside of the cell envelope is chemically reduced to Fe^{2+} by S^{2-}. The resultant S^o is oxidized to SO_4^{2-} and the electrons pass through a proton-translocating bc_1 complex to periplasmic cytochrome c and thence to cytochrome oxidase. The figure also illustrates Δp-driven reversed electron transport from S^o and from Fe^{2+} to NAD^+ through coupling sites that bring protons into the cell.

Figure 11.12 also illustrates how reversed electron flow from Fe^{2+} to NAD^+ might occur through a bc_1 complex and quinone. Reversed electron flow is probably driven by the Δp as protons enter the cell down the Δp gradient through the bc_1 complex and the NADH dehydrogenase complex.

2. Growth on pyrite

Pyrite is a cystalline ore of iron which is usually written as ferrous disulfide (FeS_2). It is actually a stable crystal of discrete $^-S-S^-$ disulfide ions (S_2^{2-}) and ferrous ion (Fe_2^{2+}). The ferrous ion and disulfide are oxidized to ferric ion and sulfuric acid, respectively, according to the following overall reaction:

$$4FeS_2 + 2H_2O + 15O_2$$
$$\longrightarrow 4Fe^{3+} + 8SO_4^{2-} + 4H^+$$

However, the oxidation is not straightforward.[23] The oxidation of FeS_2 is the result of several redox reactions (Fig. 11.12). When the bacteria are growing on iron pyrite, the Fe^{2+} is oxidized extracellularly by a complex of Fe^{3+} in the outer membrane (Fig. 11.12). The outer membrane iron complex then transfers the electrons to the periplasmic cytochrome c which transfers the electron to cytochrome oxidase. The Fe^{3+} that is formed extracellularly is recycled to Fe^{2+} by disulfide from the iron pyrite, according to the following reaction which occurs spontaneously:

$$H_2S_2 + 2Fe^{3+} \longrightarrow 2S^\circ + 2Fe^{2+} + 2H^+$$

The elemental sulfur that is produced is oxidized by the bacteria to sulfate, with the electrons travelling through a proton-translocating bc_1 complex to the periplasmic cytochrome c and from there to cytochrome oxidase.

11.4.2 A review of the energetic considerations for lithotrophic growth

ATP synthesis

As discussed in Section 3.7.1, the energy from respiratory oxidation–reduction reactions is first converted into a Δp which is then used to drive ATP synthesis via the ATP synthase. For illustrative purposes, we will consider growth on nitrite in an aerated culture as a source of energy and electrons. The E'_m for NO_3^-/NO_2^- is $+0.42$ V. The E'_m for O_2/H_2O is $+0.82$ V. The difference in potential is therefore $+0.82 - 0.42$ or $+0.40$ V. Recall that in the respiratory chain, each coupling site is associated with pairs of electrons travelling over a mid-point potential difference of approximately 0.2 V or more. The conclusion is that there is sufficient energy for a coupling site between nitrite and oxygen. Another way of examining this question is to consider the amount of energy required to synthesize an ATP. The energy required to synthesize ATP will vary with the concentrations of ATP, ADP and P_i. However, we will assume a value of 0.4 to 0.5 eV, which is a reasonable estimate. Therefore, two electrons traveling over a redox gradient of 0.4 V should give 0.8 eV of energy, which is sufficient for the synthesis of an ATP.[24]

NAD+ reduction requires reversed electron transport

In order to grow, all cells must be able to make NAD(P)H because this is a major source of electrons for reductions that occur during biosynthesis. For example, the Calvin cycle, which reduces CO_2 to the level of carbohydrate, requires two NADH per CO_2. When the cells are growing on NO_2^- and CO_2, what can they use as a source of electrons to reduce NAD^+? Clearly not CO_2 since that is already the highest oxidized form of carbon. So we are left with NO_2^-. However, a comparison of the electrode potentials points to a problem. The E'_m for NO_3^-/NO_2^- is $+0.42$ V and the E'_m for $NAD^+/NADH$ is -0.32 V. Electrons flow spontaneously only to the more electropositive acceptor. Thus, in order to make the electrons flow from NO_2^- to NAD^+, the flow must be reversed, and this requires energy. How much energy is required? The potential difference ($\Delta E'_o$) is $-0.32 - 0.42$ or -0.74 V. Thus each electron would have to be energized by 0.74 eV. The source of energy is the Δp. The inward flow of protons through the coupling sites down the Δp gradient drives electron transport in reverse (Section 3.7.1).

11.5 Summary

When it comes to inorganic metabolism, prokaryotes are versatile creatures. They can reduce oxidized forms of sulfur and nitrogen, as well as nitrogen gas, for incorporation into cell material, and they can oxidize a variety of inorganic substrates, trapping the energy released as a Δp. They can also carry out anaerobic respiration using oxidized forms of iron, manganese, nitrogen, and sulfur as electron acceptors. One can add CO_2, which is used by the methanogens, to this list of electron acceptors.

There is both an assimilatory and dissimilatory route for nitrate reduction. The assimilatory route is catalyzed by cytosolic enzymes that reduce nitrate to ammonia. The ammonia is then incorporated into amino acids using the GOGAT enzyme system. All bacteria that grow on nitrate as a source of cell nitrogen must have the assimilatory pathway.

The dissimilatory route differs in being membraneous and in producing a Δp which can be used for ATP synthesis. A common dissimilatory route reduces nitrate to N_2.

Nitrate dissimilation to N_2 is also called denitrification. Organisms that carry out denitrification are widespread. However, the enzymes for denitrification are oxygen sensitive and furthermore are formed only under anaerobic conditions or when oxygen tensions are very low. These bacteria are generally facultative anaerobes and will carry out an aerobic respiration when oxygen is available. Several bacteria are not denitrifiers, but can carry out an anaerobic respiration in which nitrate is reduced to nitrite or to ammonia by a membrane-bound nitrate reductase creating a Δp.

Most bacteria can use sulfate as a source of cell sulfur. They employ a cytosolic assimilatory pathway in which the sulfate is activated by ATP to form adenosine-3'-phosphate-5'-phosphosulphate (PAPS). The formation of PAPS raises the E_h of sulfate so that it is a better electron acceptor. The reduction of PAPS yields sulfite, which is reduced to H_2S via a soluble sulfite reductase. The H_2S does not accumulate but is immediately incorporated into O-acetylserine to form L-cysteine. The L-cysteine donates the sulfur to the other sulfur-containing compounds.

Certain strict anaerobes can use sulfate as an electron acceptor for anaerobic respiration and reduce it to H_2S. These are the sulfate reducers. Organic carbon is oxidized and the electrons are transferred to protons via an hydrogenase to form H_2. It is suggested that the H_2 diffuses into the periplasm where it is oxidized by a cytochrome c that transfers the electrons to membrane carriers. Electrons travel across the membrane to the inner surface creating a membrane potential. The protons resulting from the periplasmic oxidation of H_2 remain on the outer surface or in the periplasm, contributing towards the proton gradient. Dissimilatory sulfate reduction differs from assimilatory reduction in that APS is reduced rather than PAPS, the electron carriers are membraneous, and a Δp is created.

Nitrogen fixation is carried out by a diversity of prokaryotes, including both bacteria and archeae. These include cyanobacteria, photosynthetic bacteria, strict heterotrophic anaerobes (*Clostridium pasteuranium*,

Desulfovibrio vulgaris), and several obligate and facultative aerobes (e.g., rhizobia, *Azotobacter*, *Klebsiella*, and methanogens). The reduction of N_2 is an ATP-dependent process catalyzed by the enzyme nitrogenase. Hydrogen gas is always a by-product of nitrogen fixation. The nitrogenase is oxygen sensitive and therefore must be protected from oxygen, either in specialized cells (heterocysts) in the case of cyanobacteria, or leguminous nodules, or perhaps by an unusually high respiratory rate (*Azotobacter*), by binding to protective proteins, or by growth in an anaerobic environment (e.g., photosynthetic bacteria).

The oxidation of inorganic substances by oxygen is the source of energy for lithotrophs. Many lithotrophs are aerobic autotrophs. Electron transport takes place in the membrane and a Δp is created. The Δp is used to drive ATP synthesis and reversed electron flow so that NADH can be generated for biosynthesis.

Iron-oxidizing bacteria such as *Thiobacillus ferrooxidans* live in acid environments (around pH 2) where iron-sulfide minerals and oxygen are available. The bacteria maintain a ΔpH of around 4.5 units, by taking up cytoplasmic protons during the reduction of oxygen to water. The ΔpH drives the synthesis of ATP via a membrane ATP synthase.

Lithotrophic activities are of immense ecological significance because they are necessary for the recycling of inorganic nutrients through the biosphere. For example, consider the nitrogen cycle. All organisms use ammonia and nitrate, or amino acids, as the source of nitrogen. These are called "fixed" forms of nitrogen. Approximately 97% of all the nitrogen incorporated in living tissue comes from fixed nitrogen. However, the majority of the nitrogen on this planet is in the form of N_2 and unavailable to most organisms. Furthermore, due to the denitrification activities of bacteria living anaerobically, there is a constant drain of fixed nitrogen from the biosphere. That is to say, living systems' eventually oxidize reduced forms of nitrogen to nitrate, which is then reduced to nitrogen gas by the denitrifying bacteria. Therefore, we all depend upon the prokaryotes that reduce N_2 for our supply of fixed nitrogen.

The sulfate-reducing bacteria are responsible for much of the H_2S produced, and feed sulfide to the photosynthetic sulfur bacteria and aerobic sulfur-oxidizing bacteria. The combined activities of the sulfate reducers and the sulfide oxidizers accounts for much of the elemental sulfur deposits. Hydrogen sulfide can also have deleterious effects. Since H_2S is toxic, it can occasionally produce harmful effects on fish, waterfowl, and even plants when the soil becomes anaerobic. The latter can occur in rice paddies. The H_2S can also cause corrosion of metal pipes in anaerobic soils and waters.

Study Questions

1. What are the major distinguishing features between assimilatory and dissimilatory nitrate reduction? And assimilatory and dissimilatory sulfate reduction?

2. During assimilatory sulfate reduction O-acetylserine reacts with H_2S to form cysteine. The O-acetylserine is formed from serine and acetyl-CoA. Write a reaction showing the chemical structures suggesting how O-acetylserine might be formed from serine and acetyl-CoA.

3. Clostridia can reduce nitrogen gas to ammonia using electrons derived from pyruvate (generated from glucose using the EMP pathway). The immediate reductant for the nitrogenase is reduced ferredoxin. How is reduced ferredoxin generated from pyruvate?

4. What is the minimum number of Fe^{2+} which must be oxidized by oxygen to Fe^{3+} in order to reduce one NAD^+? Solve this problem by focusing on electron volts.

5. The ΔpH drives the synthesis of ATP in the acidophilic iron-oxidizing bacteria. How do they maintain the ΔpH?

6. What are some of the mechanisms used by nitrogen fixers to protect the nitrogenase from oxygen?

REFERENCES AND NOTES

1. Lovley, D. R. 1991. Dissimilatory Fe(III) and Mn(IV) reduction. *Microbiol. Rev.* 55:259–287.

2. Some chemolithotrophs use CO_2 as the sole or major source of carbon and are called chemolithoautotrophs.

3. Most of the reduced intracellular sulfur is H_2S and HS^-, rather than S^{2-}. This is because the pK_1 of H_2S is 7.04, the pK_2 is 11.96, and the cytoplasmic pH is usually close to 7. For example, neutrophilic bacteria have a cytoplasmic pH of approximately 7.5.

4. Thioredoxin is used in other metabolic pathways as a reductant. For example, recall that thioredoxin also reduces the nucleoside diphosphates to the deoxynucleotides (Section 9.2.5).

5. Thauer has argued that free H_2 may not be the electron carrier for lactate oxidation by sulfate. Even at very low H_2 partial pressures the reduction of H^+ by lactate is thermodyamically unfavorable. Furthermore, it has been demonstrated that H_2 does not affect lactate oxidation by sulfate by *Desulfovibrio vulgaris*, and *Desulfovibrio sapovorans* growing on lactate plus sulfate does not contain hydrogenase. Reviewed in Thauer, R. K. 1989. Energy metabolism of sulfate-reducing bacteria, pp. 397–413. In: *Autotrophic Bacteria*. H. G. Schlegel and B. Bowien (eds.). Science Tech Publishers, Madison, WI.

6. Reviewed by Postgate, J. 1987. *Nitrogen Fixation*. Edward Arnold, London.

7. Reviewed in Stacey, G., R. H. Burris, and H. J. Evans (eds). 1992. *Biological Nitrogen Fixation*. Chapman and Hall, New York, London.

8. Reviewed in Dilworth, J. J. and A. R. Glenn (eds). 1991. *Biology and Biochemistry of Nitrogen Fixation*. Elsevier. Amsterdam, New York, Oxford, Tokyo.

9. Lobo, A. L., and S. H. Zinder. 1992. Nitrogen fixation by methanogenic bacteria, pp. 191–211. In: *Biological Nitrogen Fixation*. Stacey, G., Burris, H. R., and H. J. Evans (eds.). Chapman and Hall, New York.

10. Dean, D. R., J. T. Bolin, and L. Zheng. 1993. Nitrogenase metalloclusters: Structures, organization, and synthesis. *J. Bacteriol.* 175:6737–6744.

11. There are two types of metalloclusters in the MoFe protein. These are the FeMo cofactor and the P cluster. The P cluster consists of two Fe_4S_4 clusters. Each $\alpha\beta$ dimer has an MoFe cluster and a P cluster, making two copies of each type of cluster per tetramer. One model of electron flow proposes that the electron travels from the Fe_4S_4 cluster in the Fe protein in an ATP-dependent reaction to a P cluster in the MoFe protein. From the P cluster the electron moves to the FeMo

cofactor, and from there to N_2. The precise role of ATP in electron transfer is not known, although it has been shown to bind as the Mg^{2+} chelate to the Fe protein. It is presumed that binding and/or hydrolysis of MgATP causes conformational changes in the nitrogenase that facilitate electron transfer. For example, conformational changes in the proteins might alter the environment and redox potential of the metal clusters, or bring the Fe_4S_4 cluster in the Fe protein and the P cluster in the FeMo protein into closer proximity. Proposals as to how the MoFe protein might reduce N_2 are derived from model systems. The chemical reactivity of N_2 is increased when it binds to a transition metal, especially in a complex with other ligands, such as sulfur. This has been studied in model systems where chemically synthesized metal complexes containing molybdenum bound to N_2 are reduced with artificial reductants. It is thought that the N_2 becomes activated when it binds to the molybdenum in the FeMoco cluster in nitrogenase and, while bound it is reduced to ammonia.

12. Eady, R. R. 1992. The dinitrogen-fixing bacteria, pp. 535–553. In: *The Prokaryotes*, Vol. I. A. Balows, H. G. Trüper, M. Dworkin, W. Harder and K.-H. Schleifer (eds.). Springer-Verlag, Berlin.

13. Yates, M. G., and F. O. Cambell. 1989. The role of oxygen and hydrogen in nitrogen fixation, pp. 383–416. In: *SGM Symposium*, Vol. 42, *The Nitrogen and Sulphur Cycles*. J. A. Cole and S. Ferguson (eds.). Cambridge University Press, Cambridge.

14. Haselkorn, R., and W. J. Buikema. 1992. Nitrogen fixation in cyanobacteria, pp. 166–190. In: *Biological Nitrogen Fixation*. Stacey, G., Burris, R. H., and Evans, H. J. (eds.). Chapman and Hall, New York.

15. Kelly, D. P. 1990. Energetics of chemolithotrophs, pp. 479–503. In: *The Bacteria*, Vol. XII. Krulwich, T. A. (ed.), Academic Press, New York.

16. Hooper, A. B. 1989. Biochemistry of the nitrifiying lithoautotrophic bacteria, pp. 239–265. In: *Autotrophic Bacteria*. H. G. Schlegel and B. Bowien (eds.). Springer-Verlag, Berlin.

17. Nicholls, D. G., and S. J. Ferguson. 1992. *Bioenergetics 2*. Academic Press, London.

18. Kelly, D. P. 1989. Physiology and biochemistry of unicellular sulfur bacteria, pp. 193–217. In: *Autotrophic Bacteria*. H. G. Schlegel and B. Bowien (eds.). Springer-Verlag, Berlin.

19. Kelly, D. P., W.-P. Lu, and R. K. Poole. 1993. Cytochromes in *Thiobacillus tepidarius* and the respiratory chain involved in the oxidation of thiosulphate and tetrathionate. *Arch. Microbiol.* 160:87–95.

20. Ingledew, W. J. 1990. Acidophiles, pp. 33–54. In: Microbiology of Extreme Environments. C. Edwards (ed.), McGraw-Hill Publishing Company, New York.

21. Ingledew, W. J. 1982. *Thiobacillus ferrooxidans*: The bioenergetics of an acidophilic chemolithotroph. *Biochem. Biophys. Acta* 683:89–117.

22. The amount of energy available from the oxidation of Fe^{2+} by O_2 depends upon whether the oxygen is reduced on the periplasmic side of the membrane or on the cytoplasmic side. The $E_{m,2}$ of Fe^{3+}/Fe^{2+} is usually stated to be about 0.77–0.78 V. The $E_{m,7}$ of O_2/H_2O is 0.82 V The $E_{m,2}$ of O_2/H_2O is 1.12 V. If the oxygen were reduced in the cytoplasm, then $0.82 - 0.77$ or 0.05 eV would be available per electron. If the oxygen were reduced in the periplasm then, $1.12 - 0.77$ or 0.35 eV would be available per electron. However, as pointed out by Nichols and Ferguson, periplasmic reduction of oxygen with consumption of periplasmic protons would not be helpful because a Δp would not develop. (There would be no electrogenic electron flow and no differential consumption of protons between periplasm and cytoplasm.) Bringing protons to the periplasm from the cytoplasm in order to reduce oxygen would require energy to overcome the proton concentration gradient between periplasm and cytoplasm.

23. Ehrlich, H. L., Ingledew, J. W., and J. C. Salerno. 1991. Iron- and manganese-oxidizing bacteria, pp. 147–170. In: *Variations in Autotrophic Life*. Shively, J. M. and L. L. Barton (eds.). Academic Press, New York.

24. According to the chemiosmotic theory, the actual number of ATPs made will depend upon the ratio of protons translocated per electron to the number translocated by the ATP synthase [i.e., $(H^+/e^-)/H^+/ATP$] (see Section 4.5.2).

12

C$_1$ Metabolism

Many prokaryotes can grow on C$_1$ compounds as their sole source of carbon. Some common C$_1$ compounds that support growth are:

1. Carbon dioxide (CO$_2$)

2. Methane (CH$_4$)

3. Methanol (CH$_3$OH)

4. Methylamine (CH$_3$NH$_2$)

A few strictly anaerobic prokaryotes (archaea) that use carbon dioxide as a source of cell carbon also use it as an electron acceptor, reducing it to methane, deriving ATP from the process. These are the methanogens.

12.1 Carbon dioxide fixation systems

Prokaryotes that use CO$_2$ as the sole or major carbon source are called *autotrophs* and the pathway of CO$_2$ assimilation is called autotrophic CO$_2$ fixation.[1] There are three major autotrophic CO$_2$ fixation pathways in prokaryotes. These are the *Calvin cycle*, the *acetyl-CoA pathway*, and the *reductive tricarboxylic acid pathway*. The Calvin cycle is the most prominent of the autotrophic CO$_2$ fixation systems and is found in photosynthetic eukaryotes, most photosynthetic bacteria, cyanobacteria, and chemoautotrophs. It does not occur in the archaea or in certain obligately anaerobic or microaerophilic bacteria. Instead, the acetyl-CoA pathway and

the reductive tricarboxylic acid pathway are found in these organisms. The reductive tricarboxylic acid pathway occurs in the green photosynthetic bacteria *Chlorobium* (anaerobes), in *Hydrogenobacter* (microaerophilic), in *Desulfobacter* (anaerobes), and in the archaeons *Sulfolobus* and *Thermoproteus*. The acetyl-CoA pathway is more widespread and is found in methanogenic archaea, some sulfate-reducing bacteria, and in facultative heterotrophs that synthesize acetic acid from CO$_2$ during fermentation. The latter bacteria are called *acetogens*.

12.1.1 The Calvin cycle

The Calvin cycle[2] is a pathway for making phosphoglyceraldehyde completely from CO$_2$. The pathway operates in the chloroplasts of plants and algae. However, in bacteria where there are no similar organelles, the Calvin cycle is found in the cytosol, alongside many other pathways (glycolysis, Entner–Doudoroff, pentose phosphate, citric acid cycle, and so on).

Summing the reactions of the Calvin cycle

During the Calvin cycle, three CO$_2$ molecules are reduced to phosphoglyceraldehyde, which is at the oxidation level of glyceraldehyde (C$_3$H$_6$O$_3$). Twelve electrons are required and these are provided by six NAD(P)H. [Recall that each NAD(P)H carries a hydride ion (i.e., one proton and two electrons) (Section 8.3)]. However, NAD(P)H is not a sufficiently strong reductant to reduce CO_2 without an additional source of energy. Each reduction thus requires an ATP. Therefore, six ATPs are required for the six reductions. However, the overall reaction, shown below, indicates that nine ATPs are required. The three extra ATPs are used to recycle an intermediate, ribulose-1,5-bisphosphate (RuBP) that is catalytically required for the pathway.

$$3CO_2 + 9ATP + 6NAD(P)H + 6H^+ + 5H_2O$$

$$\longrightarrow phosphoglyceraldehyde + 9ADP$$

$$+ 8P_i + 6NAD(P)^+$$

The Calvin cycle can be divided into two stages

The Calvin cycle is a complicated pathway but it will be familiar because it resembles the pentose phosphate pathway in certain key reactions (Section 8.4). It is convenient to think of the Calvin cycle as occurring in two stages:

1. Stage one is a reductive carboxylation of RuBP to form phosphoglyceraldehyde (PGALD).

2. Stage two consists of sugar rearrangements regenerating RuBP using some of the phosphoglyceraldehyde produced in stage one.

Stage 1

The first reaction is the carboxylation of RuBP forming two moles of 3-phosphoglycerate. The enzyme is called RuBP carboxylase or ribulose-1,5-bisphosphate carboxylase ("rubisco"). The reaction (and its probable mechanism) is shown in Fig. 12.1. An enolate anion probably forms that is carboxylated on the C2. Hydrolysis of the intermediate yields two moles of 3-phosphoglycerate (PGA).

The PGA is then reduced to PGALD via 1,3 bisphosphoglycerate (BPGA). These reactions also take place in glycolysis. The reactions of stage 1 can be summarized as follows:

$$3\,CO_2 + 3\,RuBP + 3\,H_2O \longrightarrow 6\,PGA$$

$$6\,PGA + 6\,ATP \longrightarrow 6\,BPGA + 6\,ADP$$

$$6\,BPGA + 6NAD(P)H + 6\,H^+ \longrightarrow 6\,PGALD$$

$$+ 6\,NAD(P)^+ + 6\,P_i$$

$$\overline{}$$

$$3\,CO_2 + 3RuBP + 3\,H_2O + 6\,ATP + 6\,NAD(P)H$$

$$+ 6\,H^+ \longrightarrow 6\,PGALD + 6\,ADP + 6\,P_i$$

$$+ 6\,NAD(P)^+$$

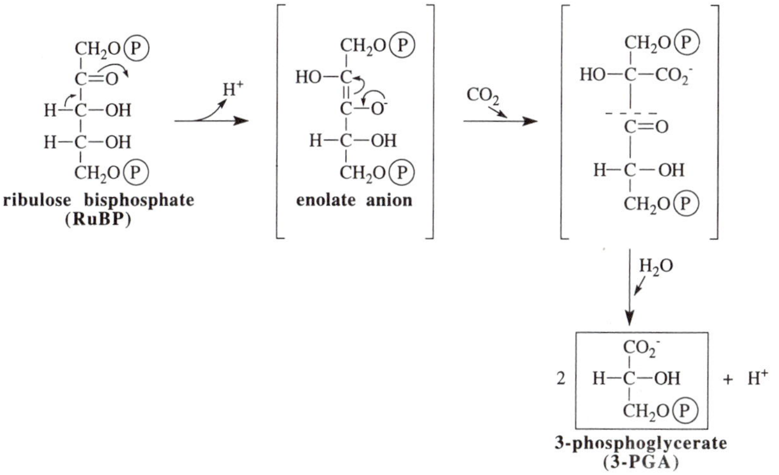

Fig. 12.1 Carboxylation of ribulose bisphosphate: the "rubisco" reaction. One possible mechanism is that the carbonyl group in RuBP attracts electrons resulting in the dissociation of a hydrogen from the carbon on the *cis*-HCOH group. Electrons then shift to form the enolate anion which becomes carboxylated at the C2. Hydrolysis of the carboxylated intermediate yields two 3-PGA molecules.

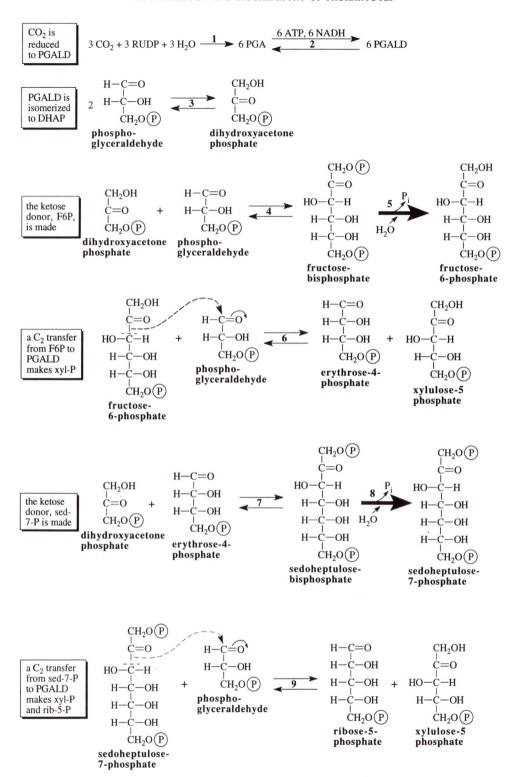

Fig. 12.2 The Calvin cycle. Enzymes: (1) ribulose-1,5-bisphosphate carboxylase; (2) 3-phosphogly-cerate kinase and triosephosphate dehydrogenase; (3) triosephosphate isomerase; (4) fructose-1,6-bisphosphate aldolase; (5) fructose-1,6-bisphosphate phosphatase; (6 and 9) transketolase; (7) sedoheptulo-se-1,7-bisphosphate aldolase; (8) sedoheptulose-1,7-bisphosphatase; (10) phosphopentose epimerase; (11) ribose phosphate isomerase; (12) phosphoribulokinase.

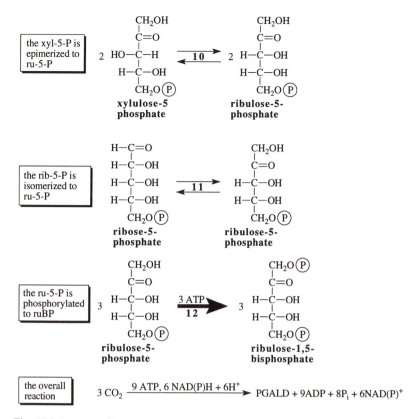

Fig. 12.2 *(continued)*

Stage 1 is summarized as reactions 1 and 2 in Fig. 12.2.

Stage 2

The whole point of the sugar rearrangements of stage 2 is to regenerate three RuBPs from five of the six phosphoglyceraldehydes made in stage 1. For reference, refer to Fig. 12.2. (The Calvin cycle is summarized without chemical structures in the discussion below. See "The relationship of the Calvin cycle to glycolysis and the pentose phosphate pathway".) Since RuBP has five carbons and PGALD has three carbons, a two-carbon fragment must be transferred to phosphoglyceraldehyde. That is to say, a transketolase reaction is required. The C_2 donor is fructose-6-phosphate which is made from PGALD (reactions 3–5, Fig. 12.2). The synthesis of fructose-6-phosphate from PGALD uses glycolytic enzymes and fructose-1,6-

bisphosphate phosphatase. One molecule of PGALD first isomerizes to dihydroxyacetone phosphate which condenses with a second molecule of PGALD to form fructose-1,6-bisphosphate. The fructose-1,6-bisphosphate is then dephosphorylated to fructose-6-phosphate by the phosphatase (reaction 5). The fructose-6-phosphate then donates a two-carbon fragment in a transketolase reaction to PGALD forming erythrose-4-phosphate and xylulose-5-phosphate (reaction 6).

The xylulose-5-phosphate is isomerized to ribulose-5-phosphate (reaction 10). The erythrose-4-phosphate is condensed with second dihydroxyacetone phosphate to form a seven-carbon ketose diphosphate, sedoheptulose-1,7-bisphosphate (reaction 7). The formation of sedoheptulose-1,7-bisphosphate is analogous to the condensation of dihydroxyacetone phosphate with PGALD to form fructose-1,6-bisphosphate, but is catalyzed by a different aldolase, sedoheptulose-

1,7-bisphosphate aldolase. (However, as discussed later, the bacterial sedoheptulose-1,7-bisphosphate aldolase and the fructose-1,6-bisphosphate aldolase may actually be a single bifunctional enzyme.) A phosphatase hydrolytically removes the phosphate from the C1 to form sedoheptulose-7-phosphate (reaction 8). This is an irreversible reaction and prevents the pentose phosphates that are formed from sedoheptulose-7-phosphate in the next step from being converted back to phosphoglyceraldehyde. The sedoheptulose-7-phosphate is a two-carbon donor in a second transketolase reaction using the last of the phosphoglyceraldehyde as an acceptor (reaction 9). The products are the pentose phosphates, ribose-5-phosphate and xylulose-5-phosphate, both of which are isomerized to ribulose-5-phosphate (reactions 10 and 11). The ribulose-5-phosphate is then phosphorylated to ribulose-1,5-bisphosphate by a ribulose-5-phosphate kinase which is unique to the Calvin cycle (reaction 12). Thus, the pentose isomerase reactions and the transketolase reactions are the same in both the Calvin cycle and the pentose phosphate pathway. The reactions of stage 2 are summarized below.

$$2\,PGALD \longrightarrow 2\,DHAP$$

$$PGALD + DHAP \longrightarrow FBP$$

$$FBP + H_2O \longrightarrow F6P + P$$

$$F6P + PGALD \longrightarrow erythrose\text{-}4\text{-}P$$
$$+ xylulose\text{-}5\text{-}P$$

$$erythrose\text{-}4\text{-}P + DHAP \longrightarrow$$
$$sedoheptulose\text{-}1,7\text{-}bisphosphate$$

$$sedoheptulose\text{-}1,7\text{-}bisphosphate + H_2O \longrightarrow$$
$$sedoheptulose\text{-}7\text{-}P + P_i$$

$$sedoheptulose\text{-}7\text{-}P + PGALD \longrightarrow$$
$$ribose\text{-}5\text{-}P + xylulose\text{-}5\text{-}P$$

$$2\ xylulose\text{-}5\text{-}P \longrightarrow 2\ ribulose\text{-}5\text{-}P$$

$$ribose\text{-}5\text{-}P \longrightarrow ribulose\text{-}5\text{-}P$$

$$3\ ribulose\text{-}5\text{-}P + 3\,ATP \longrightarrow$$
$$3\ ribulose\text{-}1,5\text{-}bisphosphate + 3\,ADP$$

$$5\,PGALD + 2\,H_2O + 3\,ATP \longrightarrow$$
$$3\ ribulose\text{-}1,5\text{-}bisphosphate + 2P_i + 3\,ADP$$

Summing stages 1 and 2, yields,

$$3CO_2 + 9ATP + 6NAD(P)H + 6H^+ + 5H_2O$$
$$\longrightarrow PGALD + 9ADP + 8P_i + 6NAD(P)^+$$

The carbon balance

Complex pathways can be seen in simpler perspective by examining the carbon balance. The carbon balance for the Calvin cycle is:

$$3\,C_1 + 3\,C_5 \longrightarrow 6\,C_3$$

$$2\,C_3 \longrightarrow C6$$

$$C_6 + C_3 \longrightarrow C_4 + C_5$$

$$C_4 + C_3 \longrightarrow C_7$$

$$C_7 + C_3 \longrightarrow 2\,C_5$$

$$3\,C_1 \longrightarrow C3$$

The carbon balance for the Calvin cycle illustrates that the pathway produces one C_3 from three C_1 molecules.

The relationship of the Calvin cycle to glycolysis and the pentose phosphate pathway

Most of the reactions of the Calvin cycle also take place in glycolysis and the pentose phosphate pathway. In Fig. 12.3, reactions 1–12 are the Calvin cycle. Reaction 1 is the carboxylation of ribulose-1,5-bisphosphate to form 3-phosphoglycerate. This reaction is unique to the Calvin cycle. The 3-phosphoglycerate enters the glycolytic pathway. Reactions 2 through 6 are reactions that take place during the reversal of glycolysis, whereby 3-phosphoglycerate is transformed into fructose-6-phosphate (see Section 8.1.) The fructose-6-phosphate enters the pentose phosphate pathway at the level of the sugar rearrangement reactions (reaction 7) (Section 8.4). Reaction 7 is the transketolase reaction which forms erythrose-4-phosphate and xylulose-5-phosphate. Now the Calvin cycle diverges from the pentose phosphate path-

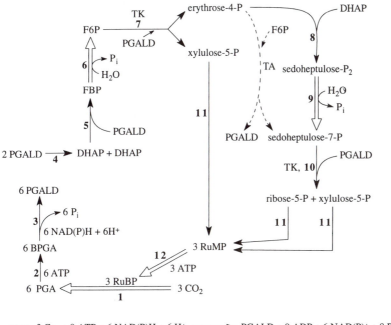

Fig. 12.3 Relationships between the Calvin cycle, glycolysis, and the pentose phosphate pathway. Reactions 1–12 are the Calvin cycle. Note that the only reactions unique to the Calvin cycle are reactions 1, 8, 9, and 12. Reaction 1 is catalyzed by ribulose-1,5-bisphosphate carboxylase.

Reactions 2-5 are glycolytic reactions. Reaction 6 is catalyzed by fructose-1,6-bisphosphatase. Reaction 7 is the transketolase reaction also found in the pentose-phosphate pathway. Reactions 8 and 9 are the sedoheptulose-bisphosphate aldolase and the sedoheptulose-1,7-bisphosphatase reactions. Reaction 10 is the transketolase reaction also found in the pentose-phosphate pathway. Reactions 11 are the pentose epimerase and isomerase reactions also present in the pentose phosphate pathway. Reaction 12 is the RuMP kinase reaction. The pentose phosphate pathway differs from the Calvin cycle only in that the pentose phosphate pathway synthesizes sedoheptulose-7-phosphate via a reversible transaldolase reaction (dotted lines) whereas the Calvin cycle synthesizes sedoheptulose-7-phosphate via an aldolase and phosphatase reaction (reactions 8 and 9). Since the phosphatase is irreversible, the Calvin cycle irreversibly converts phosphoglyceraldehyde to pentose phosphates.

way. Whereas the pentose phosphate pathway synthesizes sedoheptulose-7-phosphate from erythrose-4-phosphate via the reversible transaldolase reaction (TA), the Calvin cycle uses the aldolase (reaction 8) and the irreversible phosphatase (reaction 9) to synthesize sedoheptulose-7-phosphate. This has important consequences regarding the directionality of the Calvin cycle. Because the phosphatase (reaction 9) is an irreversible reaction, the Calvin cycle proceeds only in the direction of pentose phosphates, whereas

the pentose phosphate pathway is reversible from sedoheptulose-7-phosphate. (An irreversible Calvin cycle makes physiological sense since the sole purpose of the Calvin cycle is to regenerate ribulose-1,5-bisphosphate for the initial carboxylation reaction.) Once the sedoheptulose-7-phosphate is formed, the synthesis of ribulose-5-phosphate (reactions 10 and 11) takes place via the same reactions as in the pentose phosphate pathway. Reaction 12 of the Calvin cycle is the phosphorylation of ribulose-5-phosphate to form ribulose-1,5-bisphosphate (RuMP kinase), a second reaction that is unique to the Calvin cycle. The sedoheptulose bisphosphate aldolase and the sedoheptulose bisphosphate phosphatase also occur in the ribulose monophosphate pathway, which is a formaldehyde-fixing pathway that uses Calvin cycle reactions (Section 12.4.3).[3]

12.1.2 The acetyl-CoA pathway

Bacteria that use the acetyl-CoA pathway include methanogens, acetogenic bacteria, and most autotrophic sulfate-reducing bacteria.[4,5] This section presents a general summary of the acetyl-CoA pathway without describing the details of the individual reactions. The first part explains how acetyl-CoA is made from CO_2 and H_2, the second part explains how the acetyl-CoA is incorporated into cell material, and the third part summarizes how the acetyl-CoA pathway is used by methanogens for methanogenesis. Sections 12.1.3 and 12.1.4 describe the details of the acetyl-CoA pathway as they occur in acetogens and in methanogens.

Making acetyl-CoA from CO_2 using the acetyl-CoA pathway

The acetyl-CoA pathway can be viewed as occurring in four steps. The first step is a series of reactions that result in the reduction of CO_2 to a bound methyl group $[CH_3]$. The methyl group is then transferred to the enzyme *carbon monoxide dehydrogenase*. Carbon monoxide dehydrogenase also catalyzes the reduction of a second molecule of carbon dioxide to a carbonyl group (CO), which will become the carboxyl group of acetate. During autotrophic growth, the electrons are provided by hydrogen gas via a hydrogenase.

(1) $CO_2 + 3H_2 + H^+ \longrightarrow [CH_3] + 2H_2O$

(2) $CO_2 + H_2 \longrightarrow [CO] + H_2O$

The bound $[CH_3]$ and $[CO]$ are then condensed by carbon monoxide dehydrogenase

to form bound acetyl, $[CH_3CO]$.

(3) $[CO] + [CH_3] \longrightarrow [CH_3CO]$

The bound acetyl reacts with bound CoASH, [CoAS], to form acetyl-CoA which is released from the enzyme.

(4) $[CH_3CO] + [CoAS] \longrightarrow CH_3COSCoA$

The acetyl-CoA is converted to acetate via phosphotransacetylase and acetate kinase, (generating an ATP) and excreted, or it is assimilated into cell material. As mentioned, many anaerobic bacteria can grow autotrophically with the acetyl-CoA pathway using H_2 as the source of electrons. These bacteria must generate ATP from the Δp generated during the reduction of the CO_2 by H_2. The Δp is required for net ATP synthesis because the ATP generated during the conversion of acetyl-CoA to acetate is balanced by the ATP used for the formation of formyl-THF (Fig. 12.4).

Incorporating acetyl-CoA into cell material

For acetogens growing autotrophically, part of the acetyl-CoA that is produced must be incorporated into cell material. The glyoxylate cycle, which is generally associated with aerobic metabolism, is not present in these organisms, and therefore a different pathway must operate. As shown below, the acetyl-CoA is reductively carboxylated to pyruvate via pyruvate synthase, a ferredoxin-linked enzyme. The pyruvate can be phosphorylated to phosphoenolpyruvate using PEP synthetase (reaction 6). The phosphoenolpyruvate is used for biosynthesis in the usual way. A more detailed discussion of acetyl-CoA assimilation in methanogens is given in Section 12.1.4.

Pyruvate synthase:

Acetyl-CoA + Fd(red) + CO_2

$\longrightarrow$ pyruvate + Fd(ox) + CoASH

PEP synthetase:

Pyruvate + ATP + $H_2O \longrightarrow PEP + AMP + P_i$

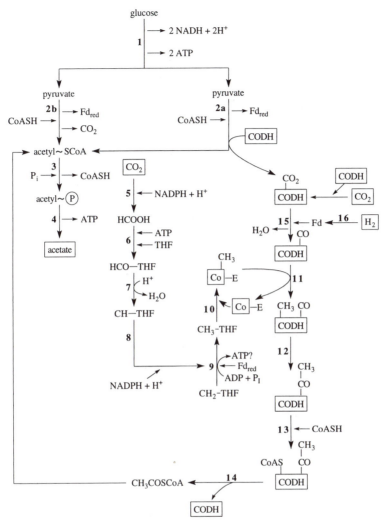

Fig. 12.4 The acetyl-CoA pathway in *Clostridium thermoaceticum*. During heterotrophic growth, the carboxyl group in some pyruvate molecules is directly transferred to CODH without being released as free CO_2. The evidence for this is that pyruvate is required in cell-free extracts for the synthesis of acetic acid from CH_3-THF even in the presence of CO_2. Furthermore, radioisotope experiments confirm that the carboxyl in acetate is derived from the carboxyl in pyruvate without going through free CO_2. However, the cells can be grown on CO_2 and H_2. Under these circumstances, the CODH reduces CO_2 to CO-CODH using hydrogenase and electrons from H_2 (reactions 15 and 16). Enzymes: (1) glycolytic; (2) pyruvate:ferredoxin oxidoreductase; (3) phosphotransacetylase; (4) acetate kinase; (5) formate dehydrogenase; (6) formyltetrahydrofolate (HCO-THF) synthetase; (7) methenyltetrahydrofolate (CH-THF) cyclohydrolase; (8) methylenetetrahydrofolate (CH₂-THF) dehydrogenase; (9) methylenetetrahydrofolate reductase; (10) methyltransferase; (11) corrinoid enzyme; (12–15) carbon monoxide dehydrogenase (CODH); (16) hydrogenase.

12.1.3 The acetyl-CoA pathway in *Clostridium thermoaceticum*

The acetyl-CoA pathway was first investigated in *C. thermoaceticum*, an acetogenic bacterium that converts one mole of glucose to three moles of acetate.[6] Under these circumstances, the production of acetate is used as an electron sink during fermentation rather than for autotrophic growth. The pathway is illustrated in Fig. 12.4. Glucose is converted to pyruvate via the Embden–

Meyerhof–Parnas pathway (reactions 1). Two of the acetates are synthesized from the decarboxylation of pyruvate using pyruvate–ferredoxin oxidoreductase, phosphotrans-acetylase, and acetate kinase (reactions 2–4). The third acetate is synthesized from CO_2 using the acetyl-CoA pathway. In the acetyl-CoA pathway, one of the carbon dioxides that is removed from pyruvate is not set free but instead becomes bound to the enzyme *carbon monoxide dehydrogenase* (CODH), where it will be used for acetate synthesis (reaction 2a). This bound CO_2 will eventually become the carbonyl group in acetyl-CoA. The second pyuvate is decarboxylated to release free CO_2 (reaction 2b). In reactions 5–9, the free CO_2 is reduced to bound methyl. The free CO_2 first becomes reduced to formate (HCOOH) by formate dehydroge-nase (reaction 5). The formate is then attached to tetrahydrofolic acid (THF) in an ATP-dependent reaction to make formyl-THF (HCO-THF) (reaction 6). (See Section 9.2.4 for a discussion of tetrahydrofolate reactions.) The formyl-THF is then de-hydrated to form methenyl-THF (CH-THF) (reaction 7). The methenyl-THF is reduced by NADPH to methylene-THF (CH_2-THF) (reaction 8). The methylene-THF is reduced by ferredoxin to methyl-THF (CH_3-THF) (reaction 9), a reaction thought to be coupled to ATP formation by a mechanism involving Δp. The methyl group is then transferred to a corrinoid enzyme, [Co]-E, which transfers the methyl group to the CO-CODH (reac-tions 10 and 11).[7] The CODH makes the acetyl moiety, [CH_3CO] (reaction 12). The CODH also has a binding site for CoASH and synthesizes acetyl-CoA from the bound intermediates (reactions 13 and 14). The CODH can also bind free CO_2 and reduce it to the level of carbon monoxide (reaction 15).

Autotrophic growth

Following the entry of CO_2 at reactions 5 and 15 (Fig. 12.4), it can be seen that the acetyl-CoA pathway allows autotrophic growth on CO_2 and H_2. Carbon dioxide is reduced to [CH_3] and [CO] (reactions 5 to 10 and reaction 15). The [CH_3] and [CO] combine with CoASH to form acetyl-CoA

(reactions 11–14). During acetogenesis an ATP can be made via substrate-level phos-phorylation from acetyl-CoA (reactions 3 and 4). However, because an ATP is required to incorporate formate (reaction 6), there can be no net ATP synthesis via substrate-level phosphorylation alone. How do organisms using the acetyl-CoA pathway during auto-trophic growth make net ATP? It is necessary to postulate that ATP is made by a chemi-osmotic mechanism coupled to the reduc-tion of CO_2 to acetate. For example, there may be an electrogenic electron flow follow-ing periplasmic oxidation of H_2 or perhaps a quinone loop or a proton pump may be operating. (Mechanisms for generating a Δp are reviewed in Sections 4.6 and 4.7.4.) It appears that homoacetogenic clostridia do indeed create a Δp during the reduction of CO_2 to acetate and that the Δp drives the synthesis of ATP via a H^+-ATP synthase.[8] But not all acetogenic bacteria use a proton circuit. *Acetobacterium woodii* couples the reduction of CO_2 to acetyl-CoA to a primary sodium ion pump.[9] The sodium ion transloca-ting step has not been identified but is thought to be either the reduction of methylene-THF to methyl-THF (reaction 9) or some later step in the synthesis of acetyl-CoA because these are highly exergonic reactions. *A. woodii* also has a sodium ion dependent ATP synthase. Thus these bacteria rely on a sodium ion current to make ATP when growing on CO_2 and H_2. A similar situation exists in *Propioni-genium modestum* where a Na^+-translocating methylmalonyl-CoA decarboxylase generates a Na^+ electrochemical potential that drives the synthesis of ATP (Section 3.8.1).

Carbon monoxide dehydrogenase

It is clear that carbon monoxide dehydro-genase is a crucial enzyme for acetogenesis, for autotrophic growth in anaerobes, and for methanogenesis from acetate. Because of the importance of this enzyme, some of the evidence for its participation in synthesizing and degrading acetyl-CoA is described next.

The CODH catalyzes the breakage and formation of the carbon–carbon bond be-tween the methyl group [CH_3] and the carbonyl group [CO] in the acetyl moiety. Evidence for this is the following exchange reaction carried out by CODH:[10]

$$^{14}CH_3\text{-}CODH + CH_3COSCoA$$

$$\longrightarrow CH_3\text{-}CODH + {}^{14}CH_3COSCoA$$

For the above reaction, the CODH was methylated with CH_3I.

The CODH catalyzes the breakage and formation of the thioester bond between CoASH and the carbonyl group of the acetyl moiety. Evidence for this is the following reaction carried out by CODH.[11] The asterisk refers to radioactive CoASH.

$$CH_3COSCoA + {}^*CoASH$$

$$\longrightarrow CH_3COS^*CoA + CoASH$$

The following exchange has also been demonstrated to be catalyzed by CO dehydrogenase:[12]

$$[1\text{-}{}^{14}C]\ \text{Acetyl-CoA} + {}^{12}CO$$

$$\longrightarrow {}^{14}CO + [1\text{-}{}^{12}C]\ \text{acetyl-CoA}$$

The three exchange reactions described above mean that the CODH is capable of disassembling acetyl-CoA into its components [CH₃], [CO], and [CoASH], and resynthesizing the molecule. It is therefore sufficient for making acetyl-CoA from the bound components.

12.1.4 The acetyl-CoA pathway in methanogens

Acetyl-CoA synthesis from CO_2 and H_2
Among the archaea are a group called methanogens that can grow autotrophically on CO_2 and H_2 (Section 1.2). During autotrophic growth, the CO_2 is first incorporated into acetyl-CoA in a pathway very similar to the acetyl-CoA pathway that exists in the bacteria (e.g., in *C. thermoaceticum*).[13-16] In other words, one molecule of CO_2 is reduced to [CO] bound to CODH. A second molecule of CO_2 is reduced to bound methyl [CH₃], which is transferred to the CODH. Then the CODH makes acetyl-CoA from the bound CO, CH₃, and CoASH.

There are, however, differences between the acetyl-CoA pathway found in the methanogenic archaea and that in the bacteria. The differences have to do with the cofactors and the utilization of formate. For reference, the structures of the cofactors are drawn in Fig. 12.5. The differences between the two pathways are:

1. The carriers for the formyl and more reduced C_1 moieties in the methanogens is not tetrahydrofolic acid (THF) but rather tetrahydromethanopterin (THMP), a molecule resembling THF (see Fig. 12.5 for a comparison of the structures).

2. The methanogens do not reduce free CO_2 to formate nor do they incorporate formate into THMP to form formyl-THMP. What the methanogens do instead is fix CO_2 on to a C_1 carrier called methanofuran (MFR) and reduce it to formyl-MFR. The formyl group is then transferred to THMP. (Recall that the acetogenic bacteria reduce free CO_2 to formate and then incorporate the formate into formyl-THF.)

Figure 12.6 is a diagram showing how the acetyl-CoA pathway is thought to operate in methanogens. (The same pathway is shown in Fig. 12.7 but with the chemical structures drawn.) The CO_2 is first condensed with methanofuran (MFR) and reduced to formyl-MFR (reaction 1). The formyl group is then transferred to tetrahydromethanopterin (THMP), to form formyl-THMP (reaction 2). Then a dehydration produces methenyl-THMP (reaction 3). The methenyl-THMP is reduced to methylene-THMP (reaction 4), which is reduced to methyl-THMP (reaction 5). The pathway diverges at the methyl level. One branch synthesizes acetyl-CoA in reactions quite similar to the synthesis of acetyl-CoA in the bacteria (Fig. 12.4), and the other branch synthesizes methane. For the synthesis of acetyl-CoA, the methyl group is transferred to a corrinoid enzyme (Fig. 12.6, 12.7, reaction 9). A second molecule of CO_2 is reduced by carbon monoxide dehydrogenase (CODH) to bound carbon monoxide [CO] (reaction 14). The CODH then catalyzes the synthesis of acetyl-CoA from [CH₃], [CO], and CoASH (reactions 11–13).

255

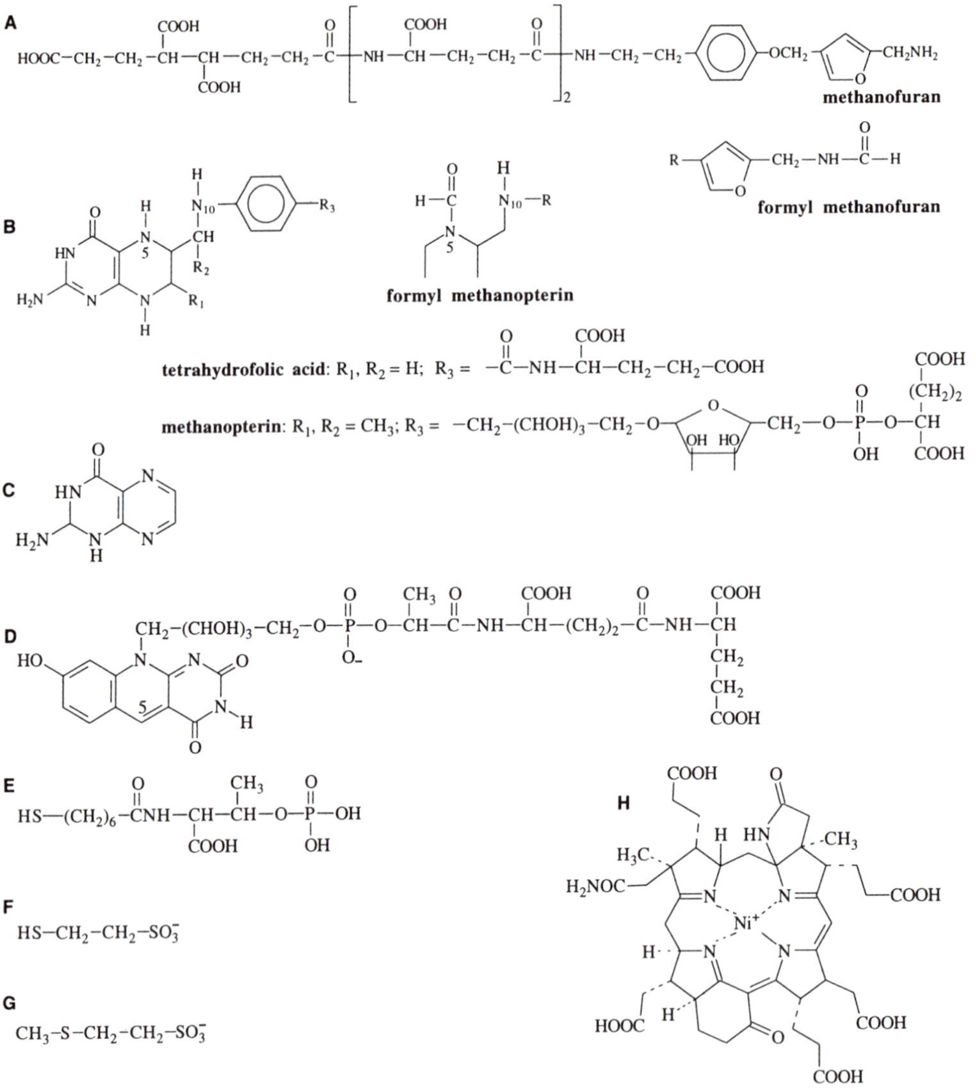

Fig. 12.5 Structures of coenzymes in the acetyl-CoA pathway and in methanogenesis. (A) Methanofuran (4-[N-(4,5,7-tricarboxyheptanoyl-γ-L-glutamyl-γ-L-glutamyl)-p-(β-aminoethyl) phenoxymethyl]-2-aminoethylfuran). The CO_2 becomes attached to the amino group on the furan and is reduced to the oxidation level of formic acid (formyl-methanofuran). (B) Tetrahydrofolic acid and methanopterin. They are both derivatives of pterin shown in (C). In formyl-methanopterin the formyl group is bound to N_5, whereas in tetrahydrofolic acid it is bound to N_{10}. (C). Pterin. (D) Oxidized coenzyme F_{420}. This is a 5-deazaflavin and carries two electrons but only one hydrogen (on the N_1 nitrogen). Compare this structure to flavins that have a nitrogen at position 5 and therefore carry two hydrogens (Section 4.2.1). (E) Factor B or HS-HTP (7-mercaptoheptanoylthreonine phosphate). (F) Coenzyme M (2-mercaptoethanesulfonic acid). (G) Methyl-coenzyme M. (H) Coenzyme F_{430}. This prosthetic group is a nickel-tetrapyrrole.

12.1.5 Methanogenesis from CO_2 and H_2

In addition to incorporating the [CH_3] into acetyl-CoA as described above, the methanogens can also reduce it to methane (CH_4) in a series of reactions not found among the bacteria. The production of methane yields ATP and is the only means of ATP formation for the methanogens. For methane synthesis the methyl group is transferred from CH_3-THMP to CoMSH to form CH_3-SCoM (Fig. 12.6, Fig. 12.7, reaction 6). The terminal

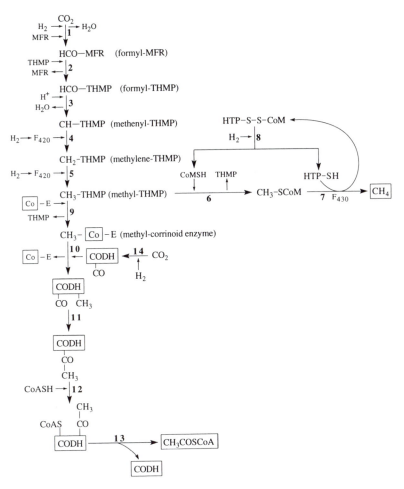

Fig. 12.6 The acetyl-CoA pathway in methanogens. There are two routes of entry of CO$_2$. In one sequence, CO$_2$ is bound to carbon monoxide dehydrogenase and is reduced to the oxidation level of carbon monoxide [CO] (reaction 14). The electron flow is from H$_2$ via a hydrogenase to ferredoxin to CODH. In a second sequence, CO$_2$ is attached to methanofuran (MF) and becomes reduced to the level of formic acid (formyl-MF) (reaction 1). The acceptor of electrons from H$_2$ in this reaction is not known. The formyl group is then transferred to a second carrier, tetrahydromethanopterin (THMP) to form formyl-THMP (reaction 2). The formyl-THMP is reduced to methyl-THMP (CH$_3$-THMP) (reactions 3–5). The CH$_3$-THMP donates the methyl group to a corrinoid enzyme for the synthesis of acetyl-CoA (reactions 9–13), or to CoMSH for the synthesis of methane (reactions 6–8). Enzymes: (1) formyl-MFR dehydrogenase (an iron–sulfur protein); (2) formyl-MFR:H$_4$MPT formyltransferase; (3) N^5,N^{10}-methenyl-H$_4$MPT cyclohydrolase; (4) N^5,N^{10}-methylene-H$_4$MPT dehydrogenase; (5) N^5,N^{10}-methylene-H$_4$MPT reductase; (6) N^5-methyl-H$_4$MPT:CoMSH methyltransferase; (7) methyl-S-CoM reductase; (8) heterodisulfide reductase; (9) methyltransferase; (10)–(13) carbon monoxide dehydrogenase. THMP is also written as H$_4$MPT.

reduction is catalyzed by the *methylreductase system* (reactions 7 and 8).[17] The methyl-reductase system has two components. One component is a methylreductase that reduces CH$_3$-SCoM to CH$_4$ and CoM-S-S-HTP. The electron donor for the methylreductase is HTP-SH.[18] A nickel-containing tetrapyrrole (F$_{430}$) is an electron carrier that is part of the methyl reductase. The second component is a FAD-containing heterodisulfide reductase that reduces CoM-S-S-HTP to CoMSH and HTP-SH.

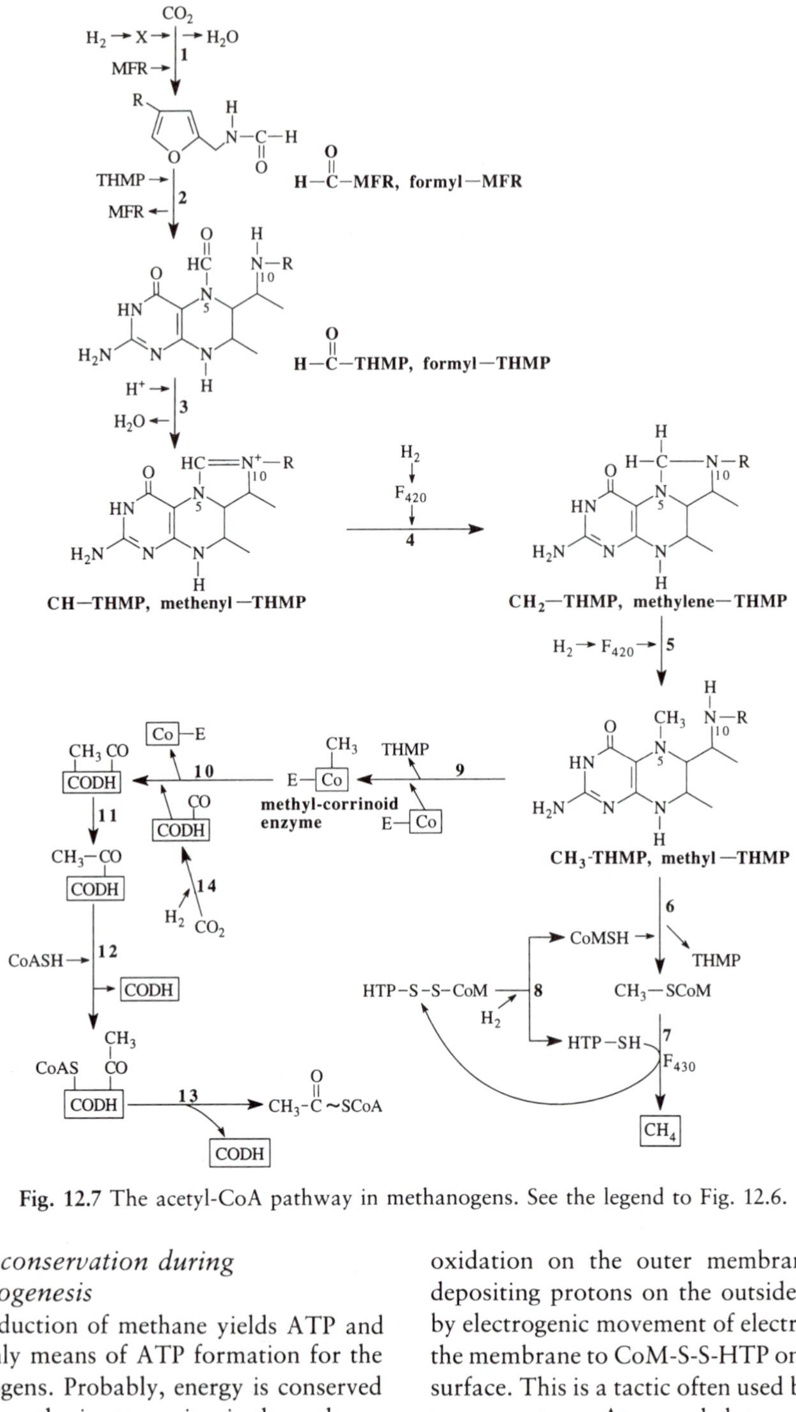

Fig. 12.7 The acetyl-CoA pathway in methanogens. See the legend to Fig. 12.6.

Energy conservation during methanogenesis

The production of methane yields ATP and is the only means of ATP formation for the methanogens. Probably, energy is conserved for ATP synthesis at two sites in the pathway of methanogenesis. It appears that electron transfer to CoM-S-S-HTP (Fig. 12.7, reaction 8) occurs in the membranes and is accompanied by the generation of a Δp which drives ATP synthesis via a H^+-translocating ATP synthase.[19] The mechanism of the generation of the Δp is not known, but could include H_2 oxidation on the outer membrane surface depositing protons on the outside, followed by electrogenic movement of electrons across the membrane to CoM-S-S-HTP on the inside surface. This is a tactic often used by bacteria to generate a Δp coupled to periplasmic oxidations. For example, recall the periplasmic oxidation of H_2 by *Wolinella succinogens* coupled to the cytoplasmic reduction of fumarate described in Section 4.7.3, or the reduction of sulfate by *Desulfovibrio* described in Section 11.2.2. There may also be a second energy-coupling site. The transfer of the methyl group from CH$_3$-THMP to CoMSH (Fig. 12.7, reaction 6) is an exergonic reaction accompanied by the extrusion of Na$^+$ in inverted vesicles, creating a primary sodium motive force. How might the sodium motive force be used? The sodium motive force probably drives ATP synthesis in whole cells, since it has been demonstrated to energize the synthesis of ATP in washed inverted vesicles made from *Methanosarcina mazei Gol*. There appear to be two different ATP synthases present, one coupled to protons and the other coupled to sodium ions.[20] It has been speculated that the sodium motive force created during methyl transfer to CoMSH may also be used to drive the formation of formyl-MFR from free CO$_2$ using H$_2$ as the reductant (Fig. 12.7, reaction 1). The latter reaction has a standard free energy change of approximately +8 kJ/mol.

Unique coenzymes in the Archaea

As discussed in Section 1.1.1, there are several unique aspects to the biochemistry of the Archaea. For example, there are several coenzymes in archaea that are not found in bacteria or eucarya. These are represented by five coenzymes used in methanogenesis. They are: methanofuran (formerly called CDR, carbon dioxide reduction enzyme), methanopterin, coenzyme M (CoSM), F$_{430}$, and HS-HTP, also known as factor B (Fig. 12.5). Coenzyme M, F$_{430}$, and HS-HTP are found only in methanogens. The other coenzymes occur in other archaea. The electron carrier, F$_{420}$, also exists in bacteria and eucarya.

12.1.6 Methanogenesis from acetate

Species of methanogens within the genera *Methanosarcina* and *Methanotrhix* can use acetate as a source of carbon and energy.[21] In fact, acetate accounts for approximately two-thirds of the biologically produced methane.[22] Those methanogens that convert acetate to methane and CO$_2$ carry out a dismutation of acetate in which the carbonyl group is oxidized to CO$_2$ and the electrons are used to reduce the methyl group to methane according to the following overall reaction:

$$CH_3CO_{2^-} + H^+ \longrightarrow CH_4 + CO_2$$

The acetate is first converted to acetyl-CoA. *Methanosarcina* uses acetate kinase and phosphotransacetylase to make acetyl-CoA from acetate.

Acetate kinase:

$$Acetate + ATP \longrightarrow acetyl\text{-}phosphate + ADP$$

Phosphotransacetylase:

$$Acetyl\text{-}phosphate + CoA$$
$$\longrightarrow acetyl\text{-}CoA + inorganic\ phosphate$$

Methanothrix uses acetyl-CoA synthetase to make acetyl-CoA.

Acetyl-CoA synthetase:

$$Acetate + ATP$$
$$\longrightarrow acetyl\text{-}AMP + pyrophosphate$$
$$Acetyl\text{-}AMP + CoA \longrightarrow acetyl\text{-}CoA + AMP$$

Reactions similar to those of the acetyl-CoA pathway in methanogens convert acetyl-CoA to methane and CO$_2$, deriving ATP from the process. Examine Fig. 12.6, beginning with acetyl-CoA. In reaction 12, acetyl-CoA combines with carbon monoxide dehydrogenase (CODH) which catalyzes the breakage of the C–S bond and the release of CoA. In reaction 11, the CODH catalyzes the breakage of the C–C bond forming the carbonyl [CO] and methyl [CH$_3$] groups. Then the carbonyl group is oxidized to CO$_2$ (reaction 13) and the methyl group is transferred to a corrinoid enzyme (reaction 10). (Note that the electrons removed from the carbonyl group are not released in hydrogen gas as implied in reaction 13 but are used to reduce the methyl group to methane.) The methyl group is transferred to tetrahydromethanopterin (THMP) (reaction 9).[23] (Acetate-grown cells of *Methanosarcina* species also contain large amounts of a derivative of THMP, tetrahydrosarcinapterin (H$_4$SPT) which has also been reported to serve as a methyl group carrier during methanogenesis from acetate.)[24] The methyl group is then transferred to coenzyme M (CoMSH) (reaction 6) and reduced to methane using the electrons derived from the oxidation of the carbonyl group (reaction 7). The electron donors for the reduction of the methyl group are HTP-SH and CH_3-S-CoM. Each of these contribute one electron from their respective sulfur atoms, and the oxidized product is the heterodisulfide, HTP-S-S-CoM. The reduction of HTP-S-S-CoM to CoMSH and HTP-SH (reaction 8) is coupled to the oxidation of the carbonyl group (reaction 13). The electron transport pathway that links the oxidation of the carbonyl group with the reduction of HTP-S-S-CoM is thought to generate ATP, presumably via a Δp.[25]

Methanogenesis from acetate is summarized below:

(7) $CH_3CO\text{-}CoA \longrightarrow [CH_3] + [CO] + CoA$

(8) $[CO] + H_2O \longrightarrow CO_2 + 2[H] +$

(9) $[CH_3] + 2[H] + ADP + P_i \longrightarrow CH_4 + ATP$

It was reported that ATP can also be made during the oxidation of [CO] to CO_2 and H_2.[26]

12.1.7 Incorporation of acetyl-CoA into cell carbon by methanogens

The glyoxylate cycle, one of the means of incorporating net acetyl-CoA into cell material, is not present in prokaryotes that use the acetyl-CoA pathway. How do they grow on the acetyl-CoA? The acetyl-CoA must be converted to phosphoenolpyruvate which feeds into an incomplete citric acid pathway and into gluconeogenesis. The enzymes to make phosphoenolpyruvate from acetyl-CoA are widespread in anaerobes. A suggested pathway for acetyl-CoA incorporation by methanogens is discussed below.

Figure 12.8 shows a proposed pathway for acetyl-CoA incorporation by methanogens.[27] The first reaction is the carboxylation of acetyl-CoA to form pyruvate, a reaction catalyzed by *pyruvate synthase*. The pyruvate is then phosphorylated to form phosphoenolpyruvate using *PEP synthase*. The phosphoenolpyruvate has two fates. Some of it enters the gluconeogenic pathway and some is carboxylated to oxaloacetate via *PEP carboxylase*. In *M. thermoautotrophicum* the oxaloacetate is reduced to α-ketoglutarate via an incomplete reductive citric acid pathway. That is to say, the reactions between citrate and α-ketoglutarate do not take place. *M. thermoautotrophicum* lacks citrate synthase and so must make its α-ketoglutarate this way. However, another methanogen, *M. barkeri*, does have citrate synthase and synthesizes α-ketoglutarate in an incomplete oxidative pathway via citrate and isocitrate. Thus, no methanogen seems to have a complete citric acid pathway.[28]

12.1.8 Using the acetyl-CoA pathway to oxidize acetate to CO_2 anaerobically

Several anaerobes can oxidize acetate to CO_2 without the involvement of the citric acid cycle. They do this by operating the acetyl-CoA pathway in reverse. These include some of the sulfate reducers that grow heterotrophically on acetate and oxidize it to CO_2.[29] To do this, they start with reactions 4 and 3 to make acetyl-CoA (Fig. 12.4) and then reverse reactions 14 through 11. This produces CH_3-[Co-E] and CO-[CODH]. In reactions 15, CO-[CODH] is oxidized to CO_2. The CH_3 is oxidized to CO_2 via reactions 10 through 5. In this way, both carbons in acetate are converted to CO_2. A Δp is made during electron transport. (However, it may be that the enzymes involved in the oxidative acetyl-CoA pathway are not identical to those of the reductive acetyl-CoA pathway that operates in autotrophs.) Some of the group II sulfate reducers (complete oxidizers) (e.g., *Desulfotomaculum acetoxidans*) use this pathway. Other group II sulfate-reducers use the reductive tricaraboxylic acid pathway to oxidize acetate (Section 12.1.6). Archaea sulfate reducers (e.g., *Archaeoglobus fulgidus*) may also use the acetyl-CoA pathway to oxidize acetate to CO_2 anaerobically.

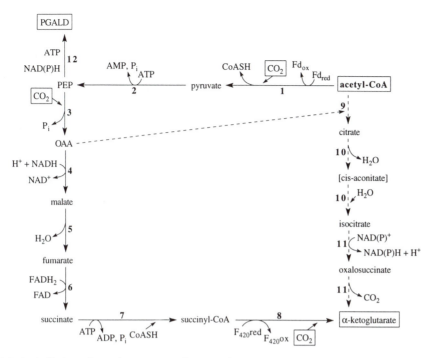

Fig. 12.8 Assimilation of acetyl-CoA into cell material in methanogens. The acetyl-CoA is carboxylated to form pyruvate which is then phosphorylated to make phosphoenolpyruvate (PEP). In *M. thermoautotrophicum*, the PEP is carboxylated to oxaloacetate (OAA) which is reduced to α-ketoglutarate, or it is reduced to phosphoglyceraldehyde. *M. barkeri* synthesizes α-ketoglutarate by a different pathway. That is to say, the oxaloacetate condenses with another acetyl-CoA to form citrate, and the citrate is then oxidized to α-ketoglutarate via aconitate, isocitrate, and oxalosuccinate (dotted lines). Enzymes: (1) pyruvate synthase; (2) PEP synthetase; (3) PEP carboxylase; (4) malate dehydrogenase; (5) fumarase; (6) fumarate reductase; (7) succinyl-CoA synthetase; (8) α-ketoglutarate synthase; (9) citrate synthase; (10) aconitase; (11) isocitrate dehydrogenase; (12) enolase, mutase, phosphoglycerate kinase, triosephosphate dehydrogenase.

12.1.9 The reductive tricarboxylic acid pathway

Bacteria that use this pathway[30] are strict anaerobes belonging to the genera *Desulfobacter* and *Chlorobium*, and the aerobic *Hydrogenobacter*.[31] In the reductive tricarboxylic acid pathway, phosphoenolpyruvate (PEP) is carboxylated to form oxaloacetate (OAA) using the enzyme *PEP carboxylase* (Fig. 12.9, reaction 1). The oxaloacetate is reduced to succinate which is derivatized to succinyl-CoA. The succinyl-CoA is carboxylated to α-ketoglutarate (the precursor to glutamate). This is a reductive carboxylation requiring reduced ferredoxin, and carried out by *α-ketoglutarate synthase*. The α-ketoglutarate is carboxylated to isocitrate which is converted to citrate. Therefore, this is really

a reversal of the citric acid pathway, substituting *fumarate reductase* for succinate dehydrogenase and *α-ketoglutarate synthase* for α-ketoglutarate dehydrogenase. Thus far, one mole of phosphoenolpyruvate has been converted to one mole of citrate. The citrate is split by a special enzyme, called *ATP-dependent citrate lyase*, to acetyl-CoA and oxaloacetate. (The ATP-dependent citrate lyase is found in eukaryotic cells but is rare in prokaryotes. Most prokaryotes use the citrate synthase reaction which proceeds only in the direction of citrate.) The oxaloacetate can be used for growth. The acetyl-CoA is used to regenerate the phosphoenolpyruvate as follows. The acetyl-CoA is reductively carboxylated to pyruvate using reduced ferredoxin and the enzyme *pyruvate synthase*. There is a thermodynamic reason why

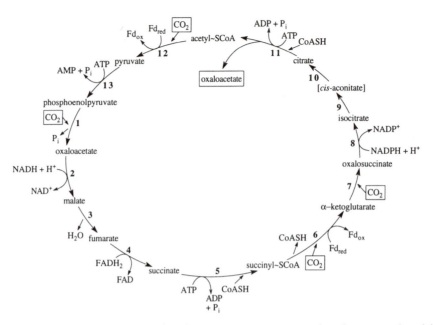

Fig. 12.9 The reductive tricarboxylic acid pathway. Enzymes: (1) PEP carboxylase; (2) malate dehydrogenase; (3) fumarase; (4) fumarate reductase; (5) succinyl-CoA synthetase (succinate thiokinase); (6) α-ketoglutarate synthase; (7, 8) isocitrate dehydrogenase; (9, 10) aconitase; (11) ATP-citrate lyase; (12) pyruvate synthase; (13) PEP synthetase.

reduced ferredoxin and not NADH is the electron donor. NADH is not a sufficiently strong reductant to reduce acetyl-CoA and CO_2 to pyruvate. Reduced ferredoxin has a potential more electronegative than that of NADH and is therefore a stronger reductant. The pyruvate is then converted to phosphoenolpyruvate using the enzyme *PEP synthetase*, thus regenerating the phosphoenolpyruvate. This discussion and Section 12.1.2 emphasize the importance of pyruvate synthase and PEP synthetase for anaerobic growth on acetate.

Note that to reverse the citric acid cycle, three new enzymes are required:

1. *Fumarate reductase* replaces succinate de-

hydrogenase. This commonly occurs in other prokaryotes that synthesize succinyl-CoA from oxaloacetate under anaerobic conditions.

2. *Alpha-ketoglutarate synthase* replaces the NAD^+-linked α-ketoglutarate dehydrogenase.

3. *ATP-dependent citrate lyase* replaces citrate synthase.

In addition, pyruvate synthase replaces pyruvate dehydrogenase, and PEP synthetase replaces pyruvate kinase.

Summary of the reductive tricarboxylic acid pathway:

$$CO_2 + PEP \longrightarrow OAA + P_i$$

$$OAA + NADH + H^+ + FADH_2 + Fd(red) + ATP + CO_2 \longrightarrow \alpha kg + NAD^+ + FAD + Fd(ox) + ADP + P_i$$

$$\alpha kg + NADPH + H^+ + CO_2 \longrightarrow citrate$$

$$Citrate + ATP + CoASH \longrightarrow OAA + acetyl\text{-}CoA + ADP + P_i$$

$$Acetyl\text{-}CoA + Fd(red) + CO_2 \longrightarrow pyruvate + Fd(ox)$$

$$Pyruvate + ATP \longrightarrow PEP + AMP + P_i$$

$$4CO_2 + 2NAD(P)H + 2H^+ + 2Fd(red) + FADH_2 + 2ATP \longrightarrow OAA + 2NAD(P)^+ + FAD + 2Fd(ox) + 2ADP + AMP + 4P_i$$

Anaerobic acetate oxidation by reversal of the reductive tricarboxylic acid pathway

As mentioned previously, some group II sulfate reducers reverse the reductive tricarboxylic acid pathway to oxidize acetate to CO_2.[32] An example is *Desulfobacter postgatei*. As shown in Fig. 12.9, by reversing reactions 11 through 6, acetyl-CoA is oxidized to CO_2. The oxaloacetate used in reaction 11 is regenerated in reactions 5 through 2. Acetyl-CoA is made from acetate by transferring the CoA from succinyl-CoA. Hence, reaction 5 is replaced by a CoA transferase.

12.2 Growth on C$_1$ compounds other than CO$_2$: The methylotrophs

Many aerobic bacteria can grow on compounds other than CO_2 that do not have carbon-carbon bonds. These bacteria are called *methylotrophs*.[33,34] Compounds used for methylotrophic growth include single carbon compounds such as methane (CH_4), methanol (CH_3OH), formaldehyde (HCHO), formate (HCOOH), and methylamine (CH_3NH_2), as well as multicarbon compounds without C–C bonds such as trimethylamine [$(CH_3)_3N$], dimethyl ether [$(CH_3)_2O$], and dimethyl carbonate ($CH_3OCOOCH_3$). These compounds appear in the natural habitat as a result of fermentations and breakdown of plant and animal products, and pesticides.[35]

The methylotrophs are divided according to whether they can also grow on multicarbon compounds. Those that cannot are called *obligate* methylotrophs. The obligate methylotrophs that grow on methanol or methylamine *but not on methane* are aerobic gram-negative bacteria, that belong to two genera, *Methylophilus* and *Methylobacillus*. (Some strains of *Methylophilus* can use glucose as a sole carbon and energy source.)[36] The obligate methylotrophs that grow on methane or methanol are called *methanotrophs*.[37] They are gram-negative and fall into five genera, *Methylomonas*, *Methylococcus*, *Methylobacter*, *Methylosinus*, and *Methylocystis*. All of the methanotrophs form extensive intracellular membranes and resting cells, either cysts or exospores. The intracellular

membranes are postulated to be involved in methane oxidation, since they are not present in methylotrophs that grow on methanol but not on methane.

Those methyltrophs that grow on *either* C$_1$ compounds (methanol or methylamine) or multicarbon compounds are called *facultative* methylotrophs. The facultative methylotrophs are found in many genera and consist of both gram-positive and gram-negative bacteria. They include species belonging to the genera *Bacillus*, *Acetobacter*, *Mycobacterium*, *Arthrobacter*, *Mycobacterium*, *Hyphomicrobium*, *Methylobacterium*, and *Nocardia*. Some species of *Mycobacterium* can grow on methane as well as on methanol or multicarbon compounds.

Methylotrophs assimilate the C$_1$ carbon source via either the ribulose-monophosphate (RuMP) pathway or the serine pathway. The RuMP pathway assimilates formaldehyde into cell material, whereas the serine pathway assimilates carbon dioxide *and* formaldehyde.[38] There also exist bacteria that grow on methanol and oxidize it to CO_2, which is assimilated via the ribulose bisphosphate (RuBP) pathway (Calvin cycle). These have been called "pseudomethylotrophs" or autotrophic methylotrophs.

12.2.1 Growth on methane

Methane is oxidized to CO_2 in a series of four reactions. The first oxidation is to methanol using a mixed-function oxidase called methane monooxygenase. There are two different methane monooxygenases, one in the membrane and one soluble. All methanotrophs have the membrane-bound enzyme. It is not yet known whether the soluble monooxygenase, which is the better characterized enzyme, is present in all methanotrophs.

$$CH_4 + NADH + H^+ + O_2$$
$$\longrightarrow CH_3OH + NAD^+ + H_2O$$

In the above reaction, one atom of oxygen is incorporated into methanol and the other

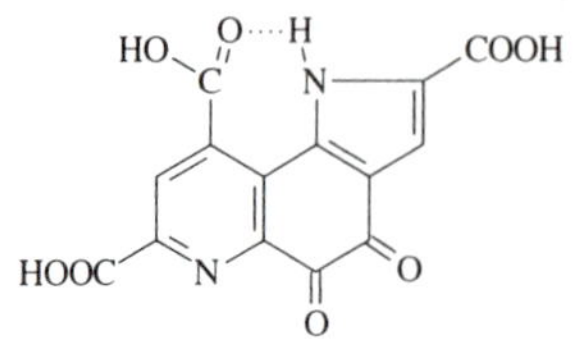

Fig. 12.10 Structure of pyrroloquinoline-quinone (PQQ). (From Gottschalk, G. 1986. *Bacterial Metabolism.* Springer-Verlag, Berlin.)

atom is reduced to water using NADH as the reductant. The second reaction is the oxidation of methanol to formaldehyde catalyzed by methanol dehydrogenase, which has as its prosthetic group a quinone called pyrroloquinoline quinone (PQQ) (Fig. 12.10).

$$CH_3OH + PQQ \longrightarrow HCHO + PQQH_2$$

In the third reaction the formaldehyde is oxidized to formate by formaldehyde dehydrogenase:

$$HCHO + NAD^+ + H_2O$$
$$\longrightarrow HCOOH + NADH + H^+$$

And the formate is oxidized to CO_2 by formate dehydrogenase:

$$HCOOH + NAD^+$$
$$\longrightarrow CO_2 + NADH + H^+$$

Because the formaldehyde and formate dehydrogenases are soluble enzymes that use NAD^+, they are probably located in the cytoplasm. Of the two NADHs produced, one is reutilized in the monooxygenase reaction and the second is fed into the respiratory chain in the usual way. However, methanol dehydrogenase is a periplasmic enzyme in gram-negative bacteria. It is thought that the methanol diffuses into the periplasm where it is oxidized by the dehydrogenase. The electrons are transferred to periplasmic cytochromes c that in turn transfer the electrons to a membrane cytochrome oxidase, which probably pumps protons out of the cell during inward flow of electrons to oxygen. A Δp is established by the inward flow of electrons, the outward pumping of protons by the cyt aa_3 oxidase, and the release of protons in the periplasm

and consumption in the cytoplasm. (See Section 4.7.2 for a description of electron transport and the generation of a Δp during methanol oxidation in *P. denitrificans*). A small number of gram-positive bacteria are also methylotrophic. However, the biochemistry of methanol oxidation may not be the same as in gram-negative bacteria.

Some of the formaldehyde produced during methane oxidation is incorporated into cell material rather than oxidized to CO_2. Depending upon the particular bacterium, one of two formaldehyde fixation pathways is used. The two pathways are the serine pathway and the ribulose-monophosphate cycle.

The serine pathway

The serine pathway produces acetyl-CoA from formaldehyde and carbon dioxide. The pathway is shown in Fig. 12.11. The formaldehyde is incorporated into glycine to form serine in a reaction catalyzed by serine hydroxymethylase (reaction 1). In this reaction, the formaldehyde is first attached to tetrahydrofolic acid to form methylene-THF. (Tetrahydrofolic reactions are described in Section 9.2.4). Methylene-THF then donates the C1 unit to glycine to form serine, regenerating THF. The serine is then converted to hydroxypyruvate via a transaminase which aminates glyoxylate, regenerating glycine (reaction 2). Hydroxypyruvate is reduced to glycerate (reaction 3) which is phosphorylated to 3-phosphoglycerate (reaction 4). The 3-phosphoglycerate is then converted to 2-phosphoglycerate (reaction 5) which is dehydrated to phosphoenolpyruvate (reaction 6). The phosphoenolpyruvate is carboxylated to oxaloacetate (reaction 7). The oxaloacetate is reduced to malate (reaction 8). Malyl-CoA synthetase then converts malate to malyl-CoA (reaction 9), which is split by malyl-CoA lyase to acetyl-Coa and glyoxylate (reaction 10). Thus, the glyoxylate is regenerated and the product is acetyl-CoA. So far, the following has happened:

$$HCHO + CO_2 + 2\,NADH + 2\,H^+$$
$$+ 2\,ATP + CoASH \longrightarrow$$
$$acetyl\text{-}CoA + 2\,NAD^+ + 2\,ADP + 2\,P_i$$

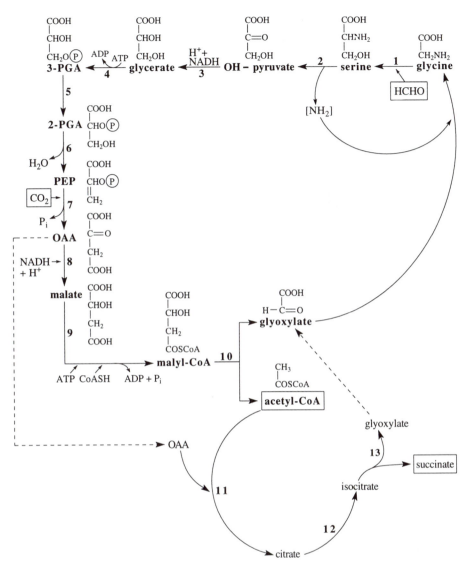

Fig. 12.11 The serine-isocitrate lyase pathway. Enzymes: (1) serine hydroxymethylase; (2) trans-aminase; (3) hydroxypyruvate reductase; (4) glycerate kinase; (5) phosphoglycerate mutase; (6) enolase; (7) PEP carboxylase; (8) malate dehydrogenase; (9) malyl-CoA synthetase; (10) malyl-CoA lyase; (11) citrate synthase; (12) *cis*-aconitase; (13) isocitrate lyase.

But how is acetyl-CoA incorporated into cell material? What happens in *some* methylotrophs is that the serine pathway goes around *a second time* to generate a *second* oxaloacetate. Then the second oxaloacetate condenses with the acetyl-CoA to form citrate (reaction 11). The citrate isomerizes to isocitrate (reaction 12). The isocitrate is cleaved by isocitrate lyase to form succinate and glyoxylate (reaction 13). The succinate can be assimilated into cell material via oxaloacetate and PEP (Section 8.12). The

second glyoxylate can be used for the second round of the serine pathway, which produces the second oxaloacetate. Therefore, the serine pathway and the isocitrate lyase pathway (called the serine–isocitrate lyase pathway) can be be described by the following overall reaction:

$$2\,HCHO + 2\,CO_2 + 3\,NADH + 3\,H^+$$
$$+ 3\,ATP \longrightarrow$$
$$succinate + 3\,NAD^+ + 3\,ADP$$
$$+ 3\,P_i + 2\,H_2O$$

Alternatively, the succinate can be converted to the second oxaloacetate via fumarate and malate, and the second glyoxylate can be converted via the serine pathway to 3-phosphoglycerate, which is assimilated into cell material.

However, only a few methylotrophs have isocitrate lyase, and it has not been established how the acetyl-CoA is converted to glyoxylate or otherwise assimilated in those strains lacking isocitrate lyase.

The reactions are summed below:

(1) $2 CH_2O + 2$ glycine $\longrightarrow 2$ serine

(2) 2 serine $+ 2$ glyoxylate $\longrightarrow 2$ glycine $+ 2$ hydroxypyruvate

(3) 2 hydroxypyruvate $+ 2 NADH + 2H^+$ $\longrightarrow 2$ glycerate $+ 2 NAD^+$

(4) 2 glycerate $+ 2 ATP \longrightarrow 2$ 3-PGA $+ 2 ADP + 2 P_i$

(5) 2 3-PGA $\longrightarrow 2$ 2-PGA

(6) 2 2-PGA $\longrightarrow 2 PEP + 2 H_2O$

(7) $2 PEP + 2 CO_2 \longrightarrow 2 OAA + 2 P_i$

(8) $OAA + NADH + H^+ \longrightarrow$ malate $+ NAD^+$

(9) malate $+ ATP + CoASH \longrightarrow$ malyl-CoA $+ ADP + P_i$

(10) malyl-CoA $\longrightarrow$ glyoxylate $+$ acetyl-CoA

(11) acetyl-CoA $+ OAA \longrightarrow$ citrate

(12) citrate $\longrightarrow$ isocitrate

(13 isocitrate $\longrightarrow$ glyoxylate $+$ succinate

$2 CH_2O + 2 CO_2 + 3 NADH + 3H^+ + 3 ATP$
$\longrightarrow$ succinate $+ 3 NAD^+ + 3 ADP$
$+ 3 P_i + 2 H_2O$

The ribulose-monophosphate cycle

There are several methanotrophs that use the ribulose-monophosphate (RuMP) pathway[39] instead of the serine pathway for formaldehyde assimilation. The pathway is shown in Fig. 12.12. It is convenient to divide the RuMP pathway into three stages. Stage 1 begins with condensation of formaldehyde with ribulose-5-phosphate to form hexulose-6-phosphate (reaction 1). The reaction is catalyzed by hexulose phosphate synthase. The hexulose phosphate is then isomerized to fructose-6-phosphate by hexulose phosphate isomerase (reaction 2). In order to synthesize a three-carbon compound (e.g., dihydroxyacetone phosphate or pyruvate) from formaldehyde (a one-carbon compound), stage 1 must be repeated three times.

In stage 2, one of the three fructose-6-phosphates is split into two C_3 compounds. There are two pathways for the cleavage, depending upon the organism. In one pathway, fructose-6-phosphate is phosphorylated to fructose-1,6-bisphosphate, which is cleaved via the fructose-1,6-bisphosphate aldolase to phosphoglyceraldehyde and dihydroxyacetone phosphate. This is called the FBPA pathway, for fructose bisphosphate aldolase. If this route is taken, then the net product of the pathway is dihydroxyacetone phosphate. In the second pathway, fructose-6-phosphate is isomerized to glucose-6-phosphate, which is oxidatively cleaved to phosphoglyceraldehyde and pyruvate using the enzymes of the Entner–Doudoroff pathway (Section 8.5). This is called the KDPGA pathway, for the 2-keto-3-deoxy-6-phosphogluconate aldolase. If this route is taken, then the net product of the pathway is pyruvate. Both pathways also produce phosphoglyceraldehyde.

Stage 3 is a sugar rearrangement stage during which the phosphoglyceraldehyde produced in stage 2 and two fructose-6-phosphates produced in stage 1 are used to regenerate the three ribulose-5-phosphates. In some bacteria the rearrangements take place using the pentose phosphate cycle enzymes (called the TA pathway, for transaldolase), whereas in other bacteria the rearrangements use the enzymes of the closely related Calvin cycle (called the SBPase pathway, for sedo-

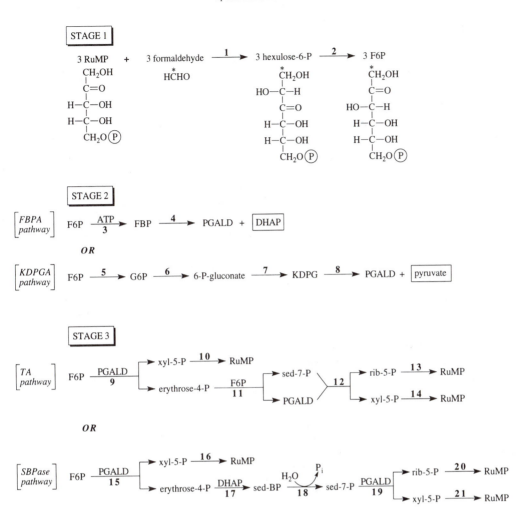

Fig. 12.12 The ribulose-monophosphate cycle. Enzymes: (1) hexulose-6-phosphate synthetase; (2) hexulose-6-phosphate isomerase; (3) phosphofructokinase; (4) fructose-1,6-bisphosphate aldolase; (5) glucose phosphate isomerase; (6) glucose-6-phosphate dehydrogenase; (7) 6-phosphogluconate dehydratase; (8) 2-keto-3-deoxy-6-phosphogluconate aldolase; (9, 12, 15, 19) transketolase; (10, 14, 16, 21) ribulose-5-phosphate epimerase; (11) transaldolase; (13, 20) ribose-5-phosphate isomerase; (17) sedoheptulose-1,7-bisphosphate aldolase; (18) sedoheptulose-1,7-bisphosphatase. 1,7- bisphosphatase. RuMP, ribulose-5-phosphate; F6P, fructose-6-phosphate; FBP, fructose-1,6-bisphosphate; PGALD, phosphoglyceraldehyde; DHAP, dihydroxyacetone phosphate; KDPG, 2-keto-3-deoxy-6-phosphogluconate; xyl-5-P, xylulose-5-phosphate; rib-5-P, ribose-5-phosphate; sed-BP, sedoheptulose-1,7-bisphosphate.

heptulose-1,7-bisphosphate aldolase). Taking into consideration the alternative pathways in stages 2 and 2, there are four different variations of the RuMP cycle that might occur in the methylotrophs. The obligate methanotrophs (*Methylococcus* and *Methylomonas*) and obligate methylotrophs (*Methylophilus*, *Methylobacillus*) that have been examined all use the KDPGA mode of cleavage and the TA pathway of sugar arrangement. The facultative methylotrophs thus far examined

use the FBPA mode of cleavage. Some use the SBPase sugar rearrangment pathway and some use the TA pathway. Use of the KDPGA and the SBPase pathways in combination has not yet been found.

12.3 Summary

There are three characterized CO$_2$ fixation pathways used for autotrophic growth: the Calvin cycle, the acetyl-CoA pathway, and the reductive tricarboxylic acid pathway. The Calvin cycle is the only one in the aerobic biosphere. It is present in green plants, algae, cyanobacteria, chemoautotrophs, and most photosynthetic bacteria. The Calvin cycle uses the transketolase and isomerization reactions of the pentose phosphate pathway and several glycolytic reactions to reduce CO_2 to phosphoglyceraldehdye. It bypasses the transaldolase reaction and synthesizes sedoheptulose-7-P via an aldolase and an irreversible phosphatase and proceeds only in the direction of pentose phosphate. The pathway has two unique enzymes, RuMP kinase and RuBP carboxylase.

The acetyl-CoA pathway is widespread among anaerobes, and occurs in both the archaea and the bacteria. However, the archaea use different coenzymes to carry the C_1 units. The pathway is used not only reductively for autotrophic growth but can also be used for anaerobic oxidation of acetate to CO_2 (e.g., by certain group II sulfate reducers). An analogous pathway is used by methanogens to form CH_4 and CO_2 from acetate. A key enzyme in the acetyl-CoA pathway is carbon monoxide dehydrogenase (CODH), which is capable of reducing CO_2 to the level of carbon monoxide and catalyzing its condensation with bound methyl and CoA to form acetyl-CoA. The bound methyl is made by reducing a second molecule of CO_2 via a separate series of enzymatic reactions and transferring the methyl group to the CODH. The reduction of CO_2 to bound methyl also takes place during methanogenesis. In methanogenesis, the bound methyl is further reduced to methane, an energy-yielding reaction, rather than transferred to CODH. Growth on acetyl-CoA requires the synthesis of phosphoenolpyruvate because the glyoxylate pathway is not present in these organisms. It involves a ferredoxin-linked carboxylation of acetyl-CoA to pyruvate catalyzed by pyruvate synthase, followed by the phosphorylation of pyruvate to phosphoenolpyruvate via the PEP synthetase. The incorporation of phosphoenolpyruvate into cell material takes place using reactions common to heterotrophs.

The third CO_2 fixation pathway has been found among the photosynthetic green sulfur bacteria, *Hydrogenobacter*, and *Desulfobacter* species. The pathway is a reductive citric acid pathway and synthesizes oxaloacetate from four moles of CO_2. There are two carboxylation reactions. One carboxylation is common, even among heterotrophs. It is the carboxylation of phosphoenolpyruvate to form oxaloacetate catalyzed by PEP carboxylase. The oxaloacetate is reduced to succinyl-CoA. The second carboxylation is found only in some strict anaerobes. It is the ferredoxin-linked carboxylation of succinyl-CoA to form α-ketoglutarate catalyzed by α-ketoglutarate synthase. The α-ketoglutarate is converted to citrate via a reversal of the reactions of the citric acid cycle, and the citrate is cleaved to oxaloacetate and acetyl-CoA via an ATP-dependent citrate lyase. The bacteria can incorporate the oxaloacetate into cell carbon. The phosphoenolpyruvate is regenerated from acetyl-CoA via its carboxylation to pyruvate using pyruvate synthase and phosphorylation of the latter to phosphoenolpyruvate using PEP synthetase. Thus, the bacteria are synthesizing oxaloacetate by carboxylating phosphoenolpyruvate, and regenerating the phosphoenolpyruvate via a reductive citric acid pathway. The pathway can also operate in the oxidative direction and oxidize acetate to CO_2. This occurs in some group II sulfate reducers.

There are many bacteria that grow aerobically on C_1 compounds such as methane or methanol. They oxidize the C_1 compounds to CO_2, deriving ATP from the respiratory pathway in the usual way, and incorporate the rest of the C_1 at the level of formaldehyde, into cell carbon. Bacteria that do this are called methylotrophs. A subclass of methylotrophs are those bacteria that are able to grow on methane. These are called methanotrophs. Two pathways for formaldehyde incorporation have been found in the bacteria. They are the serine–isocitrate lyase pathway, which incorporates both formaldehyde and carbon dioxide, and the ribulose-monophosphate cycle, which incorporates only formaldehyde.

The use of C_1 compounds is an important part of the carbon cycle. The methane produced by methanogens escapes into the aerobic atmosphere and is transformed back into carbon dioxide by aerobic methane oxidizers. The carbon dioxide is reduced to organic carbon in the aerobic environments by both photosynthetic eukaryotes and both photosynthetic and chemoautotrophic prokaryotes. It is also reduced to organic carbon anaerobically by photosynthetic prokaryotes, acetogens, and methanogens. Methanogenesis occurs in a variety of anaerobic environments including swamps and marshes, the rumen, anaerobic microenvironments in the soil, lake muds, rice paddies, and the intestine of termites. Indeed, close to 70% of the atmospheric methane is produced by methanogens. This is approximately 10^8 tons of methane per year. (The other 30% or so originates from abiogenic sources such as biomass burning and coal mines.) When one considers not only the vast amounts of methane produced by the methanogens, but also the fact that they are responsible for cycling much of the carbon into a gasous form for reutilization by the methane oxidizers and eventually the CO_2-fixing organisms, it is clear that they play a critical life-supporting role in the biosphere.

Study Questions

1. Contrast the number of ATPs required to fix three moles of CO_2 into phosphoglyceraldehyde using the Calvin cycle, the acetyl-CoA pathway, and the reductive carboxylic acid pathway.

2. Describe the role of carbon monoxide dehydrogenase in methane and carbon dioxide production from acetate, and in acetyl-CoA synthesis from CO_2.

3. What is the reason why anaerobes can carboxylate acetyl-CoA and succinyl-CoA but aerobes cannot?

NOTES AND REFERENCES

1. Autotrophic bacteria are those that use carbon dioxide as the sole source of carbon, except perhaps for some vitamins that may be required for growth. The autotrophs include the photoautotrophs and the chemoautotrophs. The prefix indicates the source of energy and the suffix the source of carbon. The *photo*autotrophs use light as their source of energy, and they include the plants, algae, and photosynthetic bacteria. The *chemo*autotrophs use inorganic chemicals as their source of energy (e.g., hydrogen gas, ammonia, nitrite, ferrous ion, and inorganic sulfur). Thus far, the only known chemoautotrophs are bacteria.

2. The properties of the Calvin cycle enzymes and their regulation are reviewed in Bowien, B. 1989. ·Molecular biology of carbon dioxide assimilation in aerobic chemolithotrophs, pp. 437–460. In: *Autotrophic Bacteria*. H. G. Schlegel and B. Bowien (eds). Science Tech Publishers, Madison, WI and Springer-Verlag, Berlin.

3. There is little information available about the bacterial aldolases and phosphatases that function in the Calvin cycle. However, it appears that the sedoheptulose-1,7-bisphosphatase and the fructose-1,6-bisphosphatase are a single bifunctional enzyme, capable of using either fructose-1,6-bisphosphate or sedoheptulose-1,7-bisphosphate as the substrate. In addition, the fructose-1,6-bisphosphate aldolase and the sedoheptulose-1,7-bisphosphate aldolase may also be a single bifunctional enzyme.

4. Reviewed by Wood, H. G., and L. G. Ljungdahl, 1991. Autotrophic character of the acetogenic bacteria, pp. 201–250. In: *Variations in Autotrophic Life*. J. M. Shively and L. L. Barton (eds). Academic Press, New York.

5. The "acetogenic bacteria" are defined as anaerobic bacteria that synthesize acetic acid solely from CO_2 and secrete the acetic acid into the media. Many of the acetogenic bacteria are facultative heterotrophs that use the acetyl-CoA pathway to reduce CO_2 to acetic acid as an electron sink, but these can also be grown autotrophically on CO_2 and H_2.

6. The early history is reviewed in Wood, H. G. 1985. Then and now. *Annu. Rev. Biochem.* 54:1–41.

7. Corrinoid enzymes are proteins containing vitamin B_{12} derivatives as the prosthetic group. Vitamin B_{12} is a cobalt-containing coenzyme.

8. Reviewed by Wood, H. G., and L. G. Ljungdahl, 1991. Autotrophic character of the acetogenic bacteria, pp. 201–250. In: *Variations in Autotrophic Life*. J. M. Shively and L. L. Barton (eds). Academic Press, New York.

9. Heise, R., V. Muller, and G. Gottschalk, 1993. Acetogenesis and ATP synthesis in *Acetobacterium woodii* are coupled via a transmembrane primary sodium ion gradient. *FEMS Microbiol. Lett.* 112:261–268.

10. Lu, W.-P., S. R. Harder, and S. W. Ragsdale, 1990. Controlled potential enzymology of methyl transfer reactions involved in acetyl-CoA synthesis by CO dehydrogenase and the corrinoid/iron–

sulfur protein from *Clostridium thermoaceticum*. *J. Biol. Chem.* **265**: 3124–3133.

11. Pezacka, E., and H. G. Wood. 1986. The autotrophic pathway of acetogenic bacteria: Role of CO dehydrogenase disulfide reductase. *J. Biol. Chem.* **261**: 1609–1615.

12. Ragsdale, S. W., and H. G. Wood. 1985. Acetate biosynthesis by acetogenic bacteria: Evidence that carbon monoxide dehydrogenase is the condensing enzyme that catalyzes the final steps of the synthesis. *J. Biol. Chem.* **260**:3970–977.

13. Fuchs, G. 1990. Alternatives to the Calvin cycle and the Krebs cycle in anaerobic bacteria: Pathways with carbonylation chemistry, pp. 13–20. In: *The Molecular Basis of Bacterial Metabolism.* G. Hauska and R. Thauer (eds). Springer-Verlag, Berlin.

14. Reviewed in Jetten, M. S. M., A. J. M. Stams, and A. J. B. Zehnder. 1992. Methanogenesis from acetate: A comparison of the acetate metabolism in *Methanothrix soehngenii* and *Methanosarcina* spp. *FEMS Microbiol. Rev.* **88**:181–198.

15. Wolfe, R. S. 1990. Novel coenzymes of archaebacteria, pp. 1–12. In: *The Molecular Basis of Bacterial Metabolism.* G. Hauska and R. Thauer (eds). Springer-Verlag, Berlin.

16. Weiss, D. S., and R. K. Thauer. 1993. Methanogenesis and the unity of biochemistry. *Cell.* **72**:819–822.

17. Olson, K. D., L. Chmurkowska-Cichowlas, C. W. McMahon, and R. S. Wolfe. 1992. Structural modifications and kinetic studies of the substrates involved in the final step of methane formation in *Methanobacterium thermoautotrophicum. J. Bacteriol.* **174**:1007–1012.

18. HTP-SH is also known as factor B or coenzyme B. It is 7-mercaptoheptanolythreonine phosphate.

19. There are several reasons for suggesting that a Δp is generated during electron transfer to the mixed disulfide: (1) *M. Barkeri* creates a Δp during catabolism of acetate; (2) the membranes of *M. thermophila* and *M. barkeri* contain electron carriers, including cytochrome b and FeS proteins; (3) over 50% of the heterodisulfide reductase is in the membranes.

20. Burkhard, B., and V. Muller. 1994. $\Delta \tilde{\mu}_{Na^+}$ drives the synthesis of ATP via an $\Delta \tilde{\mu}_{Na^+}$-translocating F_1F_0-ATP synthase in membrane vesicles of the archaeon *Methanosarcina mazei* Go1. *J. Bacteriol.* **176**:2543–2550.

21. Ferry, J. G. 1992. Methane from acetate. *J. Bacteriol.* **174**:5489–5495.

22. The methanogens are paramount in recycling organic carbon into gaseous forms of carbon in anaerobic habitats (Section 13.4). Most of the methane that enters the aerobic atmosphere is oxidized to CO_2 by the aerobic methanotrophs (Section 12.2).

23. Fischer, R., and R. K. Thauer. 1989. Methyltetrahydromethanopterin as an intermediate in methanogenesis from acetate in *Methanosarcina barkeri. Arch. Microbiol.* **151**:459–465.

24. Grahame, D. A. 1991. Catalysis of acetyl-CoA cleavage and tetrahydrosarcinapterin methylation by a carbon monoxide dehydrogenase–corrinoid enzyme complex. *J. Biol. Chem.* **266**:22227–22233.

25. There are several reasons for suggesting that a Δp is generated during the redox reaction between the carbonyl group and the mixed disulfide: (1) *M. barkeri* creates a Δp during catabolism of acetate; (2) the membranes of *M. thermophila* and *M. barkeri* contain electron carriers, including cytochrome b and FeS proteins; and (3) over 50% of the heterodisulfide reductase is in the membranes.

26. Bott, M., B. Eikmanns, and R. K. Thauer. 1986. Coupling of carbon monoxide oxidation to CO_2 and H_2 with the phosphorylation of ADP in acetate-grown *Methanosarcina barkeri. Eur. J. of Biochem.* **159**:393–398.

27. Stupperich, E., and G. Fuchs. 1984. Autotrophic synthesis of activated acetic acid from two CO_2 in *Methanobacterium thermoautotrophicum.* I. Properties of the *in vitro* system. *Arch. Microbiol.* **139**:8–13.

28. Reviewed in Danson, M. J. 1988. Archaebacteria: The comparative enzymology of their central metabolic pathways, pp. 165–231. In: *Advances in Microbial Physiology*, Vol. 29. A. H. Rose and D. W. Tempest (eds). Academic Press, New York.

29. Hansen, T. A. 1993. Carbon metabolism of sulfate-reducing bacteria, pp. 21–40. In: *The Sulfate-Reducing Bacteria: Contemporary Perspectives.* J. M. Odom and R. Singleton, Jr. (eds). Springer-Verlag, New York.

30. Reviewed by Amesz, J. 1991. Green photosynthetic bacteria and heliobacteria, pp. 99–119. In: *Variations in Autotrophic Life.* J. M. Shively and L. L. Barton (eds). Academic Press, New York.

31. Desulfobacter is a group II sulfate reducing bacterium that grows on fatty acids, especially acetate, using sulfate as an electron acceptor and producing sulfide (Section 11.2.2). *Chlorobium* is a green sulfur photoautotroph (Section 5.3). *Hydrogenobacter* is an aerobic obligately chemolithotrophic hydrogen oxidizer.

32. Hansen, T. A. 1993. Carbon metabolism of sulfate-reducing bacteria, pp. 21–40. In: *The Sulfate-Reducing Bacteria: Contemporary Perspectives.* J. M. Odom and R. Singleton, Jr. (eds). Springer-Verlag, New York.

33. Reviewed in J. Colin Murrell and H. Dalton (eds.), 1992. *Methane and Methanol Utilizers.* Plenum Press, New York, London.

34. Reviewed in Lidstrom, M. E. 1992. The aerobic methylotrophic bacteria, pp. 431–445. In: *The Prokaryotes*, Vol. 1, 2nd ed. A. Balows, H. G. Truper, M. Dworkin, W. Harder, and K.-H. Schleifer. Springer-Verlag, Berlin.

35. Methanol appears in the environment as a result of the microbial breakdown of plant products with methoxy groups (e.g., pectins and lignins). Formate is a fermentation end product excreted by fermenting bacteria. Methylamines can result from the breakdown products of some pesticides, and certain other compounds including carnitine and lecithin derivatives. Carnitine is a trimethylamine derivative that is present in many organisms and in all animal tissues, especially muscle. It functions as a carrier for fatty acids across the mitochondrial membrane into the mitochondria where the fatty acids are oxidized. Lecithin is phosphatidylcholine, which is a phospholipid containing a trimethylamine group (Section 9.1.3).

36. Green, P. N. 1992. Taxonomy of methylotrophic bacteria, pp. 23–84. In: *Methane and Methanol Utilizers.* J. C. Murrell and H. Dalton (eds). Plenum Press, New York, London.

37. The methanotrophs are responsible for oxidizing about one-half of the methane produced by methanogens. The rest of the methane escapes into the atmosphere.

38. Most of the non-methanotrophs that are obligate methylotrophs use the RuMP pathway. Within the methanotrophs, some use the RuMP pathway and some the serine pathway.

39. Dijkhuizen, L., Levering, P. R., and G. E. De Vries. 1990. The physiology and biochemistry of aerobic methanol-utilizing gram-negative and gram-positive bacteria, pp. 149–181. In: *Methane and Methanol Utilizers. Biotechnology Handbooks.* Vol. 5. J. C. Murrell and H. Dalton (eds). Plenum Press, New York, London.

13

Fermentations

Growing in the numerous anaerobic niches in the biosphere (muds, sewage, swamps, and so on) are prokaryotes that can grow indefinitely in the complete absence of oxygen, a capability that is rare among eukaryotes. (Eukaryotes require oxygen to synthesize unsaturated fatty acids and sterols.) Anaerobically growing prokaryotes re-oxidize NADH and other reduced electron carriers either by anaerobic respiration (e.g., using nitrate, sulfate, or fumarate as the electron acceptor) (Chapters 4 and 11), or they carry out a fermentation.

A *fermentation* is defined as a pathway in which NADH (or some other reduced electron acceptor that is generated by oxidation reactions in the pathway) is re-oxidized by metabolites produced by the pathway. The redox reactions occur in the cytosol rather than in the membranes and ATP is produced via substrate-level phosphorylation.

It should not be concluded that fermentation occurs only among prokaryotes. Eukaryotic microorganisms (e.g., yeast) can live fermentatively, although as mentioned, oxygen is usually necessary unless the medium is supplemented with sterols and unsaturated fatty acids. Furthermore, certain animal cells are capable of fermentation, as for example, muscle cells and human red blood cells.

Fermentations are named after the major end products they generate. For example, yeasts carry out an ethanol fermentation, and muscle cells and red blood cells carry out a lactic acid fermentation. There are many different types of fermentations carried out by microorganisms. However, the carbohydrate fermentations can be grouped into six main classes. The six major classes of carbohydrate fermentation are *lactic, ethanol, butyric, mixed acid, propionic,* and *homoacetic*. A homoacetic fermentation of glucose is described in Section 12.1.3.

13.1 Oxygen toxicity

Many anaerobic prokaryotes (strict anaerobes) are killed by even small traces of oxygen. Strict anaerobes are killed by oxygen because toxic products of oxygen reduction accumulate in the cell. The toxic products are hydroxyl radical (OH), superoxide radical (O_2^-), and hydrogen peroxide (H_2O_2).

The superoxide radical forms because oxygen is reduced by single electron steps,

$$O_2 + e^- \longrightarrow O_2^-$$

Electron carriers such as flavoproteins, quinones, and iron–sulfur proteins release small amounts of superoxide. The hydroxyl radical and hydrogen peroxide are derived from the superoxide radical.

Aerobic and aerotolerant organisms do not accumulate superoxide radicals because they have an enzyme called *superoxide dismutase* that is missing in strict anaerobes. The superoxide dismutase catalyzes the following reaction:

$$O_2^- + O_2^- + 2H^+ \longrightarrow H_2O_2 + O_2$$

Notice that one superoxide radical transfers its extra electron to the second radical, which becomes reduced to hydrogen peroxide. Strict anaerobes also lack the enzyme that converts hydrogen peroxide to water and oxygen. That enzyme is catalase:

$$H_2O_2 + H_2O_2 \longrightarrow 2H_2O + O_2$$

Catalase catalyzes the transfer of two electrons from one hydrogen peroxide molecule to the second, oxidizing the first to oxygen and reducing the second to two molecules of water. Table 13.2 shows the distribution of catalase and superoxide dismutase in aerobes and anaerobes.

13.2 Energy conservation by anaerobic bacteria

An examination of mechanisms of energy conservation and ATP production in anaerobic bacteria reveals a variety of methods that were described in previous chapters. Some of these are reviewed in Fig. 13.1. Most fermenting bacteria make most or all of their ATP via substrate-level phosphoryla-

tions and they create a Δp (needed for solute transport, motility, and so on) by reversing the membrane-bound ATPase. Many anaerobic bacteria also generate a Δp by reducing fumarate. This is an anaerobic respiration that can occur during fermentative metabolism when fumarate is produced. During fumarate reduction, NADH dehydrogenases donate electrons to fumarate via menaquinone and a Δp is created, perhaps via a quinone loop. (See Section 4.6.1 for a discussion of quinone loops.) The production of a Δp by fumarate respiration in fermenting bacteria probably spares ATP that would normally be hydrolyzed to maintain the Δp. Some anaerobes (e.g., *Wolinella succinogenes*) carry out a periplasmic oxidation of electron donors, such as H_2 or formate in the case of *W. succinogenes*, in order to create a Δp. (See Section 4.7.4). There are several other means available to anaerobic bacteria for generating a protonmotive force or a sodium-motive force. Some anaerobes and facultative anaerobes are capable of creating an electrochemical ion gradient by symport of organic acids out of the cell with protons or sodium ions. The organic acids are produced during fermentation and the energy to create the electrochemical gradient is due to the concentration gradient of the excreted organic acid (high inside). This has been demonstrated for lactate excretion (proton symport) by the lactate bacteria, and for succinate excretion (sodium ion symport) by a rumen bacterium, *Selenomonas ruminantium*. (See

Table 13.1. Balancing an acetone–butanol fermentation

Substrate and products	Yield (mol/100 mol) substrate)	Carbon (mol)	O/R value	O/R value (mol/100 mol)	Available H	Available H (mol/100 mol)
Glucose	100	600	0	—	24	2,400
Butyrate	4	16	−2	−8	20	80
Acetate	14	28	0	—	8	112
CO$_2$	221	221	+2	+442	0	—
H$_2$	135	—	−1	−135	2	270
Ethanol	7	14	−2	−14	12	84
Butanol	56	224	−4	−224	24	1,344
Acetone	22	66	−2	−44	16	352
Total		569		−425, +442		2,242

Source: Gottschalk, G. 1986. *Bacterial Metabolism.* Springer-Verlag, Berlin.

Section 3.8.3 for a discussion of lactate and succinate excretion in symport with protons and sodium ions.) *Klebsiella pneumoniae* is capable of generating an electrochemical sodium ion gradient by using the energy released from the carboxylation of oxaloacetate to pyruvate (Section 3.8.1). The decarboxylase pumps Na^+ out of the cell. The sodium potential that is created is used to drive the uptake of oxaloacetate which is used as a carbon and energy source. *Oxalobacter formigenes* creates a proton potential by catalyzing an electrogenic anion exchange coupled to the decarboxylation of oxalate to formate (Section 3.8.2).

13.3 Electron sinks

A major problem which must be addressed during fermentation is what to do with the electrons removed during oxidations. For example, how is the NADH re-oxidized? Respiring organisms do not have this problem, since the electrons travel to an exogeneous electron acceptor (e.g., oxygen or nitrate). Since fermentations usually occur in the absence of an exogenously supplied electron acceptor, the fermentation pathways themselves must produce the electron acceptors for the electrons produced during the oxidations. The electron acceptors are called "electron sinks" because they dispose of the electrons removed during the oxidations and the reduced products are excreted into the medium. Consequently, fermentations are characterized by the excretion of large quantities of reduced organic compounds such as alcohols, organic acids, and solvents. Frequently, hydrogen gas is also produced since protons are used as electron acceptors by hydrogenases. Generally, one can view fermentations in the following way, where B is the electron sink:

$$AH_2 + NAD^+ + P_i + ADP$$
$$\longrightarrow B + NADH + H^+ + ATP$$

$$B + NADH + H^+$$
$$\longrightarrow BH_2 \text{ (excreted)} + NAD^+$$

and

Hydrogenase:

$$NADH + H^+ \longrightarrow H_2 + NAD^+$$

For example, there is the lactate fermentation:

$$Glucose + 2\,NAD^+ + 2\,ADP + 2\,P_i \longrightarrow$$
$$2\,pyruvate + 2\,NADH + 2\,H^+ + 2\,ATP$$

$$2\,NADH + 2\,H^+ + 2\,pyruvate \longrightarrow 2\text{ lactate}$$
$$+ 2\,NAD^+$$

$$Glucose + 2\,ADP + 2\,P_i \longrightarrow 2\text{ lactate} + 2\,ATP$$

Or, an ethanol fermentation:

$$Glucose + 2\,NAD^+ + 2\,ADP + 2\,P_i \longrightarrow$$
$$2\,pyruvate + 2\,NADH + 2\,H^+ + 2\,ATP$$

$$2\text{ pyruvate} \longrightarrow 2\text{ acetaldehyde} + 2\,CO_2$$

$$2\text{ acetaldehyde} + 2\,NADH + 2\,H^+ \longrightarrow$$
$$2\text{ ethanol} + 2\,NAD^+$$

$$Glucose + 2\,ADP + 2\,P_i \longrightarrow$$
$$2\text{ ethanol} + 2\,CO_2 + 2\,ATP$$

13.4 The anaerobic food chain

The fermentation of amino acids, carbohydrates, purines, pyrimidines, and so on to organic acids and alcohols (acetate, ethanol, butanol, propionate, succinate, butyrate, and so on) by prokaryotes, and the conversion of these fermentation end products to CO_2, CH_4, and H_2 by the combined action of several different types of bacteria, is called the anaerobic food chain (Fig. 13.2). It takes place in anaerobic environments such as in muds, the bottom of lakes, and sewage treatment plants. This is an important part of the carbon cycle and serves to regenerate

A   Substrate level phosphorylation

1. 1,3 BPGA + ADP $\longrightarrow$ 3-PGA + ATP
2. PEP + ADP $\longrightarrow$ pyruvate + ATP
3. acetyl-P + ADP $\longrightarrow$ acetate + ATP

B   Fumarate respiration

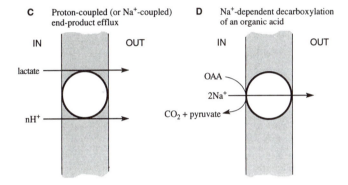

C   Proton-coupled (or Na^+-coupled)   **D** Na^+-dependent decarboxylation
       end-product efflux   of an organic acid

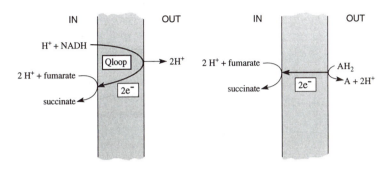

E   Electrogenic oxalate:
       formate exchange

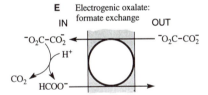

Fig. 13.1 Energy conservation in anaerobic bacteria. (A) Substrate-level phosphorylation: (1) the PGA kinase reaction; (2) the pyruvate kinase reaction; (3) the acetate kinase reaction. (B) Fumarate respiration. When the electron donor is NADH, a Q loop probably operates to translocate protons out of the cell. When the electron donor is periplasmic, proton translocation is not necessary. (C) Efflux of an organic acid coupled to protons or sodium ions, (e.g., the coupled efflux of protons and lactate by the lactate bacteria. (D) Decarboxylation of an organic acid coupled to Na^+ efflux (e.g., the decarboxylation of oxaloacetate by *Klebsiella*). (E) Electrogenic oxalate: formate exchange in *Oxalobacter*.

gaseous carbon (i.e., carbon dioxide and methane) which is reutilized by other microorganisms and plants throughout the biosphere. As illustrated in Fig. 13.2, the process can be viewed as occurring in three stages. Fermenters produce organic acids, alcohols, hydrogen gas, and carbon dioxide. These fermentation end products are oxidized to

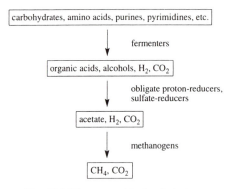

carbohydrates, amino acids, purines, pyrimidines, etc.

↓ fermenters

organic acids, alcohols, H_2, CO_2

↓ obligate proton-reducers, sulfate-reducers

acetate, H_2, CO_2

↓ methanogens

CH_4, CO_2

Fig. 13.2 The anaerobic food chain.

CO_2, H_2, and acetate by organisms that have been only partially identified. Finally, the methanogens grow on the acetate, H_2 and CO_2, converting these to CH_4 and CO_2.

13.4.1 Interspecies hydrogen transfer

Some anaerobic bacteria use protons as the major or sole electron sink. They include the *obligate proton-reducing acetogens* that oxidize butyrate, propionate, ethanol, and other compounds to acetate, H_2, and CO_2. The physiology of these organisms is not well understood. Probably the electrons travel from the organic substrate to an intermediate electron carrier such as NAD^+ to hydrogenase to H^+. Some of these oxidations are shown below.

$$CH_3CH_2CH_2COO^- + 2H_2O \longrightarrow$$
$$2CH_3COO^- + H^+ + 2H_2$$
$$\Delta G'_o = +48.1\,kJ$$
$$CH_3CH_2COO^- + 3H_2O \longrightarrow CH_3COO^-$$
$$+ HCO_3^- + H^+ + 3H_2$$
$$\Delta G'_o = +76.1\,kJ$$
$$CH_3CH_2OH + H_2O \longrightarrow CH_3COO^- + H^+$$
$$+ H_2$$
$$\Delta G'_o = +9.6\,kJ$$

Notice that the above reactions are thermodynamically unfavorable under standard conditions at pH 7. In fact, the proton-reducers live symbiotically with H_2 utilizers that keep the H_2 levels low and pull the reaction towards H_2 production. The H_2 utilizers are methanogens and sulfate reducers. (Under

conditions of sufficient sulfate, the sulfate reducers predominate over methanogens. The sulfate reducers also oxidize ethanol, lactate, and other organic acids to acetate.) The symbiotic relationship between the obligate proton reducers and the hydrogen utilizers is called a *syntrophic* association. It is also called *interspecies hydrogen transfer*. It should be pointed out that many other fermenting bacteria besides the obligate proton reducers have dehydrogenases that transfer electrons from NADH to protons and can use protons as an electron sink when grown in the presence of hydrogen gas utilizers. An example is *Ruminococcus albus*, which is discussed in Section 13.12. Hydrogenases are also coupled to the thermodynamically favored oxidation of reduced ferredoxin, as in pyruvate:ferredoxin oxidoreductase found in the clostridia and sulfate reducers (Section 7.3.2).

13.5 How to balance a fermentation

A written fermentation is said to be balanced when the hydrogens produced during the oxidations equal the hydrogens transferred to the fermentation end products. Only under these conditions is all of the NADH and reduced ferredoxin recycled to the oxidized forms. It is important to know whether a fermentation is balanced because if it is not, then the overall written reaction is incorrect. There are two methods used to balance fermentations, the O/R method and the available hydrogen method.

The O/R method

This is a bookkeeping method to keep track of the hydrogens. An oxidation/reduction (O/R) balance is computed as described later. One arbitrarily designates the O/R value for formaldehyde (CH_2O) and multiples thereof, $(CH_2O)_n$, as zero, and uses that formula as a standard with which to compare the reduction level of other molecules. The following are the steps involved in determining the O/R value of any molecule.

1. *Add or subtract water to the molecule in question to make the C/O ratio 1*. This will allow a comparison to $(CH_2O)_n$. For example, the formula for ethanol is C_2H_6O. The C/O ratio is 2. One water must be added so that the C/O ratio is 1. This changes C_2H_6O to $C_2H_8O_2$. Acetic acid is $C_2H_4O_2$. Since the C/O ratio is already one, nothing further need be done. The formula for carbon dioxide is CO_2. In order to make the C/O ratio one, one must subtract H_2O, i.e., $[CO_2-H_2O]$. The result is $C(-2H)O$.

2. *Now compare the number of hydrogens in the modified formula to $(CH_2O)_n$ which has the same number of carbons as in the modified formula*. For ethanol, $C_2H_8O_2$ is compared to $(CH_2O)_2$. There are 8H in $C_2H_8O_2$ but only 4H in $(CH_2O)_2$. Thus, ethanol has an additional 4H. For carbon dioxide, $C(-2H)O$, there are $-4H$ compared to CH_2O. For acetic acid, $C_2H_4O_2$, there are the same number of hydrogens as in $(CH_2O)_2$.

3. *Add -1 for each additional 2H and $+1$ for a decrease in 2H*. Thus, the O/R for ethanol is -2, for CO_2 it is $+2$, and for acetic acid it is 0.

Since both the oxidized and reduced fermentation end products originate from the substrate, the sum of the O/R of the products equals the O/R of the substrate, if the fermentation is balanced. For example, the O/R for glucose is 0. When one mole of glucose is fermented to two moles of ethanol and two moles of carbon dioxide, the O/R of the products is $(-2 \times 2)+(+2 \times 2)=0$. Often one simply takes the ratio $(+/-)$ of the O/R of the products when a carbohydrate is fermented. For a balanced fermentation, $+/-$ should be 1.

The available hydrogen method

Like the O/R method, this procedure is merely one of bookkeeping. It has nothing to do with the chemistry of the reactions. According to this method, one "oxidizes" the molecule to CO_2 using water to obtain the "available hydrogen." For example:

$$C_6H_{12}O_6+6H_2O \longrightarrow 24\,H+6CO_2,$$

thus glucose has 24 available H.

The available H in all of the products must add up to the available H in the starting material. In Table 13.2, the concentration of products is given per 100 moles of glucose used. The available H in the glucose is $24 \times 100=2,400$. The available H in the products adds up to 2,242. Thus, the balance is $2,400/2,242=1.07$.

13.6 Propionate fermentation using the acrylate pathway

The genus *Clostridium* comprises a heterogeneous group of bacteria consisting of

Table 13.2 The distribution of catalase and superoxide dismutase

Bacterium	Superoxide dismutase	Catalase
Aerobes or facultative anaerobes		
Escherichia coli	+	+
Pseudomonas species	+	+
Deinococcus radiodurans	+	+
Aerotolerant bacteria		
Butyribacterium rettgeri	+	−
Streptococcus faecalis	+	−
Streptococcus lactis	+	−
Strict anaerobes		
Clostridium pasteurianum	−	−
Clostridium acetobutylicum	−	−

Source: Stanier, R. Y., J. L. Ingraham, M. L. Wheelis, and P. R. Painter. 1986. *The Microbial World*. (Reprinted by permission of Prentice Hall, Englewood Cliffs, NJ.)

gram-positive, anaerobic, spore-forming bacteria that cannot use sulfate as a terminal electron acceptor. They can be isolated from anaerobic regions (or areas of low oxygen levels) in soil. The clostridia ferment organic nutrients to products that can include alcohols, organic acids, hydrogen gas, and carbon dioxide. *C. propionicum* oxidizes three moles of lactate to two moles of propionate, one mole of acetate, one mole of carbon dioxide, and 1 mole of ATP. The pathway is called the *acrylate pathway* because one of the intermediates is acrylyl-CoA. The bacteria derive ATP via a substrate level phosphorylation during the conversion of acetyl-P to acetate catalyzed by acetate kinase. Since only one acetate is made per three lactates used, the pathway yields one-third of an ATP per lactate. Growth yields are proportional to the amount of ATP produced, and it is to be expected that the growth yields for these organisms are very low. (The molar growth yield for ATP is 10.5 g cells per mole of ATP synthesized, Section 2.2.6.)

13.6.1 The fermentation pathway of C. propionicum

A molecule of lactate is oxidized to pyruvate yielding 2[H] (Fig. 13.3, reaction 1). The pyruvate is then oxidized to acetyl-CoA and

CO_2 yielding 2[H] again (reaction 2). The acetyl-CoA is converted to acetate and ATP via acetyl-P (reactions 3 and 4). During the oxidations, 4[H] are produced which must be reutilized. The electron acceptor is created from a second and third molecule of lactate (actually lactyl-CoA). The lactate acquires a CoA from propionyl-CoA (reaction 5). The lactyl-CoA is dehydrated to yield the unsaturated molecule, acrylyl-CoA (reaction 6). Each acrylyl-CoA is then reduced to propionyl-CoA using up the 4[H] (reaction 7). The fermentation is thus balanced. The propionate is produced during the CoA transfer step (reaction 5). This is catalyzed by *CoA transferase*, an enzyme that occurs in many anaerobes.

What can we learn from this pathway? The bacteria use a standard method for making ATP under anaerobic conditions. They decarboxylate pyruvate to acetyl-CoA and then, using a *phosphotransferase* (to make acetyl-P) and an *acetate kinase*, they make ATP and acetate. These reactions are widespread among fermenting bacteria. The production of acetate is presumed to be associated always with the synthesis of two moles of ATP per mole of acetate if the bacteria are growing on glucose and using the EMP pathway. One ATP is produced from acetyl-CoA and the

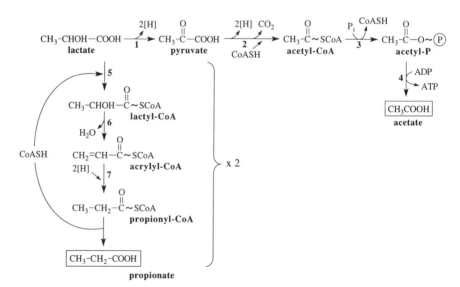

Fig. 13.3 Propionate fermentation via the acrylate pathway. Enzymes: (1) lactate dehydrogenase; (2) pyruvate–ferredoxin oxidoreductase; (3) phosphotransacetylase; (4) acetate kinase; (5) CoA transferase; (7) a dehydrogenase. Reaction 6 is not sufficiently characterized.

second ATP is produced during the production of the pyruvate in the EMP pathway. Another common reaction among fermenting bacteria is the transfer of coenzyme A from one organic molecule to another, a reaction catalyzed by *CoA transferase*. The other way of attaching a coenzyme A molecule to a carboxyl group is to transfer an AMP or a phosphate to the carboxyl group from ATP making an acyl-phosphate or an acyl-AMP, and then to displace the AMP or phosphate with CoASH. (Recall the activation of fatty acids prior to their degradation, Section 9.1.2.) However, fermenting organisms must conserve ATP. The CoA transferase reaction is one way this can be done.

13.7 Propionate fermentation using the succinate–propionate pathway

Many bacteria produce propionic acid as a product of fermentation using a pathway different from the acrylate pathway. The other pathway is called the *succinate–propionate pathway*, which yields more ATP than the acrylate pathway per mole of propionate formed. One of the organisms that utilizes this pathway, *Propionibacterium*, ferments lactate as well as hexoses to a mixture of propionate, acetate, and CO_2. *Propioni-*

bacterium is a gram-positive anaerobic, nonsporulating, nonmotile, pleomorphic rod that is part of the normal flora in the rumen of herbivores, on human skin, and in dairy products (e.g., cheese). *Propionibacterium* is used in the fermentation process that produces swiss cheese. The characteristic sharp flavor of this cheese is due to the propionate and the holes in the cheese are due to the carbon dioxide produced.

The pathway illustrated in Fig. 13.4 shows that three molecules of lactate are oxidized to pyruvate (reaction 1). This yields six electrons. Then, one pyruvate is oxidized to acetate, CO_2, and ATP, yielding two more electrons (reaction 2). We now have eight electrons to use up. The other two pyruvates are carboxylated to yield two molecules of oxaloacetate (reaction 5). The reaction is catalyzed by *methyl malonyl-CoA transcarboxylase*. The two oxaloacetates are reduced to two malates, consuming a total of four electrons (reaction 6). The two malates are dehydrated to two fumarates (reaction 7). The two fumarates are reduced via fumarate reductase to two succinates, consuming four electrons (reaction 8). The latter reduction is coupled to the generation of a Δp. The fermentation is now balanced. The two succinates are converted to two molecules of succinyl-CoA via a CoA trans-

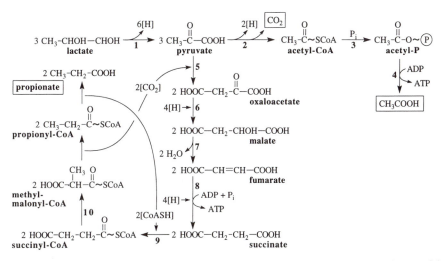

Fig. 13.4 Propionate fermentation by the succinate–propionate pathway. Enzymes: (1) lactate dehydrogenase (a flavoprotein); (2) pyruvate dehydrogenase (an NAD^+ enzyme); (3) phosphotransacetylase; (4) acetate kinase; (5) methylmalonyl-CoA–pyruvate transcarboxylase; (6) malate dehydrogenase; (7) fumarase; (8) fumarate reductase; (9) CoA transferase; (10) methylmalonyl-CoA racemase.

ferase (reaction 9). The two molecules of succinyl-CoA are isomerized to two molecules of methylmalonyl-CoA in an unusual reaction in which COSCoA moves from the α-carbon to the β-carbon in succinyl-CoA to form methylmalonyl-CoA (reaction 10). The reaction can be viewed as an exchange between adjacent carbons of an H for a COSCoA. The enzyme that carries out this reaction is *methylmalonyl-CoA racemase,* an enzyme that requires vitamin B_{12} as a cofactor. The two molecules of methylmalonyl-CoA donate the carboxyl groups to pyruvate via the transcarboxylase and in turn become propionyl-CoA (reaction 5). The propionyl-CoA donates the CoA to succinate via the CoA transferase, and becomes propionate (reaction 9). Notice the important role of transcarboxylases and CoA transferases. These enzymes allow the attachment of CO_2 and CoA to molecules without the need for ATP.

The fumarate reductase as a coupling site

When comparing Figs 13.3 and 13.4, we learn that in metabolism there is sometimes more than one route to take from point A to point B. *Propionibacterium* and other bacteria that use the succinate–propionate pathway use a circuitous route, but one which sends electrons through an energy-coupling site via the membrane-bound fumarate reductase. Electron flow to fumarate requires a quinone and presumably the Δp is generated via a redox loop involving the quinone (Section 4.6). The use of fumarate as an electron acceptor during anaerobic growth is widespread among bacteria. (See the discussion of the mixed acid fermentation in Section 13.10.) The electron donors besides lactate include NADH, H_2, formate, and glycerol-3-phosphate. The Δp which is established can be used for ATP synthesis, solute uptake, or to spare ATP that might be hydrolyzed to maintain the Δp.

The transcarboxylase reaction spares an ATP

One way to carboxylate pyruvate is to use pyruvate carboxylase and CO_2 (Section 8.8). However, this requires an ATP. *Propionibacterium* has a transcarboxylase called *methylmalonyl-CoA-pyruvate transcarboxy-*

lase that transfers a carboxyl group from methylmalonyl-CoA to pyruvate, hence an ATP is not used (Fig. 13.4, reaction 5). By using the transcarboxylase, the bacteria save energy by substituting one covalent bond for another. However, not all fermenting bacteria that produce propionate from pyruvate use the methylmalonyl-CoA-pyruvate transcarboxylase. For example, *Veillonella alcalescens* and *Propionigenum modestum* use a sodium-dependent decarboxylase to remove the carboxyl group from methylmalonyl-CoA while generating an electrochemical potential (Section 3.8). The sodium-dependent decarboxylase pumps sodium ions out of the cell generating a sodium ion potential, which can be used as a source of electrochemical energy (e.g., for solute uptake or ATP synthesis).

13.7.1 The PEP carboxytransphosphorylase of propionibacteria and its physiological significance

The reaction

Propionibacteria can produce succinate as well as propionate as an end-product of fermentation when growing on carbon sources such as glucose that enter the glycolytic pathway. This means that they must have an enzyme to carboxylate a C_3 intermediate to form the C_4 product. The C_3 intermediate that is carboxylated is phosphoenolpyruvate (an intermediate in glycolysis). The phosphoenolpyruvate is carboxylated to oxaloacetate, which is then reduced to succinate via reactions 6, 7, and 8 shown in Fig. 13.4. The enzyme that catalyzes the carboxylation of phosphoenolpyruvate is called PEP carboxytransphosphorylase and it catalyzes the following reaction:[1]

$$PEP + CO_2 + P \longrightarrow oxaloacetate + PP_i$$

During the carboxylation, a phosphoryl group is transferred from PEP to inorganic phosphate to form pyrophosphate. Pyrophosphate is a high-energy compound and propionibacteria have enzymes that phosphorylate fructose-6-phosphate to fructose-1,6-bisphosphate and serine to phosphoserine using pyrophosphate as the phosphoryl donor.

Physiological significance

The carboxylation of phosphoenolpyruvate or pyruvate to oxaloacetate and the reduction of the oxaloacetate to succinate is a widespread pathway among fermenting bacteria. (For example, see the mixed acid fermentation, Section 13.10.) These reactions were also discussed in the context of the reductive citric acid pathway (Section 8.9). The pathway from PEP or pyruvate to succinate is extremely important for anaerobes and serves three purposes: (1) the fumarate is an electron sink enabling NADH to be re-oxidized; (2) the fumarate reductase is a coupling site (i.e., a Δp is generated); and (3) the succinate can be converted to succinyl-CoA, which is required for the biosynthesis of tetrapyrroles, lysine, diaminopimeic acid, and methionine. With respect to fumarate acting as an electron sink, Gest has suggested that the carboxylation of the C_3 glycolytic intermediate and the reductive pathway from oxaloacetate to succinate may have evolved when the earth's atmosphere was still anaerobic, and served the purpose of balancing fermentations, thus sparing one of the two pyruvates derived from glucose for biosynthesis.[2] The reactions between oxaloacetate and succinate may have later become part of the oxidative citric acid cycle during the evolution of aerobic metabolism.

13.8 Acetate fermentation (acetogenesis)

As discussed in Section 12.1.3 when describing the acetogenic bacterium *Clostridium thermo-*

aceticum, some bacteria use CO_2 as an electron sink and reduce it to acetate as a fermentation end product using the acetyl-CoA pathway. This is called acetogenesis. Another acetogenic bacterium is the sulfate-reducer, *Desulfotomaculum thermobenzoicum*, when growing on pyruvate in the absence of sulfate.[3] (Another sulfate-reducer, *Desulfobulbus propionicus*, ferments pyruvate to a mixture of acetate and propionate using the succinate–propionate pathway.) The acetogenic pathway is shown in Fig. 13.5. Four pyruvate molecules are oxidatively decarboxylated to four acetyl-CoA molecules, producing eight electrons (reaction 1). The acetyl-CoA is converted to acetyl-phosphate via phosphotransacetylase (reaction 2). The acetyl-phosphate is converted to acetate and ATP via acetate kinase (reaction 3). Six of the electrons are used to reduce CO_2 to bound methyl, $[CH_3]$, (reaction 4). This requires an ATP in the acetyl-CoA pathway to attach the formic acid to the THF (Section 12.1.3). The remaining two electrons are used to reduce a second molecule of CO_2 to bound carbon monoxide, $[CO]$ (reaction 5). The $[CH_3]$, $[CO]$, and CoASH combine to yield acetyl-CoA (reaction 6). The fourth acetyl-CoA is converted to acetate with the formation of an ATP (reactions 7 and 8).

13.9 Lactate fermentation

The lactic acid bacteria are a heterogeneous group of aerotolerant anaerobes that ferment glucose to lactate as the sole or major product of fermentation. They include the genera

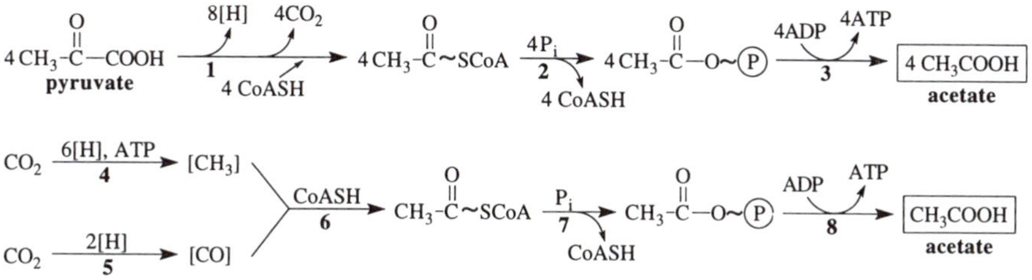

Fig. 13.5 Acetogenesis from pyruvate by *Desulfotomaculum thermobenzoicum*. Enzymes: (1) pyruvate dehydrogenase; (2, 7) phosphotransacetylase; (3, 8) acetate kinase; (4) enzymes of the acetyl-CoA pathway; (5, 6) carbon monoxide dehydrogenase.

Lactobacillus, Sporolactobacillus, Strepto-coccus, Leuconostoc, Pediococcus, and *Bifi-dobacterium.* Lactic acid bacteria are found living on the skin of animals, in the gastrointestinal tract, and in other places (e.g., mouth and throat). Some genera live in vegetation and in dairy products. Several lactic acid bacteria are medically and commercially important organisms. These include the genus *Streptococcus,* several of which are pathogenic. The lactic acid bacteria are also important in various food fermentations (e.g., the manufacture of butter, cheese, yogurt, pickles, and sauerkraut). Although they can live in the presence of air, they metabolize glucose only fermentatively, and derive most or all of their ATP from substrate level phosphorylation. Under certain growth conditions they may transport lactate out of the cell in electrogenic symport with H^+, creating a $\Delta\Psi$ (Section 3.8.3). There are two major types of lactate fermentations: *homofermentative* and *heterofermentative.* The former uses the Embden–Meyerhof–Parnas pathway (glycolysis) and the latter uses the pentose phosphate pathway. A third pathway, called the *bifidum pathway* is found in *Bifidobacterium bifidum.*

Homofermentative lactate fermentation

The homofermentative pathway produces primarily lactate. The bacteria use the glycolytic pathway to oxidize glucose to pyruvate. This nets them two ATPs per mole of glucose via the oxidation of phospho-glyceraldehyde. The NADHs produced during this oxidative step are used to reduce the pyruvate, forming lactate. The overall reaction is:

$$\text{Glucose} + 2\,\text{ADP} + 2\,\text{P}_i \longrightarrow 2\,\text{lactate} + 2\,\text{ATP}$$

Whenever lactate is produced during fermentations, it is always the result of the reduction of pyruvate. It can be assumed that one ATP is made per mole of lactate produced, since the ATP yield per pyruvate is one.

Heterofermentative lactate fermentation

The heterofermentative lactate fermentation produces lactate using the decarboxylation and isomerase reactions of the pentose

phosphate pathway (Fig. 13.6). The glucose is oxidized to ribulose-5-phosphate using glucose-6-phosphate dehydrogenase and 6-phosphogluconate dehydrogenase (reactions 1–3). Four [H] are produced in the form of two NADHs. The ribulose-5-phosphate is isomerized to xylulose-5-phosphate using the epimerase (reaction 4). Then an interesting reaction occurs during which the xylulose-5-phosphate is cleaved with the aid of inorganic phosphate to form phosphoglyceraldehyde and acetyl-phosphate (reaction 5). The enzyme that catalyzes this reaction is called *phosphoketolase* and requires thiamine pyrophosphate (TPP) as a cofactor. The phospho-glyceraldehyde is oxidized to pyruvate via reactions also found in the glycolytic pathway yielding an ATP and a third NADH, and the pyruvate is reduced to lactate by one of the 3 NADHs (reactions 9–14). The acetyl-phosphate produced in the phosphoketolase reaction is reduced to ethanol using the second and third NADH (reactions 6–8). Thus, the fermentation is balanced. The overall reaction is:

$$\text{Glucose} + \text{ADP} + \text{P}_i$$
$$\longrightarrow \text{ethanol} + \text{lactate} + \text{CO}_2 + \text{ATP}$$

Note that the heterofermentative pathway produces only one ATP per glucose in contrast to the homofermentative pathway which produces two ATPs per glucose.

Bifidum pathway

The bifidum pathway ferments two glucose to two lactates and three acetates with the production of 2.5 ATPs per glucose. The overall reaction is:

$$2 \text{ glucose} + 5 \text{ ADP} + 5 \text{ P}_i$$
$$\longrightarrow 3 \text{ acetate} + 2 \text{ lactate} + 5 \text{ ATP}$$

The ATP yields are therefore greater than for the homo- or heterofermentative pathways. The pathway uses reactions of the pentose phosphate pathway (Section 8.4) and the homofermentative pathway. Two glucose molecules are converted to two fructose-6-P, requiring two ATPs. One of the fructose-6-P molecules is cleaved by fructose-6-phosphate

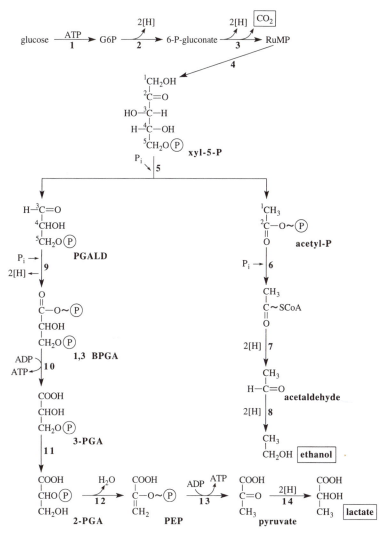

Fig. 13.6 Heterofermentative lactate fermentation. Enzymes: (1) hexokinase; (2) glucose-6-phosphate dehydrogenase; (3) 6-phosphogluconate dehydrogenase; (4) ribulose-5-phophate epimerase; (5) phospho-ketolase; (6) phosphotransacetylase; (7) acetaldehyde dehydrogenase; (8) alcohol dehydrogenase; (9) PGALD dehydrogenase; (10) PGA kinase; (11) phosphoglycerate mutase; (12) enolase; (13) pyruvate kinase; (14) lactate dehydrogenase.

phosphoketolase to erythrose-4-P and acetyl-P. The acetyl-P is converted to acetate via acetate kinase with the formation of an ATP. The erythrose-4-P reacts with the second fructose-6-P in a transaldolase reaction to form glyceraldehyde-3-P and sedoheptulose-7-P. These then react in a transketolase reaction to form xyulose-5-P and ribose-5-P. The latter isomerizes to form a second xylulose-5-P. The two xylulose-5-P are cleaved by xylulose-5-phosphate phosphoketolase to two glyceraldehyde-3-P and two acetyl-P. The

two glyceraldehyde-3-P are converted to two lactate with the production of four ATP molecules, using reactions of the homo-fermentative pathway, and the two acetyl-P are converted to two acetates with the production of two more ATP molecules. Thus, seven ATPs are produced per two glucose molecules fermented, but since two ATPs were used to make the two fructose-6-P molecules, the net gain in ATP per glucose is 5/2 or 2.5. Note that since glucose-6-P is not oxidized to 6-phosphogluconate, the acetyl-P can serve as a phosphoryl donor for ATP synthesis rather than be reduced to ethanol.

Synthesis of acetyl-CoA or acetyl-P from pyruvate by lactic acid bacteria

Not all of the pyruvate produced by lactic acid bacteria need be converted to lactate. Depending upon the species, the lactic acid bacteria may have one of three enzymes or enzyme complexes for decarboxylating pyruvate (Section 7.3.2). Streptococci have pyruvate dehydrogenase, usually found in aerobically respiring bacteria; several lactic acid bacteria are known to have pyruvate formate lyase, an enzyme also found in the Enterobacteriaceae; *L. plantarum* and *L. delbruckii* use the flavoproteins pyruvate oxidase and lactate oxidase in coupled reactions that convert two pyruvates to one acetyl-P and one lactate, as shown below:

Pyruvate oxidase:

$$Pyruvate + P_i + FAD$$

$$\longrightarrow acetyl\text{-}P + CO_2 + FADH_2$$

Lactate oxidase:

$$Pyruvate + FADH_2 \longrightarrow lactate + FAD$$

13.10 Mixed acid and butanediol fermentations

The enteric bacteria are facultative anaerobes. In the absence of oxygen several physiological changes take place as part of the adaptation to anaerobic growth. These changes are listed below.

1. Terminal reductases replace the oxidases in the electron transport chain.

2. The citric acid cycle is modified to become a reductive pathway. α-Ketoglutarate dehydrogenase and succinate dehydrogenase are missing or occur at low levels, the latter being replaced by fumarate reductase.

3. Pyruvate–formate lyase is substituted for pyruvate dehydrogenase. This means that the cells oxidize pyruvate to acetyl-CoA and formate, rather than to acetyl-CoA, CO_2, and NADH.

4. They carry out a mixed acid or butanediol fermentation.

The mixed acid and butanediol fermentations are similar in that both produce a mixture of organic acids, CO_2, H_2, and ethanol. The butanediol fermentation is distinguished by producing large amounts of 2,3-butanediol, acetoin, more CO_2 and ethanol, and less acid. The mixed acid fermenters belong to the genera *Escherichia*, *Salmonella*, and *Shigella*. All three can be pathogenic and cause intestinal infections such as dysentery, typhoid fever (*Salmonella typhi*), or food poisoning. Butanediol fermenters are *Serratia*, *Erwinia*, and *Enterobacter*.

Mixed acid fermentation

The products of the mixed acid fermentation are succinate, lactate, acetate, ethanol, formate, carbon dioxide, and hydrogen gas (Fig. 13.7). Each of these are made from one phosphoenolpyruvate and CO_2 or one pyruvate. For example, the formation of succinate is due to a carboxylation of phosphoenolpyruvate to oxaloacetate followed by two reductions to form succinate (reactions 10–13). All of the other products are formed from pyruvate. The formation of lactate from pyruvate is simply a reduction (reaction 4). Pyruvate is also decarboxylated to acetyl-CoA and formate using the pyruvate–formate lyase (reaction 3). The acetyl-CoA can be reduced to ethanol (reactions 6 and 7) or converted to acetate and ATP via acetyl-P (reactions 8 and 9). The formate can be oxidized to CO_2 and H_2 by the enzyme system formate–hydrogen lyase (reaction 5). This system actually consists of two enzymes; formate dehydrogenase and an associated hydrogenase. The formate dehydrogenase oxidizes the formate to CO_2 and reduces the hydrogenase which transfers the electrons to two protons to form hydrogen gas. *Shigella* and *Erwinia* do not contain formate–hydrogen lyase, and therefore do not produce gas.[4] Each of the pathways following phosphoenolpyruvate or pyruvate can be viewed as a metabolic branch that accepts different amounts of reducing equivalent, i.e., 0, 2[H], or 4[H], depending upon the pathway. The reducing equivalents in the different branches are: succinate (4), ethanol (4), lactate (2), acetate (0), and

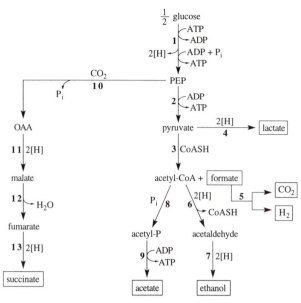

Fig. 13.7 Mixed acid fermentation. Enzymes: (1) glycolytic enzymes; (2) pyruvate kinase; (3) pyruvate–formate lyase; (4) lactate dehydrogenase; (5) formate-hydrogen lyase; (6) acetaldehyde dehydrogenase; (7) alcohol dehydrogenase; (8) phosphotransacetylase; (9) acetate kinase; (10) PEP carboxylase; (11) malate dehydrogenase; (12) fumarase; (13) fumarate reductase. Note the ATP yields: per succinate, approximately 1; per ethanol, 1; per acetate, 2; per formate, 1; per CO_2 and H_2, 1; per lactate, 1. Energy equivalent to approximately 1 ATP is conserved per succinate formed because the fumarate reductase reaction takes place in the cell membrane and generates a Δp. Note also the reducing equivalents used in the production of the end products: per succinate, 4; per ethanol, 4; per acetate, 0; per lactate, 2; per formate, 0. The number of reducing equivalents used must equal the number produced during glycolysis. Therefore, only certain ratios of end products are compatible with a balanced fermentation.

formate (0). During glycolysis, 2[H] are produced for each phosphoenolpyruvate or pyruvate formed. Therefore, in order to balance the fermentation, 2[H] must be used for each phosphoenolpyruvate or pyruvate formed. The pathways that utilize 4[H] per phosphoenolpyruvate or pyruvate spare the second phosphoenolpyruvate or pyruvate for biosynthesis. However, they may do this at the expense of an ATP. For example, the reduction of acetyl-CoA to ethanol uses 4[H], but this is at the expense of an ATP that can be formed when acetyl-CoA is converted to acetate. In this context, the production of succinate is particularly valuable. The pathway utilizes 4[H] and also includes a coupling site (fumarate reductase).

Butanediol fermentation

The butanediol fermentation is characterized by the production of 2,3-butanediol and

acetoin (Fig. 13.8). Glucose is oxidized via the glycolytic pathway to pyruvate (reactions 1). There are three fates for the pyruvate. Some of it is reduced to lactate (reaction 10), some is converted to acetyl-CoA and formate (reaction 2), and some is used for the synthesis of 2,3-butanediol (reactions 6–9). The formate is converted to CO_2 and H_2 (reaction 3), and the acetyl-CoA is reduced to ethanol (reactions 4 and 5). The first free intermediate in the butanediol pathway is α-acetolactate. This is formed by the enzyme α-acetolactate synthase which decarboxylates pyruvate in a thiamine pyrophosphate (TPP)-dependent reaction to enzyme-bound "active acetaldehyde" (reaction 6). The active acetaldehyde is transferred by the α-acetolactate synthase to pyruvate to form α-acetolactate (reaction 7). The α-acetolactate, a β-keto carboxylic acid, is decarboxylated to acetoin (reaction 8). The acetoin is reduced by NADH to 2,3-butane-

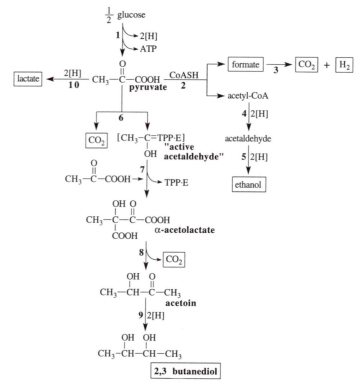

Fig. 13.8 Butanediol formation. Enzymes: (1) glycolytic enzymes; (2) pyruvate–formate lyase; (3) formate-hydrogen lyase; (4) acetaldehyde dehydrogenase; (5) alcohol dehydrogenase; (6, 7) α-acetolactate synthase; (8) α-acetolactate decarboxylase; (9) 2,3-butanediol dehydrogenase; (10) lactate dehydrogenase.

diol (reaction 9). The production of butane-diol is favored under slightly acidic conditions, and is a way for the bacteria to limit the decrease in external pH caused by the synthesis of organic acids from pyruvate.

How thiamine pyrophosphate catalyzes the decarboxylation of α-keto carboxylic acids

The decarboxylation of α-keto carboxylic acids presents a problem because there is no electron attracting carbonyl group β to the C–C bond to withdraw electrons, as there is in β-keto carboxylic acids. (See Section 8.10.2 for a description of the decarboxylation of β-keto carboxylic acids.) The problem is solved by using thiamine pyrophosphate (TPP). A proposed mechanism is illustrated in Fig. 13.9. A proton dissociates from the thiamine pyrophosphate to form a dipolar ion which is stabilized by the positive charge on the nitrogen atom. The negative center of the dipolar ion adds to the keto group of the

α-keto carboxylic acid. Then, the electron attracting $=N^{+}-$ group pulls electrons away from the C–C bond, facilitating the decarboxylation. The product is an α-ketol called "active aldehyde." The flow of electrons is reversed and the "active aldehyde" then attacks a positive center on another molecule. The product varies depending on the enzyme. As shown in Fig. 13.9, α-acetolactate synthase catalyzes the condensation of "active acetaldehyde" with pyruvate to form α-aceto-lactate, and pyruvate decarboxylase catalyzes the addition of a proton to form acetaldehyde. Pyruvate dehydrogenase and α-ketoglutarate dehydrogenase catalyze the condensation with lipoic acid to form the acyl-lipoate derivative.

13.11 Butyrate fermentation

Butyrate fermentations are carried out by the butyric acid clostridia. The clostridia are a

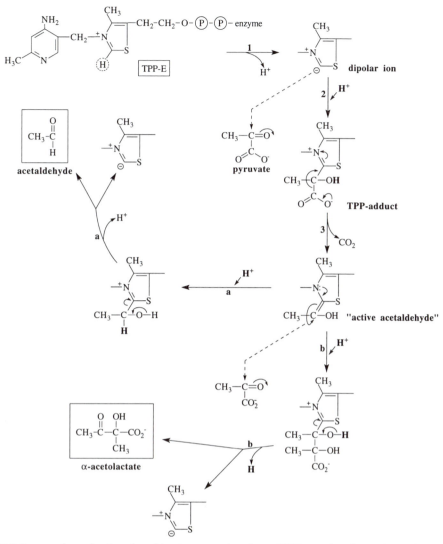

Fig. 13.9 Proposed mechanism for thiamine pyrophosphate (TPP) catalyzed reactions. Step 1: The TPP-enzyme loses a proton to form a dipolar ion. The anion is stabilized by the positive charge on the nitrogen. The anionic center is nucleophilic and can attack positive centers such as carbonyl carbons. Step 2: The dipolar ion condenses with pyruvate to form a TPP-adduct. Step 3: The electron-attracting N in the TPP facilitates the decarboxylation to form "active acetaldehyde." The "active acetaldehyde" can form acetaldehyde (step a) or α-acetolactate (step b).

heterogeneous group of anaerobic spore-forming bacteria that can be isolated from anaerobic muds, sewage, feces, or other anaerobic environments. They all are classified in the genus *Clostridium*. Some are saccharolytic (i.e., they ferment carbohydrates) and/or proteolytic. The proteolytic clostridia are important in the anaerobic decomposition of proteins, called "putrefaction." Other clostridia are more specialized and will ferment only a few substrates (e.g., ethanol, acetate, certain purines, or certain amino acids). The butyric acid clostridia ferment carbohydrates to butyric acid. The fermentation products also include hydrogen gas, carbon dioxide, and small amounts of acetate. The bacteria first oxidize the glucose to two moles of pyruvate using the Embden–Meyerhof–Parnas pathway (Fig. 13.10). This produces two NADH, plus two

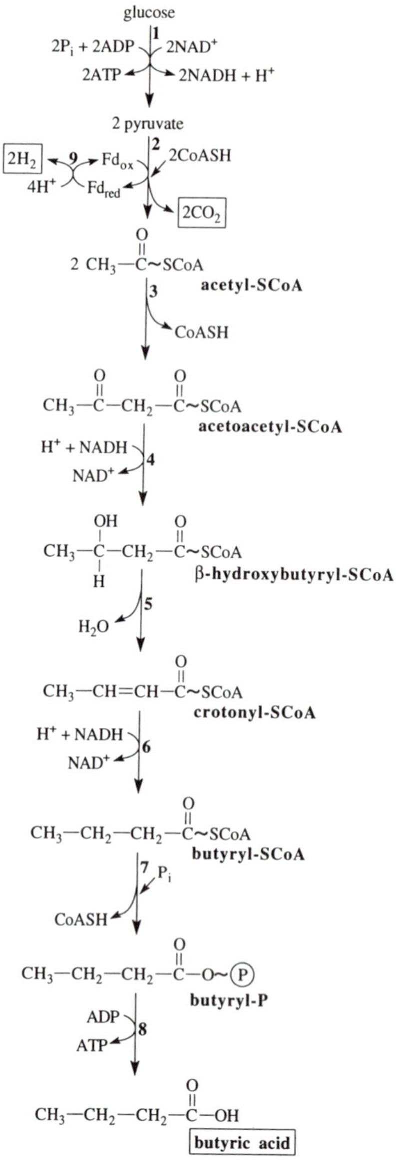

Fig. 13.10 The butyrate fermentation. The glucose is degraded via glycolysis to pyruvate which is then oxidatively decarboxylated to acetyl-CoA. Two molecules of acetyl-CoA condense to form aceto-acetyl-CoA which is reduced to butyryl-CoA. A phosphotransacetylase makes butyryl-P, and a kinase produces butyrate and ATP. Two ATPs are produced in the glycolytic pathway per glucose and one ATP from butyryl-P. Note the production of hydrogen gas as an electron sink. This actually allows the production of the third ATP from butyryl-P, rather than reducing the butyryl-SCoA to butanol in order to balance the fermentation. Enzymes: (1) glycolysis; (2) pyruvate–ferredoxin oxidoreductase; (3) acetyl-CoA acetyltransferase (thiolase); (4) β-hydroxybutyryl-CoA dehydro-

ATP molecules. The pyruvate is then decarboxylated to acetyl-CoA, CO_2 and H_2 using pyruvate–ferredoxin oxidoreductase and hydrogenase (reaction 1). The acetyl-CoA is condensed to form acetoacetyl-CoA (reaction 2) which is reduced to β-hydroxy-butyryl-CoA using one of the two NADHs (reaction 3). The β-hydroxybutyryl-CoA is reduced to butyryl-CoA using the second NADH (reactions 4 and 5). The CoASH is displaced by inorganic phosphate (reaction 6) and the butyryl-P donates the phosphoryl group to ADP to form ATP and butyrate (reaction 7). This pathway therefore utilizes three substrate level phosphorylations, the phosphoryl donors being 1,3-bisphosphogly-cerate, phosphoenolpyruvate, and butyryl-P. Note the role of the hydrogenase (reaction 9) in the hydrogen sink. (For a discussion of acetyl-CoA condensations, see Section 8.10.1)

13.11.1 Butyrate and butanol–acetone fermentation in C. acetobutylicum

Some clostridia initially make butyrate during fermentation and, when the butyrate accumulates and the pH drops (due to the butyric acid), the fermentation switches to a butanol–acetone fermentation. As we shall see, the butyrate is actually taken up by the cells and converted to butanol and acetone. The accumulation of butyric acid in media of low pH can be toxic because the undissociated form of the acid is lipophilic, and can enter the cell acting as an uncoupler and also resulting in a decrease in the ΔpH. (The pK of butyric acid is 4.82.) Butanol and acetone production by the clostridia was at one time the second largest industrial fermentation process, second only to ethanol. The solvents are now synthesized chemically.

As an example of the butyrate fermentation and the shift to the butanol–acetone fermentation, we will consider the fermentation of carbohydrates carried out by C. aceto-

genase; (5) crotonase; (6) butyryl-CoA dehydrogenase; (7) phosphotransbutyrylase; (8) butyrate kinase; (9) hydrogenase.

butylicum.[5] During exponential growth the bacteria produce butyrate, acetate, H_2, and CO_2. This is called the acidogenic phase. When the culture enters stationary phase the acids are taken up by the cells, concomitant with the fermentation of the carbohydrate, and converted to butanol, acetone, and ethanol. This is called the solventogenic phase. Pentoses are also fermented, and these are converted to fructose-6-phosphate and phosphoglyceraldehyde via the pentose phosphate pathway (Fig. 13.11). The pyruvate

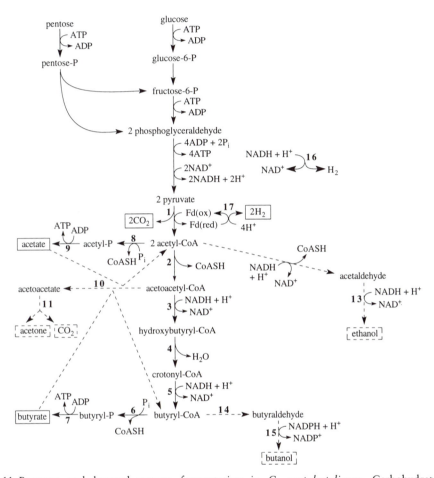

Fig. 13.11 Butyrate and butanol–acetone fermentation in *C. acetobutylicum.* Carbohydrates are oxidized to acetyl-CoA. Pentose phosphates are converted to fructose-6-phosphate and phosphoglyceraldehyde via the pentose phosphate reactions. Glucose is oxidized to pyruvate via the Embden–Meyerhof–Parnas pathway. The pyruvate is oxidized to acetyl-CoA. In the butyric acid fermentation the acetyl-CoA is converted to acetate and butyrate (solid lines). When the acetate and butyrate levels rise, they are taken up by the cells and converted to butanol and ethanol while carbohydrates continue to be fermented (dotted lines). During the butanol-acetone fermentation, the acetoacetyl-CoA donates the CoASH to butyrate and acetate, and becomes acetoacetate, which is decarboxylated to acetone. Enzymes: (1) pyruvate–ferredoxin oxidoreductase; (2) acetyl-CoA acetyltransferase (thiolase); (3) hydroxybutyryl-CoA dehydrogenase; (4) crotonase; (5) butryryl-CoA dehydrogenase; (6) phosphotransbutyrylase; (7) butyrate kinase; (8) phosphotransacetylase; (9) acetate kinase; (10) acetoacetyl-CoA:acetate/butyrate:CoA transferase; (11) acetoacetate decarboxylase; (12) acetaldehyde dehydrogenase; (13) ethanol dehydrogenase; (14) butyraldehyde dehydrogenase; (15) butanol dehydrogenase; (16) NADH–ferredoxin oxidoreductase and hydrogenase; (17) hydrogenase. (Adapted from Jones, D. T., and D. R. Woods. 1986. Acetone–butanol fermentation revisited. *Microbiol. Rev.* 50:484–524.)

formed from the sugars is oxidatively de-carboxylated to acetyl-CoA and CO2 via the pyruvate–ferredoxin oxidoreductase (reaction 1). The acetyl-CoA has two fates. Some of it is converted to butyrate and ATP as described in Fig. 13.10 (reactions 2–7). Some acetyl-CoA is also converted to acetate via acetyl-P in a reaction that yields an additional ATP (reactions 8 and 9). The ability to send electrons to hydrogen via the NADH:ferre-doxin oxidoreductase (reaction 16) allows the bacteria to produce more acetate and hence more ATP, rather than reduce acetyl-CoA to butyrate. Notice that twice as much ATP is generated per acetate produced as opposed to butyrate. However, reduction of protons by NADH is not favored energetically and is limited by increasing concentrations of hydrogen gas.

The NADH oxidoreductase can be pulled in the direction of hydrogen gas by other bacteria that utilize hydrogen in a process called interspecies hydrogen transfer, ex-plained in Section 13.4.1. In the solventogenic phase, butyrate and acetate are taken up by the cells and reduced to butanol and ethanol (dotted lines). The acids are converted to their CoA derivatives by accepting a CoA from acetoacetyl-CoA (reaction 10). The reaction is catalyzed by CoA transferase. (Recall that CoA transferase is also used in the acrylate pathway for propionate fermentation, Sec-tion 13.6.) The acetoacetate that is formed is decarboxylated to acetone and CO_2 (reaction 11). The acetyl-CoA is reduced to ethanol (reactions 12 and 13), and the butyryl-CoA is reduced to butanol (reactions 14 and 15).

The molar ratios of the fermentation end products in clostridial fermentations will vary according to the strain.[6] For example, C. acetobutylicum is an important solvent-producing strain, and when grown at pH_o below 5 produces butanol and acetone in the molar ratio of 2:1 with small amounts of isopropanol, whereas C. sporogenes and C. pasteurianum produce very little solvent. Clostridium butyricum forms butyrate and acetate in a ratio of about 2:1, whereas C. perfringens produces these acids in a ratio of 1:2, with significant amounts of ethanol and lactate.

13.12 Ruminococcus albus

Ruminococcus albus is a rumen bacterium that can ferment glucose to ethanol, acetate, H_2, and CO_2 using the glycolytic pathway (Fig. 13.12). It is a good fermentation to examine because it illustrates how growth of bacteria in mixed populations can influence fermentation end products. The fermentation is easy to understand. The glucose is first oxidized to two moles of pyruvate yielding two moles of NADH. Then each mole of pyruvate is oxidized to acetyl-CoA, CO_2, and H_2 using the pyruvate–ferredoxin oxido-reductase and hydrogenase. One of the acetyl-CoA molecules is converted to acetate allow-ing an ATP to be made. The second acetyl-CoA is reduced to ethanol using the two NADH. Thus the fermentation is balanced. The overall reaction is the conversion of one mole of glucose to one mole of ethanol, one mole of acetate, two moles of hydrogen, and two moles of carbon dioxide yielding three ATPs. Something different happens when the bacterium is grown in a co-culture with a methanogen. Growth with a methanogen shifts the fermentation in the direction of acetate with the concomitant production of more ATP. This is explained in the following way. R. albus has a hydrogenase that transfers electrons from NADH to H^+ to produce hydrogen gas. When hydrogen accumulates in the medium, the hydrogenase does not oxidize NADH because the equi-librium favors NAD^+ reduction. The NADH therefore reduces acetyl-CoA to ethanol. However, the methanogen utilizes the hydro-gen gas for methane production and keeps the hydrogen levels very low. In the presence of the methanogen, the NADH in R. albus reduces protons to hydrogen gas instead of reducing acetyl-CoA. The result is that R. albus converts the acetyl-CoA to acetate. This is also an advantage to the methanogen since methanogens can also use acetate as a carbon and energy source. The transfer of hydrogen gas from one species to another is called interspecies hydrogen transfer and is an example of nutritional synergism found among mixed populations of bacteria. (See the discussion of interspecies hydrogen transfer in Section 13.4.1.)

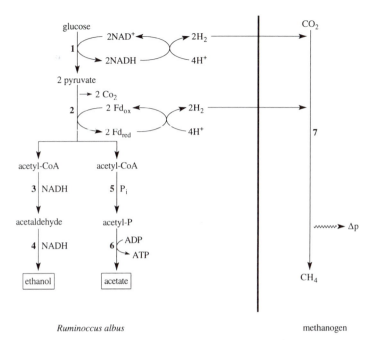

Fig. 13.12 Fermentation of glucose by *R. albus*. *R. albus* ferments glucose to a mixture of ethanol, acetate, CO_2, and H_2. Methanogens draw off the H_2 thus stimulating electron flow to H_2. The result is a shift in the fermentation end products towards acetate accompanied by a greater ATP yield. The production of H_2 by one species and its utilization by another is called interspecies hydrogen transfer. The methanogens can also utilize the acetate produced by *R. albus*.

13.13 Summary

Fermentations are cytosolic oxidation–reduction pathways where the electron acceptor is an organic compound, usually generated in the pathway.

The source of ATP in fermentative pathways is substrate-level phosphorylation. For sugar fermentations, these are the phosphoglycerate kinase, pyruvate kinase, acetate kinase, and butyrate kinase reactions. In other words, ATP is made from bisphosphoglycerate, phosphoenolpyruvate, acetyl-P, and butyryl-P. Butyryl-P is a phosphoryl donor during butyrate fermentations.

However, other means of conserving energy are available for fermenting bacteria. For example, a Δp can be created during electron flow to fumarate, the fumarate being generated during fermentation. Other means of creating a proton potential or sodium potential exist in certain groups of bacteria. These include efflux of carboxylic acids in symport with protons or sodium ions, decarboxylases that function as sodium ion pumps, and oxalate:formate exchange.

Besides making ATP, fermenting bacteria must have some place to unload the electrons removed during oxidation reactions. Of course, the reason for this is that they must regenerate the NAD$^+$, oxidized ferredoxin, and FAD to continue the fermentation. The electron acceptors are sometimes referred to as "electron sinks." During a fermentation the electron sinks are created from the carbon source. In fact, all of the electron sinks for the major carbohydrate fermentations are either pyruvate itself, or synthesized from pyruvate or phosphoenol pyruvate plus CO_2. Protons can also be used as electron sinks and many fermenting bacteria have hydrogenases that reduce protons to hydrogen gas.

The excreted end products of fermentations, including hydrogen gas, are used by other anaerobic bacteria so that an anaerobic food chain develops. At the bottom of the food chain are the methanogens that convert hydrogen gas, carbon dioxide, and acetate to CO_2 and CH_4. These are recycled to organic carbon in the biosphere as they are used by autotrophic and methanotrophic organisms as a source of carbon.

Study Questions

1. Write a fermentation balance using both the O/R and the available hydrogen method for the following:

$$C_6H_{12}O_6 \text{ (glucose)} + H_2O$$

$$\longrightarrow C_2H_4O_2 \text{ (acetate)} + C_2H_6O \text{ (ethanol)}$$

$$+ 2H_2 + 2CO_2$$

If the EMP pathway is used, what is the yield of ATP?

2. C. propionicum and Propionibacterium both carry out the following reaction:

$$3 \text{ lactate} \longrightarrow \text{acetate} + 2 \text{ propionate} + CO_2.$$

C. propionicum nets one ATP but Propionibacterium derives more. How might Propionibacterium make more ATP from the same overall reaction?

3. Write an ethanol fermentation using the Entner–Doudoroff pathway. Contrast the ATP yields with an ethanol fermentation using the EMP pathway.

4. Consider the following fermentation data for Selenomonas ruminantium[7] (in mmol) of products formed per mmol of glucose.

Lactate	0.31
Acetate	0.70
Propionate	0.36
Succinate	0.61

What is the fermentation balance using the O/R method and the available hydrogen method? What percentage of the glucose carbon is recovered in end products?

5. Consider the following data for fermentation products made by Selenomonas rumanantium in pure culture and in co-culture with Methanobacterium rumanantium (in moles per 100 moles of glucose).

Product	Selenomonas	Selenomonas + Methanobacterium
Lactate	156	68
Acetate	46	99
Propionate	27	20
Formate	4	0
Methane	0	51
CO_2	42	48

Source: Chen, M., and M. J. Wolin. 1977. Influence of CH_4 Production by Methanobacterium ruminantium on the fermentation of glucose and lactate by Selenomonas ruminantium. App. Env. Microbiol. 34:756–759.

Offer an explanation accounting for the shift from lactate to acetate in the co-culture compared to the pure culture. How might you expect this to affect the growth yields of Selenomonas?

REFERENCES AND NOTES

1. Other enzymes besides PEP carboxytrans-phosphorylase that carboxylate C_3 glycolytic intermediates to oxaloacetate are PEP carboxylase and pyruvate carboxylase (Section 8.8), and PEP carboxykinase (Section 8.12). The latter enzyme usually operates in the direction of PEP synthesis but in some anaerobes (e.g., Bacteroides) it functions to synthesize oxaloacetate.

2. Gest, H. 1983. Evolutionary roots of anoxygenic photosynthetic energy conversion, pp. 215–234. In: The Phototrophic Bacteria: Anaerobic Life in the Light. Studies in Microbiology, Vol. 4. J. G. Ormerod (ed.). Blackwell Scientific Publications, Oxford.

3. Tasaki, M., Kamagata, Y., Nakamura, K., Okamura, K., and K. Minami. 1993. Acetogenesis from pyruvate by Desulfotomaculum thermobenzoicum and differences in pyruvate metabolism among three sulfate-reducing bacteria in the absence of sulfate. FEMS Microbiol. Lett. 106:259–264.

4. Gas production is generally observed as a bubble in an inverted vial placed in the fermentation tube. The bubble is due to H_2, since CO_2 is

very soluble in water.

5. Jones, D. T., and D. R. Woods. 1986. Acetone–butanol fermentation revisited. *Microbiol. Rev.* 50:484–524.

6. Hamilton, W. A. 1988. Energy transduction in anaerobic bacteria, pp. 83–149. In: *Bacterial Energy Transduction*. C. Anthony (ed.). Academic Press, New York.

14

Homeostasis

Homeostasis refers to the ability of living organisms to maintain an approximately constant internal environment despite changes in the external milieu. As applies to the bacterial cell, this means (among other things) the ability to maintain a steady intracellular pH and a constant osmotic differential across the cell membrane, despite fluctuations in external pH and osmolarity. The regulatory mechanisms underlying homeostasis are generally unknown, although they are a subject of active research.

14.1 Maintaining a ΔpH

14.1.1 Neutrophiles, acidophiles, and alkaliphiles

Bacteria can be found growing in habitats that vary in pH from approximately pH 1–2 in acid springs to as high as pH 11 in soda lakes and alkaline soils. However, regardless of the external pH, the internal pH is usually maint   d within 1–2 units of neutrality, which is necessary to maintain viability (Table 14.1).[1-6] Thus, bacteria maintain a pH gradient (ΔpH) across the cell membrane. For example, bacteria that grow optimally between pH 1 and 4 (i.e., *acidophiles*) have an internal pH of about 6.5–7. That is to say, they maintain a ΔpH of greater than 2.5 units, inside alkaline. Those that grow optimally between pH 6 and 8 (i.e., *neutrophiles*) have

an internal pH of around 7.5–8, and can maintain a ΔpH of 0.5–1.5 units, inside alkaline. Those that grow best at pH 9 or above, often in the range of pH 10–12 [i.e., *alkaliphiles (alkalophiles)*] have an internal pH of 8.4–9. Therefore, alkaliphiles maintain a negative ΔpH of about 1.5–2 units, inside acid. (This lowers the Δp. For a discussion of this point, see Section 3.10.)

14.1.2 Demonstrating pH homeostasis

One method to demonstrate pH homeostasis is to perturb the cytoplasmic pH by changing the external pH, and then study the recovery process. For example, when *E. coli* is exposed to a rapid change in the external pH (e.g., of 1.5–2.5 units) the internal pH initially changes in the direction of the external pH. However, a recovery soon occurs as the internal pH bounces back to its initial value (Fig. 14.1). In other bacteria, pH homeostasis may be so rapid that a temporary change in the internal pH may not even be measurable.

Table 14.1 pH homeostasis in bacteria

Bacteria	pH_{out}	pH_{in}	ΔpH (in–out)
Neutrophile	6–8	7.5–8	+
Acidophile	1–4	6.5–7	+
Alkaliphile	9–12	8.4–9	−

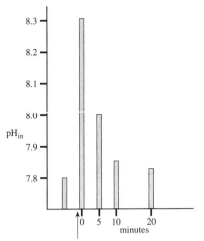

14.1.3 The mechanism of pH homeostasis

Proton pumping

One would expect that many factors influence the intracellular pH. These include the buffering capacity of the cytoplasm and metabolic reactions that produce acids and bases. However, it is generally believed that the regulation of intracellular pH is, to a large extent, a consequence of controlling the flow of protons across the cell membrane.

The reason for believing this is that the ΔpH is dissipated when the proton pumps are inhibited or when proton ionophores are added to the medium. (Recall that proton ionophores equilibrate protons across the cell membrane.) The idea is that, when the cytoplasm becomes too acid, protons are pumped out. This must be done electroneutrally (e.g., by bringing K^+ into the cell). When the cytoplasm becomes too basic, protons are brought in via exchange with outgoing K^+ or Na^+ (Fig. 14.2). This implies the existence of feedback mechanisms where the intracellular pH can signal proton pumps and antiporters. Regulation at this level is not understood. It should be pointed out that even the influx of protons depends

Fig. 14.1 pH homeostasis in *E. coli*. Growing *E. coli* cells were shifted from pH 7.2 to pH 8.3 (arrow). The difference in pH ($pH_{in} - pH_{out}$) was measured using a weak acid or a weak base as described in Chapter 3, and the pH_{in} was calculated. Immediately after shifting the cells to pH 8.3, the cytoplasmic pH rose to 8.3. However, within a few minutes the cytoplasmic pH was restored to its approximate original value. The mechanism by which *E. coli* acidifies the cytoplasm to maintain pH homeostasis is uncertain. (Data from Zilberstein, D., V. Agmon, S. Schuldiner, and E. Padan. 1984. *Escherichia coli* intracellular pH, membrane potential, and cell growth. *J. Bacteriol.* 158:246–252.)

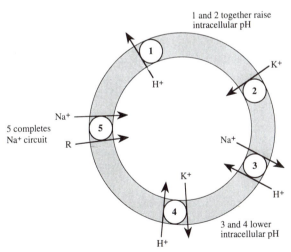

Fig. 14.2 Mechanisms of pH homeostasis. (1) Primary proton pumps create a membrane potential. (2) The uptake of K^+ dissipates the membrane potential allowing extrusion of protons via the pumps and an alkalinization of the cytoplasm. The bulk solution is kept electrically neutral because of the counterions that had neutralized the K^+. Therefore, a $\Delta\Psi$ is changed into a ΔpH, inside alkaline. (3 and 4) Cation/proton antiporters pump Na^+ and K^+ out and bring in H^+. This is suggested to be the major mechanism for acidifying the cytoplasm. (5) Sodium ion uptake systems complete the sodium circuit.

upon the outgoing proton pumps. This is because the antiporters that bring in protons in exchange for sodium or potassium ions are driven by the Δp, either by the ΔpH, outside acid, and/or by the membrane potential. The latter is the case when the inflow of protons is electrogenic [i.e., when the ratio of H^+ to Na^+ (or K^+) on the antiporter exceeds one.] Therefore, inhibition of the proton pumps should lead to a situation where $pH_{in} = pH_{out}$ (i.e., a collapse of the ΔpH). This can be tested. The two major classes of proton pumps are those coupled to respiration and the proton translocating ATPase. The former can be inhibited by respiratory poisons such as cyanide, and the latter by inhibitors of the ATPase (e.g., DCCD) or by mutation. One can also short-circuit the proton flow by using proton ionophores. The ionophores will dissipate the proton potential, thus neutralizing the pumps.[7] These experiments show that in the absence of proton pumping the ΔpH falls as the protons tend to equilibrate across the cell membrane.

The role of K^+ in maintaining a ΔpH

A problem in using the proton pumps to move protons out of the cell is that the pumps are electrogenic, and the number of protons that can be pumped out of the cell is limited by the membrane potential that develops. (See the discussion of membrane capacitance in Section 3.2.2.) Thus, in order for proton pumping to raise the intracellular pH, the excess membrane potential must be dissipated either by the influx of cations or the efflux of anions. In neutrophilic bacteria, K^+ influx dissipates the membrane potential allowing more protons to be pumped out of the cell. This means that K^+ influx is required to raise the intracellular pH.

This is seen in Fig. 14.3 where the addition of K^+ to E. coli cells suspended in media of low pH caused an immediate increase in intracellular pH, which stabilized at approximately pH 7.6. There is much more uncertainty as to how neutrophiles might decrease their intracellular pH. Potassium ions and sodium ions have been suggested to play a role in lowering the intracellular pH by bringing protons back into the cell via the H^+/K^+ and H^+/Na^+ antiporters (Fig. 14.2).

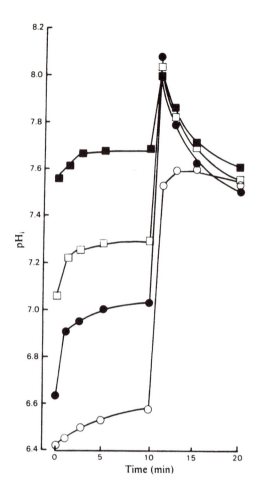

Fig. 14.3 Uptake of K^+ in E. coli is associated with pH homeostasis. Washed E. coli cells were suspended in buffer without K^+ at pH 5.3 (open circles), 6.8 (closed circles), 7.15 (open squares), or 7.6 (closed squares). The cytoplasmic pH (pH_i) was determined using the distribution of a weak acid. Glucose was added shortly after 0 min and K^+ was added at 10 min. Potassium uptake was complete within 10 min after its addition. Immediately upon the addition of K^+ the intracellular pH, which had been lowered by the extracellular pH, rose and stabilized at approximately pH 7.6. The rise in the cytoplasmic pH was due to pumping protons out of the cell in response to the depolarization of the membrane by the influx of potassium ions. The mechanism for acidification of the cytoplasm, (i.e., recovery from overshoot of the pH_{in}), is unknown. (From Kroll, R. G. and I. R. Booth. 1983. The relationship between intracellular pH, the pH gradient and potassium transport in Escherichia coli. Biochem. J. 216:709–716.)

However, the evidence for this is not as strong in neutrophiles as it is in alkaliphiles.

In alkaliphiles the Na^+/H^+ antiporter acidifies the cytoplasm

Since alkaliphiles live in a very basic medium, their main problem is keeping a cytoplasmic pH more acid than the external pH, perhaps by as much as 1.5 units. In other words, they must always be bringing protons into the cell. In contrast to research with neutrophiles, a strong case can be made for the acidification of the cytoplasm of alkaliphiles by Na^+/H^+ antiporters[8] (Fig. 14.2). When alkaliphiles are placed in a medium without Na^+ at pH 10 or 10.5, the internal pH quickly rises to the value of the external pH. However, when Na^+ is present in the external medium, the internal pH does not rise upon shifting to the more basic medium. Furthermore, mutants of alkaliphiles that cannot grow at pH values above 9 are defective in Na^+/H^+ antiporter activity. The antiporter is electrogenic $(H^+ > Na^+)$ and driven by the $\Delta\Psi$ which is generated by the primary proton pumps of the respiratory chain. The sodium ion circuit is completed when sodium ion enters the cell via Na^+/solute symporters that are also driven by the $\Delta\Psi$. The use of Na^+/solute symporters has the advantage that solute transport is driven by the sodium potential rather than the proton potential, the latter being low because of the inverted ΔpH.

pH Homeostasis in acidophiles

Acidophiles differ from other bacteria in that the external pH is several units *lower* than the cytoplasmic pH.[9] The maintenance of the large ΔpH requires an *inverted* $\Delta\Psi$ at low pH_o; otherwise proton efflux would be limited by a positive $\Delta\Psi$ as well as a low pH_o, and proton influx would be promoted by a negative $\Delta\Psi$ as well as a high pH_{in}. Accordingly, acidophilic bacteria have small membrane potentials which can be inside positive at acidic pH_o. For example, the membrane potential of *Thiobacillus ferrooxidans* is $+10$ mV at pH_o 2, and the membrane potential of *Bacillus acidocaldarius* is $+20$ to $+30$ mV at pH_o 2.5. However, acidophiles pump protons out of the cell during electron transport generating a membrane potential that is outside positive like other bacteria. How is the membrane potential inverted? It has been suggested that the maintenance of the inverted membrane potential in acidophiles is due to an *inward flux of K^+* greater than an outward flux of protons. This might be due to the electrogenic influx of K^+ catalyzed by an ATP-dependent K^+ pump known to exist. Thus, the method of maintaining a ΔpH in acidophiles and neutrophiles may be similar in relying on K^+ influx to depolarize the membrane. It has also been suggested that the *efflux* of K^+ in acidophiles may be slowed when pH_i falls, thus limiting the entry of protons against a positive $\Delta\Psi$.

Although the protons are not being pumped against a membrane potential when the external pH is acid, they are being pumped against a proton gradient. The ΔpH in acidophiles can be 4–5 units which is equivalent to 240–300 mV (i.e., 60 ΔpH). This is a large concentration gradient against which to pump protons. The energy to pump the protons is derived from aerobic respiration. However, iron-oxidizing acidophilic bacteria do not generate sufficient energy during electron transport to pump protons out of the cell because the ΔE_h between Fe^{3+}/Fe^{2+} and O_2/H_2O is very small (less than 100 mV). They appear to regulate their ΔpH by consuming cytoplasmic protons during respiration, rather than by proton pumping. This is discussed in Section 11.4.1.

14.2 Osmotic potential

When two solutions are separated by a membrane that allows the passage of water but not of solute, then the water will diffuse from the less concentrated to the more concentrated side in order to equalize the water concentration. What is happening is that the concentration of water (actually the thermodynamic activity of water) in the less concentrated solution is higher than in the concentrated solution. Thus the water is simply following its concentration gradient. The diffusion of water into the more concentrated solution is called *osmosis*. If the water were diffusing from a side with pure water, then the pressure that would have to be applied to stop the osmotic flow of water is called the *osmotic pressure*. Csonka[10] has pointed out that the term osmotic *potential* is more useful than osmotic pressure in that it emphasizes that water flows from solutions of a high osmotic potential to solutions of a low osmotic potential. The osmotic potential, π, is a function of the activity (a) of the solvent. For water, this would be:

eq. 14.1 $\pi = (RT/V_w) \ln a_w$

where R is the universal gas constant, T is the absolute temperature, V_w is the partial molal volume of water, and a_w is the activity of water. The activity of pure water is defined as one, making the osmotic potential zero. Solutes tend to lower the activity of water, therefore making the osmotic potential of solutions negative. This is because in an ideal solution (i.e., one where the interactions between solute and solvent molecules are independent of concentration), the activity of the solvent is equal to its mole fraction. For example, for water:

eq. 14.2 $a_w = n_w/(n_w + n_s)$

where n_w equals the number of water molecules and n_s equals the number of solute molecules. For dilute solutions, the osmotic potential is related to the molar concentration (moles of solute per liter), c_s, as:

eq. 14.3 $\pi = -RTc_s$

Equation 14.3[11] points out that solutions with higher concentrations of solutes have more negative osmotic potentials. Thus, water flows into these solutions. Equation 14.3 allows one to calculate the total osmolarity (osmotic concentration) of all of the solutes. The total osmolarity is defined as the sum of the concentration of solutes that give the osmotic potential, π. It is equal to $c_{total} = -\pi/RT$. It is important to point out that c represents the concentration of independent particles that contribute to the osmotic pressure. For example, c is equal to the sum of the concentration of ions produced when a salt completely ionizes in solution. The term

osmolar refers to the molarity (moles per liter of solution) of independent particles (c) that contribute to the osmotic pressure or osmotic potential. Often concentrations are expressed as molality (moles of solute per kilogram of solvent) rather than molarity, and the units of osmotic pressure are given as *osmolality*.

14.2.1 Turgor pressure and its importance for growth

Because the cytoplasm of most bacterial cells is much more concentrated in particles than is the medium in which the cells are suspended, the cytoplasm has a more negative osmotic potential than the medium, and water flows into the cell. The incoming water expands the cell membrane which exerts a pressure directed outwards against the cell wall. The pressure exerted against the cell wall is called the turgor pressure. The turgor pressure is equal to the difference in solute potential between the medium and the cytoplasm (see eq. 14.3).

eq. 14.4 $P \approx RT (c_{cells} - c_{medium})$

where P is the turgor pressure, and c is the total concentration of osmotically active solutes in the cells and in the medium.

The turgor pressures in gram-positive bacteria are about 15–20 atm and in gram-negative bacteria between 0.8 and 5 atm.[12–14] This, of course, is the reason that bacterial cell walls must be so strong. In bacteria, the tensile strength of the cell wall is due to the peptidoglycan. It is important to realize that the turgor pressure provides the force that expands the cell wall and is necessary for the growth of the wall and cell division.[15] In fact, sudden decreases in turgor pressure brought on by increasing the osmolarity of the suspending medium result in a cessation of growth accompanied by the inhibition of a variety of physiological activities (e.g., nutrient uptake and DNA synthesis). Therefore, the physiological significance of osmotic homeostasis is that it maintains an internal turgor pressure necessary for growth.

Bacteria have the capability of adjusting

their internal osmolarity to changes in external osmolarity in order to maintain cell turgor. How bacteria detect differences in external osmolarity and transfer the appropriate signals to the adaptive cellular machinery is largely unknown. The problems associated with analyzing the signaling are discussed in Section 14.2.4. Before we address these problems, we will examine the evidence for osmotic regulation and identify the molecules primarily responsible for maintaining the osmotic differential between the cytoplasm and the external medium.

14.2.2 Adaptation to high osmolarity

When cells are placed in media of high osmolarity, they increase the intracellular concentrations of certain solutes called *osmolytes*, thus ensuring that the internal osmolarity is always higher than the external, and that cell turgor is therefore maintained. The osmolytes used by bacteria are sometimes called *compatible solutes* to reflect their relative nontoxicity. Some compatible solutes are not synthesized by the cells but are accumulated intracellularly from the medium via transport. These are called *osmoprotectants*. One of the most important compatible solutes in bacteria is K^+, whose intracellular concentration is sufficiently high to be a major contributor to the internal osmolarity and hence turgor.[16] Bacteria that live in high osmolarity media have proportionally higher intracellular concentrations of K^+ because of uptake. Therefore, K^+ is important for two major aspects of homeostasis: maintenance of cell turgor and maintenance of a ΔpH (Section 14.1.3). Bacteria employ other compatible solutes in addition to K^+. The situation with regard to compatible solutes can be summarized as follows.

Solutes that increase intracellularly in many different bacteria in response to high external osmolarity include K^+, the amino acids glutamate, glutamine, and proline, the quarternary amine betaine (also called trimethylglycine or glycine betaine), and certain sugars (e.g., trehalose) which is a disaccharide of glucose.

Osmotic homeostasis in E. coli

It is possible to observe osmotic homeostasis by rapidly increasing the external osmolarity of the medium and identifying the intracellular compatible solutes that maintain turgor. As an example of osmotic homeostasis, some experiments performed with *E. coli* will be considered.[17] When *E. coli* is shifted to a medium of high osmolarity, a series of adjustments take place (Fig. 14.4). First, there is an influx of K^+ via the potassium uptake systems discussed in Chapter 15. These are thought to respond to a decrease in turgor pressure.[18] At around the same time the cells synthesize glutamic acid.[19] The idea is that the glutamic acid ionizes, providing counterions to the increased levels of K^+, and the protons are pumped out of the cell to counteract the membrane depolarization caused by the electrogenic influx of K^+. Thus, the initial major compatible solute is potassium glutamate. However, after several minutes the potassium glutamate is replaced by the newly synthesized sugar, trehalose, which then becomes the major compatible solute. This is due to the excretion of the K^+ and the catabolism or excretion of glutamate as well as the synthesis of trehalose. If proline or betaine are present in the medium, then *E. coli* transports these into the cell and replaces the trehalose or the excess K^+, indicating that *E. coli* preferentially use proline and betaine as osmotic stabilizers rather than trehalose or K^+.[20] (The K^+ is excreted and the trehalose is catabolized.) The preferential use of some osmoprotectants over others is common among bacteria. In many bacteria betaine is a preferred osmoprotectant that is readily accumulated from media of high osmolarity and can even suppress the uptake of other osmoprotectants, as well as causing the excretion of K^+ and the catabolism of trehalose, thus replacing these osmolytes.

Effect of osmolarity on transcription and on activities of enzymes

As part of adaptation to a change in media osmolarity, bacteria synthesize new enzymes or transporters which are responsible for the biosynthesis of compatible solutes or the

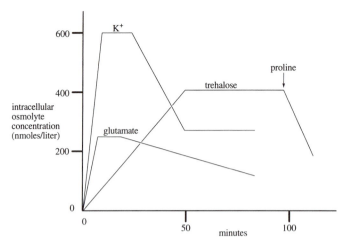

Fig. 14.4 Changes in intracellular osmolyte concentrations when *E. coli* is shifted to a medium of higher osmolarity. Growing cells were shifted to a medium containing 0.5 NaCl at time zero. The intracellular osmolyte concentrations are the differences between the concentrations in cells subjected to osmotic upshock and control cells that were not. When the cells were moved to a higher osmolarity medium, they accumulated potassium ion and synthesized glutamate. The potassium glutamate was replaced by newly synthesized trehalose. The addition of proline to the medium (arrow) caused the displacement of trehalose by proline. Proline was also capable of inducing an early efflux of K^+ if added earlier. (Data adapted from Dinnbier, U. E. Limpinsel, R. Schmid, and E. P. Bakker. 1988. Transient accumulation of potassium glutamate and its replacement by trehalose during adaptation of growing cells of *Escherichia coli* K-12 to elevated sodium chloride concentrations. *Arch. Microbiol.* 150:348–357.)

transport of these into the cells. They also activate pre-existing enzymes or transporters that synthesize compatible solutes or bring them into the cell. A few examples are listed below:

1. *E. coli* activates enzymes for the synthesis of periplasmic oligosaccharides when shifted to low osmolarity media. This is explained in Section 14.2.3.

2. *Staphylococcus aureus* activates a pre-existing proline uptake system when shifted to high osmolarity media.[21] Proline is an osmoprotectant.

3. *E. coli* and *S. typhimurium* increase the transcription of *proU*, an operon that codes for a proline (and betaine) transport system when shifted to high-osmolarity media.

4. Another set of genes whose transcription is increased in high-osmolarity media is the *kdp* operon in *Escherichia* which codes for a K^+ transport system. As mentioned, K^+ is a major osmoprotectant in most bacteria.

One can ask how bacteria sense changes in the external osmolarity and transmit the appropriate signals to the genome or to certain enzymes. As indicated earlier, the answers are largely unknown. Perhaps the best-studied signaling system that responds to changes in external osmolarity is the genes for the OmpF and OmpC porins in *Escherichia* and *Salmonella* (Section 17.7). The total amounts of OmpF and OmpC are fairly constant but the ratio changes with the osmolarity and temperature of the medium. In *high-osmolarity* media the transcription of the *ompF* gene, which codes for the *larger OmpF channel* is *repressed* and the transcription of the *ompC* gene, which codes for the *smaller OmpC channel* is *increased*. The result is a switch to a smaller porin channel in high-osmolarity media. Why bacteria should switch from one porin to the other is not clear but it probably has nothing to do with osmotic homeostasis. The smaller OmpC channel is probably an advantage in the intestinal tract where these bacteria live because of the presence of toxic molecules (e.g., bile salts).[22] The argument is that, in the intestinal tract, the osmotic pressures are higher thus favoring the smaller OmpC channels, whereas in ponds and streams where the bacteria are also found, the lower osmotic pressures favor the larger OmpF channels, which may allow more efficient uptake of nutrients. Consistent with this hypothesis is that higher temperatures, expected in the intestines of animals as opposed to habitats outside the body, also repress the transcription of *ompF*. Regardless of the physiological significance of the switch in porins, the system is of interest to us here because it is regulated in some way by osmotic pressure, and the signaling pathway from the membrane to the genome is being dissected experimentally.[23] The signaling pathway, which involves a membrane sensor protein called EnvZ, is reviewed in Chapter 17.

14.2.3 Adaptation to low osmolarity media

Thus far we have been considering the adaptation of bacteria to *high* osmolarity media, which tends to suck water out of the cytoplasm and lower the turgor pressure. As discussed above, bacteria respond to high osmolarity by raising the intracellular concentrations of compatible solutes. Many bacteria have means for adjusting to low osmolarity media, thus limiting cell turgor pressure. Not very much is known about the details. One response of bacteria to low osmolarity media is to decrease the concentration of cytoplasmic osmolytes. This might occur via specific excretion of osmolytes or their catabolism. For example, *E. coli* excretes K^+ via special transporters which may respond to cell turgor.[24]

Osmotic homeostasis in the periplasm
The periplasm is reportedly filled with a gel whose volume is still a matter of controversy (Section 1.2.4).[25] It has been suggested that gram-negative bacteria adapt to low osmolarity media by raising the osmolarity of the periplasm so that the cytoplasm never actu-

ally "sees" the low osmolarity external medium. In this way swelling of the cell membrane with concomitant compression of the periplasm are minimized, as well as the turgor pressure across the cell membrane. (The option of lowering the osmolarity of the cytoplasm by excreting solutes is limited. The cytoplasm must maintain a minimum concentration of salts and other solutes, approximately 300 mosM, to support growth.) In fact, it has been reported that the periplasm remains as a separate compartment under all conditions of external osmolarity, with an osmolarity apparently isoosmotic with that of the cytoplasm.[26] However, since the outer membrane of *Escherichia* and *Salmonella*, and by inference other gram-negative bacteria, is permeable to small molecules of molecular weight less than 600 because of the nonspecific diffusion channels formed by the porins, one can ask how bacteria maintain an osmolarity in the periplasm higher than that of the external medium. One possibility is that gram-negative bacteria synthesize and/ or accumulate periplasmic osmolytes when grown in low-osmolarity media. Indeed, many gram-negative bacteria, including *E. coli, S. typhimurium, Pseudomonas, Agrobacterium,* and *Rhizobium* synthesize negatively charged periplasmic oligosaccharides (called membrane-derived oligosaccharides, or MDO, in *E. coli* and β-glucans in other strains) when grown in media of *low* osmolarity.[27,28] (Some of the β-glucans are neutral.) The enzymes that synthesize the oligosaccharides are constitutive, therefore their *activities* are increased when the cells are grown in low osmolarity media. The increase in the synthesis of the periplasmic oligosaccharides when the bacteria are grown in media of low osmolarity has led to the suggestion that the role of the oligosaccharides is to raise the osmolarity in the periplasm. The anionic oligosaccharides would be expected to be very effective in raising the osmolarity of the periplasm because cations would accumulate in the periplasm in response to the negatively charged nonpermeable oligosaccharides.[29] However, *E. coli* mutants unable to synthesize the MDOs show no growth defects in low osmolarity media. If it is necessary to maintain a high periplasmic osmolarity when the external osmolarity is decreased, then there must exist alternative mechanisms besides the synthesis of MDOs.

14.2.4 Conceptual problems

It has been suggested that the periplasm is maintained isoosmotic with respect to the cytoplasm. If indeed this is the case, then this has important consequences for our understanding of how the cell walls of gram-negative bacteria are able to resist high turgor pressures. It is generally believed that the turgor pressure is exerted across the cell membrane not the outer membrane, and that the overlying peptidoglycan acts as a strong retainer against which the cell membrane is pressed. If the periplasm were isoosmotic with the cytoplasm, then the turgor pressure would not be across the cell membrane and against the peptidoglycan, but rather against the outer membrane. However, the outer membrane is not built to withstand high turgor pressures. If indeed the turgor pressure were exerted against the outer membrane, then one must suppose that the peptidoglycan reinforces the outer membrane by being tightly bonded to it at numerous sites. Perhaps the lipoprotein molecules that are covalently bonded to the peptidoglycan are anchored sufficently to the outer membrane via hydrophobic bonding to provide the needed stability (Section 1.2.3).[30] Another conceptual difficulty regarding a periplasm isoosmotic with the cytoplasm is that it is not clear how turgor pressure would be sensed by cell membrane proteins if there were no differential pressure across the cell membrane. (This is discussed in Section 14.2.5.) Clearly, much more needs to be learned in order to come to a better understanding of turgor pressure and osmotic regulation in the periplasm.

14.2.5 What is the nature of the signal sensed by the osmosensors?

The signals to which the putative osmosensors respond are not understood.[31,32] Indeed, the different osmosensors may not even respond to the same type of signal. Several possibilities

for osmosensor responses have been discussed, including membrane proteins that are sensitive to: (1) pressure against the peptidoglycan sacculus; (2) membrane stretch; (4) changes in the concentrations of intracellular solutes (i.e., K^+); and (5) water activity. It is important to know whether the periplasm is truly isoosmotic with the cytoplasm as discussed above. For under these conditions there should be no pressure differential across the cell membrane, and mechanisms (1) and (2) above would not apply.

It should also be understood that after the cells adapt to an osmotic upshift and begin to grow again, the osmotic differential between the cytoplasm and medium is presumably restored and, therefore, systems still activated under these circumstances cannot be responding to a change in turgor pressure or related events such as membrane stretching. They could, however, be responding to increased concentrations of specific solutes or some other parameters not dependent upon a changed turgor pressure. For example, when E. coli is shifted to a high-osmolarity medium, it continues to repress the ompF gene and stimulate the ompC gene, even though adaptation to the high-osmolarity medium has taken place. Similarly, the proU genes that specify proline and glycine betaine uptake systems in E. coli and S. typhimurium remain induced in high-osmolarity media, also implying that the sensor is not responding to a changed osmotic pressure differential on both sides of the cell membrane. On the other hand, the kdp genes for potassium ion uptake are quickly but only transiently induced by an upshift in osmolarity. This fact, plus the finding that only nonpermeable solutes induce transcription of the kdp operon, suggests that the osmosensor for these genes may indeed detect turgor pressure or something closely related, such as membrane stretch. (See Section 17.7 for a discussion of the kdp gene products.)

14.3 Summary

Bacteria must maintain a fairly constant internal pH and adjust their osmolarity in response to the external osmolarity. Very little is known about the details of how this is done. In broad outline, however, the internal pH is adjusted by using proton pumps to extrude protons, and antiporters that bring protons into the cell in exchange for sodium or potassium ions. What is not clear is precisely how the activities of the pumps and antiporters are regulated by the external pH, although it is reasonable to suggest that they respond to the internal pH, by some sort of feedback mechanism. The acidophilic iron-oxidizing bacteria represent a special problem in that there is very little energy obtainable by Fe^{2+} oxidation to pump protons out of the cell. They maintain a ΔpH by consuming cytoplasmic protons during oxygen respiration.

The reason that bacteria must maintain an osmotic differential across the cell membrane is that the resulting turgor pressure is essential for growth of the cell wall and for cell division. A variety of physiological activities come to a halt when the turgor pressure is suddenly decreased (e.g., nutrient uptake and DNA synthesis). When the external osmotic pressure is increased (causing decreased turgor), bacteria respond by increasing the concentration of internal osmolytes to raise the internal osmotic pressure and turgor. An important molecule in this regard is potassium ion which increases, in some cases transiently, in response to an increase in external osmolarity. A counterion must also increase, and in *E. coli* this is glutamate. In *E. coli*, the K^+ is subsequently replaced by trehalose. If proline or betaine are in the medium, then they replace the K^+ and trehalose.

It is clear that transcriptional changes occur when bacteria are shifted into media of different osmolarities. For example, when *E. coli* is grown in high osmolarity media, the transcription of several genes is increased. These include *proU*, the operon that codes for the uptake of two osmoprotectants, proline and betaine, and *kdp*, the operon that codes for the proteins required for the uptake of K^+. Also, when *E. coli* and *S. typhimurium* are grown in high external osmolarity, the gene for the OmpC porin is activated whereas the gene for the OmpF porin is repressed. This leads to more OmpC and less OmpF. A

protein in the cell membrane called EnvZ has a periplasmic domain and is thought to be an osmosensor (Chapter 17).

There is also osmoregulation of the periplasm in gram-negative bacteria. Gram-negative bacteria adapt to media of low osmolarity by synthesizing membrane-derived oligosaccharides (MDO) in the periplasm. The idea is that the multiple-charged anionic MDO molecules accumulate cations in the periplasm, thus raising its osmolarity. However, mutants of *E. coli* defective in MDO synthesis show no growth defects in low osmolarity media. It must be concluded that other mechanisms of osmoregulation of the periplasm must be present and that osmoregulation of the periplasm is not well understood at this time.

Study Questions

1. Na^+/H^+ and K^+/H^+ antiporters are an important way to lower the intracellular pH. What is the evidence for this in alkaliphiles?

2. No one knows for sure how acidophiles establish an inverted membrane potential. How might it affect pH homeostasis? What experiments would support the hypothesis that an electrogenic K^+ pump was responsible for the inverted membrane potential in acidophiles?

3. Upon shifting *E. coli* to a higher osmolarity, there is a transient uptake of K^+. What is the role of K^+ in this regard? What keeps the cytoplasm neutral? How is the membrane potential maintained? Is there a complication with the ΔpH? Describe the role of K^+ in pH homeostasis.

4. What is meant by turgor pressure? What is the evidence that turgor pressure is necessary for cell growth?

5. What is meant by periplasmic osmotic homeostasis? What might be the role of MDOs?

NOTES AND REFERENCES

1. Padan, E., D. Zilberstein, and S. Schuldner. 1981. pH homeostasis in bacteria. *Biochim. Biophys. Acta* 650:151–166.

2. Bakker, E. P. 1990. The role of alkali-cation transport in energy coupling of neutrophilic and acidophilic bacteria: An assessment of methods and concepts. *FEMS Microbiol. Rev.* **75**:319–334.

3. Krulwich, T. A., A. A. Guffanti, and D. Seto-Young.1990. pH homeostasis and bioenergetic work in alkalophiles. *FEMS Microbiol. Rev.* **75**:271–278.

4. Matin, A. 1990. Keeping a neutral cytoplasm; the bioenergetics of obligate acidophiles. *FEMS Microbiol. Rev.* **75**:307–318.

5. Krulwich, T. A. and D. M. Ivey. 1990. Bioenergetics in extreme environments, pp. 417–447. *The Bacteria*, Vol. XII. Krulwich, T. A. (ed.). Academic Press, New York.

6. Booth, I. R. 1985. Regulation of cytoplasmic pH in bacteria. *Microbiol. Rev.* **49**:359–378.

7. Harold, F. M., E. Pavlasova, and J. R. Baarda. 1970. A transmembrane pH gradient in *Streptococcus faecalis*: Origin, and dissipation by proton conductors and N,N′dicyclohexylcarbodiimide. *Biochim. Biophys. Acta.* **196**:235–244.

8. Reviewed in, Krulwich, T. A., A. A. Guffanti, and D. Seto-Young. 1990. pH homeostasis and bioenergetic work in alkalophiles. *FEMS Microbiol. Rev.* **75**:271–278.

9. Reviewed in, Booth, I. R. 1985. Regulation of cytoplasmic pH in bacteria. *Microbiol. Rev.* **49**:359–378.

10. Csonka, L. N. 1989. Physiological and genetic responses of bacteria to osmotic stress. *Microbiol. Rev.* **53**:121–147.

11. A derivation of eq. 14.3 can be found in the review article by Csonka (1989).[10]

12. Ingraham, J. L. 1987. Effect of temperature, pH, water activity and pressure on growth, pp. 1543–1554. In: Escherichia coli *and* Salmonella typhimurium: *Cellular and Molecular Biology.* F. C. Neidhardt, J. L. Ingraham, K. B. Low, B. Magasanik, M. Schaechter, and H.E. Umbarger (eds.). American Society for Microbiology, Washington, DC.

13. Koch, A. L., and M. F. S. Pinette. 1987. Nephelometric determination of turgor pressure in growing gram-negative bacteria. *J. Bacteriol.* **169**:3654–3668.

14. Walsby, A. E. 1986. The pressure relationships of halophilic and non-halophilic prokaryotic cells determined by using gas vesicles as pressure probes. *FEMS Microbiol. Rev.* **39**:45–49.

15. Koch, A. L. 1991. Effective growth by the simplest means: The bacterial way. *ASM News* **57**:633–637.

16. Glutamate and K^+ are present in high concentrations in most (eu)bacteria. The former is the major organic anion and the latter is the major inorganic cation. Glutamate is generally synthesized rather than accumulated from the medium, using either glutamate dehydrogenase or glutamate synthase discussed in Section 9.3.1. In *E. coli*, potassium ion is taken up by four different transport systems, including the Kdp and Trk systems discussed in Section 15.3.3. Potassium ion efflux is catalyzed by two systems, the KefB and KefC transporters.

17. Dinnbier, U. E., Limpinsel, R. Schmid, and E. P. Bakker. 1988. Transient accumulation of potassium glutamate and its replacement by trehalose during adaptation of growing cells of *Escherichia coli* K-12 to elevated sodium chloride concentrations. *Arch. Microbiol.* **150**:348–357.

18. Epstein, W. 1986. Osmoregulation by potassium transport in *Escherichia coli*. *FEMS Microbiol. Rev.* **39**:73–78.

19. McLaggan, D., T. M. Logan, D. G. Lynn, and W. Epstein. 1990. Involvement of γ-glutamyl peptides in osmoadaptation of *Escherichia coli*. *J. Bacteriol.* **172**:3631–3636.

20. Betaine is synthesized by oxidizing the hydroxyl group in choline to a carboxyl group, i.e.

$(CH_3)_3N^+–CH_2CH_2OH$ (choline)

$\longrightarrow (CH_3)_3N^+–CH_2COOH$ (betaine)

Betaine and glycinebetaine (*N,N,N*-trimethylglycine) are different names for the same molecule.

21. Townsend, D. E., and B. J. Wilkinson. 1992. Proline transport in *Staphylococcus aureus*: A high affinity system and a low-affinity system involved in osmoregulation. *J. Bacteriol.* **174**:2702–2710.

22. Nikaido, H., and M. Vaara. 1985. Molecular basis of bacterial outer membrane permeability. *Microbiol. Rev.* **49**:1–32.

23. Stock, J. B., A. J. Ninfa, and A. M. Stock. 1989. Protein phosphorylation and regulation of adaptive responses in bacteria. *Microbiol. Rev.* **53**:450–490.

24. Bakker, E. P., Booth, I. R., Dinnbier, U., Epstein, W., and A. Gajewska. 1987. Evidence for multiple potassium export systems in *Escherichia coli. J. Bacteriol.* **169**:3743–3749

25. The volume of the periplasm in *Escherichia* and *Salmonella* has been reported to be as high as 20–40% of the total cell volume by some investigators and as low as 5% of the total cell volume by others. One must conclude that there is some uncertainty regarding the size of the periplasm and that, furthermore, the periplasm of different types of bacteria may vary significantly in physical dimensions.

26. Stock, J. B., B. Rauch, and S. Roseman. 1977. Periplasmic space in *Salmonella typhimurium* and *Escherichia coli*. *J. Biol. Chem.* 252:7850–7861.

27. Miller, K. J., E. P. Kennedy, and V. N. Reinhold. 1986. Osmotic adaptation by gram-negative bacteria: Possible role for periplasmic oligosaccharides. *Science* 231:48–51.

28. Weissborn, A. C., M. K. Rumley, and E. P. Kennedy. 1992. Isolation and characterization of *Escherichia coli* mutants blocked in production of membrane-derived oligosaccharides. *J. Bacteriol.* 174:4856–4859.

29. Whenever a solution (1) of a nonpermeant ion is separated from another solution (2) (e.g., by a membrane permeable to other ions) there will be an unequal distribution of ions at equilibrium. The compartment with the nonpermeant ion will contain a higher concentration of permeant ions and hence a higher osmotic pressure (or lower osmotic potential). For the case of the periplasm and the outer membrane, suppose the nondiffusible oligosaccharide is R^- and the diffusible ions are K^+ and Cl^-. Then K^+ will passively accumulate in the periplasm as the counterion to R^-. At equilibrium, the concentration of K^+ will be greatest on the same side of the membrane as R^- (the periplasm) whereas the concentration of Cl^- will be greater on the other side (the medium). Because the K^+ can diffuse across the outer membrane and R^- cannot, a voltage potential, outside positive, develops across the membrane. This is the Donnan potential, which is really a K^+ diffusion potential. The following equations explain the relationships between the diffusible and nondiffusible ions. At equilibrium,

eq. 14.5 $[K^+]_1 = [R^-]_1 + [Cl^-]_1$ and

eq. 14.6 $[K^+]_2 = [Cl^-]_2$

Equations 14.5 and 14.6 describe electrical neutrality in the two solutions.

Because almost all the K^+ that diffuses across the membrane must be coupled with Cl^- or else the diffusion potentials would prevent further cation (or anion) flow, the diffusion kinetics can be described by a second-order rate equation:

$$k[K^+]_1[Cl^-]_1 = k[K^+]_2[Cl^-]_2,$$

where k is the rate constant for diffusion through the membrane.

k cancels out and therefore,

eq. 14.7. $[K^+]_1[Cl]_1 = [K^+]_2[Cl]_2$

But eq. 14.6 states that $[K^+]_2 = [Cl^-]_2$, therefore,

eq. 4. $[K^+]_1[Cl]_1 = [K^+]_2^2$

But eq. 14.5 states that $[Cl^-]_1 = [K^+]_1 - [R^-]_1$, therefore,

eq. 14.9 $[K^+]_1([K^+]_1 - [R^-]_1) = [K^+]_2^2$,

and rearranging,

$$[K^+]_1^2 - [K^+]_1[R^-]_1 = [K^+]_2^2$$

Thus, the concentration of K^+ on the side without R^- (side 2) is less than the concentration on the side with R^- (side 1) by $[K^+]_1[R^-]_1$. Note that if R^- is multivalent then even more K^+ will accumulate, because each negative charge on R^- is balanced by a K^+.

30. The difficulty in reconciling a periplasm isoosmotic with the cytoplasm with the fragility of the outer membrane was suggested to me by Arthur Koch.

31. Csonka, L. N. 1989. Physiological and genetic responses of bacteria to osmotic stress. *Microbiol. Rev.* 53:121–147.

32. Csonka, L. N., and A. D. Hanson. 1991. Prokaryotic osmoregulation: genetics and physiology. *Annu. Rev. Microbiol.* 45:569–606.

15

Solute Transport

Bacterial cell membranes consist in large part of a phospholipid matrix that acts as a permeability barrier blocking the diffusion of water-soluble molecules into and out of cells (Section 1.2.5). This has the advantage of allowing the bacterium to maintain an internal environment different from the external, and one conducive to growth. For example, metabolites can be maintained at an intracellular concentration that is orders of magnitude higher than the extracellular concentration. This has two important consequences: (1) the promotion of more rapid enzymatic reactions, and (2) the retention of metabolic intermediates within the cell. The lipid barrier also minimizes the passive diffusion of ions, including protons, and thus functions to maintain the electrochemical proton and sodium ion gradients that are important for driving ATP synthesis, solute transport, and other membrane activities. Since the phospholipid presents a permeability barrier, virtually everything that is not lipid soluble enters and leaves the cell through integral membrane proteins that have various names, including transporters, carriers, porters, or permeases. The amino acid sequences of a few transporters, deduced from nucleotide sequences, imply that they form multiple transmembrane loops that fold in the membrane, perhaps making an internal channel through which the solute is passed. The solute might move along amino-acid side chains that face into the channel. For example, the lactose/proton symporter from *E. coli* probably forms 12 transmembrane loops.[1] The importance of transporters is easy to demonstrate. Mutants that lack a transporter for a required nutrient will grow very poorly in low concentrations of the nutrient. In the natural environment it would not be expected that such mutants would survive. But they can be isolated during a screen for slow growers and maintained as laboratory cultures.

15.1 Reconstitution into proteoliposomes

Proteoliposomes are artificial membrane vesicles of protein and phospholipid that are of enormous value in studying solute transport, and it is instructive to describe how they are made and some of their properties. There are several methods to prepare proteoliposomes, but they are all similar and rely upon the fact that membrane proteins solubilized in detergent will integrate into phospholipid bilayers when the detergent is removed by dilution or dialysis.[2] One way to prepare phospholipid bilayers is to disperse the phospholipids in water where they spontaneously aggregate to form spherical vesicles called liposomes consisting of concentric layers of phospholipid (Fig. 15.1). The

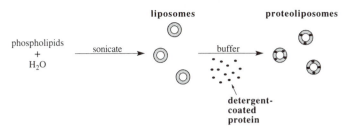

Fig. 15.1 Preparation of proteoliposomes. Phospholipids are dispersed in water and sonicated. Small vesicles form that are each surrounded by a lipid bilayer. The vesicles are mixed with detergent solubilized protein and diluted into buffer. Proteoliposomes form with the protein incorporated into the bilayer. The proteoliposomes can be loaded with substrate or ATP by including these in the dilution buffer.

liposomes are then subjected to high-frequency sound waves (sonic oscillation), which breaks them into smaller vesicles surrounded by a single phospholipid bilayer resembling the lipid bilayer found in natural membranes. Then the protein, which has been solubilized in detergent, is mixed with the sonicated phospholipids in the presence of detergent and buffer, and the suspension is diluted into buffer which lowers the concentration of detergent. The protein leaves the detergent and becomes incorporated into the phospholipid bilayer. The protein and lipid membrane vesicles that form are called proteoliposomes. One can "load" the proteoliposomes with ATP or other substrates by including these in the dilution buffer. When the proteoliposomes are incubated with solute, they catalyze uptake of the solute into the vesicles provided the appropriate transporter has been incorporated. The vesicles can be reisolated (e.g., by centrifugation or filtration) and the amount of solute taken into the vesicles can be quantified. Some examples of transporters that have been reconstituted into proteoliposomes and used to demonstrate catalyzed transport are the lactose permease from *E. coli*, the oxalate/formate antiporter from *Oxalobacter formigenes*, the Na^+/H^+ antiporter from *E. coli*, and the histidine permease from *Salmonella typhimurium*.[3-6] (Proteoliposomes are also used to study electron transport. For example, they have been important in experiments that establish that certain cytochrome oxidases are proton pumps.)

15.2 Kinetics of solute uptake

15.2.1 Transporter-mediated uptake

The existence of transporters can be revealed by the kinetics of solute uptake. If one were to add a solute (e.g., an amino acid or sugar) at different concentrations, to a bacterial suspension, and plot the initial rate of uptake into the cell as a function of the external concentration of solute, a curve such as that shown in Fig. 15.2 would be generated. Notice that the curve is a hyberbola that approaches

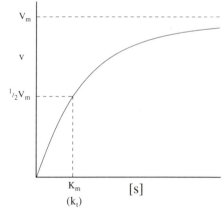

Fig. 15.2 Kinetics of transport. Bacteria are incubated with different concentrations of solute (S) and the initial rate of solute uptake is measured for each solute concentration. The rate (v) approaches a maximum (V_m) as the fraction of transporter molecules bound to solute reaches a maximum. The concentration of solute that gives $1/2 \ V_m$, the K_m (sometimes called k_t), is characteristic for each transporter.

a maximum rate. The kinetics for transporter-mediated solute uptake can be rationalized by assuming that the only significant route of entry for the solute is on a limited number of transporters. That is to say, the solute does not passively leak into the cell to any significant extent. Therefore, the rate at which the solute enters is directly proportional to the fraction of transporters that are occupied with solute. As the external concentration of solute increases, a progressively larger fraction of the transporters bind solute and the rate of transport increases to a maximum rate (V_{max}), when there are no unloaded transporters.

The concentration of solute that produces one-half the maximum initial rate of transport is called the K_m or sometimes the K_t. It is frequently assumed to be a measure of the affinity of the solute for the transporter and called the *affinity constant*. However, it will be referred to here simply as the solute concentration that gives $1/2$ V_{max}. The value of the K_m is characteristic of the transporter and can range from less than one micromolar to several hundred micromolar. The kinetics shown in Fig. 15.2 are described by the Michealis–Menten equation for enzyme catalysis which is explained in Section 6.2.1.

15.2.2 Uptake in the absence of a transporter

What if there were no transporter and the nutrient simply diffused into the cell? Slow-growing mutants have been isolated that have no functional transporter for a particular nutrient. The kinetics of uptake in these mutants reflects what one would expect in the absence of a transporter and is shown in Fig. 15.3. Note that the rates of uptake are low and they are proportional to the concentration gradient, with no saturation at very high external concentrations of solute.

15.3 Energy-dependent transport

A transporter simply facilitates the entry and exit of the solute across the membrane. At equilibrium, it does not bias the transport in

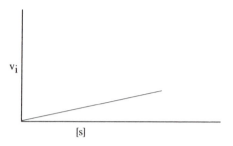

Fig. 15.3 The initial rate of uptake of solute in the absence of a transporter. In the absence of a transporter the rate of solute entry is relatively slow and does not approach a maximum, even at very high concentrations of solute (s), but is proportional to the concentration gradient.

any direction and therefore cannot, on its own, cause the accumulation of solute against a concentration gradient. However, we know that many transport mechanisms catalyze the accumulation of solutes into the cell. The internal concentration when the steady state is reached can be several orders of magnitude higher than the external concentration. Of course, this requires energy. The source of energy can be either *chemical, light,* or *electrochemical.*

Bacterial transport systems are now divided into two categories, *primary transport* and *secondary transport*. Primary transport systems are driven by an energy-producing (exergonic) metabolic event. Primary transport systems include: proton translocation driven by ATP, light, or oxidation–reduction reactions (Sections 3.7.1, 3.7.2, 3.8.4, 4.5), light-driven chloride transport (Section 3.9), sodium ion transporting decarboxylases (Section 3.8.1), the uptake of inorganic or organic solutes driven by ATP hydrolysis (described later in this chapter), and the uptake of sugars driven by phosphoryl group transfer from phosphoenolpyruvate, called the phosphotransferase (PTS) system (described later in this chapter). During transport by the phosphotransferase system, the sugar accumulates inside the cell as the phosphorylated derivative. Secondary transport systems are driven by electrochemical gradients (e.g., proton and sodium ion gradients). During secondary transport the solute moves "down" an electrochemical ion gradient, usually of protons or sodium ions.[7] Secondary transport is coupled to primary transport by the proton in the following way:

Primary transport:

$$H^+_{in} + energy \longrightarrow H^+_{out}$$

Secondary transport:

$$H^+_{out} + S_{out} \longrightarrow H^+_{in} + S_{in}$$

The distinction between primary and secondary transport is an important one because it emphasizes a central feature of the chemiosmosis theory, i.e., that the coupling between energy yielding reactions and energy requiring reactions in the membrane is via ion currents. In the example shown above, the coupling between the energy-dependent uptake of S and the primary energy-yielding reaction is via the proton current, which is most commonly used.

"Active transport" refers to primary transport during which the solute is not chemically modified, (e.g., the ATP-dependent uptake of histidine). Active transport is therefore a subclass of primary transport. (It should be pointed out that, prior to the chemiosmotic theory, the distinction between primary and secondary transport was not made, and an older definition of active transport was *any* transport that results in a concentration gradient where the solute is accumulated in a chemically unmodified form. However, the current definition of active transport restricts its usage to primary transport in which the solute does not chemically change.)

Secondary transport is catalyzed by uniporters, symporters and antiporters that use electrochemical ion gradients to accumulate solutes. Examples are illustrated in Fig. 15.4. *Uniport* refers to solute translocation in the absence of a coupling ion. For example, the uptake of K^+ down its electrochemical gradient is uniport. *Symport* refers to solute

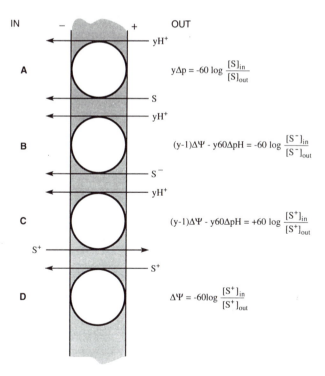

IN — + OUT

A yH^+ $y\Delta p = -60 \log \dfrac{[S]_{in}}{[S]_{out}}$

S

B yH^+ $(y-1)\Delta\Psi - y60\Delta pH = -60 \log \dfrac{[S^-]_{in}}{[S^-]_{out}}$

S^-

C yH^+ $(y-1)\Delta\Psi - y60\Delta pH = +60 \log \dfrac{[S^+]_{in}}{[S^+]_{out}}$

S^+ S^+

D $\Delta\Psi = -60\log \dfrac{[S^+]_{in}}{[S^+]_{out}}$

Fig. 15.4 Some examples of solute transport driven by the $\Delta\Psi$ and the ΔpH. (A) Symport of protons with an uncharged solute. (B) Symport of protons with a monovalent anion. (C) Antiport of protons with a monovalent cation. (D) Uniport of a monovalent cation. Another possibility, not shown, is symport of protons with cations. The ratio of protons to solute is given by y, which is assumed to have a value $\geqslant 1$.

uptake in which two solutes are carried on the carrier in the same direction. An example of this would be the uptake of a solute coupled to the uptake of one or more protons, or sodium ions. *Antiport* refers to the coupled movement of two solutes in opposite directions. For example, an exchange of H^+ for Na^+ on the same carrier is antiport.

15.3.1 Secondary transport

A general description

The way that solute transport is coupled to ion currents is that the transporter functions only when it transports both the ion and the solute in one direction (symport) or the solute in one direction and the ion in the other (antiport) (Fig. 15.4).[8] However, there are also transporters that move just an ion along its electrochemical gradient (uniport). In all of these cases, the transporter is part of the electrical circuit. This is analogous to saying that, in order for an electrical current to make a motor turn, the current must flow through the motor. Most bacteria employ both proton and sodium symporters but mainly the former.[9] However, certain bacteria (e.g., alkaliphiles, halophiles, and marine bacteria) rely more heavily on sodium symporters. The reason for this becomes clear when one considers their ecological niches. The alkaliphiles live in environments with pH values in the range of 9–11. Because their cytoplasmic pH is below 8.5, the ΔpH ($pH_{in} - pH_{out}$) is negative rather than positive. This decreases the Δp. These organisms therefore depend upon the sodium ion and the $\Delta\Psi$ for solute transport. Some of these bacteria also have flagella motors powered by a sodium potential rather than a proton potential.[10,11] Halophilic bacteria require high external NaCl concentrations (3 M to 5 M) in order to grow. They use predominantly Na^+/solute symporters. Marine bacteria also live in high concentrations of Na^+ (close to 0.5 M) and rely heavily on Na^+/solute symporters.[12] The use of ionophores can aid in identifying the ion current that is responsible for secondary transport. This is discussed in Section 3.4.

Energetics of transport

The free energy in joules required to move a mole of solute from outside of a cell to inside the cell is given by eq. 15.1, where $[S]_{in}$ is the concentration inside the cell and $[S]_{out}$ is the concentration outside the cell.

$$eq.\ 15.1\quad \Delta G = RT \ln [S]_{in}/[S]_{out}\quad J/mol$$

$$= 60 \log [S]_{in}/[S]_{out}\quad mV\ at\ 30^0C$$

For example, if $[S]_{in}/[S]_{out} = 10^3$, then $\Delta G = 17.4\,kJ/mol$ at $30°C$.[13] This means that at least $17.4\,kJ$ of work must be applied against the concentration gradient (at $30°C$) to move one mole of S into the cell when the concentration ratio, in/out is 10^3, and remains as such. This would be the case in a steady-state situation where S_{in} is used as fast as it is brought into the cell. Note that $17.4\,kJ$ does *not* refer to the energy required to move 10^3 moles of S to one side of the membrane to establish a ratio of $10^3/1$. Equation 15.1 can also be expressed as an electrical potential in volts since $(RT/F)(2.303) = 0.06\,V$ or $60\,mV$ at $30°C$. Figure 15.4 illustrates some examples of solute transport coupled to an electrochemical gradient. These are discussed next.

Some examples of solute transport coupled to an electrochemical gradient

1. Symport of an uncharged solute with protons (Fig. 15.4A)

Many solutes are transported by symport. Assume that the stoichiometric ratio of H^+/S is y. At equilibrium, the tendency of the solute to diffuse down its concentration gradient (out of the cell) is $-60 \log [S]_{in}/[S]_{out}\,mV$, and is balanced by the Δp drawing the solute in the opposite direction (into the cell) so that at $30°C$:

$$eq.\ 15.2\quad y(\Delta\Psi - 60\,\Delta pH)$$

$$= y\Delta p = -60 \log[S]_{in}/[S]_{out}$$

For example, if the ratio (y) of protons to solute transported is 1:1 and the concentration gradient S_{in}/S_{out} is 10^3, then the minimum Δp which is required to maintain that concentration gradient would be $-180\,mV$. Notice that $[S]_{in}/[S]_{out}$ is an exponential function (logarithmic function) of $y\Delta p$. For example, when y is increased to 2 the maximum concentration gradient attained at equilibrium is squared. Theoretically, very large concentration gradients can be maintained by the Δp.

2. Symport of a monovalent anion with protons (Fig. 15.4B)

The Δp can also drive the uptake of anions. However, whether or not the $\Delta\Psi$ is part of the driving force depends upon whether a net charge is transported. For example, for a monovalent anion, the ratio H^+/R^- must be greater than one in order that the $\Delta\Psi$ can be part of the driving force. The relative contributions of the $\Delta\Psi$ and the ΔpH to the driving force are as follows:

eq. 15.3 $(y-1)\Delta\Psi - y60\Delta pH$

$$= -60\log[S^-]_{in}/[S^-]_{out}$$

Equation 15.3 is the same as eq. 3.25 whose derivation can be found in Section 3.8.3. (See also eq. 3.23 in Section 3.8.3 for a multivalent anion.)

3. Antiport of a monovalent cation with H^+ (Fig. 15.4C)

An example is the proton:sodium ion antiporter, which is widespread among bacteria. It is used for pumping sodium ions out of the cell. The equation is similar to eq. 15.3 except that the sign is changed for the expression for the electrochemical potential of S^+ because it is moving out of the cell, i.e., $[S^+]_{in}/[S^+]_{out}$ is <1.

eq. 15.4 $(y-1)\Delta\Psi - y60\,\Delta pH$

$$= +60\log[S^+]_{in}/[S^+]_{out}$$

4. Electrogenic uniport of a cation (Fig. 15.4D)

For some transporters, the membrane potential alone can provide the driving force for the uptake of cations (e.g., for K^+ uptake). The energetic relationship at equilibrium is:

eq. 15.5 $\Delta\Psi = -60\log[S^+]_{in}/[S^+]_{out}$

For a divalent cation, the driving force would be $2\Delta\Psi$ since there is twice as much charge

per ion. Equation 15.5 is one form of the Nernst equation.

15.3.2 Evidence for solute/proton or solute/sodium symport

One way to demonstrate coupling of transport to proton or sodium ion influx is to measure the alkalinization of the medium (decrease in protons) or a decrease in the extracellular Na^+ concentration when bacteria or membrane vesicles are incubated with the appropriate solutes.[14] The changes in the external pH or Na^+ concentration occur as a result of symport with the solute and can be measured with H^+ or Na^+-selective electrodes. It is necessary to insure that H^+ or Na^+ influx is electroneutral and a membrane potential, which would impede further influx of cations, does not develop. Thus the experiments are done in the presence of a permeant anion (e.g., CNS^-) that serves as a counterion to protons or sodium ions, or K^+ plus valinomycin whose efflux can exchange for the incoming protons or sodium ions.[15] The experiments may be done so that the ion influx is driven by the solute concentration (i.e., symport is demonstrated rather than accumulation of the solute against a concentration gradient). An example is the experiment reported by West in 1970.[16] West suspended E. coli in dilute buffer and added the sugar, lactose, which promoted an increase in the external pH (Fig. 15.5). This implied that the lactose permease catalyzed symport of lactose with protons. Similar experiments have since been performed by other investigators demonstrating that many sugars and amino acids enter bacteria in symport with either protons or sodium ions.

15.3.3 Primary transport driven by ATP

Transport systems exist that are driven by ATP (or some phosphorylated derivative in equilibrium with ATP). These are: (1) H^+ transport [i.e., the ATP synthase (ATPase)]; (2) K^+ transport in E. coli; (3) some transport systems in gram-positive bacteria; and (4) transport systems in gram-negative bacteria

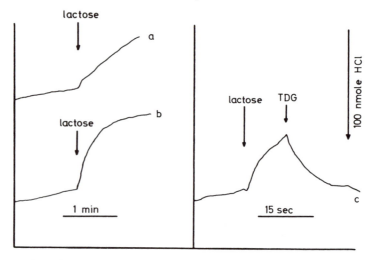

Fig. 15.5 Lactose-dependent proton uptake. Lactose was added to a suspension of *Escherichia coli* without (a) and with (b) SCN⁻ and the pH was measured with a pH meter. Lactose caused the immediate uptake of protons. Uptake was stimulated by SCN⁻ which prevented a membrane potential, inside positive, from developing. Lactose uptake was inhibited by thiodigalactoside (TDG) which is an inhibitor of the lac permease (c). These data reflect symport between lactose and protons. (From West, I. C. 1970. Lactose transport coupled to proton movements in *Escherichia coli. Biochem. Biophys. Res. Commun.* **41**:655–661.)

that use periplasmic binding proteins. (The student should review the discussion of the periplasm in Section 1.2.4.) The latter systems rely on binding proteins that combine with sugars and amino acids in the periplasm and transfer these to the actual transporters in the cell membrane. They are distinguished from all other transport systems in that they are not functional in cells that have been osmotically shocked, a treatment that makes the outer envelope permeable and causes the release of periplasmic proteins.[17] Shock-sensitive transport systems are also called *periplasmic permeases*.

Shock-sensitive transport systems
Shock-sensitive transport systems[18–22] are characteristic of gram-negative bacteria and are responsible for the transport of a wide range of solutes, including sugars, amino acids, and ions. They usually consist of a transporter which is a membrane complex consisting of four subunits (two of which are identical), and a periplasmic solute-binding protein (Fig. 15.6). Solute transport can be visualized as occurring in several sequential steps as illustrated in Fig. 15.6. This model is based upon the histidine permease system

which appears to be a typical periplasmic permease.[23,24]

Step 1. The solute enters the periplasm through an outer membrane pore (e.g., through a porin). (Consult Section 1.2.3 for a description of the outer membrane and of porins.)

Step 2. The solute binds to a specific periplasmic binding protein to form a solute:binding protein complex. The binding protein undergoes a conformational change that allows it to bind productively to the transporter in the membrane.

Step 3. The liganded binding protein binds to the transporter in the cell membrane, delivering the solute to the transporter. The membrane-bound transporter is a complex consisting of four proteins, two of which are identical.

Step 4. The transporter complex translocates the solute across the cell membrane, perhaps through a channel that opens transiently. ATP hydrolysis, catalyzed by the transporter complex, occurs at this step.

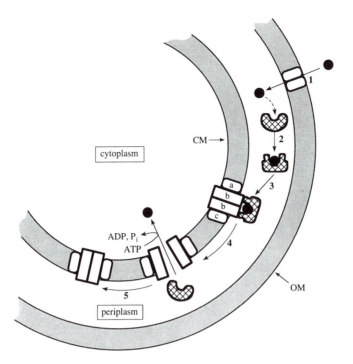

Fig. 15.6 Model for periplasmic transport. (1) The solute enters the periplasm through a pore in the outer membrane. (2) Inside the periplasm the solute binds to a binding protein which undergoes a conformational change when it binds the solute. (3) The binding protein carrying the solute binds to the transporter (a, b, c) located in the inner (cell) membrane. (4) The transporter is thought to undergo a conformational change that may result in the opening of a pore through which the solute diffuses to the cytoplasm. (5) The putative pore closes again when the binding protein is released from the permease. The transporter has a binding site for ATP which has been demonstrated to be hydrolyzed during histidine transport, and presumably during the transport of other solutes. An alternative hypothesis speculates that the binding protein triggers conformational changes in the permease that make a binding site available for the solute. The solute is then passed from one binding site to another on the permease until it is released inside the cell, rather than diffusing through a pore.

Step 5. The transporter complex returns to its unstimulated state.

In some cases the membrane-bound transporter (also called permease) interacts with more than one type of binding protein. For example, the histidine permease, besides transporting histidine, also transports arginine via the arginine–ornithine binding protein. Several different binding proteins for branched-chain amino acids use the same membrane transporter. The shock-sensitive transport systems are characterized by very high efficiencies. They are capable of maintaining concentration gradients of approximately 10^5, and K_m values for uptake are in the $0.1\,\mu M$ to $1\,\mu M$ range. This means that they can scavenge very low concentrations of solute and accumulate these to relatively high internal concentrations.

Evidence that ATP is the source of energy for histidine uptake

Direct evidence that ATP is the source of energy for the transport of histidine via a periplasmic binding protein was obtained by Bishop *et al.*[25] They extracted the proteins, including the histidine carrier proteins, from the membranes of *E. coli* and incorporated these proteins into proteoliposomes. (See Section 15.1 for a description of proteoliposomes.) The proteoliposome vesicles were sometimes loaded with ATP by including the

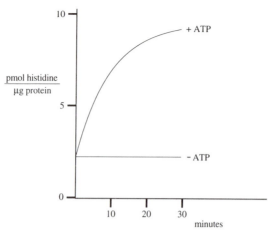

Fig. 15.7 Transport of histidine by reconstituted proteoliposomes. The histidine carrier proteins plus other membrane proteins were solubilized in detergent and incorporated into proteoliposomes in the presence or absence of ATP. Incubation mixtures contained the histidine-binding protein. Histidine was taken up by the proteoliposomes that were loaded with ATP. (Adapted from Bishop, L., R. Agbayani, Jr, S. V. Ambudkar, P. C. Maloney and G. F-Z. Ames. 1989. Reconstitution of a bacterial periplasmic permease in proteoliposomes and demonstration of ATP hydrolysis concomitant with transport. *Proc. Natl. Acad. Sci. USA* **86**:6953–6957.)

ATP in the dilution step. When the reconstituted proteoliposomes were incubated with the histidine binding protein (HisJ) and histidine, they transported histidine only when ATP was present (Fig. 15.7). The transport of histidine was not affected by the ionophores valinomycin (and K^+), nigericin, or the proton ionophore FCCP. (Ionophores and their physiological activities are described in Section 3.4.) These experiments demonstrated that histidine uptake via its periplasmic binding protein was driven by ATP and not an electrochemical potential.

ATP-driven K+ influx

Potassium ion is the principal cation in bacteria, and it plays a role not only in osmotic and pH homeostasis (Chapter 14), but also as a cofactor for many enzymes and ribosomes. Bacteria accumulate K^+ to a level several orders of magnitude higher than the external concentrations. Most of what we know about K^+ transport is derived from studies of *E. coli*.[26,27] There are two transport systems. The major route for K^+ uptake occurs via the TrK system, which is always present (constitutive) and operates at a high rate, but relatively low affinity (high K_m) for K^+. This transporter requires both a Δp and

ATP in order to function, and the reasons for the dual requirement are not clear. (It has been speculated that the energy source is actually the Δp and that ATP acts as a positive regulator.)[28] When *E. coli* is grown in media where the K^+ concentrations are very low (i.e., they limit growth), the cells synthesize a second K^+ transport system called the Kdp system, which serves to scavenge K^+. The Kdp system has a very low K_m for K^+ (2 μM) and uses ATP as the energy source. There are three structural proteins in the Kdp system, KdpA, KdpB, and KdpC, all of which are located in the cell membrane. These genes are encoded in the *kdpABC* operon. KdpB is a transmembrane protein (predicted from the nucleotide sequences suggesting several membrane spanning loops of hydrophobic domains) and catalyzes the ATP-dependent uptake of K^+ (Fig. 15.8). KdpA is exposed to the periplasm and is believed to bind to K^+, delivering it to KdpB. The reason for believing this is that mutations in the KdpA protein produce changes in the K_m for K^+ uptake, suggesting that it has the substrate binding site. The function of KdpC is not known. An increase in the external osmotic pressure induces the formation of the Kdp system, reflecting the important role for K^+

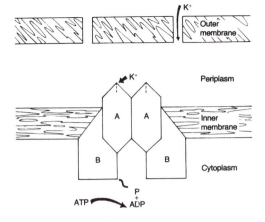

Fig. 15.8 The Kdp system for K$^+$ uptake. K$^+$ is thought to bind to KdpA in the periplasm and then be transported through the membrane via KdpB. The energy for K$^+$ transport is derived from ATP hydrolysis. The function of KdpC, which is not shown, is unknown. (From Epstein, W. and L. Laimins. 1980. Potassium transport in *Escherichia coli*: diverse systems with common control by osmotic forces. *TIBS* 5:21–23.)

as an intracellular osmolyte.[29] (See Section 14.2.2 for a discussion of K$^+$ as an osmolyte.) There are two proteins that are required for the transcription of the *kdpABC* operon. These are the KdpD protein, located in the cytoplasmic membrane, and the KdpE protein, located in the cytosol. These two proteins are part of a two-component signaling system thought to sense turgor pressure and transfer the signal to the *kdpABC* promoter. (See Section 17.8 for a discussion of this point.)

15.3.4 The phosphotransferase system

This transport system[30–32] differs from those driven by ATP or electrochemical gradients in that it catalyzes the accumulation of carbohydrates as the *phosphorylated derivatives* instead of as the free sugar. The phosphoryl donor in these transport systems is phosphoenolpyruvate (PEP), an intermediate in glycolysis (Section 8.1). Because the carbohydrate is modified (by phosphorylation), this type of transport is not referred

to as active transport but rather as group translocation. The phosphotransferase (PTS) system is characteristic of anaerobic and facultatively anaerobic bacteria, but seems to be lacking in strict aerobes. It also does not occur in eukaryotes. The overall reaction of the PTS system for glucose transport is:

$$PEP + carbohydrate_{(out)}$$

$$\longrightarrow pyruvic\ acid + carbohydrate\text{-}P_{(in)}$$

During the reaction, PEP donates a phosphoryl group to the carbohydrate which accumulates inside the cell as the phosphorylated derivative. The phosphoryl group is transferred to the carbohydrate via a consecutive series of reactions beginning with PEP as the initial donor. This is illustrated in Fig. 15.9 and described as individual steps below.

Step 1. In the soluble part of the cell, the phosphoryl group is transferred from phosphoenolpyruvate to enzyme 1 (EI).

$$PEP + E_I \longrightarrow E_I\text{-}P + pyruvate$$

Step 2. Enzyme 1 transfers the phosphoryl group to a small cytoplasmic protein called HPr.

$$EI\text{-}P + HPr \longrightarrow HPr\text{-}P + EI$$

Enzyme 1 and HPr are common to all of the PTS carbohydrate uptake systems.

Step 3. The HPr then transfers the phosphoryl group to a carbohydrate-specific permease complex in the membrane called enzyme II, which transfers the phosphoryl group to the carbohydrate during carbohydrate uptake into the cell:

$$HPr\text{-}P + EII \longrightarrow HPr + EII\text{-}P$$

$$EII\text{-}P + [CH_2O]_{out} \longrightarrow EII + [CH_2O\text{-}P]_{in}$$

Enzyme II has three domains, A, B, and C. The phosphoryl group is transferred from HPr to domain A, then to domain B, and finally to the carbohydrate in a reaction that requires domain C, which is always an

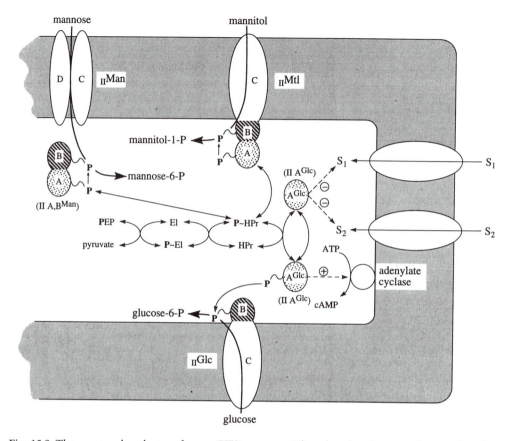

Fig. 15.9 The sugar phosphotransferase (PTS) system. The phosphoryl group is transferred via a series of proteins to the sugar. The sugar-specific carrier proteins in the membrane, $_{II}$Man, $_{II}$Mtl, $_{II}$Glc, accept the phosphoryl group from P-HPr and transfer it to the sugar-forming mannose-6-P, mannitol-1-P, and glucose-6-P. Enzyme II can be a single protein with three domains (A, B, and C as in the mannitol system), or separate proteins as in the glucose and mannose systems. The mannose carrier consists of two proteins (C and D). Note that the phosphoryl group travels from HPr to IIA to IIB to the sugar, which is translocated by IIC into the cell in an unknown manner. At some stage during translocation the sugar becomes phosphorylated. However, phosphorylation of the sugar need not take place during translocation *per se*. That is to say, the sugar may be phosphorylated on the inside surface of the cell membrane prior to its release into the cytoplasm. Also shown is the stimulation of adenylate cyclase by P-EIIA-^{Glc} (formerly called P-III^{Glc}) and the inhibition of non-PTS sugar carriers (S_1 and S_2 carriers) by EIIA^{Glc}. (Adapted from Postma, P. W., J. W. Lengeler, and G. R. Jacobon, 1993. Phosphenolpyruvate:carbohydrate phosphotransferase systems of bacteria. *Microbiol. Rev.* 57:543–594.)

integral membrane protein. This is summarized below:

$$HPr-P + EIIA \longrightarrow EIIA-P + HPr$$

$$EIIA-P + EIIB \longrightarrow EIIA + EIIB-P$$

$$EIIB-P + carbohydrate_{(out)}$$
$$\xrightarrow{EIIC} EIIB + carbohydrate-P_{(in)}$$

How EIIC brings the sugar into the cell is not understood.

Different carbohydrate uptake systems differ with respect to the number of separate proteins that constitute "enzyme II". It can be from one to four, one of which (IIC), is always membrane-bound and catalyzes the transport of the carbohydrate into the cell. For example, enzyme II for mannitol uptake (II^{Mtl} in Fig. 15.9) is a single membrane-bound protein with three domains, A, B, C. Domain C is in the membrane, whereas domains A and B project into the cytoplasm. However, in some transport systems the domains A, B, and C are on separate enzyme II proteins. Thus, enzyme II for glucose consists of two proteins, IIA and IIBC (Fig. 15.9). In this case, IIBC is membrane-bound, whereas IIA (formerly called III^{Glc}) is cytoplasmic. In a third case (EII^{Man}), EIIAB exist as a single cytoplasmic protein whereas there are two membrane-bound EII proteins, IIC and IID. In the cellobiose PTS system in *E. coli*, the enzymes II are three proteins, i.e., IIA and IIB in the cytoplasm and IIC in the membrane.

Regulation by the PTS system

It has been known for many years that when bacteria using the PTS system for the transport of glucose are presented with a choice of glucose and another carbon source, they will preferentially utilize the glucose and delay the use of the other carbon source until the glucose is depleted.[33] This is called glucose repression, or catabolite repression, and is responsible for diauxic growth, described in Section 2.2.4. The phenomenon is not restricted to glucose, since many PTS carbohydrates are used in preference over other carbon sources.[34] The discussion below applies only to the enteric bacteria. For example, PTS-mediated repression also occurs in gram-positive bacteria such as *Bacillus subtilis*, but it is not as well studied as for the enteric bacteria and the mechanism is not the same as described below.

The model

A widely accepted model for regulation by the PTS system in enteric bacteria is illustrated in Fig. 15.9. The model postulates the following.

1. IIA^{Glc} inhibits several enzymes required for carbohydrate metabolism, including certain non-PTS sugar transporters such as the lactose and melibiose transporter, the MalK protein, which is essential for the maltose transport system, and glycerol kinase. P-IIA^{Glc} is dephosphorylated to IIA^{Glc} by IICB^{Glc} during glucose transport. Therefore,

according to the model, glucose transport into the cell inhibits the above-mentioned enzymes. The inhibition by glucose of transport and metabolism of non-PTS carbohydrates is called *inducer exclusion*.

2. P-IIA^{Glc} stimulates a membrane-bound enzyme called adenylate cyclase, which makes cyclic-AMP (c-AMP), which in turn stimulates the transcription of many genes that code for catabolic enzymes. During glucose uptake by the PTS system, P-IIA^{Glc} is dephosphorylated, hence adenylate cyclase is no longer stimulated and transcription of c-AMP-dependent genes is inhibited. However, there is always a basal level of c-AMP synthesized regardless of the carbon source. This is necessary because the transcription of genes required for the metabolism of many PTS sugars also requires c-AMP.

3. The model also explains how PTS carbohydrates in addition to glucose can also depress the entry of non-PTS sugars and inhibit the expression of c-AMP-dependent genes. The uptake of PTS carbohydrates would be expected to draw phosphoryl groups away from P-HPr towards the sugars. This should decrease the phosphorylation state of IIA^{Glc}, which is phosphorylated by P-HPr. Because the phosphorylation of IIA^{Glc} by P-HPr is reversible, phosphate would be expected to flow from P-IIA^{Glc} to the PTS carbohydrates. The subsequent increase in IIA^{Glc} would be expected to inhibit the enzymes required for uptake and metabolism of non-PTS sugars, and the decrease in P-IIA^{Glc} would be expected to inhibit the adenylate cyclase.

Rationale for the model

1. The reason for believing that the PTS system is involved in the utilization of glucose-repressed carbon sources was the original finding that mutants lacking HPr or enzyme I are unable to grow on glucose-repressed carbon sources. In *E. coli*, these carbon sources include lactose, maltose, melibiose, glycerol, rhamnose, xylose, and citric acid cycle intermediates. The reason for this is that they cannot phosphorylate IIA^{Glc}, which stimulates the production of c-AMP by activating the adenylate cyclase. The requirement for EI and Hpr for growth on

non-PTS sugars is not restricted to *E. coli* or other enterics, although it is less well studied in other bacteria.

2. Because the addition of glucose to wild-type cells has the same repressive effect on growth using non-PTS sugars and citric acid cycle intermediates as does mutations in HPr or enzyme I, it was suggested that P-IIA^{Glc} is required for growth on these substances. The reasoning is that: (1) a defect in enzyme I or HPr will result in the inability to phosphorylate IIA^{Glc}; and (2) the addition of glucose to wild-type cells should result in the dephosphorylation of P-IIA^{Glc}.

3. The involvement of adenylate cyclase was implicated when it was discovered that the addition of c-AMP, the product of adenylate cyclase, could overcome the Hpr mutant phenotype with respect to growth. Since c-AMP was known to be required for the transcription of several genes, a model was postulated that stipulated a stimulation by P-IIA^{Glc} of the enzyme that synthesizes c-AMP (adenylate cyclase) (Fig. 15.9). Mutations in enzyme I or HPr, or the addition of glucose to wild-type cells, should lower the amounts of P-IIA^{Glc} and thus cause a decrease in the levels of c-AMP. Of course, it is possible that the inhibition of the adenylate cyclase is due to an increase in amounts of IIA^{Glc} rather than a decrease in levels of P-IIA^{Glc}. However, mutations that result in lowered IIA^{Glc} activity (*crr* mutants) do not relieve the inhibition of adenylate cylase by glucose. Also, it was reported several years ago that the addition of PTS carbohydrates to *E. coli* cells made permeable with toluene inhibited adenylate cyclase activity, which according to the current model can be interpreted as due to the lowering of the levels of P-IIA^{Glc} by the PTS carbohydrates.[35,36] However, it should be emphasized that the evidence for stimulation of adenylate cyclase by P-IIA^{Glc} is thus far indirect, and based primarily on genetic evidence. What is lacking is direct evidence that P-EII^{Glc} binds to the adenylate cyclase and/or stimulates its activity.

4. Mutants defective in HPr are also defective in the *uptake* of certain non-PTS sugars. The model explains this by postulating that IIA^{Glc} inhibits the carriers for these sugars. The evidence for this is abundant. Uptake in HPr mutants is restored by a mutation in the gene for IIA^{Glc} which results in lowered activity of IIA^{Glc} (*crr* mutants). Also, IIA^{Glc} is capable of binding to the lactose carrier (permease) and of inhibiting the transport of the lactose analogue (TMG) by liposomes made with the lactose permease.[37,38]

The PTS is involved in chemotaxis towards PTS sugars

In addition to functioning in sugar uptake and carbohydrate repression, the PTS system is also important for chemotaxis towards PTS sugars. Some of the evidence for this is that:

1. Mutants of *E. coli* defective in EI and HPr are usually defective in chemotaxis towards PTS sugars.

2. EII mutants defective in chemotaxis towards the sugar specifically recognized by EII also exist.

A comparison of PTS-mediated chemotaxis with the more general chemotaxis system (MCP-dependent chemotaxis)

Chemotaxis towards compounds other than PTS sugars is mediated by membrane chemosensors called MCP proteins and six cytoplasmic proteins: CheA, CheW, CheY, CheR, CheB, and CheZ (Section 17.4). The protein CheA autophosphorylates and transfers the phosphoryl group to CheY and CheB. Phosphorylated CheY (P-CheY) causes the flagella motors to turn in a clockwise rotation, which makes the bacterium tumble. When the bacterium tumbles, it swims randomly. When the MCP protein binds a chemoattractant, the autophosphorylation of CheA is inhibited. This results in a lower level of P-CheY and hence smooth swimming towards the chemoattractant. CheR and CheB are required for the adaptation phase, which involves methylation of the MCP proteins, and CheZ is necessary for dephosphorylation of CheY, which allows the bacterium to resume smooth swimming. *E. coli* mutants lacking CheB, CheR, or the MCP proteins, display normal chemotaxis to PTS sugars. This means that methylation-dependent adaptation does not occur in the PTS system. Generally it is thought that EII becomes stimulated by its PTS sugar and conveys the signal in an unknown manner to the flagellar motor causing smooth swimming, hence chemotaxis. One model postulates that, in the presence of the PTS sugar, the phosphorylation levels of EI and HPr are lowered, and that this results, via an unknown mechanism, in the lowering of the phosphorylation levels of CheA and CheY. A lowering of the phosphorylation of CheY would result in smooth swimming towards the chemoattractant, since P-CheY causes tumbling.

15.4 How to determine the source of energy for transport

Methods are available to distinguish whether bacterial transport is driven by the electrochemical proton potential (Δp) or by ATP.

The ATP synthase must be inactivated
In order to determine whether the source of energy is ATP hydrolysis or the Δp, it is necessary to isolate these sources of energy from each other and to perturb them independently (i.e., to increase and decrease the ATP and Δp levels independently of each other). Since bacteria use ATP synthase to interconvert the proton potential and ATP, one cannot vary the ATP levels and Δp independent of each other while the ATP synthase is functioning. To circumvent this problem, investigators either use mutants defective in the ATP synthase ("unc" mutants) or add inhibitors of the ATP synthase [e.g., N,N'-dicyclohexylcarbodiimide (DCCD)]. Once this is done, the ATP or the Δp can be decreased or increased independent of each other.

Perturbing the intracellular levels of ATP
How are intracellular ATP levels manipulated? Once the ATP synthase is inactivated, bacteria must rely completely on substrate-level phosphorylation to synthesize ATP, and the problem becomes one of interfering with substrate-level phosphorylation. In order to lower the levels of substrate-level phosphory-

lation, one can: (1) starve the cells of endogeneous energy reserves whose metabolism produces ATP via substrate level phosphorylation; and (2) add an inhibitor of substrate-level phosphorylation (e.g., arsenate).[39] Conversely, one can *increase* the production of ATP from substrate-level phosphorylation in starved cells by feeding them an energy source (e.g., glucose). However, one must be careful. Glucose can also stimulate respiration, and respiration can be a source of Δp. Therefore to stimulate substrate-level phosphorylation with glucose without encouraging respiration, respiration should be prevented by using a respiratory inhibitor (e.g., cyanide) or by incubating the cells anaerobically.

Perturbing the Δp
How is the Δp manipulated? One can use ionophores that collapse the proton potential.[40] (See Section 3.4 for a discussion of ionophores.) It is also possible to *increase* electrochemical potentials in de-energized cells (e.g., starved cells) or in vesicles. For example, if there is a high K^+ concentration in the cells, the addition of valinomycin will produce a potassium diffusion potential (Section 3.4). Also, the addition of certain substrates that feed electrons directly into the electron transport chain (e.g. D-lactate or succinate) will produce a proton potential without increasing substrate-level phosphorylation. A summary of some of these methods is presented in Table 15.1. However, caution

Table 15.1 Summary of data that can distinguish between energy sources for transport using whole cells

	Δp	ATP
Stimulated by:		
glucose	no*	yes
D-lactate	yes	no
Inhibited by:		
DNP[†]	yes	no[‡]
arsenate	no	yes[§]

*Respiration must be inhibited.
[†]One can use any combination of ionophores that abolishes both the ΔpH and the $\Delta \Psi$.
[‡]The ATP synthase must be inhibited.
[§]Arsenate prevents ATP formation via substrate level phosphorylation. This is the only source of ATP when the ATP synthase is not operating.

must be observed when interpreting data derived from the use of inhibitors on whole cells because of indirect effects of the inhibitors.[41] Also, it is not possible to distinguish between ATP and other high-energy molecules in equilibrium with ATP (e.g., PEP) when studying whole cells. Ideally, one should incorporate the carrier into proteoliposomes and directly test potential energy sources as described in Section 15.3.3 for the histidine permease.

15.5 A summary of bacterial transport systems

A summary of bacterial transport systems is illustrated in Fig. 15.10. These include primary transport systems (driven by chemical energy) and secondary transport systems (driven by electrochemical ion gradients). Note the role of the proton circuit for secondary transport. Most transport is in symport with protons, although several

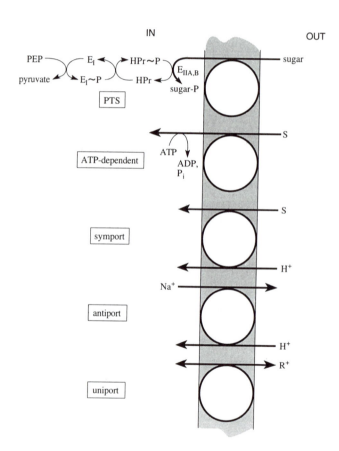

Fig. 15.10 A summary of transport systems found in the bacteria. Any single bacterium uses diverse transport systems to take up sugars, amino acids, ions, vitamins, organic acids, and so on. Many of these systems are powered by electrochemical energy (ion gradients). They catalyze symport, antiport and uniport. There are also transport systems that use chemical energy instead of electrochemical energy. An example of the latter is the PTS system, which is specific for carbohydrates, is driven by phosphoenolpyruvate, and accumulates the sugar as the phosphorylated derivative. Other chemically driven transport systems use ATP or a high-energy molecule in equilibrium with ATP. These include shock-sensitive transport systems (periplasmic permeases) that are found in gram-negative bacteria, and other chemically driven uptake systems found in both gram-negative and gram-positive bacteria.

symporters use Na^+ instead of H^+. The return of the Na^+ to the extracellar space requires a proton circuit in most bacteria.

15.6 Summary

Bacteria employ several different types of energy-dependent transport systems. A single bacterium may have representatives of all of the transport systems, except for the shock-sensitive systems which are present only in gram-negative bacteria. Transport systems may be classified as being either primary or secondary, depending on how they are coupled to the energy source. The difference between the two is that primary transport systems are directly coupled to an energy-generating reaction (e.g., ATP or PEP hydrolysis) whereas secondary transport systems are energized by an existing electrochemical gradient (e.g., a proton or sodium ion gradient), itself having been produced by energy-yielding reactions such as ATP hydrolysis or redox reactions. That is to say, secondary transport is indirectly coupled to an energy-yielding chemical reaction by ion currents. (This is one of the predictions of the chemiosmotic theory.) Active transport is defined as primary transport in which the solute accumulates in a chemically un-modified form.

ATP hydrolysis is used to drive some primary transport systems. These include osmotic shock-sensitive (periplasmic per-meases) systems present in gram-negative bacteria that transport a range of solutes similar to those systems driven by electro-chemical gradients. They consist of a peri-plasmic binding protein plus three or four membrane proteins. Other primary transport systems utilizing ATP include the ATP synthase (translocates protons) and the Kdp K^+-ATPase in E. coli (K^+ influx).

The transport of carbohydrates via the sugar–phosphotransferase (PTS) system driven by phosphoenolpyruvate is also a primary transport system. It is widespread in bacteria, being found in anaerobic and facultative anaerobes, but not in strict aerobes. However, the PTS system is absent in eukaryotes. The sugar is phosphorylated during transport and therefore this is not active transport. Some-

times this type of transport is called group translocation. Transport by the PTS system results in the inhibition of transport of non-PTS sugars and the inhibition of adenylate cyclase leading to catabolite repression by glucose. This can be explained in terms of the regulatory roles of A^{Glc} (III^{Glc}) and P-A^{Glc} (P-III^{Glc}).

Secondary transport systems require ion symport, antiport, or uniport, and make use of electrochemical ion gradients to transport a diverse range of solutes including sugars, amino acids, and ions. The coupling ions are either protons or sodium ions, depending upon the transporter. This type of transport is found in bacteria, fungi, plants, protozoa, and higher eukaryotes. It is apparently the simplest of the transport systems and consists of a single membrane-spanning protein.

The type of transport system used by a bacterium depends upon the organism, not the solute being transported. For example, Pseudomonas (a strict aerobe) actively transports glucose using symport with H^+, whereas E. coli (a facultative anaerobe) transports the same sugar using the PTS system. Furthermore, a single bacterium can use several different types of transport for the same class of compound. For example, E. coli uses the PTS system for some sugars but uses H^+ symport, Na^+ symport, or ATP for others. Harold has speculated regarding the diversity of transport systems in the bacteria, pointing out that whereas transport systems coupled to electrochemical gradients have the advantage of being simple in composi-tion, they are limited by the Δp with respect to the concentration gradients that can be attained.[42] Furthermore, they operate close to equilibrium and can theoretically be reversed if the Δp suffers a transient decrease (e.g., during starvation). On the other hand, transport systems powered by ATP or PEP are structurally more complex, but have the advantage of being driven unidirectionally by the relatively large free energies available in ATP and PEP.

Study Questions

1. Suppose the transport of X was driven by symport with H^+ in a 1:1 ratio. Assume

X is not charged, the $\Delta\Psi$ is $-120\,mV$, and the ΔpH is 1. What is the expected X_{in}/X_{out}? What would be the answer if X were X^-? If X were X^- and the ΔpH were 2?

ans. 1,000, 10, 100.

2. What is the explanation of the curious fact that mutants in the PTS system that are defective in E_1 or HPr are also defective in the transport of some sugars that do not use the PTS system?

3. What are proteoliposomes and how are they prepared?

4. Solute transport might be driven by the Δp or by ATP. Describe some experiments that could distinguish between the two sources of energy. (Hint: You will have to manipulate the ATP and Δp separately, and you might want to use proteoliposomes for some of your experiments.)

5. What is the procedure to induce "osmotic shock?" What is the cellular location of the·proteins released by osmotic shock? What are the functions of some of the proteins released by osmotic shock?

REFERENCES AND NOTES

1. Brooker, R. J. 1990. The lactose permease of *Escherichia coli. Res. Microbiol.* **141**:309–315.

2. Racker, E., B. Violand, S. O'Neal, M. Alfonzo, and J. Telford. 1979. Reconstitution, a way of biochemical research; some new approaches to membrane-bound enzymes. *Arch. of Biochem. Biophys.* **198**:470–477.

3. Vitanen, P., M. J. Newman, D. L. Foster, T. H. Wilson and H. R. Kaback. 1986. Purification, reconstitution, and characterization of the *lac* permease of *Escherichia coli*. In *Methods in Enzymology*, Vol. 125. S. Fleischer and B. Fleischer (eds). Academic Press, New York.

4. Maloney, P. C., V. Anantharam, and M. J. Allison. 1992. Measurement of the substrate dissociation constant of a solubilized membrane carrier. *J. Biol. Chem.* **267**:10531–10536.

5. Taglich, D., E. Padan, and S. Schuldiner. 1991. Overproduction and purification of a functional Na^+/H^+ antiporter coded by nhaA (ant) from *Escherichia coli. J. Biol. Chem.* **266**:11289–11294.

6. Bishop, L., R. Agbayani, Jr., S. V. Ambudkar, P. C. Maloney, and G. F.-L. Ames. 1989.

Reconstitution of a bacterial periplasmic permease in proteoliposomes and demonstraton of ATP hydrolysis concomitant with transport. *Proc. Natl. Acad. Sci. USA* **86**:6953–6957.

7. The proton gradient is created by a primary transport system. Some bacteria can create sodium gradients with a primary transport system, but most create a sodium gradient by converting a proton gradient into a sodium ion gradient using a H^+/Na^+ antiporter.

8. Wright, J. K, R. Seckler, and P. Overath. 1986. Molecular aspects of sugar:ion cotransport. *Ann. Rev. Biochem.* **55**:225–248.

9. Bacteria, yeast, and plants use primarily the proton as the coupling ion for symport, whereas animals rely on the sodium ion. This may be because the cell membranes of most bacteria, fungi, and plants create a proton potential by pumping protons out of the cell via a proton, translocating ATPase, whereas the cell membranes of animal cells create a sodium potential by pumping sodium ions out of the cell via a Na^+,K^+-ATPase which exchanges Na^+ for K^+.

10. Hirota, N., and Y. Imae. 1983. Na^+-driven flagellar motors of an alkalophilic *Bacillus* strain YN-1. *J. Biol. Chem.* **258**:10577–10581.

11. Imae, Y., and T. Atsumi. 1989. Na^+-driven bacterial flagellar motors. *J. Bioenerg. Biomemb.* **21**:705–716.

12. Maloy, S. R. 1990. Sodium-coupled cotransport, pp. 203–224. In: *The Bacteria*, Vol. XII. Krulwich, T. A. (ed.). Academic Press, New York.

13. $R = 8.3144\,J\,deg^{-1}\,mol^{-1}; T = K = 273.16 + °C$; to convert natural logarithms to log_{10} multiply by 2.303.

14. Wilson, D. M., T. Tsuchiya, and T. H. Wilson. 1986. Methods for the study of the melibiose carrier of *Escherichia coli*. In: *Methods in Enzymology* Vol. 125. S. Fleischer and B. Fleischer (eds). **125**:377–387.

15. The negative charge on SCN^- is delocalized over the three atoms of the molecule and this allows it to penetrate the lipid bilayer.

16. West, I. C. 1970. Lactose transport coupled to proton movements in *Escherichia coli. Biochem. Biophys. Res. Commun.* **41**:655–661.

17. Gram-negative cells are shocked in the following way. They are first suspended in a hypertonic solution of Tris buffer, EDTA, and sucrose. Such treatment removes much of the divalent metal cations that are holding the lipopolysaccharide together along with the lipopolysaccharide and plasmolyzes the cells. Then the cells are rapidly diluted into water or dilute $MgCl_2$ to neutralize the EDTA. This results in the release of the periplasmic proteins. The treatment inhibits

cellular functions that depend on periplasmic binding proteins (e.g., ATP-dependent transport of sugars and amino acids). Other cellular functions, including other transport systems, are retained.

18. Ames, G. F.-L. 1988. Structure and mechanism of bacterial periplasmic transport systems. *J. Bioenerg. Biomembr.* **20**:1–18.

19. Ames, G. F.-L. 1990. Energetics of periplasmic transport systems, pp. 225–245. In: *The bacteria*, Vol. 12. T. A. Krulwich (ed.). Academic Press, New York.

20. Ames, G. F.-L. 1986. Bacterial periplasmic transport systems: Structure, mechanism, and evolution. *Annu. Rev. Biochem.* **55**:397–425.

21. Ames, G. F.-L and A. K. Joshi. 1990. Energy coupling in bacterial periplasmic permeases. *J. Bacteriol.* **172**:4133–4137.

22. Furlong, C. E. 1987. Osmotic-shock-sensitive transport systems, pp. 768–796. In: Escherichia coli *and* Salmonella typhimurium: *Cellular and Molecular Biology*, Vol. 1 F. C. Neidhardt, J. L. Ingraham, K. B. Low, B. Magasanik, M. Schaechter and H. E. Umbarger (eds). ASM Press, Washington, DC.

23. Ames, G. F.-L. and H. Lecar. 1992. ATP-dependent bacterial transporters and cystic fibrosis: Analogy between channels and transporters. *FASEB* **6**:2660–2666

24. In the specific case of the histidine permease system, the transporter consists of two hydrophobic proteins, HisQ and HisM, which span the membrane, and two identical hydrophilic proteins, HisP. The HisP protein may be a peripheral membrane protein bound to the inner surface of the membrane, or it may span the membrane, being separated from the hydrophobic lipids by the HisQ and HisM proteins. The HisP protein binds ATP and is responsible for ATP hydrolysis. The periplasmic histidine binding protein is called HisJ.

25. Bishop, L., R. Agbayani, Jr., S. V. Ambudkar, P. C. Maloney, and G. F.-L. Ames. 1989. Reconstitution of a bacterial periplasmic permease in proteoliposomes and demonstration of ATP hydrolysis concomitant with transport. *Proc. Natl. Acad. Sci. USA* **86**:6953–6957.

26. Epstein, W. and L. Laimins. 1980. Potassium transport in Escherichia coli: Diverse systems with common control by osmotic forces. *TBIS* **5**:21–23.

27. Rosen, B. 1987. ATP-coupled solute transport systems pp. 760–767. In: Escherichia coli *and* Salmonella typhimurium: *Cellular and Molecular Biology* Vol. 1. Ingraham, J. L. *et al.* (eds). ASM Press, Washington, DC.

28. Stewart, L. M. D., E. P. Bakker, and I. R. Booth. 1985. Energy coupling to K$^+$ uptake via the Trk system in Escherichia coli: The role of ATP. *J. Gen. Microbiol.* **131**:77–85.

29. Epstein, W. 1986. Osmoregulation by potassium transport in Escherichia coli. *FEMS Microbiol. Rev.* **39**:73–78.

30. Saier, M. J. Jr., and A. M. Chin. 1990. Energetics of the bacterial phosphotransferase system in sugar transport and the regulation of carbon metabolism, pp. 273–299. In: *The Bacteria*, Vol. XII. T. A. Krulwich (ed.). Academic Press, New York.

31. Meadow, N. D., D. K. Fox, and S. Roseman. 1990. The bacterial phosphoenolpyruvate:glucose phosphotransferase system. *Annu. Rev. Biochem.* **59**:497–542.

32. Postma, P. W., J. W. Lengeler, and G. R. Jacobson. 1993. Phosphoenolpyruvate:carbohydrate phosphotransferase systems of bacteria. *Microbiol. Rev.* **57**:543–594.

33. Saier, M. H., Jr. 1989. Protein phosphorylation and allosteric control of inducer exclusion and catabolite repression by the bacterial phosphoenolpyruvate:sugar phosphotransferase system. *Microbiol. Rev.* **53**:109–120.

34. Inhibition of transport and metabolism of non-PTS carbohydrates such as lactose, melibiose, maltose, and glycerol (the class I PTS carbohydrates) by PTS carbohydrates is enhanced in mutants of *E. coli* that have less EI activity (leaky *ptsI* strains).

35. Harwood, J. P., C. Gazdar, C. Prasad, A. Peterkofsky, S. J. Curtis, and W. Epstein. 1976. Involvement of the glucose enzymes II of the sugar phosphotransferase system in the regulation of adenylate cyclase by glucose in Escherichia coli. *J. Biol. Chem.* **251**:2462–2468.

36. Peterkofsky, A. and C. Gazdar. 1975. Interaction of enzyme I of the phosphoenolpyruvate:sugar phosphotransferase system with adenylate cyclse of Escherichia coli. *Proc. Natl. Acad. Sci. USA* **72**:2920–2924.

37. Nelson, S.O., J.K. Wright, and P.W. Postma. 1983. The mechanism of inducer exclusion. Direct interaction between purified III^{Glc} of the phosphoenolpyruvate:sugar phosphotransferase system and the lactose carrier of Escherichia coli. *EMBO J.* **2**:715–720.

38. Osumi, T., and M.H. Saier, Jr. 1982. Regulation of lactose permease activity by the phosphoenolpyruvate:sugar phosphotransferase system: Evidence for direct binding of the glucose-specific enzyme III to the lactose permease. *Proc. Natl. Acad. Sci. USA* **79**:1457–1461.

39. Arsenate can substitute for inorganic phosphate in the synthesis of 1,3-bisphosphoglycerate. The acyl-arsenate is quickly chemically hydrolyzed to the carboxylic acid. Thus, the acyl phosphate does not form and ATP is not made.

40. As long as the ATP synthase is inhibited, the ATP levels should not be affected. However, if the ATP synthase is functioning, then collapsing the Δp will shift the equilibrium of the ATP synthase in the direction of ATP hydrolysis.

41. Ames, G. F.-L., and A. K. Joshi. 1990. Energy coupling in bacterial periplasmic permeases. *J. Bacteriol.* **172**:4133–4137.

42. Harold, F. M. 1986. *The Vital Force: A Study of Bioenergetics.* W. H. Freeman and Co.

16

Protein Export and Secretion

Many proteins synthesized by cytosolic ribosomes are destined for export to various cellular locations (cell membrane, periplasm, outer envelope, cell wall), or secretion into the medium. The traffic in exporting these proteins to their correct locations is considerable. For example, cell membranes alone contain approximately 300 different proteins. In gram-negative bacteria there may be 100 various proteins in the periplasm. The outer envelope of gram-negative bacteria is the site of perhaps 50 different proteins. In addition, both gram-negative and gram-positive bacteria also export proteins that are part of surface layers (glycocalyx) and appendages (flagella, fibrils, pili), as well as extracellular hydrolytic enzymes (e.g., proteases, lipases, nucleases, and saccharidases). The translocation process into or through the membranes is referred to as protein export. If the protein is exported to the extracellular medium, the process is called secretion. [The extracellular transport of nonproteinaceous compounds (e.g., end-products of fermentation) is called excretion.] What is it about the structure of a protein that determines whether it will be translocated and to where? What mechanisms are responsible for the translocation of exportable proteins through the cell membrane which is generally nonpermeable to proteins? What is the source of energy for protein translocation? These are important areas of research not only in prokaryotes but in eukaryotes as well, where proteins

synthesized on cytosolic ribosomes are exported to different cell compartments such as mitochondria and chloroplasts, or are secreted out of the cell via the endoplasmic reticulum. By far, most research on bacterial protein translocation through the cell membrane has been with *Escherichia coli*, which has served as a model system primarily because of the ease of genetic manipulation. Indeed, virtually all the proteins involved in *E. coli* protein translocation through the cell membrane have been purified and the genes have been cloned. This has allowed the formulation of a model that describes the general features by which most proteins are translocated through the cell membrane. It is called the *Sec system*.

16.1 The Sec system

16.1.1 The components

Before describing the model for protein export using the Sec system[1–5] (Section 16.1.2.), the components of the Sec system will be introduced. These are (1) a *leader peptide*, (2) a *chaperone protein*, and (3) a membrane-bound *translocase*.

The leader peptide
Most proteins that are translocated by bacteria are synthesized with a leader peptide (also called a leader sequence or signal

sequence) at the amino terminal end that is removed during or after translocation. The leader peptide is necessary for attachment and insertion of the protein into the cell membrane. The protein with the leader peptide is called a precursor protein or a *preprotein*. If it is secreted to the outside, it is sometimes called a *presecretory* protein. There is ample evidence that the leader peptide is necessary for the initial stages of translocation through the membrane. If the leader peptide is altered by amino acid substitutions, or deleted, then translocation is impaired or does not occur at all. Leader peptides have three regions. They are: (1) a basic region (positively charged near neutral pH) at the extreme N-terminal end, perhaps to negatively charged phospholipids; (2) a central hydrophobic region which inserts into the membrane; and (3) a C-terminal region that contains a recognition site for a peptidase that removes the leader peptide during or after translocation (Fig. 16.1). The primary amino acid

sequences of different leader peptides can vary significantly. Although the leader peptide is important for the initial stages of translocation through the membrane, it does not determine the final destination of the protein. Two reasons for believing this are:

1. There are no obvious differences in amino acid sequences between leader peptides of periplasmic and outer membrane proteins. One would expect to find such differences if the leader peptide specified the final location of the protein.

2. Exchange of leader peptide between proteins destined for two different compartments using recombinant DNA technology does not influence their final destination.

Chaperone proteins

When exportable proteins are synthesized, the nascent preprotein binds to a soluble chaperone protein upon leaving the ribosome.

a Met Lys Ala Thr Lys|Leu Val Leu Gly Ala Val Ile Leu Gly Ser Thr Leu Leu Ala GlyΔCys

b (Met) Met Ile Thr Leu Arg Lys|Leu Pro Leu Ala Val Ala Val Ala Ala Gly Val Met Ser Ala Gln Ala Met AlaΔVal

c Met Lys Ile Lys Thr Gly Ala Arg|Ile Leu Ala Leu Ser Ala Leu Thr Thr Met Met Phe Ser Ala Ser Ala Leu AlaΔLys

d Met Ser Ile Gln His Phe Arg|Val Ala Leu Ile Pro Phe Phe Ala Ala Phe Cys Leu Pro Val Phe AlaΔHis

e Met Lys Thr Lys|Leu Val Leu Gly Ala Val Ile Leu Thr Ala Gly Leu Ser Gly Ala AlaΔGlu

f Met Lys Lys Ser Leu Val Leu Lys|Ala Ser Val Ala Val Ala Thr Leu Val Pro Met Leu Ser Phe AlaΔAla

g Met Lys Lys|Leu Leu Phe Ala Ile Pro Leu Val Val Pro Phe Tyr Ser His SerΔAla

Fig. 16.1 Leader peptides of exported proteins in *E. coli*. The leader sequence consists of a basic amino terminal end that has positively charged lysine and arginine residues, followed by a hydrophobic region. The boundary between the hydrophilic basic region and the hydrophobic region is denoted with a vertical line. The cleavage site for the leader peptide peptidase is at the carboxy-terminal end (triangle). a, Lipoprotein, located in outer membrane. b, Phage lambda receptor, located in outer membrane. c, Maltose-binding protein located in periplasm. d, β-Lactamase, located in periplasm. e, Arabinose-binding protein, located in periplasm. f, fd phage major coat protein, located in cell membrane. g, fd phage minor coat protein, located in cell membrane. Note that although the leader peptides share similar features with respect to charged and hydrophobic regions, the amino acid sequences are not the same, nor are there any differences that can distinguish between leader sequences of outer membrane proteins, periplasmic proteins, and cell membrane proteins. (After Osborn, M. J. and H. C. P. Wu. 1980. Proteins of the outer membrane of gram-negative bacteria. *Ann. Rev. Microbiol.* 34:369–422. Reproduced, with permission, from the *Annual Review of Microbiology*, Vol. 34, 1980 by Annual Reviews Inc.)

The chaperone protein prevents the pre-protein from assuming a tightly folded configuration or aggregating into a complex that cannot be translocated. The signal peptide retards the folding of the preprotein, giving the chaperone protein time to bind. A major chaperone protein in E. coli is SecB. SecB binds to the mature domain (not the leader sequence) of the preprotein and prevents aggregation. SecB protein recognizes the unfolded topology of the proteins rather than specific amino-acid sequences. SecB protein also binds to SecA at the membrane site of translocation and delivers the preprotein to SecA, which is a peripheral component of the translocase. Many proteins translocate without requiring SecB and probably use other chaperone proteins. (Chaperone proteins perform other cellular functions as well.[6,7])

The SecY/E translocase

The preprotein is transferred from SecB to SecA which is attached to the membrane-bound translocase, SecY/E. It is not known how the translocase moves proteins through the membrane. For this discussion, it will be assumed that the translocase forms a hydrophilic channel for proteins. The formation of a hydrophilic channel is merely speculation at this time, but can be rationalized because many of the proteins secreted into the external medium do not have significantly hydrophobic regions that would facilitate their movement through the lipid bilayer. The translocase consists of an integral membrane protein composed of two polypeptides, SecY and SecE. The ·SecY protein spans the membrane several times and it may form a channel.

It would be misleading to imply that all proteins translocated by E. coli require the Sec proteins, although most do. The insertion of certain integral membrane proteins is not affected by mutations in the sec genes. These include the phage M13 procoat protein. Sec-independent proteins like the M13 protein may also have leader sequences, even though they do not require SecA. Thus, although a major route for protein translocation is via the Sec pathway, another pathway(s) obviously exists.

16.1.2 A model for protein secretion

Protein export can be divided into several stages that are illustrated in Fig. 16.2. Before the protein is exported, it must bind to a chaperone protein which brings the protein to the membrane-bound translocase.

Step 1. The preprotein–SecB protein complex binds to the SecA–translocase complex in the membrane.

Step 2. The preprotein binds to SecA. The SecB protein may be released at this step.

Step 3. SecA binds ATP.[8,9]

Step 4. The amino terminus of the leader peptide leaves SecA and enters the membrane. The positively charged amino-terminus binds to the negatively charged phospholipid head groups or perhaps to the SecY/E protein and remains attached to the cytoplasmic side of the membrane. The hydrophobic region of the leader sequence spontaneously inserts into the membrane forming a loop.

Step 5. The N-terminus of the leader peptide remains on the cytoplasmic side of the membrane, possibly attached to the negatively charged phospholipid head groups, while the carboxy-terminal end "flips" into the lipid bilayer as the preprotein enters the translocase. It is envisaged that a channel opens up within the translocase to accommodate the protein. During these initial stages, a small segment of the preprotein (perhaps a few kilodaltons) is translocated into the membrane. The precise mechanism by which all this occurs is not known. Limited translocation occurs in vitro using non-hydrolyzable analogs of ATP, indicating that it is the binding of ATP rather than its hydrolysis that provides the energy for translocation.[3,9]

Step 6. ATP hydrolysis takes place and SecA is released from the preprotein and the membrane.

Step 7. The rest of the protein is translocated, driven by the Δp. In agreement with this, translocation in respiring cells is immediately inhibited by uncouplers that dissipate the Δp while the intracellular ATP levels

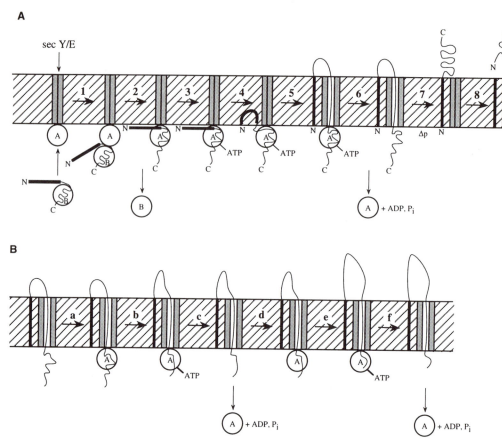

Fig. 16.2 A model for protein translocation in *E. coli*. (A) 1, The preprotein binds, either during or immediately after its synthesis, to SecB, forming a preprotein/SecB complex which moves to the SecA/SecY/E membrane complex (the translocase) and binds to SecA. 2, The preprotein is transferred to SecA and SecB is released. 3, SecA binds ATP. 4, The leader peptide leaves SecA and is inserted into the lipid bilayer as a loop. 5, The carboxy-terminus of the leader peptide "flips" to the periplasmic side and the preprotein enters the translocase channel. A short segment of the preprotein is translocated. 6, ATP is hydrolyzed and SecA is released into the cytosol. 7, Translocation continues, but now driven by the Δp alone. 8, The leader peptide is cleaved releasing the protein into the periplasm. (B) In the absence of a Δp in vitro there can be several rounds of SecA binding and release promoting ATP-dependent translocation. a, binding SecA; b, binding of ATP to SecA accompanied by the translocation of a small segment; c, hydrolysis of ATP and release of SecA; d, rebinding of SecA as the cycle continues. Symbols: A, SecA; B, SecB.

do not change. Furthermore, the direction of translocation was reversed when the polarity of the Δp was reversed in proteoliposomes translocating an outer membrane protein (proOmpA).[10]

Step 8. The leader sequence is cleaved and the translocated protein is released into the periplasm. The leader peptide must be removed from periplasmic and outer membrane proteins in order for them to leave the membrane surface, but it need

not be removed for translocation *per se* to take place. According to this model, the initiation of translocation requires the hydrolysis of one ATP. The rest of translocation is driven by the Δp and does not require ATP. However, see below.

Recycling of SecA
The initial stages of translocation, which consist of the insertion of the leader peptide into the membrane and a limited amount of

translocation (steps 1–4), require SecA and ATP binding. ATP hydrolysis releases SecA and the remainder of translocation is driven by the Δp. However, experiments have demonstrated that, in the absence of a Δp, ATP can drive translocation to completion *in vitro*.[11] This is because SecA can rebind to the portion of the protein not yet translocated and promote successive rounds of ATP-dependent translocation (Fig. 16.1B, steps a, b, c, d). For some proteins, it may be necessary to increase the SecA concentrations in order to demonstrate complete translocation in the absence of a Δp.[12] The extent of the *in vitro* requirement for the Δp apparently varies with the preprotein that is used.[13] Although SecA and ATP can be shown to drive translocation to completion in certain *in vitro* experiments in which there is no Δp, the evidence favors the conclusion that the major driving force for proton translocation *in vivo* is the Δp and not ATP. (See the description of step 7, above.) That is not to say that SecA binding and rebinding may not be important beyond the initiation stages of translocation. Cycles of SecA rebinding and dissociation may, in fact, occur *in vivo*, promoting limited translocation in conjunction with translocation driven by the Δp. One possible role for cycles of SecA binding and ATPase-dependent dissociation from the preprotein as it is being translocated might be to unfold untranslocated regions that have assumed a tight tertiary configuration that cannot be threaded through the translocase. One might expect such foldings to occur since protein secretion need not be coupled to translation and the untranslocated regions of the protein that are still in the cytosol may fold in a way not suitable for translocation.

Co-translational translocation

Polysomes translating presecretory proteins associate with membranes and it seems that some proteins might be translocated while they are being translated, although in *E. coli* most proteins are exported post-translationally. The idea is that the ribosome translating a messenger RNA for a preprotein is closely associated with a membrane translocase so that the polypeptide is threaded through the translocase during translation. Chaperones

might be necessary if the nascent polypeptide entered the cytosol before engaging the translocase.

16.2 The translocation of membrane-bound proteins

Thus far the description of protein translocation has been confined to those proteins that ultimately reside in either the periplasm or the outer membrane. The export of most of the inner membrane proteins is also Sec-dependent and occurs in a similar fashion. The question is, what keeps these proteins from being translocated through the membrane into the periplasm? The answer lies in internal hydrophobic regions of the protein that stop translocation and anchor the protein into the membrane because they bind to the lipid. Sometimes it is the signal sequence itself that anchors the protein in the membrane (Fig. 16.3A, a). These signal sequences differ from those discussed above in not being recognized by the signal peptidase and are therefore not cleaved. Other times it is an internal hydrophobic region called the "signal–membrane anchor" or "stop transfer" sequence which anchors the protein after the signal sequence has been cleaved (Fig. 16.3A, b). The "stop-transfer" regions have a stretch of 15 or more hydrophobic amino acids, which vary from one protein to another. Proteins that span the membrane several times do so via a series of alternating uncleaved signal sequences and "stop transfer" signals (Fig. 16.3A, c). (These include the membrane-bound solute transporters discussed in Chapter 15.) A model for the insertion of proteins that span the membrane multiple times is shown in Fig. 16.3B. A noncleavable signal sequence initiates insertion into the membrane. At some point, translocation stops as an internal hydrophobic "stop-transfer" signal enters the translocase. According to the model, the translocase opens laterally when the "stop-signal" enters, allowing the diffusion of the "stop-transfer" signal into the lipid bilayer. Then the channel closes. A downstream secondary signal sequence reinitiates translocation and the process is repeated until the entire protein is translocated into the membrane. This is not a well-understood process.

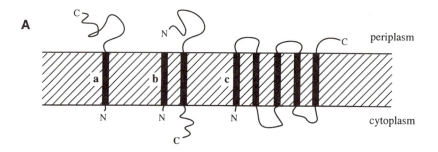

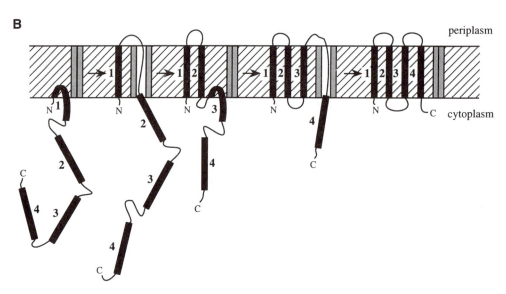

Fig. 16.3 Model for the Sec-dependent translocation of membrane proteins. (A) a, Protein anchored to the membrane by its leader peptide. b, Protein anchored to the membrane by a "stop-transfer signal." The leader sequence has been proteolytically removed and is shown in the membrane. c, Protein anchored to membrane by alternating leader peptides and stop-transfer signals. (B) Insertion of a protein with four hydrophobic domains. Domain 1 inserts into the membrane as a signal peptide. This opens the translocase channel and initiates translocation. When domain 2 (stop-transfer signal) enters the translocase, the putative channel opens laterally allowing domain 2 to diffuse laterally into the lipid matrix. The channel then closes. Domain 3 (secondary signal sequence) reinitiates transloca-tion by inserting into the lipid bilayer and reopening the channel. When domain 4 (stop-transfer signal) enters the translocase it stops translocation and moves laterally out of the translocase channel into the lipid matrix. (Adapted from Pugsley, A. P. 1993. The complete general secretory pathway in gram-negative bacteria. *Microbiol. Rev.* **57**:50–108.)

16.3 Extracellular protein secretion

Proteins that are secreted into the medium include enzymes such as proteases, nucleases, lipases, carbohydrases, toxins, and other virulence factors.[14,15] Some proteins whose secretion has been studied include immuno-globulin A proteases secreted by gram-negative pathogenic bacteria belonging to the genera *Neisseria*, *Haemophilus*, and *Serratia*, cholera and pertussis toxin secreted by *Vibrio*

cholerae and *Bordetella pertussis*, respec-tively, pilin proteins, and hemolysins (toxins that lyse red blood cells), secreted by various gram-negative bacteria. Protein secretion has been studied in gram-positive bacteria as well. These include penicillinase production, and the secretion of various proteases in *Bacillus* species. In some cases, a particular sequence of amino acids at the carboxy end seems sufficient for translocation through the outer envelope (self-promoted secretion).

However, in most instances a special secretion machinery seems to be necessary. There are two major secretion machineries, one is *sec-independent (Type I)* and the other *sec-dependent (Type II)*.

16.3.1 Sec-independent protein secretion

The *sec*-independent pathway is called the Type I system. Examples of proteins secreted by the Type I system include haemolysin by *E. coli*, proteases by *Erwinia chrysanthemi*, leucotoxin by *Pasteurella haemolytica*, cyclolysin by *Bordetella pertussis*, and alkaline protease by *Pseudomonas aeruginosa*. These proteins do not have leader sequences and do not require any of the *sec* genes for secretion.

Genes required for secretion have been identified by mutant analysis and are called *secretory genes*. The secretory genes are usually contiguous with the structural gene for the protein that is secreted. The proteins coded for by the secretory genes form a secretory apparatus. Since the carboxyl terminus of the secreted protein is required for secretion, it would appear that the secretion apparatus recognizes this region of the protein. There are three secretory proteins, HlyB, HlyD, and TolC, which are homologous in the different Type I systems. The cellular location of the secretory proteins suggest that they may form a channel through the inner and outer membrane, allowing proteins to move directly from the cytoplasm to the external medium. The mechanism of translocation is not understood.

16.3.2 Sec-dependent protein secretion

The Type II pathways are *sec*-dependent. The proteins are translocated across the cell membrane using leader sequences and the *sec* proteins in the usual way. Then, the leader sequence is proteolytically removed on the external face of the cell membrane and the proteins enter a second stage of export, specific to the Type II systems, to the extracellular milieu. *The translocation machinery and mechanism of the second stage are not understood, but it seems to be the major route*

for secretion for many proteins that are structurally unrelated. It has been speculated that the secreted proteins may move in vesicles from the periplasm to the outer membrane where they fuse with channels in the outer membrane made by the Type II secretory apparatus. Even if this were true, it does not address the problem of how the secreted proteins are selected for inclusion into the postulated vesicles.

There are several Type II systems and each can secrete more than one kind of protein. For example, a secretion machinery called the Out system exists in *Erwinia chrysanthemi* and *E. carotovora*, causative agents of soft-rot disease in plants. The enzymes secreted by the Out system are pectate lyase, exo-poly-α-D-galacturonosidase, pectin methylesterase, and cellulase. Mutants in the *out* genes have a defect in the secretion of cellulases and pectinases and accumulate these enzymes in the periplasm. Other Type II systems exist in other bacteria and are responsible for the secretion of different proteins. For example, the *pul* genes are required for the secretion of pullulanase by *Klebsiella pneumoniae*.

A resident Type II system will not secrete extracellular proteins of related bacteria when the structural genes are introduced, *even though there may be homology between the Type II proteins from the donor and recipient bacteria*. For example, there is homology in the proteins between the *out* gene products from *Erwinia* and the *pul* gene products from *Klebsiella*. Yet, when the pectate lyase gene from *Erwinia* was expressed in *Klebsiella*, the enzyme was not secreted. There can even be specificity between Out systems from two different *Erwinia* strains. For example, the *E. chrysanthemi* Out system did not secrete an extracellular pectate lyase encoded by a gene from a different species of *Erwinia*, even though both species of *Erwinia* use the Out system. The conclusion is that, when the enzyme structural gene is expressed heterogenerically, the protein products are not secreted. This is because each Type II system distinguishes not only between different secreted proteins, but also recognizes "self" proteins. The basis for this distinction is not understood. However, if the Type II system *as well as its cognate protein* are transferred to a different genus of bacterium, then protein secretion can occur. For example, when the structural gene for pullulanase as well as the adjoining secretion genes are transferred from *Klebsiella pneumoniae* to *E. coli*, pullulanase is synthesized and secreted.[16]

16.3.3 Protein secretion that does not appear to be Type I or Type II

As an example of self-promoted protein secretion that does not appear to require a secretion apparatus, we will consider the secretion of immunoglobulin A protease by the causative agent of gonorrhoeae, *Neisseria gonorrhoeae*. The protease is synthesized as a precursor protein with an N-terminal leader sequence typical for exported proteins. The leader sequence functions with the *sec* machinery and is removed by the peptidase after translocation through the inner membrane. Up to this stage, the secretion is similar to the secretion of proteins destined for the periplasm or the outer envelope. However, the immunoglobulin A protein has information in its sequence that directs it through the outer envelope to the outside. At the carboxy end of the protein, there is a helper domain that aids the protein in traversing the outer membrane where it anchors the protein. The protein is released into the medium by proteolytic cleavage (autocatalytic) at the C-terminal end, leaving the helper peptide embedded in the membrane. Some other proteins are secreted in a similar fashion, although the proteolytic cleavage may be catalyzed by a separate enzyme. How the secreted proteins cross the periplasm and insert into the outer envelope is not known. The helper domain can promote translocation of hybrid proteins across the outer membrane in *E. coli* which would not normally release the passenger protein into the medium. It would thus appear that the helper region may be sufficient for translocation through the outer envelope without the need for an accessory secretion system that has been identified thus far.

16.4 Summary

Bacteria translocate (export) proteins into the cell membrane or secrete them through the membrane into the periplasm and outer envelope of gram-negative bacteria, the cell wall of gram-positive bacteria, and into the extracellular medium. The translocation machinery and requirements for targeting to different destination sites are being studied in several bacteria, especially in *E. coli*.

The Sec system is necessary for the secretion of most proteins. A preprotein is made with a leader sequence at the amino terminal end. The role of the leader sequence is to initiate translocation by inserting into the lipid bilayer. The leader sequence also aids in preventing premature folding of the newly synthesized protein. The preprotein binds to a chaperone protein in the cytosol. An important chaperone protein is Sec B. Two roles of the chaperone protein are to prevent folding of the preprotein into a configuration that disallows translocation and to prevent the formation of protein aggregates. A third role for the chaperone protein is to deliver the preprotein to the translocase. The preprotein–chaperone protein complex binds to the translocase on the inner membrane surface and the translocation of the preprotein into the membrane is initiated. The initial stages of translocation require ATP and SecA but the remainder of translocation can be driven by the Δp. Once the preprotein has been translocated through the membrane the leader sequence is removed by a peptidase on the outer surface of the cell membrane. A few proteins (e.g., some phage proteins) are translocated independently of the Sec system, but these are a small minority.

Many proteins are secreted into the extracellular space. They must traverse the periplasm and outer envelope in gram-negative bacteria and the cell wall in gram-positive bacteria. Although a few proteins seem to be secreted without additional proteins other than the Sec proteins, most require a second secretion apparatus. There are two major secretion systems for getting through the outer envelope in gram-negative bacteria. These are called Type I and Type II. Type I is independent of the Sec proteins, and the secreted proteins seem to travel from the cytoplasm to the exterior without stopping in the periplasm. The Type II system occurs in two stages. Stage 1 is the secretion of the protein via the Sec system into the periplasm, and stage 2 is the transport from the periplasm through the outer envelope to the exterior. It is not understood how proteins destined for secretion cross the periplasm and outer envelope. A third system, which is neither Type I or Type II, also exists.

Study Questions

1. What is the experimental basis for believing that the leader peptide is necessary for translocation? What are two reasons for believing that the leader peptide is not necessary for determining the final destination of the protein?

2. What are the postulated roles for chaperone proteins in protein translocation?

3. What is the role of Sec A in protein translocation?

4. What are the functions of the different domains of the leader sequence, i.e., the basic N-terminal, middle hydrophobic, and carboxy terminus?

5. What is the evidence, based upon mutational analysis, that there are two secretory systems for proteins secreted into the medium and that one is *sec*-dependent and one is *sec*-independent.

REFERENCES AND NOTES

1. Driessen, A. J. M. 1992. Bacterial protein translocation: Kinetic and thermodynamic role of ATP and the protonmotive force. *TIBS* 17:219–223.

2. Ito, K. 1992. SecY and integral membrane components of the *Escherichia coli* protein translocation system. *Molec. Microbiol.* 6:2423–2428.

3. Wickner, W., A. J. M. Driessen, and F.-U. Hartl. 1991. The enzymology of protein translocation across the *Escherichia coli* plasma membrane. *Annu. Rev. Biochem.* 60:101–124.

4. Saier, Jr., M. H., P. K. Werner, and M. Muller. 1989. Insertion of proteins into bacterial membranes: Mechanism, characteristics, and comparisons with the eucaryotic process. *Microbiol. Rev.* 53:333–366.

5. Pugsley, A. P. 1993. The complete general secretory pathway in gram-negative bacteria. *Microbiol. Rev.* 57:50–108.

6. Chaperone proteins are important not only for protein secretion but for the folding of nonsecreted proteins into a biologically active form. For example, the GroEL and GroES proteins are cytosolic proteins required for the correct assembly of the *Rhodospirillum rubrum* enzyme ribulosebishosphate carboxylase/oxygenase (Rubisco) in *E. coli*, and for morphogenesis of certain bacteriophages. Another chaperone protein, trigger factor, was originally characterized as stabilizing proOmpa, the precursor to the *E. coli* outer membrane protein, for translocation in an *in vitro* system. However, *in vivo* studies subsequently showed that trigger factor does not promote proOmpa translocation and in fact may retard it. These experiments suggested that trigger factor is a chaperone for cell division proteins.

7. Guthrie, B. and W. Wickner. 1990. Trigger factor depletion or overproduction causes defective cell divison but does not block protein export. *J. Bacteriol.* 172:5555–5562.

8. Schiebel, A., Driessen, A. J. M., Hartl, F.-U, and W. Wickner. 1991. μ_{H^+} and ATP function at different steps of the catalytic cycle of preprotein translocase. *Cell* 64: 927–939.

9. The *in vitro* system consists of inverted *Escherichia coli* inner membrane vesicles translocating the outer membrane protein proOmpA.

10. Driessen, A. J. M. 1992. Precursor protein translocation by the *Escherichia coli* translocase is directed by the protonmotive force. *Embo J.* 11:847–853.

11. There are several ways to separate effects of the Δp from those of SecA and ATP *in vitro*. For example, membrane vesicles or proteoliposomes will not have a Δp unless one is imposed. Different methods can be used to impose a Δp. For example, the stimulation of electron transport by incubating inner membrane vesicles with NADH has been used. Vesicles are prepared from *unc⁻* cells so that the Δp and ATP are not interconvertible. A Δp has also been created by light in proteoliposomes in which bacteriorhodopsin was incorporated, or by adding reduced cytochrome c to proteoliposomes in which cytochrome c oxidase has been incorporated. Proton ionophores can be used to dissipate the Δp. SecA can be inactivated with anti-SecA antibody, and ATP levels can be experimentally manipulated.

12. Yamada, H., Matsuyama, S-ichi, Tokuda, H., and S. Mizushima. 1989. A high concentration of SecA allows proton motive force-independent translocation of a model secretory protein into *Escherichia coli* membrane vesicles. *J. Biol. Chem.* **264**:18577–18581.

13. Yamada, H., Tokuda, H., and S. Mizushima. 1989. Protonmotive force-dependent and independent protein translocation revealed by an efficient *in vitro* assay system of *Escherichia coli*. *J. Biol. Chem.* **264**:1723–1728.

14. Lory, S. 1992. Determinants of extracellular protein secretion in gram-negative bacteria. *J. Bacteriol.* **174**:3423–3428.

15. Salmond, G. P. C., and P. J. Reeves. 1993. Membrane traffic wardens and protein secretion in gram-negative bacteria. *TIBS* **18**:7–12.

16. d'Enfert, C., Ryter, A., and A. P. Pugsley. 1987. Cloning and expression in *Escherichia coli* of the *Klebsiella pneumoniae* genes for production, surface localization and secretion of the lipoprotien pullulanase. *EMBO J.* **6**:3531–3538.

17

Signaling and Behavior

Bacteria may alter cell morphology, cell metabolism, gene transcription, and cell behavior (motility and chemotaxis), in response to environmental fluctuations such as the availability of respiratory electron acceptors, the supply of carbon, nitrogen, or phosphate, or changes in the osmolarity and temperature of the medium. In this way, bacteria survive in and adapt to ever-changing environmental conditions. Underlying the adaptations are sophisticated detection systems with which the bacterium monitors the environment and transmits signals to specific intracellular targets, which can be the transcriptional machinery, enzymes, or a cellular component such as the flagellar motor. This chapter describes some of these signaling circuits and behavioral responses.

In several systems, a signal transduction pathway exists that includes a *histidine kinase* protein that receives a signal and transmits it to a partner *response regulator* protein. The response regulator protein in turn transmits the signal to the target. The signal is transmitted between the histidine kinase and the response regulator via *phosphorelay*. Specifically, the histidine kinase autophosphorylates at a histidine residue in response to a stimulus (using ATP as the phosphoryl donor), and then transfers the phosphoryl group to an aspartate residue in the partner response regulator protein. This activates the response regulator (Section 17.1.2 and Fig. 17.1). Most of the known phosphorylated

response regulators stimulate or repress the transcription of specific genes. (Exceptions are P-CheB and HP-CheY which affect the chemotaxis machinery (Section 17.4).) The signaling pathway also includes a phosphatase that dephosphorylates the response regulator, returning it to the nonstimulated state, where it once again can respond to the signal. The phosphatase may be the histidine kinase itself, the response regulator, or a separate protein. Histidine kinases may reside in the membrane or in the cytoplasm, although they are often in the membrane, whereas the response regulators are in the cytoplasm. Signaling systems that consist of a histidine kinase and a response regulator protein are called "two-component" systems, although the number of proteins in the signal transduction pathway is often more than two.

It must be emphasized that the histidine kinase need not be the first protein in the signal transduction pathway to respond to the signal. In other words, it need not be the sensor. In many systems, signals first interact with protein(s) other than the histidine kinase and the stimulus is relayed to the histidine kinase. For example, in the *E. coli* chemotaxis system described in Section 17.4, the transmembrane proteins (called chemoreceptors or MCP proteins) are sensor proteins that respond to chemoeffectors and, as a consequence, change the activity of a cytoplasmic histidine kinase (CheA). Another example where the initial receiver of the signal

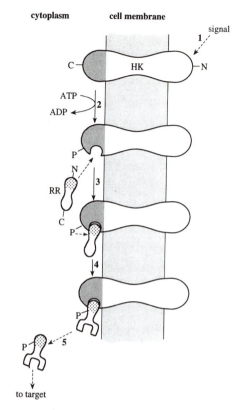

cytoplasm   cell membrane

to target

Fig. 17.1 Two-component regulatory systems. (1) A transmembrane histidine kinase (HK) is activated by a signal at its N-terminal domain. (2) The activated protein autophosphorylates in the C-terminal domain. (3) The response regulator protein (RR) binds to the C-terminal end of the histidine kinase and the phosphoryl group is transferred from the histidine kinase to the response regulator, thus activating the latter. (4) The activated response regulator leaves the histidine kinase and stimulates its target. Shaded and stippled areas of the histidine kinase and response regulator represent conserved amino acid sequences typical for the respective class of protein. The change in shape of the proteins represents a presumed conformational change. In some systems the histidine kinase is cytoplasmic and detects signals within the cytoplasm.

is not the histidine kinase occurs in the PHO regulon control system, described in Section 17.6, which is repressed by inorganic phosphate. The proteins that initially bind inorganic phosphate are in the phosphate transport system (Pts), which is believed to bind inorganic phosphate and then stimulate the enzymatic activity of the membrane-bound PhoR histidine kinase. A third (and

more complicated) example occurs in the Ntr regulon which is repressed by ammonia (Section 17.5). The ammonia levels determine the concentrations of glutamine and α-ketoglutarate via the enzymes glutamine synthetase and glutamate synthase, respectively. The α-ketoglutarate and glutamine, in turn, influence the activity of a bifunctional enzyme, uridydyl transferase-uridyl-removing (UT-UR) enzyme that modifies a signal transduction protein (P_{II}). P_{II} regulates the activity of a cytoplasmic histidine kinase (N_{II}). In this case the histidine kinase is indeed far removed from the initial signal, ammonia.

This chapter also discusses the Fnr protein. Although the Fnr protein is not part of a two-component regulatory system, it is discussed in this chapter because it functions along with two-component regulatory systems (the Arc system and the NarL/NarX/NarQ system) during the metabolic adaptations that accompany the shift from aerobic to anaerobic growth conditions.

17.1 An introduction to two-component signaling systems

Two-component signaling systems[1-3] have been discovered in many bacteria, both gram-negative and gram-positive, and have been implicated in a wide range of physiological responses. These include nitrogen assimilation, outer membrane porin synthesis, chemotaxis in *E. coli* and *S. typhimurium*, nitrogen fixation in *Klebsiella* and *Rhizobium*, sporulation in *Bacillus*, oxygen regulation of gene expression in *E. coli*, and the uptake of carboxylic acids in *Rhizobium* and *Salmonella*. It is clearly a widespread and important signal transduction system that enables bacteria to adapt to changes in the external milieu.

17.1.1 Components

The components of the "two-component" systems include:

1. A histidine kinase (HK) which is sometimes called a sensor/kinase. The histidine kinase

receives a signal and autophosphorylates at a histidine residue:

$$HK + ATP \xrightarrow{\text{signal}} HK\text{-}P + ADP$$

2. A partner response regulator (RR). The partner response regulator is activated by HK-P and sends a signal to its target, e.g., the genome:

$$RR + HK\text{-}P \longrightarrow RR\text{-}P \ (\text{active}) + HK$$

3. A phosphatase. The phosphatase inactivates the RR-P. The phosphatase may be the histidine kinase, the response regulator, or a separate protein.

$$RR\text{-}P + H_2O \longrightarrow RR + P_i$$

17.1.2 Signal transduction in the two-component systems

Figure 17.1 illustrates a simplified model that summarizes signal transduction in two-component systems. The model will be modified later to accommodate differences between the various signaling systems. In the example shown in Fig. 17.1, the histidine kinase (HK) is depicted as a transmembrane protein composed of three domains (i.e., an N-terminal domain that is presumed to be at the outer surface of the cell membrane and to bind to a signaling ligand in the periplasm, a hydrophobic domain that is transmembrane, and a conserved C-terminal domain that is cytoplasmic). Some histidine kinases (e.g., CheA and NR_{II}) are cytoplasmic proteins (Sections 17.4.4 and 17.5.1). Signal transduction as depicted in Fig. 17.1 can be conveniently thought of as occurring in three steps.

Step 1. In response to a stimulus at the N-terminal domain, the histidine kinase autophosphorylates (usually) at the C-terminal domain. The phosphoryl donor is ATP.

Step 2. The phosphoryl group is transferred from the histidine kinase to its partner response regulator protein. All of the response regulator proteins are related in

having conserved amino acid sequences in the N-terminal domain (usually) that may bind to the conserved C-terminal region of the histidine kinases.

Step 3. The response regulator becomes activated on being phosphorylated and changes the activity of its target. The effect is usually the stimulation or repression of gene transcription. As we shall see, some response regulator proteins (e.g., NarL) do both (Section 17.3). Other response regulators may have targets other than the genome. For example, in chemotaxis, the phosphorylated derivatives of the response regulators CheY and CheB change the direction of the flagella motors and the extent of methylation of the chemoreceptor proteins in the membrane, respectively (Section 17.4.4). The response regulator proteins differ at their C-terminal domains and this probably confers specificity with regard to their targets and activities.

Histidine kinases are sometimes bifunctional enzymes that can act either as kinases or as phosphatases when stimulated. When stimulation of the histidine kinase promotes phosphatase activity, this results in the *dephosphorylation* of the response regulator and hence a *repression* of transcription. (In the absence of the signal, the kinase phosphorylates the response regulator and transcription is stimulated.) An example where the signal results in stimulating the phosphatase activity of the histidine kinase, and therefore represses gene transcription, is the repression of the Ntr regulon by ammonia. In the presence of excess ammonia, the histidine kinase, NRII, acts as a phosphatase rather than as a kinase and inactivates the response regulator, NR_I, which in its phosphorylated form is a positive transcription factor (Section 17.5). Another example may be the repression of the PHO regulon by inorganic phosphate. The histidine kinase, PhoR, appears to respond to excess inorganic phosphate by dephosphorylating the response regulator, PhoB (Section 17.6). However, as explained in Sections 17.5 and 17.6, NRII and PhoR do not themselves bind the ammonia or phosphate, respectively.

Amino acid sequences define histidine kinases and response regulator proteins

The histidine kinases are defined by a conserved sequence of about 200 amino acids at the terminal end. The C-terminal domain is the site of the conserved histidine residue that becomes phosphorylated in response to a stimulus (Fig. 17.2). As indicated in Fig. 17.1, most of the known histidine kinases are transmembrane. The conserved C-terminal end is in the cytoplasm where it interacts with the response regulator protein. The N-terminal end may be exposed on the extracellular membrane surface (the periplasm in gram-negative bacteria). The amino acid sequence at the N-terminal domain varies with the different histidine kinases presumably because they respond to different stimuli. The response to the stimulus within the N-terminal region causes the enzyme to autophosphorylate the conserved histidine residue in the cytoplasmic domain (Fig. 17.1).

The response regulator proteins are defined by a conserved amino-terminal domain of about 100 amino acids (Fig. 17.3). The conserved amino-terminal end of the response regulator protein is thought to interact with the conserved carboxy end of the histidine kinase and become phosphorylated at an aspartate residue. The phosphate is eventually removed by a phosphatase, which may be the histidine kinase, the response regulator protein, or perhaps a third protein. Thus far, phosphorylation and dephosphorylation have been demonstrated for only a few of the histidine kinases and response regulator

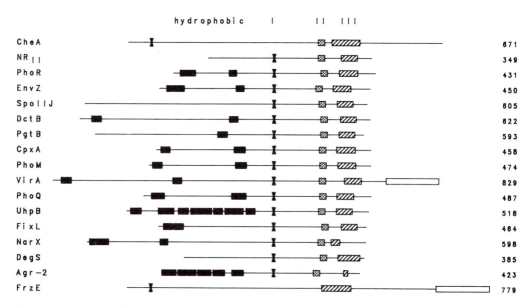

Fig. 17.2 Structures of the histidine protein kinases. The histidine protein kinases are part of the two-component regulatory systems. They autophosphorylate at a histidine residue and then transfer the phosphate to a response regulator protein. Most of the histidine kinases are believed to be transmembrane proteins with an extracytoplasmic amino terminus that responds to stimuli. Exceptions are CheA, NR$_{II}$, and FrzE, which are cytoplasmic. Hydrophobic domains that presumably span the membrane are indicated by the black boxes at the amino-terminal end. The cytoplasmic domain that includes the phosphorylated histidine residue is indicated by a filled X (domain I). Regions II and III (stippled and hatched-filled boxes) represent regions in the carboxy-terminal domain where certain amino acids appear with high frequency at specific locations in the sequence. These are called conserved regions. For example, when the amino acid sequences are lined up for comparison, position 43 may be a glycine in all of the proteins (totally conserved), whereas position 44 may be arginine in e.g., 60% of the proteins, and so on (partially conserved). The empty boxes at the extreme carboxy ends are homologous to response regulator domains at their amino-terminal ends. (From Stock, J. B., A. J. Ninfa, and A. M. Stock. 1989. Protein phosphorylation and regulation of adaptive responses in bacteria. *Microbiol. Rev.* 53:450–490.)

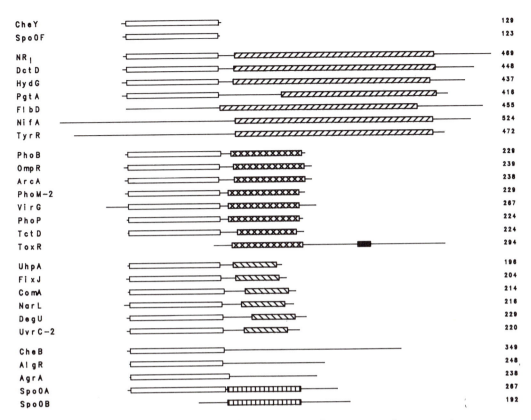

Fig. 17.3 Structures of the response regulator proteins. Response regulator proteins are cytoplasmic proteins that are phosphorylated, or presumed to be phosphorylated, by the histidine kinase proteins. The phosphorylated regulator proteins transmit the signal to the genome or to some other cellular machinery, (e.g., the flagellar motor). The amino-terminal region of the response regulator protein has conserved amino acids (open boxes). Approximately 20–30% of the amino acids are identical at corresponding posititions when the sequences are aligned. This is the region thought to interact with the carboxy domain of the histidine kinase, and to become phosphorylated. Other conserved regions are indicated by cross-hatched boxes. These are in the carboxy-terminal domain that is thought to interact with target molecules. For example, NR_I, DctD, NifA share a homologous carboxy-terminal domain that is thought to interact with one of the E. coli sigma factors, sigma 54. NR_I and NifA have a common carboxy-terminal region that is thought to bind to DNA. One of the proteins, ToxR, spans the membrane, and the hydrophobic region is indicated by the filled box. The numbers to the right are the lengths of the protein in amino acid residues predicted from the nucleotide sequences. (From Stock, J. B., A. J. Ninfa, and A. M. Stock. 1989. Protein phosphorylation and regulation of adaptive responses in bacteria. *Microbiol. Rev.* 53:450–490.)

proteins. At present, these activities are inferred for the remainder of the proteins if the amino acid sequences place these proteins in the established classes of the known histidine kinases and response regulators. Since the carboxy ends of the histidine kinases and the amino ends of the different response regulator proteins are conserved, it is theoretically possible that a histidine kinase in one signaling system may activate a response regulator of a different system. This possible interaction between different signaling systems has been called "cross-talk."[4]

17.2 Adaptive responses by facultative anaerobes to anaerobiosis

A shift from an aerobic to an anaerobic atmosphere results in extensive changes in the metabolism of facultative anaerobes. These adaptive responses,[5] which have stimulated

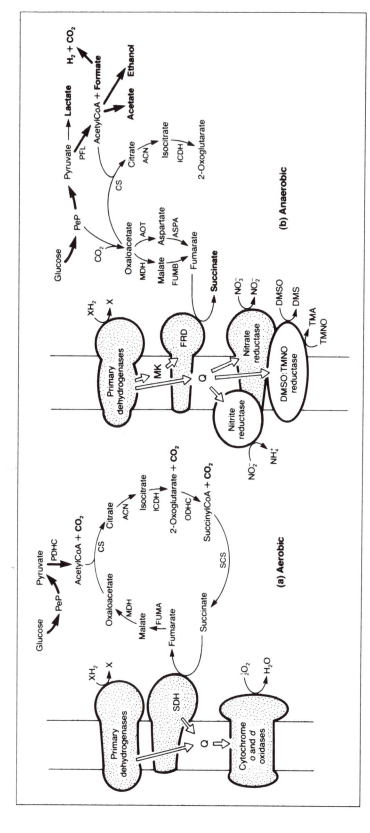

Fig. 17.4 A comparison of respiratory systems and carbon metabolism in *E. coli*. (a) An oxidative citric acid cycle feeds electrons into succinic dehydrogenase (SDH) and NADH dehydrogenase (a primary dehydrogenase). (b) The citric acid pathway is non-cyclic and reductive because the α-ketoglutarate dehydrogenase activity is absent (or severely diminished). The electron transport chain has been altered so that electrons flow to acceptors other than oxygen. When nitrate is present, nitrate reductase (NR) is synthesized. Nitrate represses the synthesis of fumarate reductase (FRD) and DMSO:TMNO reductase. In the absence of an electron acceptor for anaerobic respiration, the major energy-yielding pathways are fermentative.(From Spiro, S., and J. R. Guest. 1991. Adaptive responses to oxygen limitation in *Escherichia coli*. *TIBS* 16:310–314.).

much interest, are described below. The adaptive responses to anaerobiosis are regulated by two systems [i.e., the Arc system (two-component) and the Fnr system (not two-component)]. The metabolic changes will be discussed first followed by a description of the Arc and Fnr systems. (A third system, one that regulates the response to nitrate, is discussed in Section 17.3.)

17.2.1 Metabolic changes accompanying shift to anaerobiosis

During aerobic respiration, the citric acid pathway is cyclic and oxidative (Fig. 17.4a). However, in the absence of oxygen several important changes take place (Fig. 17.4 b). These include the replacement of succinate dehydrogenase by fumarate reductase and the repression of the synthesis of α-ketoglutarate dehydrogenase. The result of these enzymatic changes is the conversion of the oxidative citric acid cycle into a reductive noncyclic pathway. Additionally, in E. coli, acetyl-CoA is no longer made by pyruvate dehydrogenase but rather by pyruvate–formate lyase. This is an advantage under anaerobic conditions because it decreases the amount of NADH which must be reoxidized. The acetyl-CoA thus formed is converted to acetate and ethanol, and the formate is converted to hydrogen gas and carbon dioxide. There is also a decrease in synthesis of other enzymes used during aerobic growth (e.g those of the glyoxylate cycle and fatty acid oxidation.)

Depending upon the presence of particular electron acceptors, major changes also take place in the respiratory pathway.[6] Facultative anaerobes such as E. coli carry out aerobic respiration in the presence of oxygen, but in the absence of oxygen they carry out anaerobic respiration using either nitrate, fumarate, or some other electron acceptor (e.g., TMAO or DMSO). (TMAO and DMSO are formed naturally in nature and presumably exist in the intestine where E. coli grows.) There is a hierarchy of electron acceptors that are used, oxygen being the most preferred with nitrate second, followed by fumarate and the other electron acceptors. The hierarchy parallels the maximum work that can

be done when electrons travel over the electrode potential gradient to the terminal acceptor. The work is proportional to the difference in electrode potential between the electron acceptor and donor. For example, the maximum work that can be done is greatest when oxygen is the electron acceptor $(E'_m = +0.82\,V)$, less when nitrate is the electron acceptor $(E'_m = +0.42\,V)$, and least when fumarate is the electron acceptor $(E'_m = +0.03\,V)$. As described in Section 4.4, the electrons are passed to these terminal electron acceptors from reduced quinone. For example, ubiquinone $(E'_m = +0.1\,V)$ transfers electrons from the reductant to either the cytochrome oxidase or the nitrate reductase module. Menaquinone $(E'_m = -0.074\,V)$, rather than ubiquinone, transfers electrons to the fumarate reductase complex. E. coli determines which electron acceptors will be used, in part by regulating the transcription of genes coding for the electron acceptors. Thus, in the presence of oxygen, the genes for nitrate reductase, fumarate reductase, and the other reductases are repressed, and therefore only aerobic respiration can take place. In the absence of oxygen but in the presence of nitrate, nitrate reductase genes are induced (by nitrate) but the genes for fumarate reductase and the other reductases are repressed by nitrate. Therefore, nitrate respiration takes place. In the absence of both oxygen and nitrate there is no longer any represssion of fumarate reductase and so fumarate respiraton takes place. Changes also take place within the aerobic respiratory pathway. When oxygen levels are high, E. coli uses cytochrome o, encoded by the cyo operon, as the terminal oxidase. When oxygen becomes limiting (and during stationary phase), cytochrome o is replaced by cytochrome d, encoded by the cyd operon. One advantage to this is that cytochrome d has a higher affinity for oxygen $(K_m = 0.23–0.38\,\mu M)$ than has cytochrome o $(K_m = 1.4–2.9\,\mu M)$.

Thus, oxygen represses the synthesis of the anaerobic reductases ensuring that oxygen is used as the electron acceptor in air. Nitrate induces the synthesis of nitrate reductase and represses the synthesis of the other reductases, ensuring that nitrate is used as the electron acceptor in the presence of nitrate anaerobically. In the absence of any exogenously supplied electron acceptor, *E. coli* relies on fermentation as its major source of ATP.

17.2.2 Regulatory systems

Three systems for the regulation of gene expression by oxygen or nitrate have been characterized in *E. coli*. Two of these are responsible for regulation by oxygen, and the third mediates regulation by nitrate. They are:

1. The Arc system (two-component). This represses the transcription of several "aerobic" genes and stimulates transcription of the gene for cytochrome d oxidase.

2. The Fnr system (not a two-component system). This stimulates the transcription of several "anaerobic" genes and represses the transcription of some "aerobic" genes.

3. The NarL/NarX/NarQ system (two component). This stimulates the transcription of the nitrate reductase genes and represses the transcription of the other reductase genes.

The Arc system ("aerobic respiration control")
The Arc system ("aerobic respiration control") consists of a histidine kinase, ArcB, and a partner response regulator, ArcA. Phosphorylated ArcA is responsible for:

1. Anaerobic repression of the genes for:
   Citric acid cycle enzymes
   Glyoxylate cycle enzymes
   Several dehydrogenases for aerobic growth (e.g., pyruvate dehydrogenase)
   Fatty acid oxidation enzymes
   Cytochrome o oxidase

2. Induction in low oxygen of the genes for:
   Cytochrome d oxidase and cobalamin synthesis.[9]

Interestingly, ArcA also activates the transcription of the mating system genes (i.e., mutations in *arc* A produce a defect in the synthesis of F-pili).[7]

It appears that the ArcB protein monitors the level of oxygen indirectly and that when oxygen levels are sufficiently low, the ArcB

protein autophosphorylates using ATP as the phosphoryl donor, and then transfers the phosphoryl group to ArcA (Fig. 17.5).[8] How does ArcB detect changes in the levels of oxygen? Apparently, it is not oxygen itself that is the signaling molecule. This conclusion has been reached from two lines of evidence. The level of expression of the *sdh* operon (succinate dehydrogenase) varies with the mid-point potential of the electron acceptor. The level of expression is highest with oxygen, lower with nitrate, and lowest with fumarate. This parallels the mid-point potential of the electron acceptor, which is most positive for oxygen, less positive for nitrate, and least positive for fumarate. The ArcB protein may respond to the ratio of the oxidized to reduced forms of an electron transport carrier (e.g., flavoprotein, quinone, cytochrome) rather than to fumarate, nitrate, and oxygen *per se*. One way to test this hypothesis is to delete the cytochrome o and d genes and measure the expression of genes under the control of the Arc system. Deletion of the cytochrome oxidase genes would be expected to increase the ratio of reduced to oxidized forms of electron carriers. If a high ratio of reduced over oxidized form of an electron transport carrier signaled the ArcB protein that oxygen was absent, then deletion of the cytochrome oxidase genes would be expected to mimic the absence of oxygen. This experiment was done using *cyo-lacZ* and *cyd-lacZ* fusions as probes to monitor expression of the cyo and cyd genes.[10,11] When both the *cyo* and *cyd* genes were deleted and the cells were grown in air, *cyo-lacZ* expression was lowered and *cyd-lacZ* expression increased, as if oxygen were absent.[12]

The implication of these experiments is that the reduced form of an electron transport carrier signals ArcB which in turn converts ArcA into a repressor for *cyo* and an inducer for *cyd* (Fig. 17.5).

The Fnr system
When *E. coli* is shifted from aerobic to anaerobic growth, a number of genes are induced while others are repressed (Fig. 17.4). The *fnr* gene codes for a protein (Fnr) that acts as a positive regulator of transcription

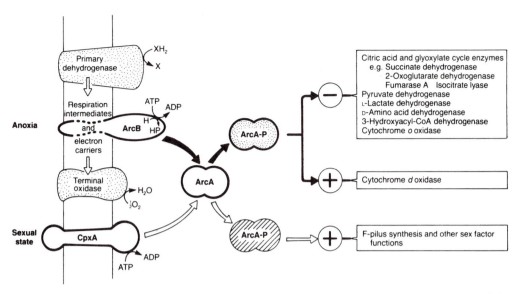

Fig. 17.5 The ArcA/ArcB regulatory system in *E. coli*. ArcB is a membrane protein activated by anoxia, perhaps by a reduced form of an electron carrier. The model postulates that the activated form of ArcB becomes phosphorylated and then phosphorylates ArcA, which then becomes a repressor of aerobically expressed enzymes and an inducer of cytochrome d oxidase. Recently it was reported that ArcA-P also activates the genes for cobalamin synthesis (*cob* genes) in *Salmonella typhimurium*. ArcA also responds to CpxA, a membrane-bound sensor protein that is necessary for the synthesis of the F-pilus and other sex factor functions. (From Spiro, S. and J. R. Guest. 1991. Adaptive responses to oxygen limitation in *Escherichia coli*. *TIBS* **16**:310–314.)

for many genes that are expressed only during anaerobic growth, and a repressor for certain genes that are expressed only during aerobic growth (Fig. 17.6).[13] Genes whose expression requires Fnr include those coding for the anaerobic respiratory enzymes fumarate reductase and nitrate reductase, as well as several other enzymes, including formate dehydrogenase II that is part of formate–hydrogen lyase (a complex of formate dehydrogenase II and hydrogenase). Mutations in the *fnr* gene result in an inability to grow anaerobically on fumarate or nitrate as electron acceptors.

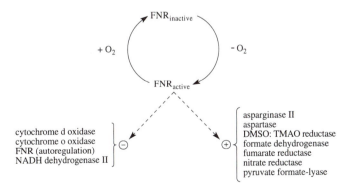

Fig. 17.6 Fnr is a transcription regulator protein during anaerobic growth. In the absence of oxygen FNR becomes activated to become an inducer for many anaerobically expressed genes and a repressor for certain aerobically expressed genes. It has been speculated that Fnr might become activated by the reduction of bound ferric ion causing a conformational change in the protein. This is not an example of a two-component regulatory system.

There is no evidence that Fnr is phosphorylated or is part of a two-component regulatory system. It is discussed here in the context of a protein involved in aerobic/anaerobic regulation of gene expression.

Studies have shown that the Fnr protein is present in comparable amounts in both aerobically and anerobically growing cells, and it is believed to be largely in an inactive state during aerobic growth.[14,15] It is not known how Fnr becomes activated during anaerobic growth. Perhaps a reductant is generated that either directly or indirectly promotes a conformational change in the protein, or a small molecule that accumulates during anaerobiosis might activate the protein. Possible targets for a reductant are sulfhydryl groups and iron atoms that are part of the purified protein.[16]

An interesting parallel exists between the Fnr protein and another transcriptional regulator, the cAMP receptor protein, Crp. The Crp protein is a positive transcriptional regulator for catabolite-sensitive genes (i.e., those genes repressed by glucose). Crp, in response to binding to c-AMP, binds to specific sites on the promoter region of the target gene to activate transcription. A comparison between the nucleotide-derived amino acid sequences of Fnr and Crp reveals that the structure of Fnr is very similar to Crp, including the presence of a DNA-binding and a nucleotide-binding domain (although numerous attempts to show specific binding of nucleotides to Fnr have failed). This has led to the suggestion that Fnr and Crp are examples of a family of proteins that regulate transcription at the promoter region of target genes.

In summary, ArcA and Fnr are activated by anaerobiosis and regulate the enzymological changes that accompany the shift from aerobic to anaerobic growth. Activated Arc A (ArcA-P) represses several aerobically expressed genes, including pyruvate dehydrogenase, succinate dehydrogenase, α-ketoglutarate dehydrogenase, and cytochrome o oxidase (Fig. 17.5). Activated Fnr is required for the induction of many genes during anaerobic growth, including pyruvate–formate lyase, fumarate reductase, and nitrate reductase, and represses several genes normally expressed during aerobic growth, including cytochrome o oxidase (compare Fig. 17.5 with Fig. 17.6).

17.3 The NarL/NarX/NarQ system

When *E. coli* is given a choice of electron acceptors under anaerobic conditions, one of them being nitrate, it will utilize the nitrate. The reason that nitrate is preferentially used as an electron acceptor during anaerobic respiration is that it induces the synthesis of nitrate reductase and represses the synthesis of the other reductases (e.g., fumarate reductase and DMSO/TMAO reductase). This may be advantageous because nitrate has a more positive redox potential than the alternative electron acceptors, and therefore more energy is potentially available from electron transport when nitrate is the electron acceptor.

The regulation of gene transcription by nitrate is mediated by NarL, NarX, and NarQ (Fig. 17.7). Nucleotide-derived amino acid sequence analysis and analogy with other signaling systems have placed NarL and NarX in the two-component regulatory system. NarX is a membrane-bound sensor/kinase protein, and NarL is a cytoplasmic regulator protein (Figs 17.2 and 17.3). The gene *narQ* was recently discovered and, by analogy with other two-component systems, is thought to be a membrane-bound sensor/kinase. How were these genes discovered? Originally mutants that were defective in nitrate repression of the fumarate reductase gene, *frdA*, were isolated. (The actual screen was to look for mutant colonies in which nitrate did not repress the expression of the fusion gene *frdA-LacZ*.) The mutations were found to be in two genes, called *narL* and *narX*. Whereas *narL* was definitely required for nitrate repression or induction of the nitrate-regulated genes, mutations in *narX* caused only a partial loss of nitrate regulation. It therefore appeared that *narX* was dispensable. Another gene was later discovered, *narQ*, that was also a sensor for nitrate reductase.[17] *E. coli* can use either NarX or NarQ as the sensor when one or the other is inactivated. Cells must be mutated in both *narX* and *narQ* in order to observe

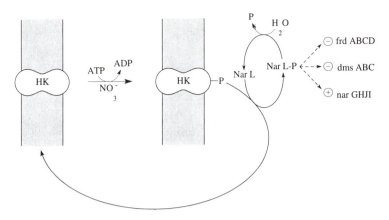

Fig. 17.7 A model for nitrate control of anaerobic gene expression. The membrane-bound hexo-kinase sensor proteins are probably NarQ and NarX. It is proposed that, in the presence of external nitrate, they autophosphorylate and transfer the phosphoryl group to a cytoplasmic regulator protein which is NarL. Phosphorylated NarL is thought to be a transcriptional activator for the nitrate reductase genes (*narGHJI*) and a transcriptional inhibitor for the fumarate reductase genes (*frdABCD*) and the genes for DMSO/TMAO reductase (*dmsABC*). Evidence for the phosphorylation pathway is based upon the resemblance in nucleotide sequences of *narQ* and *narX* with known histidine kinases and *narL* with known regulator proteins. HK, histidine kinase.

the loss of nitrate-dependent repression of fumarate reductase and induction of nitrate reductase. According to the model, NarX and NarQ are activated when they are stimulated in the presence of excess nitrate. The activated NarX and NarQ autophosphorylate and then transfer the phosphoryl group to NarL. P-NarL is a transcription regulator that stimulates transcription of the gene for nitrate reductase and represses genes for other reductases. Recently a second response regulator protein, NarP, has been discovered which responds to both NarX and NarQ.[18]

17.4 Chemotaxis

Chemotaxis[19–22] refers to the ability of bacteria to move along a concentration gradient towards a chemical attractant (positive chemotaxis), or away from a chemical repellent (negative chemotaxis). The attractants and repellents are called *chemoeffectors*. Most research concerning chemotaxis has employed flagellated, swimming bacteria, and the discussion that follows is restricted to these. The student should refer to Section 1.2.1 for a description of bacterial flagella.

17.4.1 Bacteria measure changes in concentration over time

One might think that bacteria swimming along a chemical concentration gradient are detecting the spatial gradient itself, but this is not the case. Calculations indicate that, because of the small size of bacteria, the difference in concentration of a chemoeffector between both ends of the cell would be too small to be measured accurately.[23] What the bacterium actually measures is the absolute concentration of chemoeffector and then compares it to the concentration previously measured. In other words, *bacteria measure changes in concentration over time*. They actually "remember" the previous concentration. If the bacterium finds that it is in a higher concentration of an attractant or a lower concentration of repellent than at a previous time, it continues to move in that direction. If the bacterium finds that it is in a lower concentration of attractant or a higher concentration of repellent than at a previous time, it moves in a *randomly* different direction. The propensity to swim randomly is fundamental to chemotaxis as will be made clear in Section 17.4.2.

Some bacteria swim randomly because they periodically tumble, whereas other bacteria swim randomly without tumbling. Chemotaxis in bacteria that tumble is understood best, and will be discussed first. Random swimming without tumbling is discussed in Section 17.4.7.

17.4.2 Tumbling

E. coli and many other bacteria swim along a path for a few seconds and then tumble. Very importantly, when they recover from the tumble, they swim in a *randomly* different direction. If nothing affected the frequency of tumbles and swimming, then bacteria would just swim in random directions. This is the situation when they are not responding to chemoeffectors. However, *attractants decrease the frequency of tumbles whereas repellents increase the frequency of tumbles.*[24,25] This is illustrated in Fig. 17.8. If a bacterium moves into an area where the concentration of chemoattractant is higher than at a previous

moment, the cell detects the higher concentration (because of increased binding to chemoreceptors), tumbles less frequently, and continues to move in that direction. Likewise, if the cell swims into an area where the concentration of repellent is higher, then tumbling increases and the cell changes its direction and moves away from the repellent. But why does the cell tumble?

E. coli and *S. typhimurium* are peritrichously flagellated bacteria in which the flagella are wrapped around each other in a helical bundle that extends to the rear of the swimming cell (Fig. 17.9). The flagella are usually in the conformation of a left-handed helix. When the flagella rotates counterclockwise (viewed from the tip of the flagella toward the cell), they remain in a bundle as a helical wave moves from the proximal to the distal portion of the flagella. This pushes the bacterium forward. However, when the flagella rotate clockwise, they unwind (fly apart) and there is no longer coordinated movement. The result is tumbling. Restoration of the original direction of rotation causes the flagella to come together again and the bacterium to resume swimming, but in a randomly different direction. Thus, the repellents cause the flagellar motor to reverse and the cells to tumble, whereas attractants decrease the frequency of reversal and promote smooth swimming. The tumbling and smooth swimming responses to changes in chemoeffector concentrations can be seen by observing the swimming of individual cells in the microscope. It is also possible to monitor the change in direction of flagellar rotation by tethering the bacteria to a glass slide by a flagellum. One way to do this is to use an antibody to the flagellum that sticks to both the flagellum and the glass slide.[26] The cells are grown in minimal media with glucose as the carbon source. Under these conditions, the average number of flagella per cell is reduced and cells tethered by a single flagellum can be observed. When the flagellar motor rotates, the cell rotates because the flagellum is fixed to the slide. It is then possible to infer that the addition of attractants causes the flagella to rotate counterclockwise and the addition of repellents causes clockwise rotation.[27]

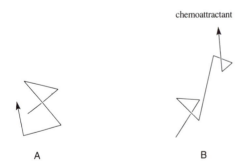

chemoattractant

A   B

Fig. 17.8 Chemoeffectors bias random swimming. (A) The swimming pattern is smooth swimming for a short period of time interspersed with brief periods of tumbling. After tumbling, the bacterium resumes swimming in a randomly different direction. (B) A chemoattractant decreases the frequency of tumbling thus prolonging the periods of smooth swimming. The result is that the bacterium swims in the direction of higher concentrations of chemoattractant. A chemorepellent increases the frequency of tumbling causing the bacterium to swim away from the repellent.

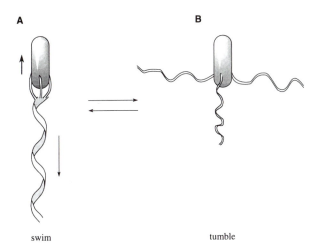

Fig. 17.9 Flagella fly apart during tumbling. In *E. coli* the flagella are peritrichously located and form a trailing bundle while the cell is swimming. (A) The filaments are in a left-handed helix. When the motors rotate counterclockwise a helical wave travels proximal to distal pushing the cell forward. (B) When the motors rotate clockwise the filaments undergo a transition to a right-handed wave form causing them to fly apart and the cells to tumble. (Adapted from MacNab, R. M. 1987. Motility and chemotaxis, pp. 732–759. In: Escherichia coli and Salmonella typhimurium. F. C. Neidhardt, J. L. Ingraham, K. B. Low, B. Magasanik, M. Schaechter and H. E. Umbarger (eds). American Society for Microbiology, Washington, DC.)

17.4.3 Adaptation

An important aspect of the chemotactic response is *adaptation*. Bacteria adapt (i.e., become desensitized), to a chemoeffector so that after a short period of time (seconds to minutes), they no longer respond to it at that concentration, but may respond to a higher concentration. In fact, one could say that adaptation is the way that the bacterium "remembers" the previous concentration of chemoeffector. Why is an adaptation circuit built into the signaling pathway? Adaptation makes good sense. When a bacterium adapts to a chemoattractant, it resumes tumbling at the nonstimulated frequency and will stay in the area, but if it randomly swims into an area of higher concentration of the chemoattractant, tumbling will again be suppressed. Consequently, the bacterium will remain in the area of the higher concentration of chemoattractant. Adaptation to repellents is also easily rationalized. If a bacterium did not adapt to the repellent, it would continue to tumble at the stimulated rate and be trapped in the area.

17.4.4 Proteins required for chemotaxis

Chemotaxis employs complex regulatory circuits involving both cytoplasmic and membrane proteins. A description of the proteins and their functions will be given first, followed by an explanation of the regulatory circuits.

Cytoplasmic proteins involved in chemotaxis

Mutants of *E. coli* and *Salmonella typhimurium* have been isolated that fail to show chemotaxis. These mutants have resulted in the identification of six genes required for chemotaxis. The genes are *cheA*, *cheB*, *cheR*, *cheW*, *cheY*, and *cheZ*. Deletions of any one of these genes prevents chemotaxis without affecting motility. The Che proteins are cytoplasmic proteins that are part of a signal transduction pathway between the attractant or repellent, and the flagellar motor switch.[28,29] CheA belongs to the class of signaling proteins called *histidine kinases*. It has conserved amino acid sequences at the carboxy-terminal domain similar to the other histidine kinases, is phosphorylated at a histidine residue, and it transfers the phosphoryl group to a response regulator protein (Fig. 17.1). CheY and CheB are *response regulator proteins*. They have amino-terminal domains similar to the other response regulators, and undergo phosphorylation and dephosphorylation.

Membrane proteins involved in chemotaxis

Chemoreceptor proteins for the chemoeffectors are built into the bacterial cell membrane. The chemoreceptor proteins are also called *receptor-transducer proteins*, or simply *transducer proteins*. The receptor-transducer proteins are thought to loop across the membrane with a periplasmic domain separating two hydrophobic membrane-spanning regions that connect to cytoplasmic domains. This is similar to the configuration illustrated in Fig. 16.3b for a protein anchored to the membrane by a stop-transfer signal and a leader peptide, which for the case of the receptor-transducer protein would be uncleaved. The cytoplasmic domain interacts with cytoplasmic components such as the CheW and CheA proteins, whereas the periplasmic domain interacts with chemoeffectors or periplasmic proteins to which the chemoeffectors are bound. The cytoplasmic domain of the receptor-transducers also contains glutamate residues that become methylated and demethylated as part of the adaptation response described in Section 17.4.5. Therefore, the receptor-transducers are also called *methyl-accepting chemotaxis proteins* (MCPs). *E. coli* has four different MCPs. These are the *Tsr* protein (taxis to serine and away from repellents), *Tar* (taxis to aspartate and away from repellents), *Trg* (taxis to ribose, glucose, and galactose), and *Tap* (taxis-associated protein that responds to dipetides).[30] Some chemoeffectors, primarily certain sugars, first bind to periplasmic binding proteins, also called primary chemotaxis receptors, and the sugar-bound periplasmic binding proteins interact with specific receptor-transducers in the membrane. Under these circumstances, the receptor-transducer (MCP) acts as a *secondary* chemoreceptor rather than a primary chemoreceptor. These are the same periplasmic binding proteins that bind sugars and interact with membrane-bound transporters in the shock-sensitive permease systems described in Section 15.3.3. They therefore function in both chemotaxis and solute transport. However, the majority of periplasmic binding proteins in *E. coli* function only in transport, not in chemotaxis.

17.4.5 A model for chemotaxis

Stimulation

The cytoplasmic domains of the receptor-transducer proteins (Tar, Tsr, Trg, or Tap) control the autophosphorylation activity of CheA in an activity that requires CheW (Fig. 17.10):

$$CheA + ATP \longrightarrow CheA\text{-}P + ADP + P_i$$

CheA-P is a protein kinase that phosphorylates CheY:

$$CheA\text{-}P + CheY \longrightarrow CheA + CheY\text{-}P$$

CheY-P has an autophosphatase activity which is considerably enhanced by CheZ:[31]

$$CheY\text{-}P + H_2O \xrightarrow{\ CheZ\ } CheY + P_i$$

In the nonstimulated state there is a steady state level of CheY-P that is determined by the rate of phosphorylation of CheY and the rate of dephosphorylation of CheY-P.

CheY-P is a response regulator protein. It binds to the flagellar motor switch (FliM, FliN, FliG) and reverses the motor so that it rotates in a CW direction causing the cell to tumble. In the nonstimulated cell, there is a certain intermediate level of CheY-P that supports normal run–tumble behavior.

When an attractant or repellent stimulate a receptor-transducer protein, a signal is sent to the cytoplasmic region of the receptor-transducer protein. The signal is presumably a conformational change in the periplasmic region of the receptor-transducer protein that is propogated to the cytoplasmic region of the receptor-transducer. An attractant *decreases* the rate of autophosphorylation

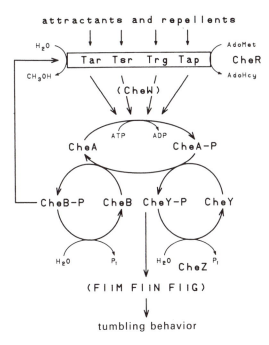

attractants and repellents

tumbling behavior

Fig. 17.10 A model for signal transduction during chemotaxis in *E. coli* and *S. typhimurium*. Attractants bind to the receptor-transducer proteins (Tar, Tsr, and so on) in the membrane. The receptors also mediate the response to certain repellents (leucine, indole, acetate, Co^{2+}, and Ni^{2+}). This activates a CheW-dependent change in the rate of autophosphorylation of CheA and the proteins that are phosphorylated by CheA (CheB and CheY). Attractants reduce the rate of autophosphorylaton and repellents increase the rate of autophosphorylation. The phosphoryl group is transferred from CheA-P to CheY and CheB. CheY-P interacts with the switch proteins (FliM, FliN, FliG) to cause clockwise rotation of the motor and tumbling. Thus repellents increase tumbling because they increase the level of CheY-P, and attractants reduce tumbling because they reduce the level of CheY-P. CheB-P is a methylesterase whose activity results in demethylation of the receptor-transducer proteins. As CheB-P goes up, methylation goes down. Thus, repellents reduce the level of methylation because they increase the level of CheB-P and attractants increase methylation because they reduce the levels of CheB-P. CheB-P autodephosphorylates. CheR is a methyltransferase that methylates the receptor-transducer proteins. The more highly methylated receptor-transducer protein does not transmit the chemoattractant signal to CheA. Hence adaptation to a chemoattractant is due to increased methylation of the receptor-transducer proteins. Adaptation to a repellent is due to undermethylation of

of CheA and a repellent *increases* the rate of autophosphorylation. Thus, attractants promote CCW (counterclockwise) rotation and smooth swimming, whereas repellents promote CW rotation and tumbling.

Adaptation
CheA-P also phosphorylates a second response regulator protein, CheB:

$$CheA\text{-}P + CheB \longrightarrow CheA + CheB\text{-}P$$

CheB-P is the active form of a methylesterase that removes methyl groups from receptor-transducer proteins. There is also a methyl-transferase, CheR, that adds methyl groups. As a result of the stimulation of the receptor-transducer protein by the chemo-attractant, the levels of CheB-P also decrease because there is less of the phosphoryl donor, CheA-P. When phosphorylation of CheB is inhibited by binding of the attractant to the receptor-transducer, the removal of methyl groups is slowed relative to the addition of methyl groups and *all* the receptor-transducer proteins *become more highly methylated*. The more highly methylated receptor-transducer protein no longer responds to its bound attractant, hence autophosphorylation of CheA is restored. (Another way of saying this is that the methylation cancels the attractant-induced signal.) As a result, the pre-stimulus levels of CheY-P and CheB-P are restored and the cell resumes the pre-stimulus run–tumble behavior, despite the presence of bound attractant. As long as the attractant is bound to the receptor-transducer protein, it remains highly methylated and the cell is said to be *adapted* to that particular concentration of attractant. However, the *other* (non-occupied)

the receptor-transducer protein. CheZ is thought to be a phosphatase that removes the phosphate from CheY-P. The relative activities of CheA-P and CheZ determine the level of CheY-P, hence the frequency of tumbling. Theoretically, changes in the activity of CheZ could also alter the frequency of tumbling. (From Stock, J. B., A. J. Ninfa, and A. M. Stock. 1989. Protein phosphorylation and regulation of adaptive responses in bacteria. *Microbiol. Rev.* 53:450–490.)

receptor-transducer proteins quickly lose their extra methyl groups so that the cell is adapted only to the specific attractant. If the receptor-transducer protein is not fully saturated with bound attractant, then the cells will respond to the attractant when it swims into areas of higher concentration of attractant. However, as the receptor-transducer binds more chemoattractant, the average number of methyl groups per transducer is increased two- to four-fold, thus balancing the increased binding of attractant and resulting in adaptation to the higher concentration of chemoattractant.[32] When the chemoattractant is removed, the overmethylated transducer stimulates CheA autophosphorylation. This results in increased CheY-P and consequent tumbling. However, the methylesterase CheB-P also increases and returns the transducer to the methylation level of the pre-stimulus state and normal run–tumble behavior. Adaptation to a repellent differs from adaptation to an attractant in that during repellent adaptation, the activated CheB-P *demethylates* the transducer that responds to the repellent.

Some supporting evidence in favor of the model

The phenotypes of chemotaxis mutants support the model. This is summarized in Table 17.1. For example, *cheW*, *cheA*, and *cheY* mutants do not tumble, whereas *cheB* and *cheZ* mutants tumble constantly. This is in agreement with the model since it stipulates that CheW and CheA promote the phosphorylation of CheY, and CheY-P causes the motor to reverse and tumbling to take place. CheB competes with CheY for the phosphoryl group on CheA-P and would therefore be expected to promote smooth swimming. Hence a mutation in *cheB* should make the cells tumble. CheZ dephosphorylates CheY and therefore a mutation in *cheZ* should

result in high levels of CheY-P and therefore increased tumbling. There are also biochemical data to support the model. When an attractant binds to the extracellular domain of the chemoreceptor protein, several events occur: (1) the rate of phosphorylation of CheY decreases; (2) the rate of demethylation is reduced; (3) the rate of methylation is increased.

17.4.6 Mechanism of repellent action

Repellents for *E. coli* and *S. typhimurium* include indole, glycerol, ethylene glycol, phenol (for *S. typhimurium* only), organic acids (e.g., formic, acetic, benzoic, salicylic), alcohols, Co^{2+} and Ni^{2+} (for *E. coli* only), and hydrophobic amino acids (leucine, isoleucine, tryptophan, valine). In general, the mechanism of action of organic repellents is not well understood, except that the response requires the receptor-transducer (MCP) proteins. Yet, there is no direct evidence for specific binding between an organic repellent and a specific MCP. In some cases (e.g., glycerol, ethylene glycol, and aliphatic alcohols), there is not even specificity regarding which receptor-transducer protein is required. This has been learned by using mutants lacking one or more of the receptor-transducer proteins and showing that any of the remaining receptor-transducer proteins can mediate the repellent response. However, in other instances there is specificity. For example, the response to leucine, isoleucine, and valine requires the Tsr protein, and the response to the cations Co^{2+} and Ni^{2+} requires Tar. Because much higher concentrations of organic repellents (in the millimolar range) are usually required than for attractants (in the micromolar range), it has been suggested that the organic repellents (which

Table 17.1 Effects of chemotaxis mutants

Gene	Mutant phenotype	Rationale for mutant phenotype
cheY⁻	Smooth swimming	CheY-P is required for tumbling
cheA⁻, *cheW⁻*	Smooth swimming	CheA-P and CheW phosphorylate CheY
cheB⁻	Tumbling	CheB competes with CheY for the phosphoryl group from CheA-P
cheZ⁻	Tumbling	CheZ dephosphorylates ChY-P

all have some hydrophobic character as well as polar groups) do not actually bind to the receptor-transducer proteins. They are thought to act indirectly by perturbing the membrane at the site of the receptor-transducer proteins. For example, they might make the membrane more fluid, which could result in the alteration of receptor-transducer activity. Recently, a study was carried out to investigate the action of repellents on the membrane fluidity of *E. coli*.[33] It was concluded that one could *not* account for repellent activity by changes in membrane fluidity, suggesting that repellents may indeed bind to receptor-transducer proteins, however, with low affinity, and in some cases low specificity. In summary, the response to repellents *does* involve the receptor-transducer proteins (MCP proteins), but the mechanism of how they affect the activities of the receptor-transducer proteins is not understood. However, there is more known about the repellent activity of weak organic acids. These are believed to affect the activities of the receptor-transducer proteins Tsr and Tap by lowering the intracellular pH.

17.4.7 Bacteria that do not tumble

Not all bacteria tumble, yet they are capable of chemotaxis. For example, *Rhodobacter sphaeroides* and *Rhizobium meliloti* rotate their flagella only in a clockwise direction and do not tumble. Yet they swim randomly and show chemotaxis. *R. meliloti* has 5–10 peritrichously located flagella that form a bundle when the cell swims. Swimming consists of straight runs interrupted by very quick turns. Cells tethered to a microscope slide by a single flagellum (the other flagella were mechanically sheared off the cell) rotated only clockwise with brief intermittent stops. It was suggested that the runs of swimming cells were due to the clockwise rotation of most of the flagella in the bundles, and that when a sufficient number of flagella stopped rotating, the cell turned, perhaps because of single rotating flagella situated unevenly on the sides of the rod-shaped cell.[34] Chemoattractants presented to *R. meliloti* prolonged clockwise rotation of the flagella,

hence extending the motility in the direction of higher concentrations of chemoattractant.

R. sphaeroides has a single medially (subpolar) located flagellum that rotates in one direction only (clockwise), pushing the cell forward. The flagellum stops rotating intermittently, and when rotation resumes, the bacterium swims randomly in a different direction. It was speculated that Brownian motion turns *R. sphaeroides* cells, which have stopped rotating, randomly in a different direction.[35] *R. sphaeroides* is attracted to organic acids, sugars and polyols, glutamate, ammonia, and certain cations (e.g., K^+ and Rb^+). As discussed in Section 17.4.8, chemotaxis in *R. sphaeroides* does not involve the MCP proteins, which are lacking in this bacterium, or the phosphotransferase system. Metabolism of the chemoattractant appears necessary for chemotaxis because, if metabolism is blocked, e.g., by a specific inhibitor, then chemotaxis is also inhibited. There is very little known about the metabolism-dependent chemotactic system in *R. sphaeroides* and nothing about its relationship to the MCP-dependent system of the enteric bacteria. Since *R. sphaeroides* changes its direction as a result of stopping rather than tumbling, it is reasonable to suppose that an increased concentration of chemoattractant might decrease the frequency of stops (the equivalent of the suppression of tumbling), thus ensuring that the cells continue to swim in the direction of the chemoattractant. A decreased concentration of chemoattractant might increase stopping frequency, and hence promote changes in the direction of swimming. This, in fact, is the case. When tethered cells were examined, it was observed that an increase in attractant concentration caused a decrease in stopping frequency which persisted for many minutes, whereas a reduction in the concentration of attractant caused the cells to stop rotating for between 10 and 30 sec, after which the cells resumed the pre-stimulus stopping frequency. Since the cells adapted very slowly to an increased concentration of attractant (up to 60 min) but adapted within 30 sec to a step-down in attractant concentration, it has been suggested that perhaps cell accumulation in response to a chemoattractant gradient by *R.*

sphaeroides is dependent only upon sensing a reduction and not an increase in the concentration of chemoattractant. Some metabolites, especially weak organic acids, cause an increase in the speed of swimming, a response called *chemokinesis*. Many compounds that cause chemokinesis also cause chemotaxis. However, the two phenomena are separable because: (1) most amino acids and sugars that are chemoattractants do not cause chemokinesis; and (2) metabolism is required for chemotaxis but not for chemokinesis. (It appears that transport across the cell membrane may be involved in chemokinetic signalling.) The increased rate of swimming in response to certain metabolites lasts for a long time, i.e., there is no rapid adaptation. It has been pointed out that in the absence of adaptation, chemokinesis should result in the spreading of the population rather than its accumulation. (Section 17.4.3.)

Bacteria with polar flagella (e.g., *Caulobacter*) change direction by reversing the direction of flagellar rotation. For example, in organisms with polar flagella (flagella at one pole or both poles, either single or in bundles) the reversal of the flagellar motor causes the cells to reverse the direction of swimming rather than tumble. Because they do not back up in an absolutely straight line, and/or do not swim forward in a straight line, the direction of forward swim is random to the original forward direction. Brownian motion may also contribute towards changing the direction of swimming.

17.4.8 Taxis that does not involve the MCP proteins

There have been reports that at least two bacteria, *Azospirillum brasilense* and *Rhodobacter sphaeroides*, lack a methylation-dependent pathway for chemotaxis towards organic compounds.[36,37] In *Azospirillum brasilense*, it appears that the strongest chemoattractants are electron donors to the aerobic redox chain. This, plus the fact that chemoattraction towards oxygen is so strong in this organism that it masks responses to other chemoattractants, has led to the suggestion that the signaling system involves the respiratory chain. It is reasonable to suggest that the signaling system in these cases (and in phototaxis) may involve the redox level of one of the electron carriers or the proton-motive force.

Another class of chemotaxis sensors are related to the phosphotransferase system (PTS system) for sugar uptake.[38] (See Section 15.5.3 for a description of the PTS system.) Chemotaxis towards the PTS sugars does not involve the methylation system since even mutants with defects in the MCP proteins will chemotax towards PTS sugars, and furthermore, PTS sugars do not promote methylation of the MCP proteins. Instead of the MCP proteins, the membrane-bound permease, enzyme II, of the phosphotransferase sugar transport system functions as a chemotaxis sensor for PTS sugars. In agreement with this conclusion, mutants defective in enzyme II (or enzyme I, or HPr) are defective in chemotaxis towards PTS sugars. Mutants lacking CheA, CheW, and CheY do not show chemotaxis to PTS sugars, suggesting that the signal from the PTS sugars interacts with these proteins. However, the signal transduction pathway is not understood.

17.5 Nitrogen assimilation: The Ntr regulon

Bacteria use inorganic and organic nitrogenenous compounds as a source of nitrogen for growth. The inorganic nitrogen is in the form of ammonia, nitrate, or dinitrogen gas, and the organic nitrogen is in the form of amino acids, urea, and other nitrogenous compounds. Ultimately, all of these nitrogenous compounds are either reduced or catabolized by the bacteria to ammonia, which is assimilated via the glutamine synthetase reaction. When the ammonia supply is adequate, then ammonia is the preferred source of nitrogen because it represses genes required for the assimilation of other nitrogenous compounds. However, when ammonia in the growth medium becomes limiting, then genes required for ammonia production from external nitrogen sources, as well as the gene for glutamine synthetase, are induced. The genes that are regulated by the ammonia supply belong to the Ntr regulon.[39–47] A *regulon* is a set of operons controlled by a common regulator, and for the *Ntr regulon*, the regulator is a response regulator protein called NR$_I$. The Ntr regulon consists of many genes, including the genes for glutamine synthetase, NR$_{II}$ (histidine kinase) and NR$_I$ (response regulator), which are in an operon called *glnALG*,[42] the genes for the nitrogenase system in *Klebsiella* (*nif* genes), the genes for the degradation of urea in *Klebsiella*, the genes for periplasmic binding proteins required for amino acid transport (e.g., for glutamine (*glnH*), arginine, lysine, and ornithine (*argT*), and histidine (*hisY*) in *Salmonella*), the genes for histidine degradation (*hut*) in *Klebsiella*, and the genes for proline degradation (*put*) in *Salmonella*.[43] It should not be concluded automatically that, if a gene system is under Ntr regulation in one genus of bacterium, it is also under Ntr regulation in all bacteria. This is clearly not the case. For example, the *hut* genes are not regulated by ammonia levels in *Salmonella typhimurium*, although they are so regulated in *Klebsiella pneumoniae*.

One can consider glutamine as the intracellular signal that informs the cell whether the ammonia levels are adequate. To understand this, let us review the biochemistry of ammonia assimilation (Sections 9.3.1 and 17.5.1). When ammonia levels are high in the growth medium, the ammonia is incorporated into glutamate by the NADPH-dependent reductive amination of α-ketoglutarate catalyzed by the enzyme glutamate dehydrogenase. Ammonia is also incorporated into glutamine in an ATP-dependent reaction with glutamate catalyzed by glutamine synthetase (GS). These reactions are crucial because glutamate supplies nitrogen to approximately 85% of the nitrogen-containing molecules in the cell and glutamine supplies the remaining 15%. However, when the supply of ammonia is limiting (e.g., when cells derive ammonia from organic nitrogen compounds in the growth medium, or from the reduction of nitrate or dinitrogen gas), the glutamate dehydrogenase reaction does not significantly contribute to glutamate

synthesis (Section 9.3.1). Instead, glutamate is made by transferring an amino group from the amide nitrogen of glutamine to α-ketoglutarate in a reaction catalyzed by glutamate synthase. The result of the glutamate synthase reaction is two molecules of glutamate, one of which is recycled back to glutamine using glutamine synthetase. Therefore, when the cells are limited for ammonia, glutamine is the immediate source of nitrogen for glutamate. As described below, the cell uses glutamine as an indicator of nitrogen supply, and adjusts the transcription of the Ntr regulon and the activity of glutamine synthetase accordingly.

17.5.1 A model for the regulation of the Ntr regulon

This is a complex signaling pathway.[44] It will be explained at two levels with one giving a more detailed explanation than the other.

The less detailed explanation
Transcription of the Ntr regulon is inhibited by excess ammonia and the inhibition is relieved when the ammonia concentration falls to below 1 mM. It works in the following way. When the ammonia concentration is shifted from a high concentration to below 1 mM, the intracellular concentration of glutamine falls. This is because the level of glutamine synthetase, the enzyme that synthesizes glutamine from ammonia, glutamate, and ATP, is low in cells grown in the presence of high ammonia. Additionally, glutamate dehydrogenase, the enzyme that synthesizes glutamate from α-ketoglutarate, ammonia, and NADPH, has a high Km for ammonia (approximately 1 mM) and is therefore relatively inactive when the ammonia concentrations fall below 1 mM (Section 9.3.1, Fig. 9.16). When the glutamine levels drop, the concentration of α-ketoglutarate increases because it is the glutamine that donates an amino group to α-ketoglutarate via the glutamate synthase reaction when the ammonia concentrations are low (Fig. 9.16). The increased amounts of α-ketoglutarate lower the concentration of a protein, P$_{II}$, by stimulating its conversion to P$_{II}$-UMP. (The

α-ketoglutarate stimulates an enzyme called uridydyl transferase that adds UMP to P_{II}, converting it to P_{II}-UMP.) *The decrease in P_{II} stimulates transcription of the Ntr regulon.* The reason why lowering the amount of P_{II} stimulates transcription of the Ntr regulon is that P_{II} activates the phosphatase activity of the histidine kinase, NR_{II}, and the phosphatase activity of NR_{II} inactivates the response regulator (positive transcription factor), NR_I-P. When the the levels of P_{II} decrease, NR_{II} acts as a kinase rather than as a phosphatase, and phosphorylates NR_I, which then activates transcription. Thus, the sequence of events is as follows: (1) low ammonia results in an increase in the concentrations of α-ketoglutarate because its conversion to glutamine is slowed; (2) the α-ketoglutarate lowers the concentration of P_{II} by stimulating its conversion to P_{II}-UMP; (3) when the P_{II} levels fall, NR_{II} no

longer acts as a phosphatase but rather as a kinase and phosphorylates NR_I; and (4) P-NR_I increases and stimulates transcription. When ammonia levels are high, then transcription is repressed because glutamine stimulates the conversion of P_{II}-UMP to P_{II} and P_{II} stimulates the phosphatase activity of NR_{II}, which results in a decrease in P-NR_I. These events are described in more detail below and summarized in Fig. 17.11.

The more detailed explanation
The central operon involved is called the *glnALG* operon which encodes three genes (Fig. 17.11A). They are: (1) *glnA*, the gene for glutamine synthetase; (2) *glnL*, the gene for NR_{II}, which is a bifunctional histidine kinase/phosphatase whose substrate is N_RI; and (3) *glnG*, the gene for NR_I, a response regulator. Under nitrogen excess (high ammonia), the

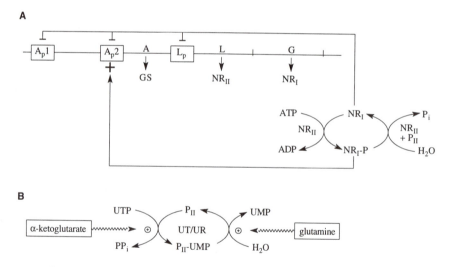

Fig. 17.11 Model for transcriptional regulation of *glnALG*. The protein products of the operon (are glutamine synthetase (GS), a histidine protein kinase/phosphatase (NR_{II}) and a response regulator (NR_I). (A). Under ammonia-limiting conditions, NR_{II} phosphorylates NR_I, forming NR_I-P. The increased levels of NR_I-P stimulate a large increase in transcription rate from Ap2, which uses sigma 54 RNA polymerase. As the levels of NR_I-P increase, other Ntr operons are also stimulated. Under conditions of excess ammonia, NR_{II} is stimulated by P_{II} to dephosphorylate NR_I-P. As a consequence, transcription from Ap_2 stops and sigma 70 RNA polymerase transcribes from promoters Ap1 and Lp at a low rate. Transcription from these promoters is further regulated by repression by NR_I. (B). The levels of P_{II} are increased by glutamine (signaling high ammonia) and decreased by α-ketoglutarate (signaling low ammonia). A bifunctional enzyme, uridylyl transferase/uridydyl removal (UT/UR), is stimulated by glutamine to remove UMP from P_{II}-UMP and stimulated by α-ketoglutarate to add UMP to P_{II}. The enzyme adds a total of four UMP moieties to each of four subunits of P_{II}. Thus, high glutamine (high ammonia) increases the levels of P_{II}, which stimulates the dephosphorylation of NR_I-P and represses transcription, whereas high α-ketoglutarate (low ammonia) stimulates the addition of UMP to P_{II}, thus lowering the levels of P_{II} and activates transcription.

operon is transcribed at a low level from the *glnAp1* and *glnLp* promoters by sigma 70 RNA polymerase. The small amount of transcription from *glnAp1* is sufficient to guarantee the synthesis of enough glutamine synthetase to meet the cell's needs for glutamine when the ammonia concentrations are high. Part of the reason that only a small amount of transcription takes place from these promoters is that transcription from them is repressed by NR_I (Fig. 17.11 A).[45] Under nitrogen-limiting conditions, the *glnALG* operon is transcribed at a high frequency from the *glnAp2* promoter by the sigma 54 RNA polymerase. Transcription from the *glnAp2* promoter requires the phosphorylated form of the response regulator NR_1 (i.e., NR_1-P).[46] The protein kinase, NR_{II}, is itself regulated by another protein, called P_{II}. When P_{II} levels are high, NR_{II} acts like a phosphatase rather than a kinase, and the levels of P-NR_1 drop. Hence, P_{II} inhibits transcription of the *glnALG* operon. Under conditions of excess ammonia, P_{II} is generated from P_{II}-UMP because glutamine stimulates the removal of UMP from P_{II}-UMP catalyzed by the bifunctional enzyme uridylyl transferase-uridylyl removal (UT/UR) (Fig. 17.11B). When ammonia levels fall, the P_{II} is converted to P_{II}-UMP in a reaction catalyzed by UT/UR and stimulated by α-ketoglutarate. Under these circumstances, NR_{II} acts as a kinase rather than as a phosphatase and transcription is stimulated. In summary then:

1. High ammonia leads to high PII because glutamine stimulates the removal of UMP from P_{II}-UMP.

2. High P_{II} leads to phosphatase activity of NR_{II}.

3. Phosphatase activity of NR_{II} leads to dephosphorylation of P-NR_I.

4. Dephosphorylation of P-NR_I leads to a lowering of transcription of the glnALG operon, since P-NR_I is a positive transcription factor.

A key enzyme in the signal transduction pathway is the bifunctional enzyme UT/UR that either adds or removes UMP from P_{II}. It can be considered a sensor protein that responds to the α-ketoglutarate/glutamine ratio and modifies P_{II}, which in turn regulates the activity of the histidine kinase/phosphatase (NR_{II}). Alternatively, one might consider glutamine synthetase to be the sensor because it responds to the ammonia supply and determines the levels of glutamine and α-ketoglutarate, which signal the UT/UR.

Regulation of the other operons in the Ntr regulon

It has been estimated that the level of NRI in ammonia repressed cells is about five molecules per cell, which when phosphorylated under conditions of ammonia starvation will activate transcription of *glnALG*. When the levels of NR_1-P become sufficiently high (around 70 molecules per cell) due to increased transcription of *glnALG*, it also activates transcription of the other operons in the Ntr regulon system.

The stimulation of some of the Ntr operons (e.g., *glnALG*) by P-NRI is direct. However, in other instances, P-NR_1 activates the transcription of a gene encoding a second positive regulator, which activates transcription of the target promoters. For example, in *Klebsiella pneumoniae*, NR_1-P activates the transcription of *nifAL*. The product of *nifA* is a positive regulator of the *nif* genes. [The product of *nifL* inactivates the *nifA* product in response to oxygen and ammonia, thus making certain that the *nif* genes (except for *nifAL*) are expressed only during anaerobiosis and ammonia starvation.] Transcription of the *nif* genes also relies on sigma 54. In *Klebsiella* species, but not *Salmonella typhimurium*, P-NR1 also activates the transcription of *nac*, which encodes a positive regulator of the *hut* operons that encode enzymes for the catabolism of histidine as a source of nitrogen. The *hut* genes use a sigma 70 RNA polymerase, rather than the sigma 54 polymerase.

17.5.2 Effect of P_{II}-UMP and P_{II} on glutamine synthetase activity

When ammonia levels are low, not only is transcription of the gene for glutamine

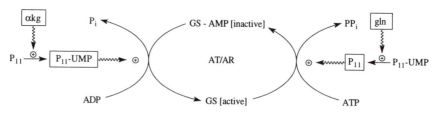

Fig. 17.12 Adenylylation of glutamine synthetase regulates its activity. When glutamine (gln) levels are high (high ammonia), the enzyme is inactivated by adenylylation catalyzed by the bifunctional enzyme adenylyl transferase/adenylyl removal (AT/AR), which is stimulated by P_{II}, itself produced from P_{II}-UMP in the presence of high glutamine. When the α-ketoglutarate (αkg) levels are high (low ammonia), the glutamine synthetase is activated by deadenylylation via AR/AT stimulated by P_{II}(UMP), which is produced from P_{II} in the presence of high α-ketoglutarate. Thus the ratio gln/αkg regulates the activity of glutamine synthetase by changing the ratio P_{II}/P_{II}-UMP. A total of 12 AMP moieties can be added to each of 12 subunits of glutamine synthetase.

synthetase stimulated, but the activity of the existing enzyme is also stimulated. This is because P_{II}-UMP, which is made when ammonia levels are low, indirectly stimulates glutamine synthetase activity. This makes sense because it is the glutamine synthetase that is responsible for the assimilation of low concentrations of ammonia. To understand how P_{II}-UMP does this, we have to discuss how glutamine synthetase is regulated in the first place. When ammonia concentrations are high, glutamine synthetase activity is low because it has been adenylylated by an enzyme called adenylyl transferase (AT), which itself is activated by P_{II} (Fig. 17.12). [The adenylylated enzyme is further inactivated because it is susceptible to feedback inhibition by a variety of nitrogenous compounds that depend upon glutamine for their synthesis (e.g., glucosamine-6-phosphate, carbamoyl phosphate, CTP, AMP, tryptophan, and histidine).] However, adenylyl transferase is a bifunctional enzyme and in the presence of P_{II}-UMP, it removes the AMP via a phosphorolysis reaction. Therefore, low ammonia stimulates glutamine synthetase by increasing the levels of P_{II}-UMP, and high ammonia inhibits glutamine synthetase by increasing the levels of P_{II}. Note that just as in the regulation of transcription, P_{II} affects a separate enzyme that activates or inactivates a component. For example, in transcription, P_{II} stimulates the phosphatase activity of NR_{II}, which results in the inactivation of the transcription regulator, NR_{I}. In the regulation of glutamine synthetase activity,

P_{II} activates AT which in turn inactivates glutamine synthetase, whereas P_{II}-UMP inactivates AT, which in turn leads to activation of glutamine synthetase. Table 17.2 summarizes an overview of the regulation of transcription and glutamine synthetase activities by ammonia and P_{II}.

17.6 The PhO regulon

Bacteria have evolved a signaling system to induce the formation of phosphate assimilation pathways when the supply of phosphate becomes limiting. When inorganic phosphate is in excess, it is transported into the cell by a low-affinity transporter called Pit. Under low inorganic phosphate conditions, *E. coli* stimulates the transcription of at least 31 genes (most of them in operons) involved in phosphate assimilation, including: genes encoding a periplasmic alkaline phosphatase that can generate phosphate from organic

Table 17.2 Regulation of transcription of *glnALG* and glutamine synthetase

NH₃	P_{II}/P_{II}-UMP	Transcription*	Glutamine† synthetase
High	High	Low	Inactive
Low	Low	High	Active

*Indirectly repressed by P_{II}.
†Indirectly inactivated by P_{II}. Indirectly activated by P_{II}-UMP.

phosphate esters (*phoA*); an outer membrane porin channel for anions, including phosphate (*phoE*); a high-affinity inner-membrane phosphate-uptake system and a protein required for phosphate repression (*pstSCAB-phoU*); a histidine kinase and response regulator (*phoBR*); 14 genes for phosphonate[47] uptake and breakdown (*phn*CDEFGHIJKLMNOP); genes for glyceraldehyde-3-phosphate uptake; and a gene encoding a phosphodiesterase that hydrolyzes glycerophosphoryl diesters (deacylated phospholipids) (*ugpBAECQ*). All of these genes (except for *pit*) are repressed by phosphate and are in the PHO regulon. The PHO regulon is controlled by PhoR, which appears to be a histidine kinase, and a response regulator, PhoB. The phosphorylated form of PhoB (i.e., PhoB-P) activates the transcription of the genes in the PHO regulon. The Pho promoters use the sigma 70 RNA polymerase.

17.6.1 The signal transduction pathway

A model for the regulation of the PHO regulon by Pi is shown in Fig. 17.13.[48] The components required for regulation are: (1) PstS, a periplasmic Pi binding protein; (2) PstABC, integral membrane proteins required for P_i uptake (PstB is the permease); (3) PhoU; (4) PhoR; and (5) PhoB. The model proposes that P_i binds to PstS and that the P_i-bound PstS binds to PstABC. A "repressor complex" is postulated that consists of P_i bound to PstS, PstABC, PhoU, and PhoR. PhoR is thought to be a histidine kinase/phosphatase bifunctional enzyme. In the "repressor complex" PhoR is suggested to be a monomer with phosphatase activity and to maintain PhoB in its dephosphorylated (inactive) state. The model further proposes that, when phosphate becomes limiting, PhoR is released from the "repressor complex" and functions as a

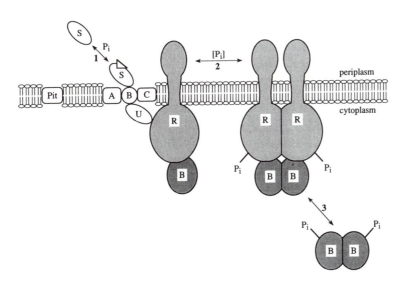

Fig. 17.13 Model for the regulation of the PHO regulon. Inorganic phosphate in the periplasm binds to the phosphate binding protein, PstS. (S), The P_i-PstS binds to the phosphate transporter, PstABC, in the cell membrane. A "repressor complex" forms between P_i-PstS, PstABC, PhoU, and PhoR. PhoR in the repressor complex acts as a phosphatase and maintains PhoB in the dephosphorylated state. When P_i becomes limiting, PhoR is released from the repressor complex, autophosphorylates, and then phosphorylates PhoB. Phosphorylated PhoB is a positive transcription regulator for the PHO regulon. PhoR as a phosphatase is depicted as a monomer and PhoR as a kinase is drawn as a dimer, although this has not been demonstrated. Unshaded: Pit, low-affinity P_i transporter; Pst, high-affinity P_i-specific transporters: S, P_i binding protein; A, C, integral membrane proteins; B, P_i permease; U, negative regulator PhoU. Shaded: B, response regulator PhoB; R, P_i sensor PhoR. (From Wanner, B. L. 1993. Gene regulation by phosphate in enteric bacteria. *J. Cell. Biochem.* 51:47–54.)

histidine kinase, which is depicted in Fig. 17.13 as a dimer. The dimer form of PhoR autophosphorylates and activates PhoB by transferring to it the phosphoryl group. In agreement with this model, Pst and PhoU are required for repression of the PHO regulon, but not for its activation. However, it must be added that the interaction of the Pst system and PhoU with PhoR to cause P$_i$ repression is not really understood.

17.7 Regulation of porin synthesis

When growing in higher osmolarity or high temperature, *E. coli* increases the synthesis of the slightly smaller porin channel, OmpC, relative to the larger OmpF channel. For example, when *E. coli* is grown at 37°C and in the presence of 1% NaCl (which is the temperature and approximate osmolarity of the intestine), only OmpC is made. When growing at lower temperatures and at osmolarities that approximate the conditions in lakes and streams, OmpF is preferentially made. The changes in porin composition of the outer membrane do not change the intracellular osmotic pressure, and therefore are not part of a homeostatic response to osmotic pressure changes. One can rationalize the change in porins by assuming that the smaller OmpC channel is advantageous in the intestinal tract because it may retard the inflow of toxic substances such as bile salts, whereas in lower osmolarity environments, such as would be experienced outside of the body, the larger OmpF channel might be an advantage to increase inward diffusion of dilute nutrients.[49]

17.7.1 Porin synthesis is regulated by a two-component system

EnvZ is an inner membrane histidine protein kinase that has been postulated to function also as an osmotic sensor. The response regulator is a cytoplasmic protein called OmpR. One model proposes that increased external osmolarity activates EnvZ so that it phosphorylates OmpR (Fig. 17.14). However, it must be emphasized that there is very little known concerning the signals to which EnvZ

responds when it phosphorylates or dephosphorylates OmpR. The model proposes that high levels of P-OmpR repress *ompF* and stimulate *ompC* by binding to low-affinity sites upstream from the respective promoters, and that low levels of P-OmpR stimulate transcription of *ompF* by binding to high-affinity sites upstream from the *ompF* promoter. Thus, when the external osmotic pressure is raised, the levels of OmpR-P should increase and repress *ompF* while stimulating *ompC*. Several lines of evidence support the model:[50–52] (1) it has been demonstrated *in vitro* that EnvZ can accept a phosphoryl group from ATP and in turn transfer it to OmpR forming OmpR-P; (2) it is also known that when *E. coli* is shifted to a medium of high osmolarity, the levels of OmpR-P increase; (3) in a mutant of *E. coli* that has elevated levels of OmpR-P, *ompF* is repressed and *ompC* is expressed all of the time (constitutively); and (4) OmpR-P stimulates transcription from *ompF* and *ompC* promoters.

Regulation of translation by micF RNA
There appears to be an additional mode of inhibition of *ompF* expression. Another regulatory gene, *micF*, is also stimulated by high concentrations of OmpR-P and produces an RNA molecule that inhibits *ompF* mRNA translation.[53] When multiple copies of *micF* are introduced on a high copy-number plasmid there is inhibition of *in vivo* synthesis of OmpF. However, deletion of the *micF* gene does not result in any change in the synthesis of OmpF under high or low osmolarity conditions. Therefore, it must be concluded that there is insufficient information regarding its physiological role.

17.8 Regulation of the kdpABC operon

When *E. coli* is placed in a medium of high osmolarity, it responds by synthesizing a K$^+$ uptake system, called the KdpABC transporter, discussed in Section 15.3.3. The role of K$^+$ uptake in pH and osmotic homeostasis is discussed in Chapter 14. There are two regulatory genes for the *kdp*ABC operon,

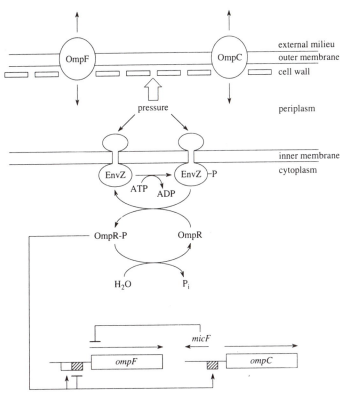

Fig. 17.14 A model to explain the regulation of porin synthesis. EnvZ is a transmembrane histidine kinase/phosphatase whose substrate is the cytoplasmic response regulator OmpR. To explain how a shift to high osmolarity media represses *ompF* and stimulates *ompC*, it is assumed that high osmotic pressures activate (not necessarily directly) the kinase activity of EnvZ. The resulting high levels of OmpR-P repress transcription of *ompF* and stimulate transcription of *ompC*. OmpR-P can effect these changes by binding to DNA sequences upstream of the respective promoters. A third gene, *micF*, is transcribed from the *ompC* promoter but in the opposite direction. Its transcription is also activated by OmpR-P. The gene *micF* codes for an RNA transcript that is complementary to the 5′ end of the *ompF* mRNA (i.e., the region where protein translation is initiated) and blocks translation. MicF RNA is therefore "antisense" RNA because it binds to the *ompF* RNA transcript that is translated into protein (i.e., the "sense" RNA). It may be that an additional control of *ompF* expression is the inhibition of translation of *ompF* mRNA by *micF* RNA. When the external osmotic pressures are lower, the concentrations of OmpR-P are too low to bind to the DNA sites for repression of *ompF* and stimulation of *ompC*. However, low levels of OmpR-P can stimulate the transcription of *ompF* because of high-affinity binding sites for OmpR-P upstream of the *ompF* promoter. (From Stock, J. B., A. J. Ninfa, and A. M. Stock. 1989. Protein phosphorylation and regulation of adaptive responses in bacteria. *Microbiol. Rev.* 53:450–490.)

kdpD and *kdpE,* both of which are required for expression of the *kdpABC* genes. The model is that the KdpD protein, which is located in the inner (cell) membrane, senses a change in turgor pressure and transmits the signal to KdpE which activates the transcription of the *kdpABC* operon. (However, see Sections 14.2.4 and 14.2.5 for a discussion of whether the pressure differential due to an osmotic gradient is across the cell membrane.) Recent evidence suggests that the KdpD and KdpE proteins are part of a two-component regulatory system, with KdpD being a sensory protein kinase that autophosphorylates, and KdpE being a response regulator protein that is phosphorylated by KdpD. The phosphorylated form of KdpD is thought to activate the *kdpABC* promoter.[54,55]

17.9 Summary

Bacteria can sense environmental and cytoplasmic signals and transmit information to the genome or to other parts of the cell to elicit a response. Many of the signaling systems are called "two-component" signaling systems because they have a histidine kinase (component 1) that causes the phosphorylation of a response regulator (component 2). However, as was discussed, other proteins may also be involved in signal transduction.

All of the "two-component" systems have four functional components: (1) a sensor that receives the signal; (2) an autophosphorylating histidine protein kinase; (3) a partner response regulator that is the substrate for the histidine kinase; and (4) a phosphatase that removes the phosphate from the regulator. As discussed below, some of the proteins can be bifunctional and carry out two of the activities. During signaling, phosphate (actually the phosphoryl group) travels from ATP to the histidine kinase to the response regulator to water (the phosphatase reaction). The phosphorylated form of the response regulator is the active state and transmits the signal to the genome (stimulation or inhibition of transcription), to the flagellar motor (reversal of turning direction), or to an enzyme. Although many signals cause an increase in phosphorylation of the response regulator, some (e.g., chemoattractants, excess inorganic phosphate, excess ammonia) cause a decrease in phosphorylation (dephosphorylation) and therefore an inactivation of the response regulator.

All of the histidine protein kinases have conserved domains that bear a startling similarity to each other, and are the basis for classification of these proteins. These conserved domains are in the carboxy-terminus of the protein and are the site of the conserved histidine residue that is phosphorylated. It should be emphasized that, at this time, enzymatic activity has been demonstrated for only a few of the kinases. The response regulators have conserved amino-terminal domains that are believed to interact with the carboxy-terminus of the kinase. For some of the response regulators, it has been shown that the phosphoryl group is transferred from the histidine to the conserved aspartate residue in the amino-terminus of the response regulator.

Responses that use a two-component regulatory system include the Che system in chemotaxis, the ArcA/ArcB system for oxygen regulation of gene expression, the NarL/X/Q system for nitrate regulation, the PHO system for phosphate assimilation, the Ntr system for nitrogen assimilation, the EnvZ system for porin gene expression, and the KdpABC system for K^+ uptake. Many other signaling systems use proteins that can be classified on the basis of amino acid sequence as histidine kinases and response regulators, and are therefore two-component systems. These include *Bacillus* sporulation, flagellar synthesis, *Agrobacterium* virulence, and *Salmonella* virulence. However, not all adaptive responses use a two-component regulatory system. For example, the Fnr system, which regulates anaerobic gene expression, is not a two-component regulatory system.

Study Questions

1. Describe the components of two-component regulatory systems and how they work. In your answer, explain what is meant by "cross-talk."

2. What criteria must be established in order to characterize a protein as a kinase or regulator protein in a two-component system?

3. What is Fnr and what is its role? Why is it not believed to be part of a two-component regulatory system?

4. It has been suggested that Fnr might be activated by some reduced molecule in the cell that rises in concentration in the absence of oxygen. How might you test whether the redox levels of electron transport carriers are involved?

5. Describe the signaling pathway for chemotaxis in *E. coli*.

6. What causes *E. coli* to swim randomly? Is this the mechanism for all bacteria?

Explain the differences. How is random swimming related to chemotaxis?

7. What is the relationship between adaptation and methylation of MCPs?

8. In the Arc and Nar systems, which proteins are thought to be the sensor/kinase proteins and which ones the regulator proteins? What is the evidence for this?

9. What is the phenotype of an *fnr⁻* mutant? An *arc⁻* mutant?

10. Describe the role that P_{II} plays in transcriptional regulation of the *glnALG* operon and in the regulation of activity of glutamine synthetase. How are the levels of P_{II} and P_{II}-UMP regulated?

REFERENCE AND NOTES

1. Saier, M. H., Jr. 1993. Introduction: Protein phosphorylation and signal transduction in bacteria. *J. Cell. Biochem.* 51:1–6.

2. Stock, J. B., A. J. Ninfa, and A. M. Stock. 1989. Protein phosphorylation and regulation of adaptive responses in bacteria. *Microbiol. Rev.* 53:450–490.

3. Parkinson, J. S., and E. C. Kofoid. 1992. Communication modules in bacterial signaling proteins. *Annu. Rev. Genet.* 26:71–112.

4. Wanner, B. L. 1992. Is cross regulation by phosphorylation of two-component respons regulator proteins important in bacteria? *J. Bacteriol.* 174:2053–2058.

5. Spiro, S. and J. R. Guest. 1991. Adaptive responses to oxygen limitation in *Escherichia coli*. *TIBS* 16:310–314

6. Gunsalus, R. P. 1992. Control of electron flow in *Escherichia coli*: Coordinated transcription of respiratory pathway genes. *J. Bacteriol.* 174:7069–7074.

7. ArcA also responds to a second sensor/kinase protein (i.e., CpxA) which is necessary for the production of the F-pilus in donor strains of E. coli. The CpxA gene was originally discovered as a mutation that reduced the efficiency of DNA transfer as a consequence of reduced F-plasmid *tra* gene expression. It is now known that the CpxA protein is an inner membrane protein whose amino acid sequence places it in the class of sensor/kinase proteins (Fig. 17.2).

8. Iuchi, S., and E. C. C. Lin. 1992. Purification and phosphorylation of the Arc regulatory compo-nents of *Escherichia coli*. *J. Bacteriol.* 174:5617–5623.

9. Anderson, D. I. 1992. Involvement of the Arc system in redox regulation of the cob operon in *Salmonella typhimurium*. *Mol. Microbiol.* 6:1491–1494.

10. Gene fusions are valuable probes to monitor the expression of genes of interest. The fused gene has the promoter region of the target gene but not the promoter for the *lacZ* gene. Expression of the fused gene is therefore under control of the promoter region of the target gene. The fusions produce a hybrid protein whose amino-terminal end is derived from the target gene and the carboxy-terminal end from β-galactosidase. The hybrid protein has β-galactosidase activity. Therefore, an assay for β-galactosidase is a measure of the expression of the target gene. Thus, one can measure the expression of virtually any gene simply by constructing the proper gene fusion and performing an assay for β-galactosidase. One can construct gene fusions *in vitro* or *in vivo*. In vitro construction involves using restriction endonucleases to cut out a portion of the gene with its promotor region from a plasmid containing the cloned DNA, and ligating it to a *lacZ* gene, without its promotor or ribosome binding site, in a second plasmid. The plasmid containing the fused gene is then introduced into the bacterium and transformants are selected on the basis of being resistant to an antibiotic-resistant marker on the plasmid, and the production of β-galactosidase. This method was used by Iuchi *et al. In vivo* construction of gene fusions can also be performed. In this case, one uses Tn5 transposons fused to a promoterless *lacZ* gene. The transposon is introduced into the bacterium where it can recombine with the bacterial chromosome. Cells harboring the transposon are selected using the antibiotic-resistance marker on the transposon. Many of the strains have the transposon inserted into the host bacterial genes in the proper orientation and frame so that galactosidase production is under the control of the promoter of the interrupted gene. Insertion of the transposon into a gene interrupts the gene so that the normal gene product is not made. The gene is identified by mutant analysis.

11. Iuchi, S., V. Chepuri, H.-A. Fu, R. B. Gennis, and E. C. C. Lin. 1990. Requirement for terminal cytochromes in generation of the aerobic signal for the *arc* regulatory system in *Escherichia coli*: Study utilizing deletions and lac fusions of *cyo* and *cyd*. *J. Bacteriol.* 172:6020.

12. The control experiments were done to show that, in an *arc⁺ cyo⁺ cyd⁺* backround, anaerobiosis repressed *cyo-lacZ* and induced *cyd-lacZ* expression, and that a mutation in the *arc* genes prevented the repression of *cyo-lacZ* and lowered the expression of *cyd-lacZ*. These experiments showed that the fusion genes were regulated by

the availability of oxygen, and that the Arc system was responsible for the regulation.

13. Spiro, S., and J. R. Guest. 1990. FNR and its role in oxygen-regulated gene expression in *Escherichia coli. FEMS Microbiol. Rev.* **75**:399–428.

14. Jones, H. M., and R. P. Gunsalus. 1987. Regulation of *Escherichia coli* fumarate reductase (*frd*ABCD) operon expression by respiratory electron acceptors and the *fnr* gene product. *J. Bacteriol.* **169**:3340–3349.

15. Unden, G., and J. R. Guest. 1985. Isolation and characterization of the Fnr protein, the transcriptional regulator of anaerobic electron transport in *Escherichia coli. Eur. J. Biochem.* **146**:193–199.

16. Spiro, S., and J. R. Guest. 1990. FNR and its role in oxygen-regulated gene expression in *Escherichia coli. FEMS Microbiol. Rev.* **75**:399–428.

17. Chiang, R. C., R. Cavicchioli, and R. P. Gunsalus. 1992. Identification and characterization of narQ, a second nitrate sensor for nitrate-dependent gene regulation in *Escherichia coli. Mol. Microbiol.* **6**:1913–1923.

18. Rabin, R. S., and V. Stewart. 1993. Dual response regulators (NarL and NarP) interact with dual sensors (NarX and NarQ) to control nitrate- and nitrite-regulated gene expression in *Escherichia coli* K-12. *J. Bacteriol.* **175**:3259–3268.

19. Reviewed in Macnab, R. M. 1987. Motility and chemotaxis, pp. 723–759. In: Escherichia coli *and* Salmonella typhimurium: *Cellular and Molecular Biology*, Vol. 1. F. C. Neidhardt, J. L. Ingraham, K. B. Low, B. Magasanik, M. Schaechter and H. E. Umbarger (eds.). ASM Press, Washington, DC.

20. Reviewed in Stock, J. B, A. J. Ninfa, and A. M. Stock. 1989. Protein phosphorylation and regulation of adaptive responses in bacteria. *Microbiol. Rev.* **53**:450–490.

21. Reviewed in Manson, M. D. 1992. Bacterial motility and chemotaxis, pp. 277–346. In: *Advances in Microbial Physiology*, Vol. 33. A. H. Rose (ed.). Academic Press., New York

22. Park, C., D. P. Dutton, and G. L. Hazelbauer. 1990. Effects of glutamines and glutamates at sites of covalent modification of a methyl-accepting transducer. *J. Bacteriol.* **172**:7179–7187.

23. Macnab, R. M., and D.E. Koshland, Jr. 1972. The gradient-sensing mechanism in bacterial chemotaxis. *Proc. Natl. Acad. Sci. USA* **69**:2509–2512.

24. Berg, H. C., and D. A. Brown. 1972. Chemotaxis in *Escherichia coli* analysed by three-dimensional tracking. *Nature.* **239**:500–504.

25. Macnab, R. M., and D. E. Koshland, Jr. 1972. The gradient-sensing mechanism in bacterial chemotaxis. *Proc. Natl. Acad. Sci. USA.* **69**:2509–2512.

26. Silverman, M., and M. Simon. 1974. Flagellar rotation and the mechanism of bacterial motility. *Nature.* **249**:73–74.

27. Larsen, S. H., R. W. Reader, E. N. Kort, W.-W. Tso, and J. Adler. 1974. Changes in direction of flagellar rotation is the basis of the chemotactic response in *Escherichia coli. Nature.* **249**:74–77.

28. Iino, T., Y. Komeda, K. Kutsukake, R. M. Macnab, P. Matsumura, J. S. Parkinson, M. I. Simon,, and S. Yamaguchi, 1988. New unified nomenclature for the flagellar genes of *Escherichia coli* and *Salmonella typhimurium. Microbiol. Rev.* **52**:533–535.

29. Three proteins called FliG, FliM, and FliN, coded for by *fliG*, *fliM*, and *fliN* are the switch proteins. These proteins are not in the basal body and seem to be peripheral (rather than integral) cell membrane proteins closely associated with the basal body. A description of the flagellar motor is given in Section 1.2.1. Mutants that completely lack these proteins do not make flagella (Fla⁻ phenotype) even though these proteins are not part of the basal body, hook, or filament. Missense mutations in these genes cause a Fla⁻ phenotype, a Che⁻ phenotype (CW or CCW-biased), or paralyzed flagella (Mot phenotype), depending upon the mutation. It therefore appears that these proteins are necessary for basal body synthesis and switching, and perhaps are involved in energy coupling. Two other proteins should be mentioned, even though they are not chemotaxis proteins. The MotA and MotB proteins are necessary for motor function. Mutations in these genes cause paralyzed flagella. The proteins are integral membrane proteins that are thought to form a ring around the S and P rings of the basal body, and couple the proton potential to flagellar rotation. The MotA protein is a proton channel and the MotB protein may mediate the interaction of MotA with the basal body.

30. The MCP proteins are truly remarkable with respect to the different classes of signals that they transduce. Tsr responds to the chemoattractant L-serine. It also responds to the repellents L-leucine, indole, and low external pH. Tsr also responds to weak organic acids such as acetate that act as repellents because they lower the internal pH. In addition, the Tsr protein is a thermoreceptor mediating taxis towards warmer temperatures up to 37°C. The Tar protein in *E. coli* responds to the chemoattractant L-aspartate as well as to maltose

bound to the periplasmic maltose binding protein. Tar also detects the repellents Co^{2+} and Ni^{2+}. In mutants lacking Tsr, the Tar protein mediates taxis towards higher temperatures. The Trg protein responds to the chemoattractants ribose, glucose, and galactose when they are bound to their respective periplasmic binding proteins. The Tap protein (present in *E. coli* but not *S. typhimurium*) mediates taxis towards a variety of dipeptides when they are bound to DPP, a periplasmic dipeptide binding protein. Trg and Tap from *E. coli* serve as repellent receptors for phenol. Trg is also an attractant receptor, whereas Tap is a repellent receptor for weak organic acids.

31. Lukat, G. S., and J. B. Stock. 1993. Response regulation in bacterial chemotaxis. *J. Cell. Biochem.* 51:41–46.

32. Presumably the binding of chemoattractant induces a conformational change in the receptor-transducer protein exposing additional methylating sites on the receptor-transducer protein, thus increasing the level of methylation.

33. Eisenbach, M., Constantinou, C., Aloni, H., and M. Shinitzky. 1990. Repellents for *Escherichia coli* operate neither by changing membrane fluidity nor by being sensed by periplasmic receptors during chemotaxis. *J. Bacteriol.* 172:5218–5224.

34. Gotz, R., and R. Schmitt. 1987. *Rhizobium meliloti* swims by unidirectional, intermittent rotation of right-handed flagellar helices. *J. Bacteriol.* 169:3146–3150.

35. Armitage, J. P., and R. M. MacNab. 1987. Unidirectional, intermittent rotation of the flagellum of *Rhodobacter sphaeroides*. *J. Bacteriol.* 169:514–518.

36. Sockett, R. E., J. P. Armitage, and M. C. W. Evans. 1987. Methylation-independent and methylation-dependent chemotaxis in *Rhodobacter sphaeroides* and *Rhodospirillum rubrum*. *J. Bacteriol.* 169:5808–5814.

37. Zhulin, I. G., and J. P. Armitage. 1993. Motility, chemokinesis, and methylation-independent chemotaxis in *Azospirillum brasilense*. *J. Bacteriol.* 175:952–958.

38. Reviewed in Titgemeyer, F. 1993. Signal transduction in chemotaxis mediated by the bacterial phosphotransferase system. *J. Cell. Biochem.* 51:69–74.

39. Magasanik, B. 1993. The regulation of nitrogen utilization in enteric bacteria. *J. Cell. Biochem.* 51:34–40.

40. Magasanik, B. 1988. Reversible phosphorylation of an enhancer binding protein regulates the transcription of bacterial nitrogen utilization genes. *TIBS* 13:475–479.

41. Magasanik, B. 1993. The regulation of nitrogen utilization in enteric bacteria. *J. Cell. Biochem.* 51:34–40.

42. The operon is called *glnALG* in *Escherichia coli* and *glnABC* in *Salmonella typhimurium*.

43. Magasanik, B. 1982. Genetic control of nitrogen assimilation in bacteria. *Ann. Rev. Gen.* 16:135–168.

44. Reviewed in Magasanik, B., and F. C. Neidhardt. 1987. Regulation of carbon and nitrogen utilization, pp. 1318–1325. In: Escherichia coli *and* Salmonella typhimurium, *Cellular and Molecular Biology*, Vol. 2. F. C. Neidhardt, J. L. Ingraham, K. B. Low, B. Magasanik, M. Schaechter and H. E. Umbarger (eds.). ASM Press, Washington, D. C.

45. MacNeil, T., G. P. Roberts, D. MacNeil, and B. Tyler. 1982. The products of *gln*L and *gln*G are bifunctional regulatory proteins. *Mol. Gen. Genet.* 188:325–333.

46. The reason for the requirement of NRI-P for transcription is that the sigma 54 polymerase binds to the glnAp₂ promoter but cannot by itself initiate transcription. It is unable to do this because it cannot form an open complex (i.e., it cannot "melt" the DNA double helix around the promoter site to gain access to the transcription start site). However, NRI-P binds to sites on the DNA 100–130 base pairs upstream from the promoter region (called "enhancer sites"), and interacts with the RNA polymerase to form the open complex and thus initiate transcription.

47. Phosphonates are organophosphates in which there is a direct carbon–phosphorus bond rather than a phosphate ester linkage.

48. Wanner, B. L. 1993. Gene regulation by phosphate in enteric bacteria. *J. Cell. Biochem.* 51:47–54.

49. Nikaido, H., and M. Vaara. 1987. Outer membrane. In: Escherichia coli *and* Salmonella typhimurium, *Cellular and Molecular Biology*. F. C. Neidhardt, J. L. Ingraham, K. B. Low, B. Nagasanik, M. Schaechter and H. E. Umbarger (eds.). ASM Press, Washington, DC.

50. Waukau, J., and S. Forst. 1992. Molecular analysis of the signaling pathway between EnvZ and OmpR in *Escherichia coli*. *J. Bacteriol.* 174:1522–1527.

51. Aiba, H., T. Mizuno, and S. Mizushima. 1989. Transfer of phosphoryl group between two regulatory proteins involved in osmoregulatory expression of the *omp*F and *omp*C genes in *Escherichia coli*. *J. Biol. Chem.* 264:8563–8567.

52. Aiba, H., F. Nakasai, S. Mizushima, and T. Mizuno. 1989. Evidence for the physiological importance of the phosphotransfer between two regulatory components, EnvZ and OmpR, in

osmoregulation in *Escherichia coli. J. Biol. Chem.* **264**:14090–14094.

53. Mizuno, T., M.-Y. Chou, and M. Inouye. 1984. A uniqe mechanism regulating gene expression: translational inhibition by a complementary RNA transcript (micRNA). *Proc. Natl. Acad. Sci. USA* **81**:1966–1970.

54. Sugiura, A., K. Nakashima, K. T. Tanaka, and T. Mizuno. 1992. Clarification of the structural and functional features of the osmoregulated kdp operon of *Escherichia coli. Molec. Microbiol.* **6**:1769–1776.

55. Nakashima, K., H. Sugiura, H. Momoi, and T. Mizuno. 1992. Phosphotransfer signal transduction between two regulatory factors involved in the osmoregulated kdp operon in *Escherichia coli. Molec. Microbiol.* **6**:1777–1784.

Index

365